AF767569

Human Biomonitoring II

Hermann Fromme

Human Biomonitoring II

Levels and Toxicokinetics of Environmental and Industrial Chemicals in Blood, Urine and Breast Milk

Springer

Hermann Fromme
Institute and Outpatient Clinic for
Occupational, Social and Environmental
Medicine
Ludwig-Maximilians-University Munich
Munich, Germany

ISBN 978-3-032-10035-1   ISBN 978-3-032-10036-8 (eBook)
https://doi.org/10.1007/978-3-032-10036-8

© The Editor(s) (if applicable) and The Author(s), under exclusive license to Springer Nature Switzerland AG 2026

This work is subject to copyright. All rights are solely and exclusively licensed by the Publisher, whether the whole or part of the material is concerned, specifically the rights of translation, reprinting, reuse of illustrations, recitation, broadcasting, reproduction on microfilms or in any other physical way, and transmission or information storage and retrieval, electronic adaptation, computer software, or by similar or dissimilar methodology now known or hereafter developed.
The use of general descriptive names, registered names, trademarks, service marks, etc. in this publication does not imply, even in the absence of a specific statement, that such names are exempt from the relevant protective laws and regulations and therefore free for general use.
The publisher, the authors and the editors are safe to assume that the advice and information in this book are believed to be true and accurate at the date of publication. Neither the publisher nor the authors or the editors give a warranty, expressed or implied, with respect to the material contained herein or for any errors or omissions that may have been made. The publisher remains neutral with regard to jurisdictional claims in published maps and institutional affiliations.

This Springer imprint is published by the registered company Springer Nature Switzerland AG
The registered company address is: Gewerbestrasse 11, 6330 Cham, Switzerland

If disposing of this product, please recycle the paper.

We have only the world that we can bring forth with others, and only love helps bring it forth.
(H. Maturana, F. Varela)

Preface

Dear reader,

We are constantly influenced by a diverse range of environmental factors, each affecting us in distinct ways. Among these, physical, chemical, and biological pollutants—alongside social and psychosocial influences—deserve special attention due to their potential for long-term health impacts. Increasingly, our environment is recognized as a complex and dynamic system of interrelated influences in which we are embedded. This complexity becomes apparent when considering that nearly 100,000 chemicals are used worldwide on a large scale, exposing us to many of them—sometimes at significant concentrations—in our daily lives. Importantly, this system is neither static nor uniform; it is in a state of constant variation. New chemicals and influences are regularly introduced, while others diminish in relevance or disappear entirely. Early identification of these changes is essential, enabling us to implement effective preventive measures.

Human biomonitoring is an essential method for evaluating exposure to environmental chemicals. It involves measuring chemicals and their metabolites in human samples and analyzing the body's biological and biochemical reactions to these substances. Human biomonitoring is part of a comprehensive system that tracks, assesses, and understands environmental effects on human health. It complements traditional environmental monitoring by providing data on internal exposures rather than just external factors. By measuring the overall body burden of substances from all potential exposure sources, biomonitoring gives a complete picture of individual and population exposures. Its results can be directly connected to health-based guidelines, enabling informed assessments of potential health risks.

This book aims to give a complete overview of current knowledge about the contamination of human samples—such as blood, urine, and breast milk—by various chemicals or groups of chemicals. It offers detailed summaries of the toxicokinetic properties (absorption, distribution, metabolism, and excretion) of each substance, which are vital for properly interpreting and applying biomonitoring data. Important health aspects are also covered, along with options for health assessment. The main focus is on substances that are commonly found in humans, are especially concerning due to their health impact, or for which limited information is currently available. Additionally, the book highlights gaps where reliable data on exposure and risk are missing, pointing out areas where new challenges may arise.

Munich, Germany
2025

Prof. Dr. Hermann Fromme

Acknowledgments I want to dedicate this book to my entire family with all my heart for our exciting life together and your constant support. It brings me so much joy and happiness. Thank you for every moment.

Contents

About the Author

Prof. Dr. Hermann Fromme is a professor at the Institute and Clinic for Occupational, Social, and Environmental Medicine at University Hospital, Ludwig-Maximilians-University Munich (LMU), Germany. He started his career studying social sciences at the University of Göttingen and earned his MD degree from the Freie Universität Berlin. After holding various management roles in the environmental health departments of the Berlin Senate Department for Social Affairs and Health, he served as the head of the Institute for Environmental Analysis and Human Toxicology in Berlin. Later, Prof. Fromme led the Department of Chemical Safety and Toxicology at the Bavarian Health and Food Safety Authority. His work mainly focuses on identifying chemicals in environmental media and humans, as well as conducting health-related risk assessments. This work has resulted in numerous scientific articles, particularly in the fields of exposure science, indoor air contaminants, human biomonitoring, health effects, and regulatory issues related to pollutants.

About the Author

[illegible]

Chapter 1
Pesticides

Abstract The focus of this chapter is on pesticides, which play a significant role in controlling pests in agriculture, crop production, and crop protection. Nevertheless, it is important to keep in mind that pesticides can lead to contamination of environmental media, bioaccumulate in the food chain, and pose potential health risks to nontarget organisms, as well as to humans. Therefore, this chapter provides a brief description of the chemical and physical characteristics and usage patterns of various pesticides, their primary health risks, and toxicokinetic properties, along with references to further scientific literature. The primary objective is to present a detailed overview of their occurrence in human matrices, such as blood, urine, and breast milk. In this context, detailed information is provided on organophosphate pesticides, carbamates and dithiocarbamates, glyphosate and glufosinate, pyrethroids and pyrethrins, neonicotinoids, toxaphenes, and the insect repellent N,N-diethyl-meta-toluamide (DEET).

Keywords Pesticides · Organophosphates · Carbamates · Glyphosate · Pyrethroids · Neonicotinoids · Toxaphenes

Ecosystems, along with humans, are exposed to various natural and chemical substances due to the rapid advancement of our technological capabilities. This also pertains to a large group of pesticides. A significant issue related to pesticides is the increasing development of resistance to individual substances, which leads to further changes in application practices and, most importantly, to the ongoing creation of new substances (Rivero et al. 2010). In a review, Umetsu and Shirai (2020) conclude that at least 105 chemical pesticides have been introduced during the past decade or are currently in development, and that global pesticide development has been gradually shifting from synthetic chemical pesticides to biological pesticides. This may mark the beginning of future strategies for initiating and implementing transition processes in this area as well.

Pesticides play a significant role in controlling pests in agriculture, crop production and distribution, and crop protection. This is particularly important due to the growing world population, as it helps safeguard the food supply. Another critical application

© The Author(s), under exclusive license to Springer Nature Switzerland AG 2026
H. Fromme, *Human Biomonitoring II*, https://doi.org/10.1007/978-3-032-10036-8_1

area is the prevention and control of so-called vector-borne diseases, in which an infectious agent is transmitted to humans by an infected animal (e.g., mosquitoes, ticks, flies, lice). Vector-borne diseases are among the leading causes of illness and mortality worldwide, especially in tropical and subtropical regions. They account for more than 17% of all infectious diseases, causing more than 700,000 deaths annually (WHO 2020). The most prevalent diseases include malaria, dengue fever, chikungunya fever, Zika virus fever, yellow fever, and West Nile fever.

In addition, it must always be kept in mind that uncontrolled and high-dose use of pesticides, in particular, can lead to contamination of water and soil, bioaccumulation in the food chain, and thus pose a permanent risk to nontarget organisms, the sensitive balances in ecosystems in general, and to humans (Sharma et al. 2019; Pathak et al. 2022). In principle, targeted use limited to what is necessary is essential in all application areas to minimize ecological and health impacts.

According to the U.S. Environmental Protection Agency (US EPA 2023c), pesticide law defines a "pesticide" as:

- Any substance or mixture of substances intended for preventing, destroying, repelling, or mitigating any pest.
- Any substance or mixture of substances intended for use as a plant regulator, defoliant, or desiccant.
- Any nitrogen stabilizer (any substance or mixture of substances intended to prevent or hinder nitrification, denitrification, ammonia volatilization, or urease production through action upon soil bacteria).

A fundamental distinction can be made between active and so-called inert ingredients. In the latter case, these substances do not have a direct pesticidal effect but can improve the product's applicability and effectiveness. Although they are labeled as inert, they can still possess toxic potential. An active ingredient is a chemical in a pesticide product that controls pests.

One common way of categorizing pesticides is to use them according to the target organisms they are intended to control (EEA 2023):

- Herbicides: weeds, algae
- Insecticides: insects
- Acaricides: mites
- Fungicides: fungi
- Bactericides: bacteria
- Nematicides: nematodes
- Molluscicides: snails, molluscs
- Avicides: birds
- Wood preservatives: mainly fungi and insects
- Antifoulants: mainly barnacles (e.g., on ship hulls or other undersea surfaces)
- Antimicrobial: microbiological organisms in general, such as bacteria and viruses.

Figure 1.1 presents a simplified diagram categorizing pesticides based on their chemical characteristics, emphasizing organic and synthetic pesticides, and highlights the main groups along with their individual representatives. Data on the human body burden are specifically available for individual organic substances and are discussed in more detail in the following chapters. As illustrated in Fig. 1.2, five pesticide groups are particularly significant in today's market, representing 20% (neonicotinoids), 19% (pyrethroids), 16% (diamides), 14% (organophosphates), and 6% (carbamates) of total pesticide sales worldwide in 2019 (Ahamad and Kumar 2023).

Additionally, sales figures for Europe refer to data at the beginning of the trade chain, while no information on the use of pesticide products is available. In 28 European countries (excluding the UK), sales of pesticide products increased from 282,000 tons in 2012 to 356,000 tons in 2021, with significant differences among individual countries (EUROSTAT 2023). For example, in 2021, Spain sold about 76,000 tons, whereas the three Scandinavian countries sold only about 7000 tons combined. Of the total, approximately 121,000 tons consist of herbicides, haulm destructors, and moss killers; 155,000 tons are fungicides and bactericides; and 50,000 tons are insecticides and acaricides. In 2012, pesticide usage in the US at the producer level totaled over 540,000 tons, with roughly 91% of herbicides, 82% of fungicides, and 57% of insecticides utilized in the agricultural sector alone (US EPA 2017b).

1.1 Organochlorines

Organochlorine pesticides are chlorinated hydrocarbons that have been used as insecticides for many decades. The chemical structures of some essential compounds are shown in Fig. 1.3. The prominent representatives of this class of substances, for which ample exposure data are available, include hexachlorobenzene (HCB), hexachlorocyclohexanes (HCH), and dichlorodiphenyltrichloroethane (DDT). These are highly persistent lipophilic compounds that accumulate in the environment and the food chain. Even though their prevalence is declining, high-fat animal meat remains the primary source of human exposure. Since 2001, these substances have not been approved as plant protection products in the EU.

HCH isomers are volatile and relatively persistent substances in the atmosphere, expected to be capable of long-range transport (ATSDR 2023a). HCH has been widely used as an insecticide since the 1940s. Its use has been banned, with only lindane (γ-HCH) remaining permitted for topical treatment of scabies and head lice. The technical mixture contains approximately 60–70% α-HCH, 5–12% β-HCH, 10–15% γ-HCH, 6–10% δ-HCH, and 3–4% ε-HCH isomers (ATSDR 2023a).

Approximately 2.8 million tons of DDT have been produced worldwide, with about 50% manufactured in the USA (UNEP 2015). Technical-grade DDT is a complex mixture consisting of 65–80% p,p'-DDT, the primary insecticide component, along with smaller proportions of o,p'-DDT (around 15–21%), p,p'-DDE (dichlorodiphenyldichloroethene), and p,p'-DDD (dichlorodiphenyldichloroethane) (ATSDR 2019). DDT has been extensively used as an insecticide since 1938. In

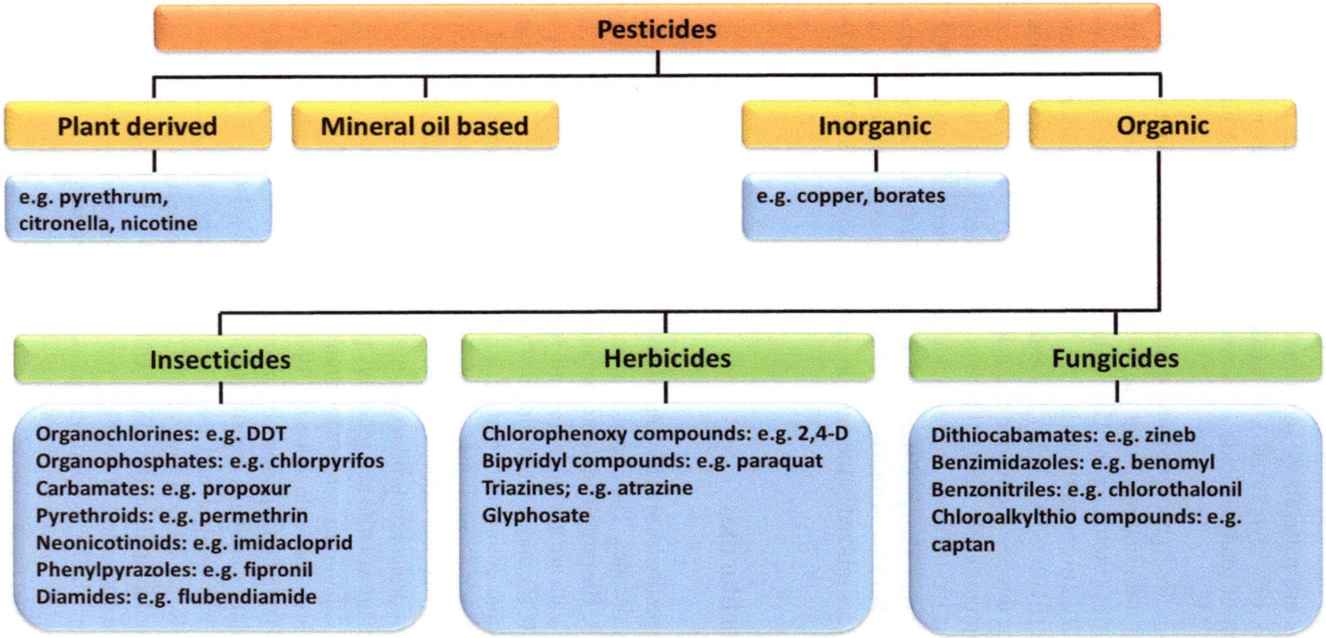

Fig. 1.1 Major classes of pesticides (modified from Alengebawy et al. 2021)

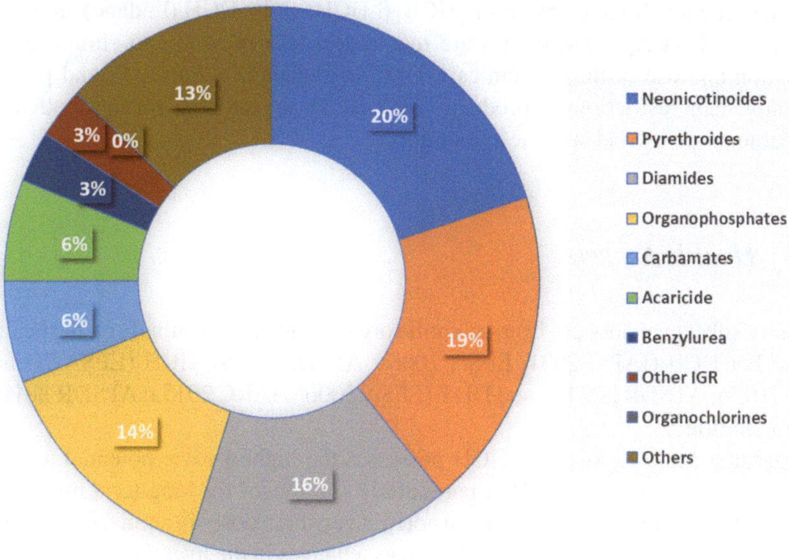

Fig. 1.2 Share of individual pesticide classes in the global sales market in 2019 (IGR: insect growth regulator) (modified from Ahamad and Kumar 2023)

Fig. 1.3 Chemical structure of **a** DDT, **b** hexachlorobenzene (HCB), **c** γ-hexachlorocyclohexane (γ-HCH)

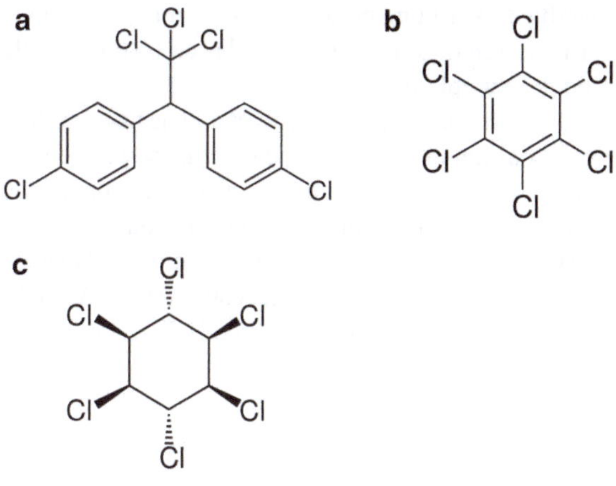

Europe and the US, DDT has been banned for decades; only in developing countries is it still used to control vector-borne diseases such as malaria (ATSDR 2019).

HCB is a white crystalline solid that is practically insoluble in water (ATSDR 2015). It was introduced in North America during the 1940s and used as a fumigant, fungicide, and in pyrotechnic materials until it was banned. Although it has not been produced for decades, it can still appear as an impurity during the production of chlorinated solvents and pesticides.

Under the Stockholm Convention, HCB, β-HCH, and γ-HCH (lindane) are listed in Annex A. Consequently, parties are required to take measures to eliminate the production and use of these chemicals. DDT is included in Annex B, and parties must implement restrictions on production and use, taking into account any applicable acceptable purposes and specific exemptions.

1.1.1 Health Aspects

Extensive substance dossiers from scientific organizations are available on the health effects. For HCB (IARC 2001; EFSA 2006a; ATSDR 2015), HCH (EFSA 2005a; IARC 2015a; ATSDR 2023a), and DDT (EFSA 2006b; IARC 2015a; ATSDR 2019).

Here, in brief:

Regarding acute exposure, γ-HCH possesses the highest toxic potential among all HCH isomers, while β-HCH is particularly significant for long-term exposure. Effects on the nervous system have been noted at very high doses in animal studies and in cases of poisoning. In experimental animals, changes in the liver, immune system, reproduction, and hematological parameters have been associated with long-term exposure. Concerning γ-HCH, the primary effects involve the immune system and development. Research on its carcinogenic potential remains inconclusive, partly due to frequent mixed exposure to other organochlorine compounds. Lindane is classified as a human carcinogen (Group 1) by the WHO's International Agency for Research on Cancer (IARC), based on epidemiological evidence related to non-Hodgkin lymphoma.

DDT primarily targets the liver and nervous system, and it also impacts reproduction. Additionally, the immune system and fetal development have been identified as other target areas. Based on findings from animal studies, IARC has classified DDT as probably carcinogenic to humans (Group 2A).

HCB has various effects on the organism, primarily targeting the liver, skin, bones, thyroid, immune system, and nervous system (ATSDR 2015). Based on animal experimental results, IARC has classified HCB as possibly carcinogenic to humans (Group 2B).

1.1.2 Toxicokinetics

Technical-grade HCH is well absorbed in the gastrointestinal tract and is primarily distributed in fatty organs. It can cross the placental barrier and be present in breast milk. The main metabolites are glucuronidated and sulfated chlorophenols, which are excreted in urine. The half-life of β-HCH in humans is approximately seven years, while that of γ-HCH ranges from 18 to 163 h.

DDT is rapidly absorbed by the organism after oral exposure and likely also through inhalation within a few hours. However, in animal experiments, oral absorption efficiency was significantly enhanced by the simultaneous administration of digestible oil (ATSDR 2019). Once distributed throughout the organism, DDT is stored in tissues in proportion to their lipid content and is also found in breast milk. DDT and its metabolites can cross the placenta (ATSDR 2019). p,p′-DDT and o,p′-DDT are metabolized in the body into p,p′-DDD and o,p′-DDD, while p,p′-DDT is additionally broken down into DDE (see Fig. 1.4). DDE is more lipophilic and is excreted less readily than the parent compound. The elimination half-lives are 5 years for p,p′-DDT and 10 years for p,p′-DDE. The primary routes of excretion include transfer to breast milk and excretion as DDA (2,2-bis[4-chlorophenyl]acetic acid) conjugates in urine (ATSDR 2019).

Hexachlorobenzene is minimally absorbed after inhalation and only slightly absorbed in the gastrointestinal tract following oral exposure. It is slowly metabolized in mammals by microsomal enzymes in the liver, kidneys, lungs, and intestines into pentachlorophenol, pentachlorobenzene, and other substances (ATSDR 2015). HCB accumulates in fatty tissues and breast milk and can cross the placental barrier. Generally, humans excrete HCH isomers and their metabolites in urine, breast milk, sweat, and semen (ATSDR 2015). It is primarily excreted unchanged in feces, with smaller amounts found in urine and breast milk (ATSDR 2015).

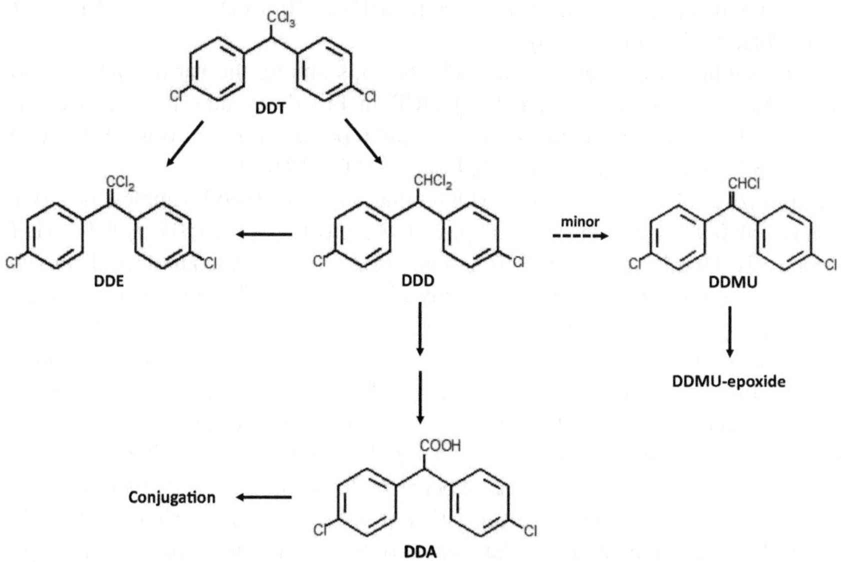

Fig. 1.4 Metabolism of dichlorodiphenyltrichloroethane (DDT) (according to ATSDR 2019) (*DDT* dichlorodiphenyl trichloroethane, *DDE* dichlorodiphenyl dichloroethane, *DDD* tetrachloro diphenylethane, *DDA* 2,2-bis[4-chlorophenyl]acetic acid, *DDMU* 1-chloro-2,2-bis(4-chlorophenyl)ethylene)

1.1.3 Occurrence in the Blood

Numerous studies on organochlorine pesticides (OCPs) in blood are available. A compilation of the more recent studies, divided into adults, children, and adolescents, as well as umbilical cord blood, is presented in Table 1.1. The highest levels for adults have been reported in recent years for all OCPs by Svarcova et al. (2019) from the Czech Republic. In a representative study from Germany involving 1135 children aged 6–11 years, geometric mean values were lower than in earlier studies from Denmark and the USA.

Levels of OCP were lower among women with a history of breastfeeding and higher for those experiencing their first pregnancy (Reid et al. 2013). Additionally, concentrations were higher in children whose mothers breastfed (Bandow et al. 2020). Overall, several authors describe an increase in OCP concentrations with age (Reid et al. 2013; Mørck et al. 2014; Thomas et al. 2017). Studies conducted in Australia and the USA report a time trend of internal load. In total, 183 pooled samples of human blood serum were collected in Australia during 2002/03, 2006/07, 2008/09, 2010/11, and 2012/13, demonstrating a decreasing temporal trend for all organochlorines, except for an increase in concentrations observed between 2006/07 and 2008/09 (Thomas et al. 2017). Utilizing data from six cycles of the US National Health and Nutrition Examination Survey, which involved between 1973 and 2322 participants, β-HCH, HCB, and p,p′-DDE in all age groups showed a consistently similar downward trend from 2005–2006 to 2015–2016 (Li et al. 2022). The results are graphically illustrated in Fig. 1.5.

Some studies have examined the relationships among the various investigative media. Mørck et al. (2014) analyzed OCP in blood samples from 116 Danish schoolchildren and their mothers (N: 143) and reported children-to-maternal serum ratios of 3.5 for p,p′-DDE, 0.9 for HCB, and 1.0 for β-HCH.

Bergonzi et al. (2009) studied 70 maternal and cord blood samples, observing medians of 0.96 and 0.25 μg/l for p,p′-DDE, and 0.16 and 0.05 μg/l for HCB, respectively. Highly significant correlations were found between umbilical cord and maternal serum levels, with mean concentration ratios (cord serum/maternal serum) of 0.25 for p,p′-DDE and 0.29 for HCB. The study's mean placenta-to-maternal serum ratios were 0.57 and 1.06. In a Slovak study, the mean HCB cord blood to maternal blood concentration ratio on a lipid basis, utilizing 1027 paired samples, was 0.25 (Murínová et al. 2017). Moreover, Santos et al. (2022) found median cord blood to maternal blood ratios of 1.3 for p,p′-DDE and 0.8 for β-HCH. A Spanish study examined 155 samples collected between 2004 and 2008 and reported median lipid-based cord and maternal serum ratios of 0.7 for p,p′-DDE, HCB, and β-HCH (García-Villarino et al. 2020). In the context of the study of maternal serum samples in the first trimester and placenta and fetal samples after miscarriage, significantly higher concentrations of p,p′-DDE, HCB, and β-HCH were found in fetal tissues compared to wet weight in Norway, relative to maternal serum or placenta (Björvang et al. 2021).

Table 1.1 Geometric mean concentration of organochlorine pesticides in blood in ng/g lipid (in brackets in µg/l)

References	N	Age (years)	HCB	p,p′-DDT	p,p′-DDE	p,p′-DDD	α-HCH	β-HCH	γ-HCH	Country, year
Adults										
Xu et al. (2022)	38[d]	19–40	– (0.05[c])	– (0.09[c])	–	–	–	–	–	Norway, 2019
Santos et al. (2022)	139	20–39	–	–	– (0.131)	–	–	– (0.121)	– (0.175)	Brazil, 2017/18
Björvang et al. (2021)	20	25–40	– (0.026 [b])	– (0.017[b])	– (0.213[b])	–	– (< LOQ[b])	– (0.025[b])	– (< LOQ[b])	Sweden, 2015/16
Li et al. (2022)	1989	> 16	8.88	2.96	199	–	–	3.29	–	USA, 2015/16
Svarcova et al. (2019)	38		29.6[b]	13.5[b]	207[b]	3.84[b]	2.65[b]	3.97[b]	1.02[b]	Czech Republic, 2015
Palaniswamy et al. (2025)	102	33[c]	–	< LOQ	11.8	< LOQ	–	–	–	Finland, 2012–14
Thomas et al. (2017)	2400[e]		5.6	4.3	129	–	–	5.0	–	Australia, 2012/13
Mørck et al. (2014)	143	31–52	14[b]	3.0[b]	124[b]	–	–	2.0[b]	–	Denmark, 2011
Lee et al. (2019)	1874	18–70	– (0.009)	– (0.052)	– (0.041)	– (0.005)	–	–	–	South Korea, 2009–11
Reid et al. (2013)	157	> 18	– (0.06)	– (0.07)	– (0.68)	– (0.01)	– (< 0.06)	– (0.10)	– (< 0.09)	Australia, 2009–11
Haines et al. (2017)	1665	20–79	– (0.053)	– (<0.05)	– (0.75)	–	–	– (0.031)	– (< 0.01)	Canada, 2007–09

(continued)

Table 1.1 (continued)

References	N	Age (years)	HCB	p,p′-DDT	p,p′-DDE	p,p′-DDD	α-HCH	β-HCH	γ-HCH	Country, year
García-Villarino et al. (2020)	155	19–42	75.1[b]	19.53[b]	255[b]	0.36[b]	–	25.0[b]	2.44[b]	Spain, 2004–08
Children/adolescents										
Schoeters et al. (2022)	415	14–15	7.6	1.8	41	–	–	1.1	< LOQ	Belgium, 2017/18
Pineda et al. (2013)	1082	11–21	– (0.042)	– (< 0.015)	– (0.11)		– (< 0.020)	– (< 0.015)	– (< 0.020)	Sweden, 2016/17
Bandow et al. (2020)	1135[a]	6–11	9.7 (< 0.07)	2.2 (< 0.02)	30 (0.158)	1.9 (< 0.02)	2.8 (< 0.03)	2.4 (< 0.02)	5.5 (< 0.05)	Germany, 2014–17
Mørck et al. (2014)	116	6–11	14[b]	3.0[b]	38[b]	–	–	< LOD	–	Denmark, 2011
CDC	588	12–19	13.3	< 7.8	105	–	–	< 7.8	< 7.8	USA, 2003/04
Cord blood										
Santos et al. (2022)	139		–	– (0.099)	– (0.150)	– (0.066)	–	– (0.071)	– (0.157)	Brazil, 2017/18
Palaniswamy et al. (2015)	102		–	< LOQ	13.1	< LOQ	–	–	–	Finland, 2012–14
García-Villarino et al. (2020)	229		50.3[b]	33.4[b]	179[b]	0.63[b]	–	17.4[b]	5.01[b]	Spain, 2004–08

[a] Number for γ-HCH 1028; [b] median; [c] mean; [d] in four pools; [e] in 24 pools; *LOQ* limit of quantification

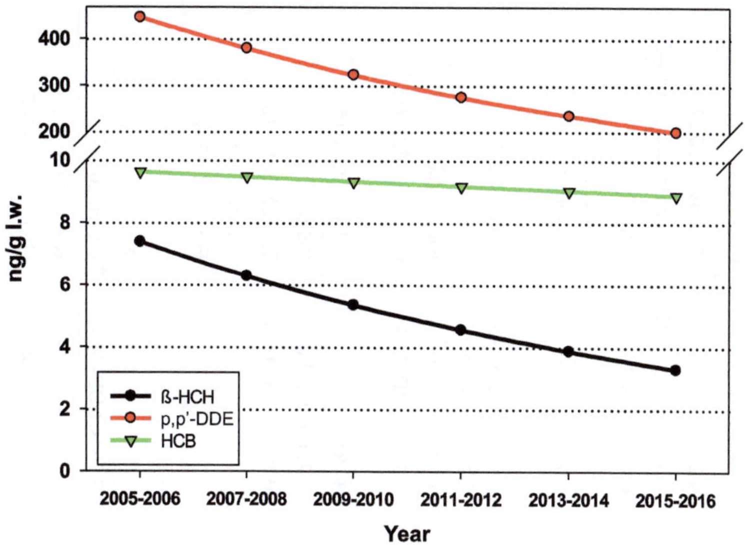

Fig. 1.5 Temporal trend of organochlorine pesticides in serum samples in the US (according to Li et al. 2022)

1.1.4 Occurrence in Breast Milk

Many studies on organochlorine pesticides in breast milk have been conducted. A compilation of more recent studies is provided in Table 1.2. The significant differences in results between countries and within populations are striking, primarily due to varying amounts of active ingredients used and the duration of their application. For example, very high concentrations of p,p′-DDE, β-HCH, and HCB have been reported in breast milk from western Iran (Shahmoradi et al. 2019). Extremely high p,p′-DDE concentrations are also noted in India, while β-HCH and HCB levels do not exhibit such wide variations (Bawa et al. 2018). The lowest median concentrations have been found in Poland and Turkey (Witczak et al. 2021; Agus et al. 2022).

Various studies have shown that dietary factors, such as an increased consumption of fatty foods, can influence the concentrations found in breast milk. Besides temporary and spatial differences in levels, fluctuations in concentrations during breastfeeding must be taken into account. For instance, Witczak et al. (2021) gathered breast milk samples from 96 Polish mothers at 16 different time points during the infant period, showing decreasing mean concentrations of p,p′-DDE and β-HCH from 0.12 and 3.19 ng/g lw at less than 7 days to only 0.036 and 0.37 ng/g lw at 12 months. Similarly, Du et al. (2016) collected breast milk from 16 Australian mothers at 2, 5, 9, and 12 months into breastfeeding, reporting a decline in p,p'-DDE concentrations during this period, which decreased from 71 ng/g lw at 2 months to 22 ng/g lw at 12 months.

Table 1.2 Median (range) concentrations of organochlorine pesticides in breast milk in ng/g lw

References	N	HCB	*p,p'*-DDE	β-HCH	γ-HCH	Country, year
Nermo et al. (2025)	120	–	16.9 (6.36–46.8[a])	1.04 (0.56–2.85[a])	0.23 (0.08–2.47[a])	Norway, 2020/21
Parizek et al. (2023)	231	49.8 (4.1–25.6[a])	70.2 (21–329[a])	4.42 (< 0.05–16.7[a])	< 0.05 (< 0.05–1.05[a])	Czech Republic, 2019–21
Dwivedi et al. (2023)	130	–	790 (< –1300)	1108 (10–8500)	663 (30–1900)	India, 2020
Kuang et al. (2020)	55	22.2 (< LOD-146)	50.0 (< LOD-414)	< LOD (< LOD-303)	–	China, 2018–19
Shahmoradi et al. (2019)	50	337[c] (30–4120)	1430[c] (80–7230)	854[c] (15–5870)	170 (10–2740)	Iran, 2017
Bawa et al. (2018)	150	–	407[c] (< LOD-3772) 345[c] (< LOD-3213)	37[c] (< LOD-475) 31[c] (< LOD-305)	9.04[c] (< LOD-145) 4.8[c] (< LOD-209)	India, 2015/16[d]
EL-Saeid et al. (2021)	50	117[c] (57–248)	80[c] (17.2–135)	70[c] (35.8–146)	14.0[c] (9.2–24.0)	Saudi Arabia
Torres-Moreno et al. (2023)	49	–	259 (0.05–1800)	< LOD (< LOD-0.05)	0.43 (< LOD-2.14)	Colombia, 2014/15
Witczak et al. (2021)	96	–	0.14[bc]	2.88[bc]	3–30[bc]	Poland, 2013–15
Aerts et al. (2019)	206	5.5 (< 2.0–16.6)	37.0 (7.8–256)	2.4 (< 2.0–46.3)	< 2.0 (< 2.0–4.0)	Belgium, 2014
Chen et al. (2018)	68	–	9.2 (0.35–44.6)	0.12 (< 0.04–1.33)	0.10 (< 0.03–0.62)	Taiwan, 2013–16
Agus et al. (2022)	48	0.02[c] (< –0.77)	0.84[c] (< –14.2)	0.16[c] (< –2.7)	–	Turkey, 2013
Antignac et al. (2016)	42	10.3 (4.0–68.1)	60.0 (7.0–1069)	15.1 (2.4–195)	–	French, 2011–14
Antignac et al. (2016)	42	10.3 (4.0–68.1)	60.0 (7.0–1069)	15.1 (2.4–195)	–	French, 2011–14
Du et al. (2016)	99	–	53[c] (5.7–278)	< LOD (< LOD-65)	–	Australia
Song et al. (2013)	48	–	334[c] (76–840)	175[c] (37–478)	7.76 (< LOD-24.4)	China, 2012
Lee et al. (2013)	206	6.2 (< LOD-56.1)	91.7 (< LOD-375)	20.5 (< LOD-107)	–	Korea, 2011
Raab et al. (2013)	516	16 (3–140)	63 (8.0–2330)	6.0 (< 2.0–1520)	< 1.0 (< 1.0–34)	Germany, 2007/08

[a] 5th and 95th percentiles of one sampling site; [b] collected in 2nd month; [c] mean; [d] two districts; *LOD* limit of determination

1.2 Organophosphate Pesticides

The synthetically produced organophosphate pesticides (OPPs) are either ester, amide, or thiol derivatives of phosphoric and phosphorothioic acids. The general chemical structure is illustrated in Fig. 1.6, where the R1 and R2 moieties typically consist of alkyl or aryl groups, which may be bonded directly or via an oxygen or sulfur atom. X denotes one of a wide variety of substituted or branched aliphatic, aromatic, or heterocyclic groups linked to the phosphorus atom via a labile bond (COT 1999). Organophosphates contain a phosphorus atom that is usually bonded to two oxygen atoms, forming a carbonyl group that is crucial for the effects of organophosphates (Araújo et al. 2023).

Over 100 individual substances in this class exhibit considerable variation in their chemical, physical, and biological properties. Most are highly reactive, slightly soluble in water, have low vapor pressure, and are comparatively low in volatility, with the exception of dichlorvos.

Table 1.3 summarizes the most commonly used OPPs. For instance, chlorpyrifos has been used as a pesticide in both agricultural and nonagricultural areas of the US since 1965, primarily on corn, soybeans, fruit trees, walnut trees, Brussels sprouts, blueberries, broccoli, and cauliflower. Additionally, it is applied on golf courses, ornamental plants, wood, and homes to control mosquitoes, cockroaches, and ants.

Between 1998 and 2014, OPPs accounted for nearly half of all insecticides used worldwide, and chlorpyrifos is among the most widely used (Mora-Gutiérrez et al. 2021). In the US, the most commonly used OPPs include acephate, chlorpyrifos, malathion, naled, phorate, dicrotophos, phosmet, dimethoate, terbufos, ethoprophos, and tetrachlorvinphos (US EPA 2017b). The amount of OPPs used in the US has declined by more than 70% since 2000, falling from approximately 32 million kilograms to 9.1 million kilograms in 2012, and the usage of OPPs as a percentage of total insecticide use has decreased from 71% in 2000 to 33% in 2012 (US EPA 2017b).

1.2.1 Health Aspects

In several human and animal experiments, OPPs inhibit acetylcholinesterase (AChE) activity by forming a covalent bond at the active site. The phosphorylated AChE from OPP is usually readily reactivated by oxime (COT 1999). Acetylcholine is a common

Fig. 1.6 General structure of organophosphates (left: phosphate compounds; right: phosphorothioate compounds) (according to COT 1999)

Table 1.3 Technical important organophosphate pesticides (OPPs)

Name	CAS No.	Chemical name
Acephate	30560-19-1	N-(Methoxy-methylsulfanylphosphoryl)acetamide
Azamethiphos	35575-96-3	S-[(6-Chloro-2-oxo[1,3]oxazolo[4,5-b]pyridin-3(2H)-yl)methyl] O,O-dimethyl phosphorothioate
Azinphos-Methyl	86-50-0	O,O-Dimethyl S-[(4-oxo-1,2,3-benzotriazin-3(4H)-yl)methyl] phosphorodithioate
Chlorpyrifos	2921-88-2	O,O-Diethyl O-3,5,6-trichloropyridin-2-yl phosphorothioate
Chlorpyrifos-Methyl	5598-13-0	O,O-Dimethyl 0-3,5,6-trichloro-2-pyridyl phosphorothioate
Diazinon	333-41-5	O,O-Diethyl O-[4-methyl-6-(propane-2-yl)pyrimidin-2-yl] phosphorothioate
Dichlorvos	62-73-7	2,2-Dichlorovinyl dimethyl phosphate
Dimethoate	60-51-5	O,O-Dimethyl S-[2-(methylamino)-2-oxoethyl] dithiophosphate
Ethion	563-12-2	O,O,O′,O′-Tetraethyl S,S′-methylene bis(phosphorodithioate)
Fenitrothion	122-14-5	O,O-Dimethyl O-(3-methyl-4-nitrophenyl) phosphorothioate
Fenthion	55-38-9	O, O-dimethyl-O-(4-methylthion-m-tolyl) phosphorothioate
Malathion	121-75-5	Diethyl 2-[(dimethoxyphosphorothioyl)sulfanyl]butanedioate
Methyl-parathion	298-00-0	O,O-Dimethyl O-4-nitrophenyl phosphorothioate
Parathion	56-38-2	O,O-Diethyl O-(4-nitrophenyl) phosphorothioate
Phosmet	732-11-6	O, O-Dimethyl phthalimidomethyl S-phosphorodithioate
Terbufos	13071-79-9	S-((tert-butylthio)methyl) O,O-diethyl phosphorodithioate
Tetrachlovinphos	961-11-5	(Z)-2-chloro-1-(2,4,5-trichlorophenyl) vinyl dimethylphosphate

neurotransmitter released after nerve stimulation to transmit the nerve impulse to a neuron or effector organ.

Acute and long-term exposure has been linked to various dose-dependent clinical signs such as nausea, vomiting, abdominal cramps, diarrhea, excessive salivation, headache, weakness, difficulty breathing, vision impairment, convulsions, central nervous system depression, paralysis, coma, and respiratory failure. Most studies on acute poisoning with OPPs indicate that polymorphism in metabolic enzymes accounts for inter-individual variability in effects. Depressions in plasma pseudocholinesterase and red blood cell acetylcholinesterase enzyme activities serve as biochemical indicators of exposure (EFSA 2019; US EPA 2023a). In addition to typical exposure history and clinical symptoms, a 20–30% reduction in AChE may suggest significant OPP poisoning. In contrast, in severe cases, the enzyme is typically depressed by 80–90% of normal levels.

Some OPPs, such as chlorpyrifos, monocrotophos, mipafox, trichlorfon, tri-o-cresyl phosphate, trichloronate, leptophos, EPN (O-ethyl O-4-nitrophenyl phenylphosphonothioate), and methamidophos, may also cause organophosphorus-induced delayed neuropathy (OPIDN), which results from the inhibition of neuropathy target esterase (NTE). This condition can lead to tingling and loss of sensitivity in the hands and feet, followed by progressive muscle weakness in both the upper and lower limbs, as well as ataxia (EFSA 2019; Song et al. 2021). NTE is an integral membrane protein found in neural tissue (Sepahi et al. 2023). Since NTE activity in nervous tissue correlates with its activity in lymphocytes, measuring NTE in lymphocytes can serve as a bioindicator (DFG 2008).

In addition, there is increasing evidence that low-dose chronic exposure of humans to OPPs might increase the risk of neurodevelopmental, attention deficit, and cognitive disorders in children (Antonangeli et al. 2023; Jokanović et al. 2023).

1.2.2 Toxicokinetics

The scientific literature has reported toxicokinetic characteristics of individual substances (WHO 1986a; ATSDR 1997; Egeghy et al. 2011; IARC 2015b).

Here is a brief summary. Generally, after ingestion, inhalation, or topical application, OPPs are rapidly absorbed by mammals; however, absorption is less efficient or delayed through the skin compared to oral exposure. For example, in experimental settings with humans, only around 3 and 7% of the applied chlorpyrifos and malathion doses were absorbed (IARC 2015b). OPPs distribute uniformly after absorption in experimental animals, with no evidence of accumulation.

OPPs follow similar metabolic pathways across species (Casida and Quistad 2004). For example, the oxidation and hydrolytic biotransformation of malathion are key enzymatic metabolic pathways. Following oral exposure, biotransformation of this substance occurs primarily in the liver and to a lesser extent in the small intestine. Malathion metabolites and their glucuronide or sulfate conjugates are mainly excreted in urine.

Overall, OPPs undergo rapid metabolism, resulting in a range of group-specific dialkylphosphate (DAP) and substance-specific metabolites that can be used for biomonitoring. Figure 1.7 illustrates an example of chlorpyrifos metabolism. Chlorpyrifos is metabolized by esterases into trichloro-2-pyridinol (TCPy), which is converted to DETP or bioactivated to chlorpyrifos oxon via cytochrome P-450, and then rapidly converted to DEP through TCPy. In contrast, diazinon is metabolized quickly into short-lived diazoxon or directly to 2-isopropyl-4-methyl-6-hydroxypyrimidine (IMPY), DETP, and DEP.

1.2.3 Occurrence in Urine

Table 1.4 lists examples of nonspecific dialkyl phosphates and specific metabolites and their parent substances that are suitable and applied for human biomonitoring.

Table 1.5 summarizes findings from recent studies on DAP metabolites in urine, categorized by adults and children/adolescents, along with specific OPP metabolites (see Table 1.6). DMP, DMTP, and DEP are the metabolites most commonly identified

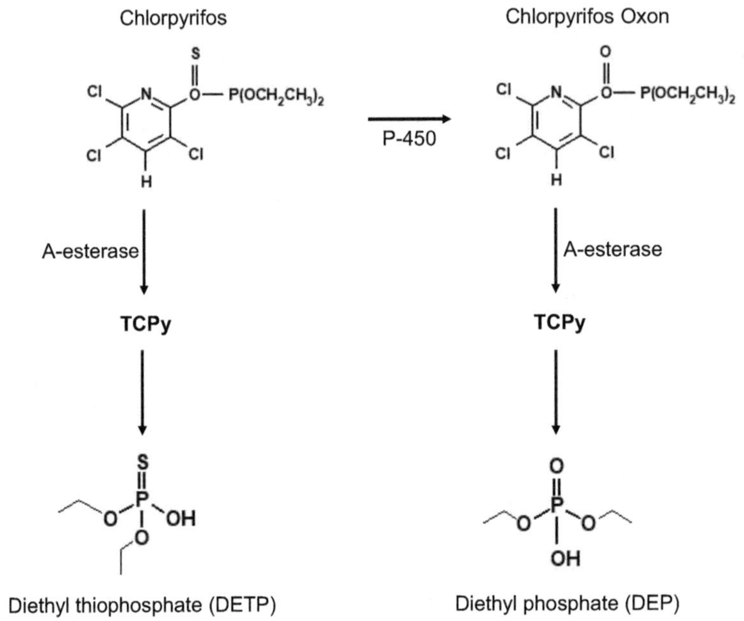

Fig. 1.7 Simplified scheme of chlorpyrifos metabolism (adapted from ATSDR 1997)

with higher median values. Notable differences exist in the concentration levels of individual substances due to variations in application ranges and quantities. Very high median values were reported for adults in an Australian study conducted in 2012/13, for adults in Israel in 2011, and for children in Japan in 2012/13. Regarding specific metabolites, TCPy and IMPY stand out, with the highest median values reported for Australian preschoolers in 2014/15.

Li and Kannan (2018) analyzed 322 adult urine samples collected from eight countries between 2006 and 2014. Median PNP concentrations were highest in China at 5.0 μg/l, followed by 4.7 μg/l in India, 2.8 μg/l in Korea, 2.6 μg/l in Vietnam, 2.1 μg/l in Japan, 1.7 μg/l in Saudi Arabia, and 1.6 μg/l in the US and Greece.

Using pooled samples from the Australian population, age was significantly associated with concentrations of DMDTP, DMTP, DETP, TCPY, and PNP, displaying a curvilinear relationship, with higher concentrations observed in younger individuals and those over 60 years old (Heffernan et al. 2016). A Spanish study involving 222 children aged 3–5 years found significantly higher levels of DEP, DETP, and DEDTP compared to children in the 6–11 year age group (Hernández et al. 2019). In a Chinese study, urinary DAP metabolites were analyzed in 436 mothers, along with 236 children aged 12 months and 237 children aged 24 months (Wang et al. 2017). Overall, concentrations in children were higher than those in their mothers, showing an increase in DAP concentrations from 12 to 24 months of age.

Some studies analyzing time trends are available. During cycles of the national Canadian Health Measures Survey (CHMS), median concentrations of DMP and

Table 1.4 Examples of organophosphate pesticides and their metabolites (extended according to Danish EPA 2016; HC 2021)

Metabolite name	Acronym	CAS-No	Parent pesticide
Unspecific metabolites			
Dimethyl phosphate	DMP	813-79-5	Azinphos-methyl, chlorpyrifos-methyl, dichlorvos, dicrotophos, dimethoate, fenthion, fenitrothion, malathion, methidathion phosmet, pirimiphos-methyl, temephos, tetrachlorvinphos
Dimethyl thiophosphate	DMTP	1112-38-5	Azinphos-methyl, chlorpyrifos-methyl, dimethoate, fenitrothion, fenthion, isazaphos-methyl, malathion, methidathion, oxydemeton-methyl, phosmet, pirimiphos-methyl, temephos
Dimethyl dithiophosphate	DMDTP	756-80-9	Azinphos-methyl, dimethoat, malathion, methidathion, phosmet
Diethyl phosphate	DEP	598-02-7	Chlorethoxyphos, chlorpyrifos, coumaphos, diazinon, disulfoton, ethion, parathion, phorate, sulfotepp, terbufos
Diethyl thiophosphate	DETP	2465-65-8	Chlorethoxyphos, chlorpyrifos, coumaphos, diazinon, disulfoton, ethion, parathion, phorate, terbufos
Diethyl dithiophosphate	DEDTP	298-06-6	Disulfoton, ethion, phorate, sulfotepp. terbufos
Specific metabolites			
3,5,6-trichloro-2-pyridinol	TCPy	6515-38-4	Chlorpyrifos

(continued)

Table 1.4 (continued)

Metabolite name	Acronym	CAS-No	Parent pesticide
2-isopropyl-4-methyl-6-hydroxypyrimidine	IMPY	2814-20-2	Diazinon
Malathion dicarboxylic acid	MDA	1190-28-9	Malathion
2-diethylamino-6-methyl pyrimidine-4-ol	DEAMPY	42487-72-9	Pirimiphos-methyl
4-nitrophenol	4-NP	100-02-7	Parathion, methylparathion

DEP decreased from 3.5 and 2.8 μg/l in 2009/11 to 1.6 and 2.1 μg/l in 2016/17, and further to 1.4 and 2.1 μg/l in 2018/19 (HC 2021). A detailed trend analysis of the data, based on geometric mean values, showed a significant decrease of 40% for DMP and a nonsignificant decrease of about 8% for DEP (Pollock et al. 2021). For the three most commonly detected DAPs, the time course of the 95th percentile in urine based on the representative NHANES study in the USA is shown in Fig. 1.8. It is clear that concentrations of DMTP have decreased significantly, DMP levels are declining again after an increase in the mid-2000s, and DEP is found at relatively constant levels in urine samples.

A Chinese study collected between 240 and 482 urine samples each year from 2018 to 2022, reporting a slight downward trend in TCPy, from a median of 2.92 μg/l in 2018 to 1.81 μg/l in 2022 (Jia et al. 2024). The median value of 4-PNP in urine increased from 4.47 μg/l in 2018 to 6.16 μg/l in 2020, then steadily decreased to 4.05 μg/l in this study. In contrast, young adults in Sweden (between 195 and 314 individuals) aged 17–21 years were studied in 2000, 2004, 2009, 2013, and 2017, revealing medians (95th percentiles) of 0.82 μg/l (3.5 μg/l), 0.84 μg/l (4.01 μg/l), 1.41 μg/l (5.21 μg/l), 1.11 μg/l (5.32 μg/l), and 0.92 μg/l (6.54 μg/l) for TCPy, respectively (Norén et al. 2020). An increasing temporal trend of 1.7% per year was found for this biomarker.

Some studies describe the differences between adults and children or mother–child pairs. DEP and DETP levels in mothers were statistically significantly higher than in children (Cequier et al. 2017). In Slovenia, the OPP metabolites DEAMPY, IMPY, MDA, and 4-PNP were statistically significantly higher in children than in their mothers (between 3 and 24%), except for TCPy, which was 26% lower (Bravo et al. 2020). In Norway, significant correlations were observed between mothers and children (6–12 years old) for all DAP (Cequier et al. 2017). In this study, the median concentrations in maternal urine samples were higher compared to those in children, while the 95th percentile was higher in children. Moreover, a Danish study indicated a tendency for higher DAP concentrations in children than in their mothers (Danish EPA 2016). The European Human Biomonitoring Initiative (HBM4EU) Aligned Studies summarize the results of urine samples collected from 3,576 children aged 6–12 years and 4,102 adults aged 20–39 years between 2014 and 2021 in four countries. TCPy geometric means (95th percentiles) of 0.89 μg/g crea (4.25 μg/g crea) in adults and 2.43 μg/g crea (10.8 μg/g crea) in children were reported (Govarts et al. 2023).

Table 1.5 Median concentrations of dialkyl organophosphate metabolites in urine in μg/l

References	N	Age (years)	DMP	DMTP	DMDTP	DEP	DETP	DEDTP	Country
Adults, general population									
Li et al. (2025)	425	≥ 20	4.27	3.42	< LOQ	2.92	1.12	< LOQ	China, 2018/19
HC (2021)	2526	3–79	1.4	1.0	0.10	2.1	0.27	< 0.07	Canada, 2018/19
Fišerová et al. (2021)	20	9–15	10.9	1.7	–	< 0.2	0.63	< 0.2	South Africa, 2018
Baumert et al. (2022)	330[a]	18–39	4.5	1.0	0.4	3.7	2.5	0.3	Thailand, 2017–19
CDC (2023)		> 6	1.3	0.64	0.11	2.2	0.12	< 0.1	USA, 2017/18
Heffernan et al. (2016)	24[d]	> 0	13.6[f]	10.6[f]	0.74[f]	6.18[f]	1.84[f]	< 0.5[f]	Australia, 2012/13
Nishihama et al. (2021)	4575	32[c]	2.9	3.3	< LOD	2.8	< LOD	< LOD	Japan, 2011–14
Wang et al. (2017)	436	28[c]	9.9	0.78	–	5.2	0.78	–	China, 2011–13
Cequier et al. (2017)	48	32–56	2.6	5.5	< 1.1	5.3	1.0	< 0.3	Norway, 2012
Abdeen et al. (2016)	148	16–43	3.7[f]	8.6[f]	0.4[f]	2.9[f]	0.9[f]	0.02[f]	Palestine, 2010–12
Berman et al. (2013)	120	41[c]	14.4[f]	9.5[f]	0.4[f]	2.1[f]	0.5[f]	0.03[f]	Israel, 2011
Danish EPA (2016)	145	31–52	2.3	3.1	< 0.3	4.0	< 0.3	< 0.3	Denmark, 2011
Forde et al. (2015)	150	18–44	1.4	0.8	< 0.3	1.5	< 0.3	< 0.3	Caribbean Islands, 2008–11
Kokkinaki et al. (2014)	86	30[c]	7.1	–	–	4.6	2.1	0.4	Greece, 2008/09
Children, adolescents									
Oya et al. (2017)	116	1.5–1.8	3.4	4.0	1.0	6.7	0.6	< 0.07	Japan, 2015

(continued)

Table 1.5 (continued)

References	N	Age (years)	DMP	DMTP	DMDTP	DEP	DETP	DEDTP	Country
Li et al. (2019)	20[e]	0–5	6.1	5.3	0.39	1.5	1.8	0.02	Australia, 2014/15
Osaka et al. (2016)	223	3	14.3	5.5	–	5.3	0.55	–	Japan, 2012/13
Wang et al. (2017)	235	1	4.3	0.9	–	2.4	0.6	–	China, 2011–13
Wang et al. (2017)	237	2	9.1	0.8	–	4.0	1.0	–	China, 2011–13
Cequier et al. (2017)	54	6–12	< 1.1	5.3	< 0.6	3.8	0.43	< 0.1	Norway, 2012
Danish EPA (2016)	144	6–11	4.0	3.2	< 0.3	4.8	< 0.3	< 0.3	Denmark, 2011
Hernández et al. (2019)	222	3–11	0.27 f	–	–	0.11[f]	0.09[f]	0.05[f]	Spain, 2010
Mat Sutris et al. (2016)	180	7–12	2.15	< 0.1	0.4	11.6	< 0.1	< 0.1	Malaysia
Bennett et al. (2022)	237[b]	2–5	7.5	4.1	0.3	3.0	0.4	< 0.05	USA, 2006–17
Myridakis et al. (2016)	500	4	< 0.24	4.0	< 0.05	2.0	1.1	< 0.03	Greece, 2007/08

[a] In early pregnancy; [b] group with typical development; [c] mean; [d] 24 pools of 2400 individuals; [e] 20 pools of 400 individual samples; [f] geometric mean; *LOD* limit of detection, *LOQ* limit of quantification

An attempt to identify the sources of internal exposure has been made in only a few studies. Thus, current or past work in agriculture, the husband's occupation as a farmer, and the family member's occupation as a farmer were not linked to increased dialkyl phosphate metabolite levels (Abdeen et al. 2016). Moreover, the risk of exposure to DEP and DMTP was higher in children whose parents reported using any pesticide at work (Babina et al. 2012). In Japan, 3-year-old children living with individuals employed in agriculture, pest control, and garden centers showed higher urinary DMP concentrations than those not living with individuals engaged in such activities (Osaka et al. 2016).

Bradman et al. (2015) collected urine samples over 16 consecutive days from children who consumed conventionally grown food for 4 days, organic food for 7 days, and then conventionally grown food for 5 days. The adjusted geometric

Table 1.6 Median concentrations of specific OPP metabolites in urine in μg/l

References	N	Age (years)	DEAMPY	MDA	4-PNP	IMPY	TCPy	Country, year
Deng et al. (2024)	159	1–6	–	–	1.97	–	0.91	China, 2022
Castiello et al. (2023)		14–18	–	0.31	–	0.27	< 0.13	Spain, 2017–19
Fiserová et al. (2021)	20	9–15	–	1.15	< 0.2	0.20	5.32	South Africa, 2018
Schoeters et al. (2022)	415	14–15	–	–	–	–	4.45	Belgium, 2017/18
Bustamante et al. (2025)	399	7–9	0.60	–	–	< 0.06	1.50	Poland, 2014–18
HC (2019)	2422	3–79	–	< 0.19	–	–	1.30	Canada, 2014/15
Bravo et al. (2019)	199	7	3.0	< 0.07	1.2	< 0.01	0.36	Italy, 2014/15
Li et al. (2019)	20[c]	0–5	–	< 0.06	1.1	12	9.7	Australia, 2014/15
Bravo et al. (2020)	164	7–8	0.30	< 0.07	0.84	< 0.01	0.06	Slovenia, 2014/15
Bravo et al. (2020)	168	39[a]	< 0.02	< 0.07	0.61	< 0.01	0.18	Slovenia, 2014/15
Bravo et al. (2022)	42	49[a]	1.1	< 0.07	1.8	< 0.01	2.4	Spain
Gao et al. (2022)	25	–	–	0.96[a]	–	0.08	0.45	Belgium
Gari et al. (2018)	80	–	0.81	< 0.07	1.3	< 0.01	2.2	Spain
CDC (2023)	2648	> 6	–	–	0.61	< 0.1	–	USA, 2013/14
Heffernan et al. (2016)	24[b]	> 0	–	1.0[d]	1.76[d]	0.38[d]	23.0[d]	Australia, 2012/13
Davis et al. (2013)	55	–	1.02[a]	1.18[a]	1.40[a]	0.75[a]	1.87[a]	USA, 2012
Bliznashka et al. (2023)	284	23[a]	–	–	18.7[e]	–	3.2[e]	Bangladesh, 2008–11
Li and Kannan (2018)	322		–	0.4	2.7	0.3	0.2	Eight countries worldwide, 2006–14
Li et al. (2020)	619	18–44	–	0.22	0.64	2.7	0.60	USA, 2007–09

[a] mean; [b] 24 pools of 2400 individuals; [c] 20 pools of 400 individual samples; [d] geometric mean; [e] in μg/g creatinine

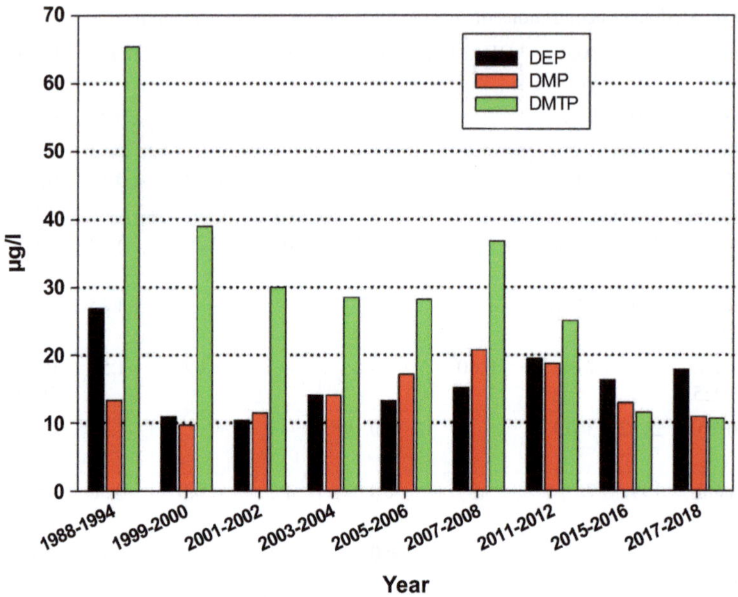

Fig. 1.8 Time-trend of the concentrations (95th percentiles) of three alkylphosphate pesticides in urine in µg/l (Adapted from Clune et al. 2012 and CDC 2023)

mean concentrations of DAP during the organic phase were significantly lower for all children. In a Norwegian study, DAP was positively correlated with fruit but not vegetable consumption in adults (Cequier et al. 2017). Within the Japan Environment and Children's Study, a nationally representative birth cohort study, 4575 maternal urine samples were analyzed, and the months of urine sampling, consumption of apples, and maternal body mass index were essential factors for predicting urinary DAP concentrations (Nishihama et al. 2021).

In a Spanish study of 125 children aged 6–11 years, multivariate regression analysis revealed that age, vegetable consumption, and residential use of pesticides were predictors of exposure for TCPy and IMPY. Meanwhile, location and vegetable consumption were associated with DEP concentrations (Roca et al. 2014).

1.2.4 Occurrence in Breast Milk

So far, few studies on the presence of OPPs in breast milk have been published in the scientific literature. Most of these involve chlorpyrifos, and often, the concentrations are at or below the detection limit, and only in areas where pesticides are known to be used are the levels somewhat higher.

A total of 43 samples were obtained from 21 women living in an urban area and 13 from an agricultural region in California during 2002–2004 and 2007/08 (Weldon et al. 2011). Chlorpyrifos and chlorpyrifos-methyl were detected in all samples and in 67 and 23% of the samples, respectively, with medians of 0.025 ng/g milk and 4.0 ng/g in the urban area, and 0.028 ng/g and < 0.0005 ng/g in the agricultural region. No significant differences were found between the two sampling sites. Fonofos, disulfoton, and diazinon were detected only in some samples, with mean LODs ranging from 0.0001 to 0.006 ng/g milk. In another study from California, pesticides were analyzed in 79 breast milk samples collected 35–42 days after birth (Pedersen et al. 2021). The mean values (maxima) in this study were as follows: azinphos methyl 1.86 μg/l (34.2 μg/l), malathion 0.05 μg/l (0.85 μg/l), oxydemeton methyl 0.03 μg/l (0.82 μg/l), chlorpyrifos 0.03 μg/l (0.38 μg/l), diazinon 0.02 μg/l (0.31 μg/l), and chlorpyrifos-methyl 0.02 μg/l (0.58 μg/l).

Thus far, results have been reported from two studies in the Netherlands conducted between 2011 and 15 ($n = 120$) and in northwest Spain from 2016 to 18 ($n = 81$) (Čechová et al. 2017; Sanguos et al. 2023). For chlorpyrifos, the mean values were 0.63 and 2.8 ng/g lw. In the Spanish study, the mean levels (interquartile ranges) for diazinon were 12 ng/g lw (0.002–5.8 ng/g lw), for parathion 5.9 ng/g lw (0.02–3.8 ng/g lw), and for fenthion 0.16 ng/g lw (0.002–0.075 ng/g lw).

In Sri Lanka in 2005/6, breast milk samples were collected from mothers living in a farming community at the end of the pesticide spray season ($N = 41$) and just before the spray season ($n = 25$) and analyzed for chlorpyrifos, dimethoate, diazinon, and fenthion (Samarawickrema et al. 2008). Only 0.632 μg/l of dimethoate was detected during the spray season in one sample, while all other pesticides were below the detection limit of 0.5 μg/l. In contrast, mean concentrations in colostrum ($n = 25$) and mature milk samples ($N = 8$) were 85.9 and 26.7 ng/g lw for dimethoate, 48.0 and 745 ng/g lw for ethion, and 4.3 and 37.3 ng/g lw for chlorpyrifos, respectively, in breast milk from a rural area in India in 2006 (Srivastava et al. 2011). In another study from India, 53 samples were analyzed in 2011/12, yielding mean chlorpyrifos values of 84.1 ng/g lw (Bedi et al. 2013). Naksen et al. (2016) analyzed 30 breast milk samples collected from participants who lived and worked in an agricultural community in Fang District, Thailand, in 2013, where OP pesticides have been used intensively. Of the 11 OP pesticides, only 5 were detectable. Chlorpyrifos was found in 73% of samples at levels ranging from < 0.22–0.46 μg/l, malathion in 23% at < 0.12–0.35 μg/l, ethion in 20% at < LOD-0.14 μg/l, dicrotophos in 3.3% at < LOD-0.14 μg/l, and dimethoate in 3.3% at < LOD-0.12 μg/l. Higher levels were reported by Brahmand et al. (2019), who discovered a mean chlorpyrifos value of 2.1 μg/l (0.2–6.5 μg/l) in 61 Iranian breast milk samples in 2017.

In a study conducted in 2018/19, milk samples from 40 refugees in Lebanon were tested for various pesticides. Chlorpyrifos was detected in approximately 1% of the samples, exceeding the limit of quantification of 5 μg/l, with a mean value of 2.1 μg/l in the positive samples (Smadi et al. 2019).

In southern Brazil, bromophos, chlorpyrifos, fenitrothion, and malathion were not detected above the limit of 3 ng/g lw in 20 samples collected 3–8 weeks after birth (Viera et al. 2022).

1.3 Carbamates and Dithiocarbamates

Carbamates are N-substituted esters of carbamic acid and chemically represent N-methylcarbamic acid esters. The general chemical structure is illustrated in Fig. 1.9. Three classes can be distinguished based on their chemical structure. While carbamates used as insecticides have a methyl group as R1, they are generally stable, with low vapor pressure and low water solubility. In contrast, the R1 in carbamate herbicides (and sprout inhibitors) and fungicides features an aromatic group and a benzimidazole group, respectively (WHO 1986b; Mdeni et al. 2022). R2 is an aromatic or aliphatic moiety.

Since the introduction of carbaryl in 1958, the share of carbamates in the total quantity of insecticides produced has steadily increased. Many carbamates act systemically, meaning they are distributed throughout the plant via transport systems, allowing control of even difficult-to-reach pests on both the shoot and root of the plant being protected (ATSDR 2023b). This property is valuable for use as a soil insecticide and nematicide. Technically important carbamate insecticides are listed in Table 1.7.

Furthermore, dithiocarbamates (DTC), such as ziram, maneb, and mancozeb, are commonly used as fungicides in vineyards, potato farms, and the floricultural industry. These compounds contain a carbamate group and two sulfur atoms in their molecules (see Fig. 1.10) (Stadler et al. 2022). The usage of mancozeb and ziram has remained steady over the past decade, while maneb has been phased out in the U.S., and mancozeb lost its EU approval as an active ingredient in plant protection products effective 2021 (Stadler et al. 2022). For instance, mancozeb is the common name for manganese ethylenebis (dithiocarbamate) (polymeric) complex with a zinc salt (EFSA 2020).

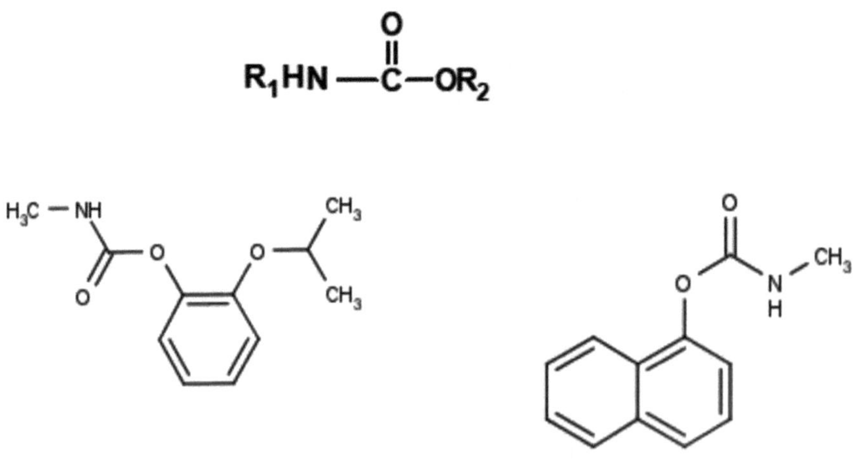

Fig. 1.9 General structure of carbamate insecticides (above) and propoxur (bottom left) and carbaryl (bottom right)

Table 1.7 Typical carbamates and their metabolites

Pesticide	CAS No.	Metabolite	Abbreviation
Aldicarb	116-06-3		
Aldicarb sulfone	1646-88-4		
Aldicarb sulfoxide	1646-87-3		
Bendiocarb	22781-23-3	2,2-dimethylbenzo[1,3]dioxol-4-ol	DMBDO
Benfuracarb	82560-54-1	Carbofuranphenol (2,2-dimethyl-2,3-dihydrobenzofuran-7-ol)	CFP
Carbaryl	63-25-2	1-naphthol	1-Nap
Carbofuran	1563-66-2	Carbofuranphenol (2,2-dimethyl-2,3-dihydrobenzofuran-7-ol)	CFP
Carbosulfan		Carbofuranphenol (2,2-dimethyl-2,3-dihydrobenzofuran-7-ol)	CFP
Formetanate hydrochloride	23422-53-9		
Furathiocarb	65907-30-4	Carbofuranphenol (2,2-dimethyl-2,3-dihydrobenzofuran-7-ol)	CFP
Methiocarb	2032-65-7	3,5-dimethyl-4-(methylthio)phenol	DMMTP
Methomyl	16752-77-5		
Oxamyl	23135-22-0		
Pirimicarb	23103-98-2	2-amino-5,6-dimethyl-pyrimidine-4-ol; 2-(Dimethylamino)-4-hydroxy-5,6-dimethylpyrimidine	ADMP; DMADMP
Propoxur	114-26-1	2-isopropoxyphenol	2-IPP
Thiodicarb	59669-26-0		

Modified from Denghel and Göen (2018)

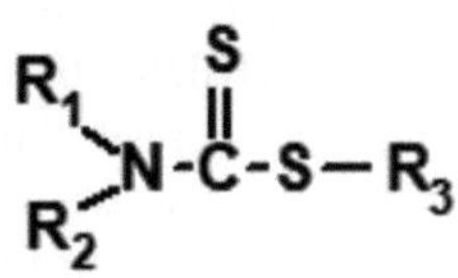

Fig. 1.10 General structure of dithiocarbamates

1.3.1 Health Effects

Carbamates' mode of action involves the reversible inhibition of the enzyme acetylcholinesterase (AChE) at the postsynaptic membrane, resulting in an accumulation of the neurotransmitter acetylcholine (Araújo et al. 2023). However, unlike organophosphate insecticides, carbamylation is unstable, leading to a faster regeneration of AChE, which results in lower toxicity compared to OPPs (US EPA 2023b).

The most well-studied and significant effects are the neurological effects mediated by the inhibition of AChE (ATSDR 2023b). Poisoning has been reported, particularly in workplaces, such as when used in spray form indoors, with symptoms starting within a few minutes of exposure and lasting for several hours. Additionally, results from subchronic and chronic animal toxicity studies have identified several more effects on the kidneys, liver, or spleen, as well as decreased body weight gain. Nevertheless, AChE inhibition remains a sensitive effect for all studied carbamate insecticides (ATSDR 2023b). The US EPA concluded that a common mechanism of toxicity exists for N-methyl carbamates and that dose addition is a reasonable and appropriate approach for estimating the cumulative risks of this group (ATSDR 2023b). Only aldicarb and carbaryl were classified in 1987 in Group 3 (not classifiable as to carcinogenicity to humans), while others were not classified (IARC 2012).

In rats, some DTCs induce dose-dependent adverse effects on reproduction and endocrine structures and functions at high doses, thereby reducing reproductive capacity (WHO 1988). Furthermore, elevated levels of DTC impact the thyroid gland in rats. In long-term oral studies, the thyroid (effects on thyroid hormones, thyroid hypertrophy, and hyperplasia) is the primary target of mancozeb in rodents, and the substance is classified for developmental toxicity as Reproduction 1B, based on malformations observed in rat developmental toxicity studies (EFSA 2020). Additionally, for certain DTCs like disulfiram and diethyldithiocarbamate, the application, for instance, as an anti-cancer agent for infectious diseases and against substance abuse, is being discussed (Kaul et al. 2021).

1.3.2 Toxicokinetics

The toxicokinetic characteristics of carbamates are compiled in some reviews (WHO 1986b; ATSDR 2023b; US EPA 2023b). In brief, carbamates are well absorbed into the organism through inhalation and oral exposure, while dermal absorption is limited. They are widely distributed in organs and tissues and reach the nervous system due to their lipophilicity. The first step in the metabolism of carbamates is the hydrolysis of carbamic acid. Additionally, oxidation processes by mixed functional oxidases (MFO) have been described, resulting in various reaction products because of the different functional groups. Therefore, the metabolism of carbaryl leads to at least 15 different compounds in mammals through a range of oxidative and hydrolytic reactions. Elimination half-lives of several hours have been observed in animal studies.

Overall, DTCs can be absorbed by the body through the skin, mucous membranes, and respiratory system, as well as rapidly from the gastrointestinal tract. However, metal-complexed alkylene bisdithiocarbamates are poorly absorbed from both the gastrointestinal tract and skin (WHO 1988). The DTC mancozeb, for instance, is partially (50%) but rapidly (3–6 h) absorbed after a single oral administration in rats, and is widely distributed, with the thyroid exhibiting the highest levels that are

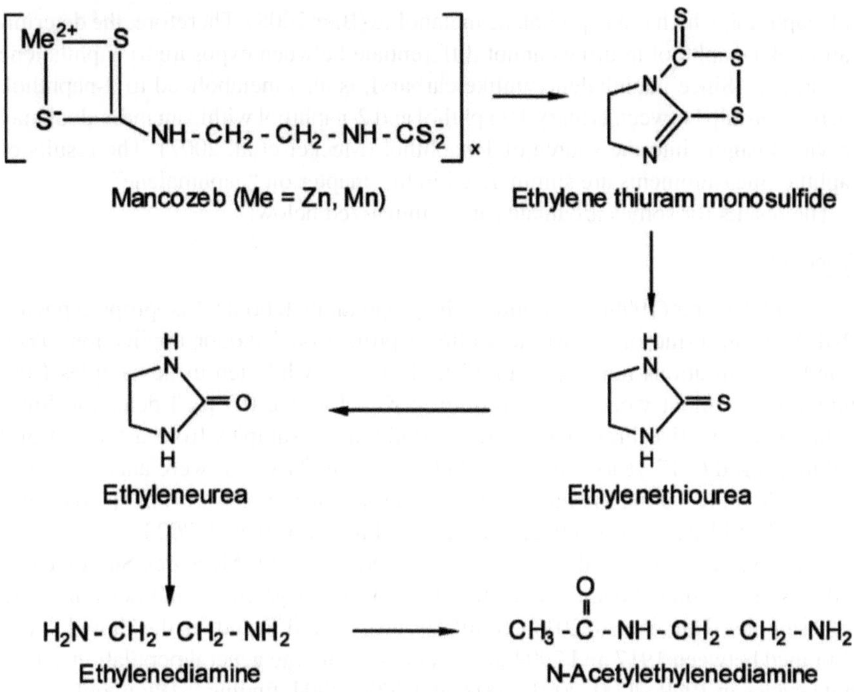

Fig. 1.11 Metabolism of mancozep in mammals (from Colosio et al. 2002)

quickly excreted, mainly via urine and feces (EFSA 2019). Mancozeb undergoes extensive metabolism in experimental animals through hydrolysis and oxidation, leading to the main urinary metabolites: ethylenethiourea (ETU), ethylenurea (EU), ethylenediamine (EDA), and N-acetyl EDA (see Fig. 1.11) (EFSA 2020).

To date, only one study on toxicokinetics in humans has been published. In this study, the elimination kinetics of DDHP and MDHP were examined in five volunteers who received a single oral dose of pirimicarb (Sams et al. 2010). Total MDHP and DDHP recovered over 48 h accounted for 74% (range 32–123%) of the administered dose, with elimination half-lives ranging from 2.8 to 4.6 h.

1.3.3 Occurrence of Carbamates in Urine

Similar to organophosphate pesticides, there are both specific and nonspecific metabolites of carbamate. For instance, the particular metabolite 2-isopropoxyphenol (2-IPP) can be detected in urine in the case of propoxur, while the metabolite carbofuranphenol can be produced from the substances benfuracarb, carbofuran, carbosulfan, and furathiocarb. This is also true for carbaryl, whose most predominant metabolite is 1-naphthol, which is a naphthalene metabolite (Barr 2008). Therefore, the determination of 1-naphthol in urine cannot differentiate between exposure to naphthalene or carbaryl. Since naphthalene, unlike carbaryl, is also metabolized to 2-naphthol, the relationship between urinary 1-naphthol and 2-naphthol within an individual may provide insights into the source of 1-naphthol (Meeker et al. 2007). The results of naphthol measurements are summarized in the chapter on "naphthalene".

The results for some carbamates are summarized below:

Propoxur

Hardt and Angerer (1999b) determined the propoxur metabolite 2-isopropoxyphenol (2-IPP) in the urine of six workers after a professional indoor application. They found concentrations ranging from 37 to 170 μg/l, while ten urine samples from individuals without workplace exposure were below the 0.5 μg/l detection limit. In the representative French ESTEBAN study, urine samples from a total of 500 children (aged 6–17 years) and 899 adults (aged 18–74 years) were analyzed from 2014 to 2016, revealing medians (95th percentiles) of < 0.05 μg/l (0.30 μg/l) and < 0.05 μg/l (0.11 μg/l), respectively, for 2-IPP (Tagne-Fotso et al. 2023).

2-IPP was analyzed in the 2009–2011 Canadian Health Measures Survey cycle and was only found above the detection limit of 0.05 μg/l in a small portion of the 2560 samples (Haines et al. 2017). Similarly, the NHANES study in the United States examined between 1917 and 2994 urine samples from the general population across three cycles in 1999–2000, 2001–2002, and 2003–2004, finding 2-IPP in only a few samples exceeding the detection limits of 1.1, 0.4, and 0.4 μg/l, respectively (CDC 2023).

2-IPP was detected in 13% of 150 random samples collected from 10 Caribbean countries between 2011 and 2018, with a median of < 0.05 μg/l (0.025–0.71 μg/l) (Forde et al. 2015). Urine samples were obtained from 11 children aged 4–51 months and 18 adults in China, revealing mean (95th percentile) 2-IPP levels of 0.01 μg/l (0.04 μg/l) and 0.19 μg/l (0.59 μg/l), respectively (Lu et al. 2015b).

Berman et al. (2016) studied 42 adults from northern Israel between 2013 and 2014, detecting 2-IPP above 0.5 μg/l (LOD) in 29% of samples, with a geometric mean of 0.72 μg/g crea and a 95th percentile of 3.5 μg/g crea.

Carbofuran

Petropoulou et al. (2006) studied ten occupationally unexposed smokers and nonsmokers in Greece and found no traces of carbofuranphenol-3-keto, carbofuran-3-keto, 3-hydroxycarbofuran, or carbofuranphenol (CFP) above the detection limits of 0.04–0.08 μg/l. In contrast, concentrations ranging from 39.7 to 113 μg/l for CFP and 1.22–2.53 μg/l for carbofuranphenol-3-keto were reported in five urine samples from farmers. The carbofuran metabolite 3-hydroxycarbofuran was detected above the detection limit of 10 nmol/l in only two of 130 urine samples from healthy UK children aged 1.0–4.2 years (Jones et al. 2014). The representative French ESTEBAN study examined urine from 500 children (ages 6–17) and 899 adults (ages 18–74) between 2014 and 2016. CFP was detected above the limit of 5.0 μg/l in only 2% of the samples in each case (Tagne-Fotso et al. 2023).

CFP was also analyzed in the 2009–2011 Canadian Health Measures Survey cycle and was found above the detection limit of 0.05 μg/l in only a few of the 2557 samples (Haines et al. 2017). Similarly, in the NHANES study in the USA, 1917 to 2994 urine samples from the general population were analyzed across three cycles: 1999–2000, 2001–2002, and 2003–2004. CFP was detected in only a few samples above the detection limit of 0.4 μg/l (CDC 2023). In a study of 150 random urine samples from 10 Caribbean countries, CFP was found above the detection limit of 0.1 μg/l in only one sample from Belize (Forde et al. 2015).

Higher concentrations are partly reported in Asia compared to Europe and the Americas. For example, Hung et al. (2019) analyzed 30 urine samples from Taiwanese individuals and detected CFP in 25% of the samples, with a median of 3.4 μg/l (maximum: 34.5 μg/l). Zhang et al. (2022) found lower medians (maximum) of 0.9 μg/l (395 μg/l) in 1100 mothers in China from 2009 to 2016 and 0.4 μg/l (163 μg/l) as well as 0.32 μg/l (8239 μg/l) in their three- and seven-year-old children, respectively. Urine samples from 11 children aged 4–51 months and 18 adults were collected in China, revealing median (95th percentile) CFP levels of 0.19 μg/l (14 μg/l) and 0.17 μg/l (2.32 μg/l), respectively (Lu et al. 2015b).

Methiocarb

Methiocarb sulfoxide is specific to methiocarb and was detected above the detection limit of 10 nmol/l in only four of 130 urine samples from healthy UK children aged 1.0–4.2 years (Jones et al. 2014).

Pirimicarb

Hardt and Angerer (1999a) identified the three primicarb metabolites DDHP (2-dimethylamino-5,6-dimethyl-4-hydroxypyrimidines), MDHP (2-methylamino-5,6-dimethyl-4-hydroxypyrimidines), and ADHP (2-amino-5,6-dimethyl-4-hydroxypyrimidines) in the urine of seven workers after the professional application of pirimicarb. Concentrations ranged from 0.9 to 2.2 μg/l for DDHP, 26.1–60.0 μg/l for MDHP, and 4.2–17.3 μg/l for ADHP. In contrast, all ten urine samples from individuals without occupational exposure were below the detection limits of 0.5 μg/l (DDHP), 1.0 μg/l (MDHP), and 4.0 μg/l (ADHP). Conversely, in the UK, MDHP was detected in 40% of urine samples from 95 healthy children aged 1.0–4.2 years, with a 95th percentile of 20 μg/l and a maximum of 99 μg/l (Jones et al. 2014).

1.3.4 Occurrence of DTC Metabolites in Urine

Thus far, only measurements of the nonspecific mancozep, manep, and thiram metabolite ethylenethiourea (ETU), along with the propinep metabolite propylenethiourea (PTU), are available in the scientific literature. Propineb is a compound belonging to the group of zinc compounds with DTC. Regarding ETU, it is important to consider that it can be a relevant impurity of up to 0.3% in the product mancozep and therefore also appears directly in the environment (EFSA 2020).

Colosio et al. (2002) found no ETU levels above the LOQ of 0.5 μg/g crea when analyzing 13 subjects without occupational exposure in Italy. Between 38 and 51 unexposed adults were studied across four European countries. Medians of < 0.5 μg/g crea were observed for ETU in Finland, < 0.5 μg/g crea in Italy, 0.7 μg/g crea in the Netherlands, and < 0.5 μg/g crea in both of the Bulgarian studies (Fustinoni et al. 2008). The 90th percentiles ranged from 2.9 to 3.8 μg/g creatinine at all but one study site, from 2.9 to 3.8 μg/g crea, with only a notably high 90th percentile of 80.2 μg/g crea in one of the Bulgarian sites. In the Spanish INMA cohort, a total of 201 children aged 14–18 years were studied between 2017 and 2019 at two sites, with an overall median of 0.15 μg/l ETU (< 0.24–19.4 μg/l) reported (Castiello et al. 2023). Norén et al. (2020) analyzed between 149 and 254 urine samples from Swedish adults at five time points between 2000 and 2017, observing an increase in ETU medians from 0.29 to 0.40 μg/l from 2000 to 2004, followed by a continuous decrease to 0.34 μg/l, then to 0.25 μg/l, and finally to 0.18 μg/l in 2017. The 95th percentile exhibited a similar trend. In 2017, ETU concentrations varied from < 0.02 μg/l to a maximum of 7.3 μg/l in this study. PTU was not detected above the LOD of 0.1 μg/l in 95% of the samples.

In the CHAMACOS study, urine was collected from approximately 450 women at each time point between 1999 and 2001 in an area heavily dominated by agriculture in California (Castorina et al. 2010). ETU was detected in only 24 and 8% of samples exceeding the limit of detection (LOD) of 0.1 μg/l in this study, with 95th percentiles at 1.5 and 0.4 μg/l, respectively. In 2011, medians of 1.7 and 0.72 μg/g crea (0.38–6.9 μg/g crea) were reported for two groups of 15 individuals each who had not been occupationally exposed, in Florida, US (Runkle et al. 2013). Urine samples were obtained from 129 non-farmworkers in 2013 and from 78 non-farmworkers in 2013 in the US, where ETU and PTU were found with means of 5.0 μg/g crea and 6.0 μg/g crea in 2012, as well as 0.40 μg/g crea and 0.43 μg/g crea in 2013 (Arcury et al. 2018a). Moreover, the urine samples of 55 Hispanic non-farmworkers collected in 2012 and 28 collected in 2013 showed ETU medians of 1.4 and 0.9 μg/l, respectively, while PTU was not detected above the LOQ of 0.2 μg/l (Arcury et al. 2018b). In the 2003–2008 NHANES, a nationally representative survey in the US, ETU was analyzed in 4613 adult urine samples, with a median of 0.15 μg/l found and 8% of samples above the LOD of 0.07 μg/l (Stadler et al. 2022). This study observed similar findings with a median of 0.15 μg/l in 2708 children and adolescents aged 6–18 years old.

In Costa Rica, a total of 140 children aged 6–9 years living near banana plantations were studied in 2007. ETU medians of 1.2 μg/l (< 0.1–11.0 μg/l) were found in a single measurement, and 1.6 μg/l (< 0.1–34.5 μg/l) in repeated measurements (van Wendel de Joode et al. 2016). In the ISA (Infantes y Salud Ambiental) study, a prospective birth cohort study of pregnant women and their infants living near banana plantations in Costa Rica, higher median levels of 3.35 μg/l (0.81–127 μg/l) were found during pregnancy from 2010 to 2013 (Mora et al. 2020). In 2011, Handal et al. (2016) examined the urine of 10 pregnant women in Ecuador who had not worked on farms for five years and reported medians of 5.24 μg/l (< 0.1–23.1 μg/l) for ETU and 0.73 μg/l (< 0.1–6.3 μg/l) for PTU.

1.3.5 Occurrence in Breast Milk

Overall, carbamates in breast milk have been studied only to a limited extent. When carbamates are found in breast milk samples, their concentrations appear to be low, but the range of values shows greater variation.

For example, Weldon et al. (2011) collected 21 breast milk samples from mothers aged 30–40 years in an urban area and 13 samples from an agricultural region in California between 2007 and 2008. They found propoxur in 67 and 8% of these samples, exceeding the median detection limits of 0.96 pg/g and 1.7 pg/g milk, respectively. The median (range) of the urban samples was 4.3 pg/g (< 0.59–127 pg/g). In this study, bendiocarb was detected in only 19% of the samples from the urban area, with a median detection limit of 0.7 pg/g, but was absent in the agricultural samples. In another study conducted in California, pesticides were analyzed in 79 breast milk samples collected 35–42 days after birth (Pedersen et al. 2021). Carbaryl and oxamyl were present above the LOQ in 88% of the samples, with a mean value of 0.38 μg/l (0.011–9.37 μg/l), and in 23% of the samples with a mean of 0.28 μg/l (0.137–5.10 μg/l), respectively. In 10 breast milk samples collected in 2010–2011 in Atlanta, USA, bendiocarb was detected in 10% of the samples with a maximum of 0.11 μg/l, while fenobucarb was not detected (Chen et al. 2014).

Du et al. (2017) analyzed 40 Australian breast milk samples from 2013 to 2015 and could not find carbamates like methiocarb, pirimicarb, and Indoxacarb above the detection limits of 0.2–2.0 μg/l milk.

1.4 Glyphosate and Glufosinate

Glyphosate (Gly; N-(phosphonomethyl) glycine; CAS No. 1071-83-6) is a colorless and odorless crystalline powder under environmental conditions. The structural formula of Gly is shown in Fig. 1.12.

For decades, glyphosate products have been used as systemic herbicides in agricultural and residential settings. They are applied against emerging weeds in all crops and used as foliar sprays for desiccation in cereals and oilseeds (IARC 2015b). Gly inhibits a 5-enolpyruvyl shikimate-3-P synthase, which is essential in plants and some microorganisms but is not expressed in animals and humans (Rodríguez-Gil et al. 2021). Initially, the use was limited to post-harvest applications, but the development of glyphosate-resistant crops has broadened its applications. One major US manufacturer has announced that it will cease selling it for nonagricultural use in 2023.

Glyphosate is marketed as an isopropylamine, ammonium, or sodium salt in water-soluble concentrates or granules (IARC 2015b). Commercially used glyphosate-based herbicides are complex mixtures. Quantitatively relevant

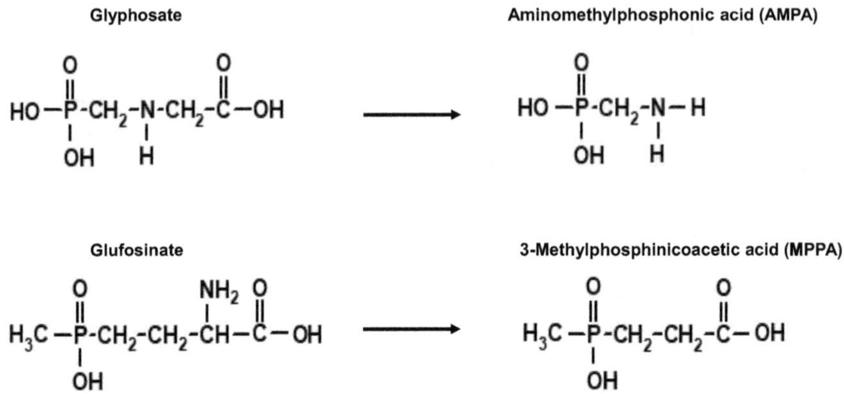

Fig. 1.12 Chemical Structures of glyphosate and glufosinate and their hydrolysis products (adapted from Kazui et al. 2014)

impurities may include formaldehyde, N-nitrosoglyphosate, and N-nitroso-N-phosphonomethylglycine. The finished product may contain various nonionic surfactants, such as polyethyloxylated tallow amine (POEA, CAS No: 24911-53-5) (Tush et al. 2013). POEA enhances glyphosate's herbicidal efficacy by facilitating its penetration of the plant cuticle. There is increasing evidence that POEA has toxic effects, particularly on aquatic systems and soil organisms (e.g., Mesnage et al. 2022; Mikó and Hettyey 2023).

In 1974, the total use of Gly in the USA as an active ingredient was approximately 0.635 million kg. By 2014, total annual glyphosate use had risen to around 125 million kg, with agriculture accounting for about 90% of that usage (Benbrook 2016). Globally, glyphosate use increased more than 12-fold, from approximately 67 million kilograms in 1995 to 826 million kilograms in 2014.

Gly has limited potential to persist in the environment and is inactivated by sorption to soil particles and microbial degradation in soils and water (ATSDR 2020). The primary degradation product in the environment and in humans is aminomethylphosphonic acid (AMPA, CAS No. 1066-51-9). A thorough overview of ecotoxicology and environmental fate is provided by Rodríguez-Gil et al. (2021).

Glufosinate (Gluf) is the ISO common name for 4-[hydroxy(methyl)phosphinoyl]-DL-homoalanin, but glufosinate ammonium salt (CAS No. 77182-82-2) is utilized in the formulated product (EFSA 2005a, b). The structural formula of Glu is illustrated in Fig. 1.12. Glufosinate ammonium consists of a mixture of the D and L enantiomers and acts as a nonselective herbicide by inhibiting glutamine synthetase in plants. It has been widely employed in agriculture to manage glyphosate-resistant weeds (Ashley-Martin et al. 2023).

1.4.1 Health Aspects

Animal experimental data indicate a low level of acute toxicity for Gly. Previous results on toxicity in mammals—except humans—after long-term exposure showed effects such as body weight impairment, reduced food consumption or efficiency, and increased alkaline phosphatase activity (ATSDR 2020).

The carcinogenic effects of glyphosate have been discussed intensely and controversially in recent times. In 2015, the WHO's International Agency for Research on Cancer classified glyphosate as a probable carcinogen to humans (group 2A) (IARC 2015b). Some rodent studies indicated a positive trend in the incidence of renal tubule carcinomas, adenomas, as well as liver and pancreatic adenomas. Regarding epidemiological studies, IARC concluded that case–control studies on non-Hodgkin lymphoma (NHL) showed no effects or statistically insignificant effects, or they were not adjusted for exposure to other pesticides. More recently, in the large Agricultural Health Study (AHS), a prospective cohort study with 54,251 applicators, no association was found between glyphosate exposure and solid tumors (Andreotti et al. 2018). Additionally, an international consortium that pooled three studies from Norway, France, and the United States, which including data from 316,270 farmers accumulating 3,574,815 person-years at risk, found no significant association between Hodgkin lymphoma and exposure to glyphosate (Kim et al. 2023). Once again, pooled data from 10 case–control studies participating in the International Lymphoma Epidemiology Consortium, which included 9,229 cases and 9626 controls from North America, the European Union, and Australia, found no association with all NHL, but an association with follicular lymphoma that was limited to short-duration use (De Roos et al. 2022).

In contrast to the IARC assessment, the Joint Food and Agriculture Organization FAO/WHO Meeting on Pesticide Residues found that Gly was not carcinogenic in rats and only at very high doses, not relevant to human exposure, if any (JMPR 2016). Given the absence of genotoxicity by the oral route in mammals and considering the epidemiological evidence from occupational exposures, they concluded that glyphosate was unlikely to pose a carcinogenic risk to humans. This view was later shared by the US Environmental Protection Agency (US EPA), the European Food Safety Authority (EFSA), the European Chemicals Agency (ECHA), and Health Canada (CARC 2015; US EPA 2017a; EFSA 2015, 2023; ECHA 2022; HC 2017). Tarazona et al. (2017) and Benbrook (2019) gave a detailed comparison of these differences.

In 2015, EFSA established an ADI (acceptable daily intake) of 0.5 mg/kg bw based on developmental toxicity (EFSA 2015). The U.S. Environmental Protection Agency and the ATSDR have determined, based on chronic intake, a Reference Dose (RfD) and a Minimal Risk Level (MRL) of 1.0 mg/kg bw, respectively (US EPA 2017a; ATSDR 2020).

Primary information is available for glufosinate ammonium salt. For this compound, EFSA (2005a, b) summarizes the toxicological and health effect data as follows: The main long-term effects observed in experimental animals included changes in hematological and biochemical parameters, increased kidney weight, and no evidence of genotoxicity or carcinogenicity, direct impact on reproductive performance or fertility, or neurotoxicity. Nevertheless, glufosinate ammonium induced pre- and post-implantation losses and was classified by ECHA in category 2 (suspected human reproductive toxicant) and category 3, R62 (possible risk of impaired fertility). Several human poisoning cases due to (suicidal) ingestions of glufosinate ammonium have been described in the literature, presenting dose-dependent symptoms such as nausea, vomiting, diarrhea, abdominal pain, tremors, hypotonia, bradycardia or tachycardia, muscle weakness, drowsiness, loss of consciousness and coma, convulsions, and respiratory arrest (JECFA 2012).

1.4.2 Toxicokinetics

After oral administration, Gly is rapidly absorbed at an average of about 27% of the applied dose in animal experiments (ECHA 2022). Excretion remains largely unchanged through the urine in the first 48 h and is nearly complete within 72–168 h or via the feces. In humans and mammals, Gly undergoes little effective metabolism. Accumulation has not been observed in laboratory animals. Uptake following dermal exposure is estimated to be low, and after inhalation, absorption appears possible but relies on limited data (ATSDR 2020).

In the first study, 12 healthy volunteers consumed a test meal with known Gly and AMPA concentrations (Zoller et al. 2020). Of the ingested Gly dose of 193 μg (median), 0.91% was excreted as Gly, with a median half-life of 9.2 h. For AMPA, 23% of the 1.6 μg dose was excreted in urine, assuming that only limited glyphosate metabolism to AMPA occurred. If all the excreted AMPA was a glyphosate metabolite, this corresponds to 0.3% of the glyphosate dose on a molar basis. Additionally, Faniband et al. (2021) studied Gly excretion in three healthy Swedish subjects (one female and two males) who received a single dose of 50% of the ADI. Gly excretion followed first-order kinetics and exhibited two-phase excretion, with a half-life of 6–9 h in the rapid phase and 18–33 h in the slower phase. The total dose recovered as unchanged Gly in the urine samples of volunteers was 1.1, 1.9, and 6.3%, while AMPA was found to be only 0.01–0.04% of the total Gly dose.

Glufosinate ammonium is rapidly absorbed, though poorly, across various animal species, with an average absorption rate of about 10%. It is widely distributed, with the highest concentrations found in the liver, kidneys, and testes, and shows no signs of accumulation (EFSA 2005a, b). The primary metabolite is 3-methylphosphinicopropionic acid (3-MPPA). In a case involving a 65-year-old man who orally ingested a product containing 60 g of glufosinate ammonium, blood concentrations were continuously monitored over two days and urinary excretion was measured (Hirose et al. 1999). The result indicated an elimination half-life of 9.6 h.

1.4.3 Occurrence of Glyphosate in Urine

Gillezeau et al. (2019), Conolly et al. (2020), and Olejade et al. (2022) provide a comprehensive overview of urinary Gly concentrations in individuals exposed both occupationally and non-occupationally.

Hoppe (2013) analyzed 182 samples from 18 European countries and found Gly and AMPA levels above the limit of quantitation of 0.15 μg/l in 44 and 36% of samples, respectively. The highest mean Gly values were 0.82 μg/l in Malta, while the lowest were 0.09 μg/l in Bulgaria, Macedonia, and Switzerland. Table 1.8 summarizes the results of a recent study on adult urinary concentrations. Very high median levels of 3.25 μg/l were reported by Parvez et al. (2018) in 2015/16 in the USA and 1.19 μg/l by Grau et al. (2022) in a study conducted in France from 2018 to 2020. For AMPA, median and mean concentrations are comparable, ranging from 0.06 to 0.30 μg/l.

Table 1.9 summarizes the results of a more recent study on urine concentrations found in children and adolescents. The mean and median concentrations ranged from < 0.05 to 0.28 in the more recent studies, with only Fagan et al. (2020) reporting a median of 1.03 μg/l for 16 participants in the United States. AMPA concentrations are comparable to those of Gly.

Results on gender differences are inconsistent. While some studies from Germany, Slovenia, and France found higher concentrations in boys/men compared to girls/women (Conrad et al. 2017; Stajnko et al. 2020; Lemke et al. 2021; Grau et al. 2022), a contrasting picture emerged in Portugal and the USA (Ferreira et al. 2021; Ospina et al. 2022).

In 2011, Kundsen et al. (2017) examined the urine of 13 Danish schoolchildren and found higher concentrations than in their mothers' urine. Buekers et al. (2022a,b) also note that children show higher concentrations (95th percentile: 0.48 μg/l) compared to adults (95th percentile: 0.29 μg/l) in their summary review of four European studies within the HBM4EU program. The age-stratified analysis of NHANES 2013–2014 participants revealed a downward trend in urinary geometric mean concentration at younger ages, from 0.52 μg/l for the 6–11 year age group to 0.48 μg/l for those aged 12–19. In contrast, concentrations showed a significant upward trend from 0.37 μg/l for individuals aged 20–59 to 0.46 μg/l for those 60 years and older (Ospina et al. 2022).

To date, few studies have explored the temporal evolution of internal load. In the US, the mean Gly level rose significantly among a group of older adults from 0.024 μg/l in 1993–1996 to 0.314 μg/l in 2014–2016, while the mean AMPA levels increased from 0.008 μg/l in 1993–1996 to 0.285 μg/l in 2014–2016 (Mills et al. 2017). Conrad et al. (2017) examined 24-h urine samples from 40 German students aged 20–29 years at ten different time points between 2001 and 2015. The fraction of Gly levels at or above the limit of quantification (LOQ) peaked in 2012 and 2013 (57.5 and 56.4%), following a continuous increase of 10% in 2001, and then declined

Table 1.8 Median (range) glyphosate and AMPA concentrations in urine of adults in µg/l

References	N	Age (range)	Glyphosate	AMPA	Country, year
Ben Khadda et al. (2025)	50	38.4[e]	< 0.05 (< 0.05–0.71)	< 0.05	Marocco
Junqué et al. (2023)	11	–	– (0.14–1.74)	– (0.10–1.84)	Spain
Connolly et al. (2022)	54[a] 53[b]	29–57 26–54	< 0.05 (< 0.05–0.17) < 0.05 (< 0.05–0.17)	0.06 (< 0.05–4.12) 0.12 (< 0.05–2.86)	Ireland, 2019/20
Grau et al. (2022)	6848	0.5–94	1.19[e] (< 0.01–7.36)	–	France, 2018–20
Mellor et al. (2024)	725	≥ 18	0.36	0.37	USA, 2016–23
Schütze et al. (2021)	50[c]	–	0.39 (< 0.2–4.09)	–	USA, 2017–19
Faniband et al. (2021)	196	18–19	< 0.1 (< 0.1–3.39)	< 0.1 (< 0.1–0.99)	Sweden, 2017
Fagan et al. (2020)	7	36–52	0.26[e] (< 0.05–0.82)	0.27[e] (< 0.05–1.33)	USA, 2017
Lucia et al. (2023)	338	45–66	0.1	0.04	USA, 2016
Silver et al. (2021)	247	18–40	0.5 (< 0.2–2.73)	0.30 (< 0.2–1.85)	Puerto Rico
Parvez et al. (2018)	71	18–40	3.25 (< 0.5–7.2)	–	USA, 2015/16
Ruiz et al (2021)	97	25–45	0.12 (< 0.1–1.31)	0.13 (0.1–2.38)	Spain, 2015
Mills et al. (2017)	100	–	0.31 (0.24–0.39)	0.29 (0.22–0.35)	USA, 2014–16
Dereumeaux et al. (2022)	60	23–74	–	0.07[e] (< 0.05–0.37[f])	France, 2014–16
Conrad et al. (2017)	40	20–28	< 0.1 (< 0.1–0.57)	< 0.1 (> 0.1–0.41)	Germany, 2015
Nomura et al. (2020)	24	31–63	< 0.1 (< 0.1–1.99)	–	Japan, 2015
McGuire et al. (2016)	41	29[e]	0.28 (< 0.02–1.93)	0.30 (> 0.03–1.33)	USA, 2014/15
Eaton et al. (2022)	227	18–40	0.44 (< LOD-3.45)	< LOD (< LOD-15.9)	Puerto Rico, 2012–17
Ospina et al. (2022)	2310	6–80	0.39 (< 0.2–8.13)	–	USA, 2013–14
Campbell et al. (2022)	125[d]	–	< 0.2 (< 0.2–1.25)	< 0.2	Australia, 2013
Soukup et al. (2020)	312	18–80	< 0.05 (< 0.05–1.36)	< 0.09 (< 0.09–1.53)	Germany, 2012/13

(continued)

Table 1.8 (continued)

References	N	Age (range)	Glyphosate	AMPA	Country, year
Lesseur et al. (2022)	163	18–44	0.23 (< 0.04–4.13)	0.14 (<0.04–11.69)	USA, 2010–12
Ashley Martin et al. (2023)	1829	–	0.23 (< 0.08–7.8)	0.12 (< 0.09–5.7)	Canada, 2008–11
Knudsen et al. 2017	13	31–52	1.28[e] (0.49–3.22)	–	Denmark

[a] males; [b] females; [c] subgroup with organic diet; [d] 125 pools representing 1875 individuals; [e] mean; [f] 95th percentile; *LOD* limit of determination

in 2014 and 2015 to 32.5 and 40.0%, respectively. The medians (95th percentiles) rose from < 0.1 μg/l (0.12 μg/l) in 2001 to a peak of 0.11 μg/l (1.25 μg/l) in 2013, before falling to < 0.1 μg/l (0.45 μg/l) in 2015.

In Japan, morning urine from a total of 150 children aged 4–6 years was analyzed in 2006, 2011, and 2015, and significantly increasing Gly medians (95th percentiles) were found from < 0.05 μg/l (0.25 μg/l) in 2006 to < 0.05 μg/l (0.50 μg/l) in 2015 (Nomura et al. 2022). The values above the detection limit increased from 18 to 50%.

The sources of internal Gly contamination cannot always be substantiated. While an association with agricultural use in close proximity was observed in studies from Denmark, Portugal, and Morocco (Knudsen et al. 2017; Ferreira et al. 2021; Berni et al. 2023), this was not noted in a Slovenian study (Stajnko et al. 2020). Some studies identified a link between organic food (Fagan et al. 2020; Grau et al. 2022) or fasting diets (Ospina et al. 2022) and lower urinary Gly concentrations; however, Lemke et al. (2021) found no significant difference between vegetarian and regular diets in subjects aged 3–17 years. For instance, during a six-day organic diet, Fagan et al. (2020) reported a 71 and 77% decrease in Gly and AMPA concentrations among 16 subjects. Additionally, Lucia et al. (2023) found that frequent organic eaters in Southern California had significantly lower Gly and AMPA levels, though this was not apparent after adjusting for covariates.

1.4.4 Occurrence of Glufosinate in Urine

Data on internal exposure to Gluf or its primary metabolite are limited.

In examining 30 blood samples from pregnant women at term and 39 from nonpregnant women in a Canadian hospital, Gluf was not detected in any samples. Meanwhile, 3-MPPA was found in 100% of serum samples from pregnant women, with a mean of 120 μg/l (range 21.9–417 μg/l), and in 100% of cord serum samples, with a mean of 57.2 μg/l (range 8.8–193 μg/l) (Aris and Leblanc 2011).

Table 1.9 Median (range) glyphosate and AMPA concentrations in the urine of children and adolescents in μg/l

References	N	Age (range)	Glyphosate	AMPA	Country, year
Conolly et al. (2022)	54	3–19	< 0.05 (< 0.05.–2.48)	0.09 (< 0.05–7.24)	Ireland, 2019/20
Ferreira et al. (2021)	41	2–13	1.77[a] (0.87–4.35)	–	Portugal, 2018–19
Stajnko et al. (2020)	149	7–10	< 0.1 (< 0.1–0,39)	0.10 (< 0.1–0.45)	Slovenia, 2018[e]
Stajnko et al. (2020)	97	12–15	< 0.1 (< 0.1–0.33)	< 0.1 (< 0.1–0.44)	Slovenia, 2018[e]
Fagan et al. (2020)	9	4–15	1.03 (< 0.05–6.22)	0.79 (< 0.05–1.96)	USA, 2017
Makris et al. (2022)	177	10–11	< 0.19 (< 0.19–3.18)	0.18 (< 0.1–1.44)	Cyprus, 2016/17
Lemke et al. (2021)	2144	3–17	0.1 (< 0.1–11.1)	< 0.1 (< 0.1–13.1)	Germany, 2015–17
Dereumeaux et al. (2022)	61	6–17	–	0.1[a] (< 0.05–0.6[d])	France, 2014–16
Nomura et al. (2022)	50	4–6	< 0.03 (< 0.03–0.61)	–	Japan, 2015
LANUV (2016)	250	2–6	0.14 (< 0.1–3.70)	0.13 (< 0.1–2.21)	Germany, 2014–15
Ospina et al. (2022)	337	6–11	0.52 (0.44–0.61)[c]	–	USA, 2013–14
Trasande et al. (2020)	108	0–8	0.28[a] (0.11–2.13)	–	USA
Knudsen et al. (2017)	145	6–11	1.96[b] (< 0.08–3.31)	–	Denmark
Curwin et al. (2007)	51	< 16	2.7[b] (0.1–9.4)	–	USA, 2001

[a] Arithmetic mean; [b] geometric mean; [c] 95% confidence interval; [d] 95th percentile; [e] sampling January to March

Ashley-Martin et al. (2023) AM collected 1855 single spot urine samples from women in their first trimester of pregnancy as part of the Canadian Maternal-Infant Research on Environmental Chemicals (MIREC) study conducted from 2008 to 2011. They found that only 1 and 6% of samples had medians (maxima) of < 0.08 μg/l (6.1 μg/l) and < 0.08 μg/l (4.0 μg/l), respectively, for Gluf and 3-MPPA. In another Canadian study, 3-MPPA was detected in only six of 35 samples, with concentrations reaching a maximum of 0.96 μg/l, while Gluf remained below the detection limit of 0.08 μg/l in all samples (Bienvenu et al. 2021). In 20 archived urine samples from two cities in the USA, Gluf could not be detected above the method detection limit of 0.12 μg/l (Li and Kannan 2022).

1.4.5 Occurrence in Breast Milk

Studies on breast milk for Gly or AMPA have been conducted relatively infrequently. In 2014, a report published on the Moms Across America website garnered considerable attention. In the report, using a commercially available enzyme-linked immunosorbent assay (ELISA), Gly was detected in 10 breast milk samples from the United States, with concentrations ranging from < 75 to 166 μg/l (three of 10 samples exceeded the limit of quantitation) (Honeycutt and Rowlands 2014). Additionally, a more recent study by Camiccia et al. (2022) found Gly in all 67 breast milk samples from an agricultural region in Brazil, with a mean value of 1.45 μg/l. A commercially available ELISA was also utilized in this study. Given the complex protein- and lipid-rich matrix of breast milk, this measurement method, initially developed for aqueous media, must be viewed critically. Thus, highly sensitive, matrix-optimized, and validated sample preparation and analytical procedures are essential for valid analytics. In the USA, using a liquid chromatography-tandem mass spectrometry method, Gly and AMPA were not detected above the quantification limit of 1 μg/l in any of the 41 breast milk samples (McGuire et al. 2016). Moreover, Steinborn et al. (2016) also investigated 114 German breast milk samples collected in 2015, employing liquid chromatography-tandem mass spectrometry (LC–MS/MS) and gas chromatography-tandem mass spectrometry (GC–MS/MS). Gly, above the detection limits of 0.5 μg/l (first method) and 1.0 μg/l (second method), was not found in any sample.

1.5 Pyrethroids/Pyrethrins

Pyrethrins are among the oldest known natural insecticides, used indoors since the early nineteenth century. They are extracted and purified from the dried flower heads of various agriculturally grown chrysanthemum species (e.g., Chrysanthemum cinerariaefolium). Pyrethrins comprise six insecticidally active esters of chrysanthemic acid or pyrethric acid, along with the cyclic ketoalcohols pyrethrolone, cinerolone, and jasmolone (ATSDR 2003). The main constituents of the extract are the pyrethrins, making up approximately 40%, followed by cinerins at 10% and jasmolins at 5%. To enhance their application and insecticidal properties, pyrethroids were synthesized from naturally occurring pyrethrins.

Pyrethroids are organic compounds that have been produced on a large scale since the 1980s by modifying and introducing various functional groups based on the structure of natural pyrethrins. Depending on their properties (e.g., duration of action, stability, volatility), pyrethroids can be classified as short-term or long-term acting. These substances are present as mixtures of enantiomers. Generally, a distinction exists between type I pyrethroids, which lack an α-cyano moiety at the phenyl alcohol position (e.g., allethrin, bifenthrin, bioresmethrin, resmethrin, tefluthrin, tetramethrin, d-phenothrin, and permethrin), and type II pyrethroids, which contain α-cyano

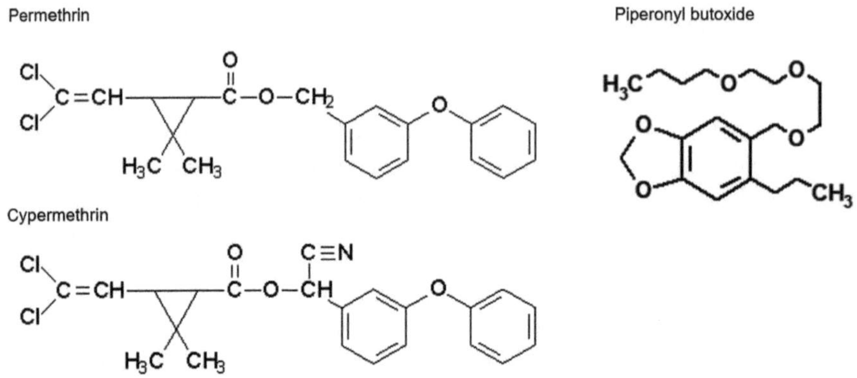

Fig. 1.13 Chemical structure of permethrin (type I pyrethroid), cypermethrin (type II pyrethroid), and piperonyl butoxide

substitution (e.g., cyfluthrin, cyhalothrin, cypermethrin, deltamethrin, fenvalerate, fenpropathrin, flumethrin, fluvalinate, and tralomethrin) (Ahamad and Kumar 2023). Figure 1.13 shows the chemical structures of two typical type I and type II pyrethroids, along with piperonyl butoxide. Piperonyl butoxide is a substance derived from the natural compound safrole, which acts as a synergist and enhances the action of pyrethrins.

Pyrethroids are commonly used to manage pests and nuisances. They are applied by professional pest controllers and by homeowners as household products containing insecticides, such as insect sprays, gels, strips, dusters, and electric vaporizers. Additionally, they serve as moth and beetle repellents for treating home textiles. Furthermore, they have been used in products designed to control insect pests in crops (Morgan 2012).

Because vector control has been crucial in preventing, controlling, and eliminating various vector-borne diseases, the WHO implemented a specific strategy for insecticide use that includes insecticide-treated nets (ITNs), residual spraying, space spraying, and larviciding. Van den Berg et al. (2021) concluded that in the African region, pyrethroids accounted for 90% of all vector control insecticide use, while in the Asia–Pacific and Latin American and Caribbean regions, they represented 59%. The global pyrethroids market reached a value of USD 3.58 billion in 2022, with the Asia Pacific being the largest user of pyrethroid insecticides, particularly in China, India, Malaysia, Thailand, and Indonesia (EMR 2023).

1.5.1 Health Aspects

Pyrethrins and pyrethroids act as neurotropic contact poisons and repellents. Despite their strong insecticidal activity and significant fish toxicity, they are only moderately toxic to mammals after oral ingestion (Chrustek et al. 2018). However, there are considerable differences in the potency of individual pyrethroids. The mechanism of action is thought to influence the sodium permeability of the nerve membrane, leading to the permanent opening of sodium channels and, consequently, prolonged depolarization (Soderlund 2012). The effects of type I pyrethroids are typically characterized by the so-called "T-syndrome," which usually includes tremors, hyperexcitability, ataxia, and convulsions. In contrast, type II pyrethroids cause persistent nerve impulses and may reduce the resting membrane potential to such an extent that no action potential is generated. In this case, the typical poisoning pattern is known as the "CS syndrome," characterized by choreatic-athetotic mixed hyperkineses, along with salivation and tonic–clonic seizures (Zhu et al. 2020).

The primary effects of acute and chronic toxicity observed in mammal experiments influence the nervous system (ATSDR 2003). The US EPA has indicated that pyrethroids are unlikely to impact the human hormone system, fertility, reproductive capability, or the immune system (summary in Burns and Pastoor 2018). Two comprehensive reviews of human studies by Burns and Pastoor (2018) and Burns and LaKind (2022) are accessible.

Animal studies showed lung and liver tumors at very high doses. However, the underlying mechanism is of minor relevance to humans. In a review, the US EPA concluded, based on human studies, that there is insufficient evidence for childhood brain tumors and childhood leukemia, as well as no evidence for breast cancer (US EPA 2019). Additionally, for other endpoints such as neurodevelopmental, neurobehavioral, and neurocognitive effects in children, birth effects, and male reproductive health outcomes, there is insufficient evidence to conclude that there is a clear associative or causal relationship with pyrethroid exposure.

1.5.2 Toxicokinetics

Animal and in vitro studies indicate low dermal absorption of 0.2–1.8% in healthy, uninjured skin (ATSDR 2003; Buchholz et al. 2021). For instance, Khemiri et al. (2018) discovered that 0.12–0.08% of the administered dose of cyhalothrin was recovered in the urine of six volunteers who were dermally dosed. This phenomenon has also been observed with dermal application in medical treatment contexts. Results from animal studies and test subjects show absorption rates after oral intake of between 40 and 60%, and there also seems to be significant absorption through inhalation (ATSDR 2003).

In mammalian organisms, pyrethroids are converted in the intestinal mucosa through hydrolytic cleavage and oxidative processes into numerous metabolites, which are primarily excreted in urine. The initial detoxification step appears to be the cleavage of the central ester bond (EFSA 2022). PBA-based metabolites and conjugates with glucuronic acid, glycine, or sulfate are the predominant metabolites in the pyrethroid metabolic pathway. Notably, the overall rate of metabolism is lower for the (+)-cis isomers than for the (+)-trans isomers. Table 1.10 provides a brief overview of typical pyrethroid metabolites and their parent compounds.

Table 1.10 Typical pyrethroid metabolites and their parent compounds (adapted from Sabbioni et al. 2022)

Name	Abbreviation	CAS No.	Structure	Parent compound
3-phenoxybenzoic acid	3PBA	3739-38-6	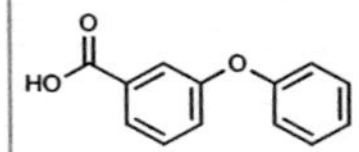	Cyhalothrin, cypermethrin, deltamethrin, fenpropathrin, permethrin, tralomethrin
4-fluoro-3-phenoxybenzoic acid	4F3PBA	77279-89-1	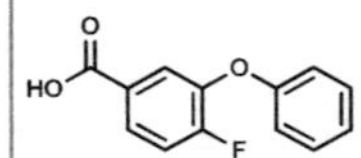	Cyfluthrin
cis-(2,2-dibromovinyl)-2,2-dimethylcyclopropanecarboxylic acid	cis-DBCA	63597-73-9	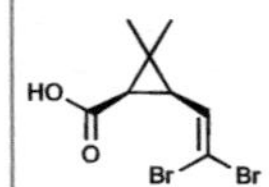	Deltamethrin
cis-3-(2,2-dichlorovinyl)-2,2-dimethylcyclopropane-1-carboxylic acid	cis-DCCA	59042-49-8	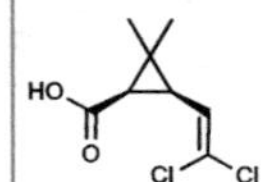	Permethrin, cypermethrin, cyfluthrin
trans-3-(2,2-dichlorovinyl)-2,2-dimethylcyclopropane-1-carboxylic acid	trans-DCCA	59042-50-1	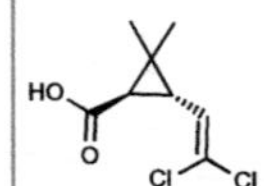	Permethrin, cypermethrin, cyfluthrin

(continued)

Table 1.10 (continued)

Name	Abbreviation	CAS No.	Structure	Parent compound
trans-Chrysanthemumdicarboxylic acid	trans-CDCA	72120-98-0	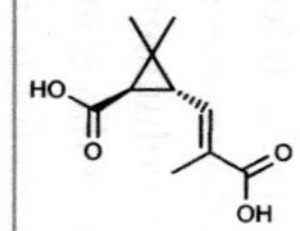	Phenothrin, allethrin, prallethrin, imiprothrin, resmethrin, tetramethrin
cis-3-(2-chloro-3,3,3-trifluoroprop-1-en-1-yl)-2,2-dimethylcyclopropanecarboxylic acid	ClF3CA	72748-35-7	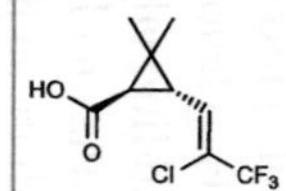	Cyhalothrin

First, Woollen et al. (1992) analyzed the individual time courses of cypermethrin metabolites in the urine of six volunteers over a 120-h period following a single oral dose of cypermethrin. Based on ClF3CA metabolite measurements, the mean absorption of cypermethrin was estimated to be 36% of the administered dose. In another study, seven volunteers ingested a single dose of lambda-cyhalothrin, with blood and urine samples collected over 72 and 82 h, respectively, and the metabolites ClF3CA and 3-PBA were analyzed (Khemiri et al. 2017). The elimination phase exhibited a rapid decay, showing mean half-lives of approximately 5.3 h for ClF3CA and 6.4 h for 3-PBA. On average, 21% of lambda-cyhalothrin doses were excreted in urine as CFMP, compared to 30% as 3-PBA. Additionally, six volunteers received one dose of cypermethrin (Ratelle et al. 2015a). Plasma levels of three metabolites peaked on average around 5–7 h post-dosing, with mean half-lives for trans-DCCA, cis-DCCA, and 3-PBA of 5.1, 6.9, and 9.2 h, respectively. The same research group exposed six volunteers to permethrin, with peak plasma concentration reached after about 7 h. The mean half-life of the three metabolites ranged between 4.5 and 5.7 h (Ratelle et al. 2015b). Over the 84 h following treatment, 43–46% of the administered molar dose was excreted in urine as trans-DCCA. After the single oral administration of deltamethrin to five volunteers, DBVA, and to a lesser extent 3-PBA, were detected with mean half-lives of 3.6 and 7.1 h, respectively (Sams and Jones 2010). In this study, the mean recovery of urinary DBCA was expressed as a proportion of the administered deltamethrin dose over the initial 24-h period post-dose, which was 45%.

1.5.3 Occurrence in Urine

Tables 1.11 and 1.12 summarize the occurrence of pyrethroid metabolites in the urine of adults and children/adolescents. Several reviews of urinary concentrations in the general population have been published (e.g., Tang et al. 2018; Li and Kannan 2018; Sabbioni et al. 2022). Comparable median levels were found in urine samples from adults in Europe, North America, and the Caribbean, while higher levels were detected in East Asia and Australia. The median levels were significantly higher, particularly for 3-PBA, trans-DCCA, and cis-DBCA. For instance, Hung et al. (2019) reported medians of 8.5 μg/l (maximum: 23.6 μg/l) for 3-PBA and 2.8 μg/l (maximum: 71.3 μg/l) for cis-DCCA following the application of a new analytical method. In contrast, the medians observed in a least-developed city in central China (Han et al. 2017) were significantly lower than those from the study in Taiwan. The type and quantity of pyrethroid used likely account for these differences. This finding is supported by the analysis of Li and Kannan (2018), who analyzed 322 adult urine samples collected from eight countries between 2006 and 2014. Median 3-PBA concentrations were highest in Vietnam at 2.1 μg/l, followed by numbers of 1.0 μg/l in Korea, 0.9 μg/l in India, 0.7 μg/l in the US, 0.6 μg/l in China and Saudi Arabia, and 0.5 μg/l in Greece and Japan. Again, in a recent review, Lehmler et al. (2022)

concluded that factors contributing to higher 3-PBA levels in the general population primarily included dietary exposures and residential and agricultural pyrethroid applications.

The European Human Biomonitoring Initiative (HBM4EU) Aligned Studies present a summary of the results from urine samples collected from 3,576 children aged 6–12 years and 4,102 adults aged 20–39 years between 2014 and 2021 across 4–6 countries (Govarts et al. 2023). The 95th percentiles for children and adults are 5.4 and 2.0 μg/g crea for 3-PBA, 0.12 and 0.11 μg/g crea for 4F3PBA, 4.01 and 1.12 μg/g crea for trans-DCCA, and 4.34 and 1.72 μg/g crea for cis-DBCA, respectively.

Other metabolites have received limited investigation. For instance, Makris et al. (2022) quantified CIF3CA, the metabolite of cyhalothrin, with a median of 0.07 μg/l (0.06–1.66 μg/l) in 177 urine samples from children in Cyprus (Makris et al. 2022). Trans-CDCA, a nonspecific metabolite of various pyrethroids, was found in 92% of the samples with a median of 0.14 μg/l (< 0.02–28 μg/l) in the urine of Japanese children (Osaka et al. 2016).

Only for trans-DCCA, higher concentrations were found in females compared to males, but not for other metabolites, in the study by Li and Kannan (2018). The same gender differences were observed in the studies by Barr et al. (2010) in the representative US National Health and Nutrition Examination Survey 1999–2002 and by Osaka et al. (2016) in Japan. In urine samples collected from eight countries, cis-DCCA concentrations were highest in the age group of ≥ 50 years, followed by 21–49 years and ≤ 20 years (Li and Kannan 2018). Conversely, children reported significantly higher urinary concentrations of 3-PBA compared to adolescents and adults in the USA (Barr et al. 2010). Sulc et al. (2022) also found significantly higher concentrations of 3-PBA and the sum of cis- and trans-DCCA in children (4–15 years old) compared to adults (age: 31–54 years) in a study of 110 parent–child pairs in the Czech Republic. Govarts et al. (2023) also describe this relationship in summarizing European data.

A Chinese study collecting 240–482 urine samples annually between 2018 and 2022 reported 3PBA's consistent downward trend from a median of 1.19 μg/l in 2018 to 0.74 μg/l in 2022, while the medians of cis-DCCA and trans-DCCA did not change significantly (Jia et al. 2024).

1.5.4 Occurrence in Breast Milk

So far, few results on pyrethroid exposure in breast milk have been published in the scientific literature. Table 1.13 shows the mean values and ranges. Overall, the results indicate a wide range of variation, and depending on the predominantly used substance, there are also differences in exposure among individual pyrethroids.

The highest concentrations are reported in South Africa and Mozambique, where malaria control is prevalent, and pyrethroids have been used extensively along with other pesticides (Bouwman et al. 2006; Feo et al. 2012).

Table 1.11 Median (range) pyrethroid metabolite concentrations in urine of adults or the general population samples in μg/l

References	N	Age (years)	3-PBA	4F-3PBA	trans-DCCA	cis-DCCA	cis-DBCA	Country, year
Jia et al. (2024)	240	35[b]	0.97 (< 0.03–105)	< 0.02 (< 0.02–6.15)	1.1 (< 0.03–114)	0.34 (< 0.06–57)	–	China, 2022
Li et al. (2025)	425	≥ 20	0.44	< LOQ	0.72	–	–	China, 2018/19
Sulc et al. (2022)	110	31–54	0.16 (0.91[d]) 0.19 (1.18[d])	– –	0.56 (3.16[d]) 0.30 (1.77[d])		– –	Czech Republic, 2020 (winter and summer)
Radwan et al. (2022)	450	20–45	0.30 (< 0.1–2.1[d])	–	0.36 (< 0.1–2.9[d])	0.21 (< 0.01–1.3[d])	0.14 (< 0.1–2.2[d])	Poland, 2017–19
CHMS (2019)	2719	> 6	0.46 (9.[d])	< 0.006 (0.08[d])	0.23 (–)	0.15 (–)	0.02 (0.18[d])	Canada, 2016/17
Schettgen et al. (2016)	38	26–58	0.22 (< 0.01–4.0)	0.04 (< 0.01–4.9)	0.17 (< 0.01–3.8)	0.08 (< 0.01–2.3)	0.04 (< 0.01–0.36)	Germany, 2016
Tagne-Fotso et al. (2023)	900	18–74	0.72 (3.2[d])	< 0.015 (0.07[d])	0.17 (1.5[d])	0.25 (1.5[d])	0.69 (4.4[d])	France, 2014–16
Bravo et al. (2020)	168	> 18	0.24 (< 0.02–12)	< 0.02 (< 0.02–0.73)	–	–	–	Slovenia, 2014/15
CDC (2022)	2622	> 6	0.63 (7.1[d])	< 0.1 (0.8[d])	< 0.6 (5.2[d])	–	–	USA, 2013/14
Han et al. (2017)	136	60[b]	0.74 (< 0.1–5.93)	–	0.42 (< 0.1–2.75)	1.03 (< 0.1–5.6)	–	China, 2013/14

(continued)

Table 1.11 (continued)

References	N	Age (years)	3-PBA	4F-3PBA	trans-DCCA	cis-DCCA	cis-DBCA	Country, year
Heffernan et al. (2016)	24[c]	> 0	1.21[a] (0.51–4.0)	< 0.1	1.89[a] (1.1–5.5)	–	1.25[a] (0.68–2.3)	Australia, 2012/13
Hung et al. (2019)	30	adults	8.52 (< 0.1–23.6)	1.25 (< 0.1–27.9)	10.2 (< 0.1–34.6)	2.7 (< 0.1–71.3)	4.4 (< 0.1–24.5)	Taiwan
Davis et al. (2013)	55	> 18	1.52[b] (0.07–8.5)	0.84[b] (0.09–2.25)	6.34[b] (1.1–28.2)	3.4[b] (1.4–11.0)	< 0.5	USA, 2012
Gari et al. (2018)	80	adults	1.8 (< 0.02–15)	0.08 (< 0.02–0.34)	–	–	–	Spain
Lewis et al. (2014)	54	18–40	< 0.1 (< 0.1–11.3)	< 0.1	< 0.6 (< 0.6–7.6)	–	< 0.5	Puerto Rico, 2010–12
Dalsager et al. (2019)	948	20–45	0.24 (< 0.03–29)	< 0.2 (< 0.2–2.9)	< 0.4 (< 0.4–21)	< 0.5 (< 0.5–20)	< 0.5 (< 0.5–2.0)	Denmark, 2010–12
Dewailly et al. (2014)[a]	295	27[b]	0.54 (0.03–24.0)	0.02 (< 0.008–0.40)	0.40 (< 0.01–14.0)	0.17 (< 0.007–5.0)	0.03 (< 0.006–9.7)	Caribbean, 2008–11
Li et al. (2020)	619	18–44	0.17 (< 0.001–24.0)	0.008 (< 0.001–4.9)	0.06 (< 0.004–37.2)	0.09 (< 0.008–34.3)	< 0.01 (< 0.01–7.3)	USA, 2007–09

[a] Geometric mean; [b] arithmetic mean; [c] 24 pools of 100 samples; [d] 95th percentile; *LOQ* limit of quantification

Table 1.12 Median (range) pyrethroid metabolite concentrations in urine samples of children/adolescents in µg/l

References	N	Age (years)	3-PBA	4F-3PBA	trans-DCCA	cis-DCCA	cis-DBCA	Country, year
Deng et al. (2024)	159	1–6	0.71 (< 0.02–16.7)	< 0.01 (< 0.01–0.31	1.08 (< 0.02–18.2)	0.25 (< 0.03–9.28)	–	China, 2022
Sulc et al. (2022)	110	4–15	0.56 (2.25[e]) 0.38 (1.57[e])	– –	1.45 (6.66[e]) 0.73 (3.23[e])		– –	Czech Republic, 2020 (winter and summer)
Castiello et al. (2023)	151	14–18	< 0.12 (< 0.12–0.72)	–	1.35 (< 0.17–9.0)		–	Spain, 2017–19
Göl et al. (2023)	186	3–6	0.07 (< 0.03–4.88)[b]	< 0.05	< 0.03		–	Turkey, 2017/18
Makris et al. (2022)	177	10–12	1.93 (0.34–20.0)	–	0.93 (0.10–10.8)	0.61 (0.06–9.6)	0.61 (0.11–6.1)	Cyprus, 2016/17
Bustamante et al. (2025)	399	7–9	0.44 (< 0.08–24.5)	< 0.10 (< 0.10–2.6)	–	–	–	Poland, 2014–19
Ueyama et al. (2022)	1075	16–23 months	0.98 (< 0.04–28.2)	< 0.04 (< 0.04–1.62)	0.51 (< 0.04–121)		< 0.04 (< 0.04–2.17)	Japan, 2015/16
Tagne-Fotso et al. (2023)	499	6–17	1.03 (7.3[e])	< (0.09[e])	0.18 (1.1[e])	0.32 (1.9[e])	1.16 (5.7[e])	France, 2014–16
Bravo et al. (2019)	199	7	0.56 (< 0.02–36)	< 0.02 (< 0.02–1.3)	–	–	–	Italy, 2014/15
Bravo et al. (2020)	164	7–8	0.40 (< 0.02–12)	< 0.02 (< 0.02–0.53)	–	–	–	Slovenia, 2014/15
Fiserová et al. (2021)	20	9–15	0.53 (0.17–2.27)	4.43 (1.2–27)	0.52 (< 0.2–3.1)	–	–	South Africa

(continued)

Table 1.12 (continued)

References	N	Age (years)	3-PBA	4F-3PBA	trans-DCCA	cis-DCCA	cis-DBCA	Country, year
Li et al. (2019)	20[c]	0–5	2.3[a] (0.45–5.3)	0.09[a] (< 0.01–1.0)	2.2[a] (0.38–7.9)	–	< 1.3 (< 1.3–7.5)	Australia, 2014/15
Osaka et al. (2016)	223	3	0.01 (0.33[d]–25.3)	–	0.72 (< 0.02[d]–48.7)			Japan, 2012/13
Roca et al. (2014)	125	6–11	< 0.8 (< 0.8–64.4)	< 0.2	< 0.4 (< 0.4–82.8)	< 0.4 (< 0.4–13.2)	< 0.8 (< 0.8–8.6)	Spain, 2010

[a] Geometric mean; [b] gravity adjusted; [c] 20 pools of 400 samples; [d] 5th percentile; [e] 95th percentile

Table 1.13 Mean (range) concentrations of pyrethroid pesticides in breast milk in ng/g lw and μg/l

References	N	Permethrin	Deltamethrin	Cyfluthrin	Cypermethrin	Bifenthrin	λ-Cyhalothrin	Tetramethrin	Country, year
In ng/g lw									
Sanguos et al. (2023)	83	7.5 (0.06–75)[b]	4.8 (0.051–142)[b]	2.3 (0.023–46)[b]	7.6 (0.82–83)[b]	–	–	–	Spain, 2016–18
Anand et al. (2021)	56	63	< LOD	< LOD	< LOD	196	108	–	India, semi-urban, 2013/14
	25	< LOD	< LOD	< LOD	< LOD	2250	< LOD	–	India, urban, 2016
Feo et al. (2012)		79 (10–230)	< LOD	80 (11–220)	54 (3.3–160)	6.5 (1.1–36)	140 (16–440)	80 (6.7–230)	Mozambique, 2002
Bouwman et al. (2006)	152	186–1343[a]	< LOD-258[a]	< LOD-1307[a]	< LOD-127[a]	–	–	–	South Africa, 2002
Čechová et al. (2017)	120	0.57 (0.06–1.41 [c])	0.37 (0.17–0.84[c])	0.43 (0.22–0.64[c])	–	0.21 (0.15–0.31[c])	0.15 (0.04–0.53)[c]	< LOD	Netherlands, 2011–15
Corcellas et al. (2012)	8	0.36 (0.03–1.7)	0.14 (< LOD-0.57)	–	1.44 (0.61–2.89)	2.8 (0.21–6.71)	0.29 (0.05–0.78)	0.45 (< LOD-0.96)	Brazil, 2009/10
	27	0.91 (< LOD-5.12)	0.23 (< LOD-1.9)	–	4.25 (4.7–16.4)	0.02 (< LOD-0.48)	1.06 (0.05–4.49)	2.2 (< LOD-14.8)	Colombia, 2011
	6	2.2 (1.3–3.2)	0.42 (0.09–0.71)	–	0.4 (0.09–1.0)	0.19 (> LOD-0.5)	0.31 (0.04–0.85)	1.28 (< LOD-4.5)	Spain, 2009
In μg/l									
Kuang et al. (2020)	55	< 0.5–1.0[d]	–	< 1.4–3.9[d]	–	< 0.76	< 0.9–2.0[d]	–	China, 2018/19

(continued)

Table 1.13 (continued)

References	N	Permethrin	Deltamethrin	Cyfluthrin	Cypermethrin	Bifenthrin	λ-Cyhalothrin	Tetramethrin	Country, year
Anand et al. (2021)	56	2 (< LOD-54)	< LOD	< LOD	< LOD	7 (< LOD-119)	4 (< LOD-45)	–	India, semi-urban, 2013/14
	25	< LOD	< LOD	< LOD	< LOD	77 (< –420)	< LOD	–	India, urban, 2016
Bouwman and Kylin (2009)		57 (< LOD-114)	8.4 (< LOD-83)	42 (< LOD-459)	4.2 (< LOD-28)	–	–	–	South Africa
Bouwman et al. (2006)	152	7.3–48.4 [a]	< 0.35–8.4[a]	< 0.18–41.7[a]	< 0.34–4.2[a]	–	–	–	South Africa, 2002

[a] Range of means from primipara and multipara of three towns; [b] interquartile range; [c] 5th and 95th percentiles; [d] range of detection limit; LOD: limit of detection

Additionally, very high fat-related results are reported from an agricultural semi-urban area in Bengal, India, while in samples from mothers in nearby Kolkata, all but bifenthrin were below the limit of quantification (Anand et al. 2021). The high levels of bifenthrin observed in certain study areas are particularly striking, as was also observed in a rural region of Colombia (Corcellas et al. 2012). Low mean concentrations were detected in the Netherlands and urban areas of Brazil and Colombia, whereas slightly higher mean levels were reported in Spain. Conversely, none of the four investigated pyrethroids could be found above the limit of quantification of the analytical method in 55 breast milk samples obtained from mothers living in Jinhua City, China (Kuang et al. 2020).

Results of piperonyl butoxide have only been published in one study to date, which analyzed 120 samples from the Netherlands; the substance was detected in 85% of the samples, with a mean value of 0.65 ng/g lw (5th–95th percentiles: 0.09–2.13 ng/g lw) (Čechová et al. 2017).

1.6 Neonicotinoids

Neonicotinoids (Neonics) can be classified into three chemical groups: N-nitroguanidines (e.g., imidacloprid, thiamethoxam, clothianidin, and dinotefuran), nitromethylenes (e.g., nitenpyram), and N-cyanoamidines (e.g., acetamiprid and thiacloprid) (Goulson 2013). Additionally, a subdivision based on the development time or market launch (first to fourth generation) is possible. Table 1.14 compiles data about the Neonics, including chemical structure, CAS Number, and observed metabolites. It is estimated that over 600 Neonics or Neonic-like substances have been synthesized to date (Thompson et al. 2020).

Neonics are systemically acting pesticides that are rapidly absorbed by plants after application and distributed throughout all parts of the plant, including pollen and nectar (Simon-Delso 2015). They are potent agonists of the nicotinic acetylcholine receptors in insects' central nervous systems, leading to nervous stimulation at low concentrations but causing receptor blockage, paralysis, and death at higher concentrations (Goulson 2013).

After their introduction in the early 1990s, neonics quickly became one of the most widely used classes of insecticides. In this context, they were originally developed as an alternative to organophosphates and carbamates (Cimino et al. 2017). They serve various purposes in agriculture and for domestic and companion animal use. Neonicotinoids are currently registered for application on a wide range of crops, including citrus fruits, oilseed crops (e.g., cotton), cucurbit vegetables, fruiting vegetables, pome fruits, stone fruits, cereal grains, tree nuts, root and tuber vegetables, leafy vegetables, legume vegetables, and bulb vegetables (CDPR 2018).

Table 1.14 Description of different neonicotinoids (adapted from Thompson et al. 2020; Lewis et al. 2016; Li and Kannan 2022)

Parent substance	Abbreviation	CAS Number		Formula	Year introduced	Metabolite (in humans)
First generation						
Imidacloprid	IMI	138261-41-3	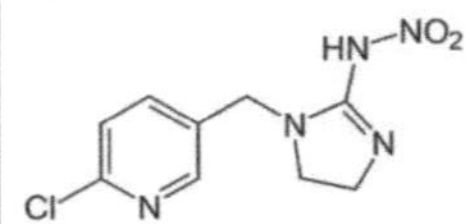	$C_9H_{10}ClN_5O_2$	1991	5-hydroxy-imidacloprid (5-OH-IMI), desnitro imidacloprid (DN-IMI), IMI-olefin, IMI-urea
Acetamiprid	ACE	135410-20-7	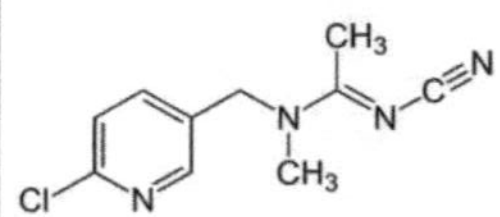	$C_{10}H_{11}ClN_4$	1995	N-desmethyl-acetamiprid (DME-ACE)
Nitenpyram	NIT	150824-47-8	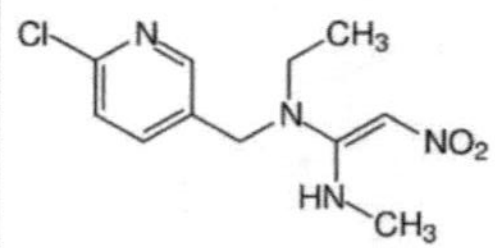	$C_{11}H_{15}ClN_4O_2$	1995	
Thiacloprid	THI	111988-49-9	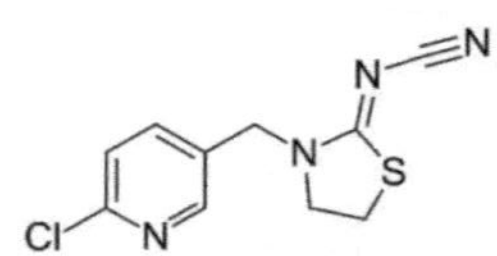	$C_{10}H_9ClN_4S$	2001	4-OH-THI, THI-amide

(continued)

Table 1.14 (continued)

Parent substance	Abbreviation	CAS Number		Formula	Year introduced	Metabolite (in humans)
Second generation						
Thiamethoxam	THX	153719-23-4	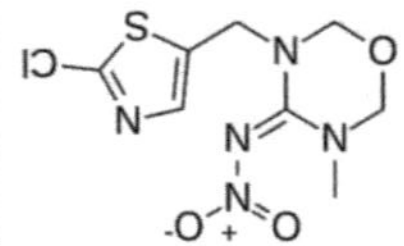	$C_8H_{10}ClN_5O_3S$	1998	N-desmethyl-thiamethoxam (DME-THX),
Clothianidin	CLO	210880-92-5	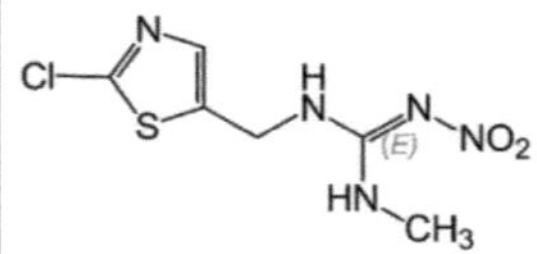	$C_6H_8ClN_5O_2S$	2001	N-desmethyl-clothianidin (DME-CLO), CLO-urea
Third generation						
Dinotefuran	DIN	165252-70-0	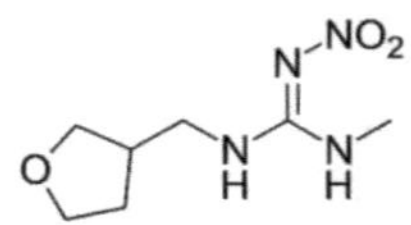	$C_7H_{14}N_4O_3$	2002	

(continued)

Table 1.14 (continued)

Parent substance	Abbreviation	CAS Number		Formula	Year introduced	Metabolite (in humans)
Fourth generation						
Cycloxaprid	CYC	1203791-41-6	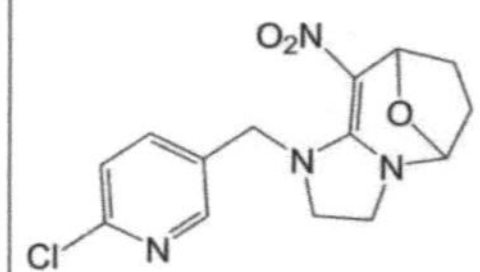	$C_{14}H_{15}ClN_4O_3$		
Guadipyr	GUA	1376342-13-0	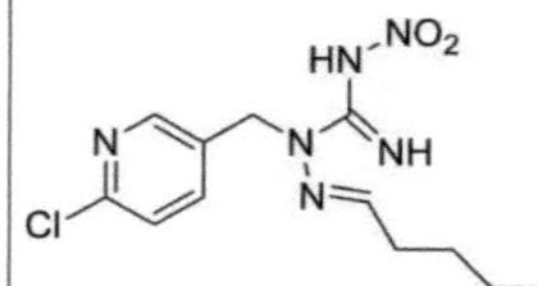	$C_{12}H_{17}ClN_6O_2$	2008	
Imidaclothiz	IMID	105843-36-5	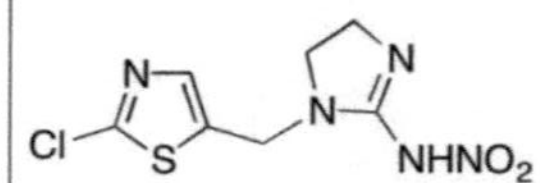	$C_7H_8ClN_5O_2S$		
Paichongding	PCD	948994-16-9	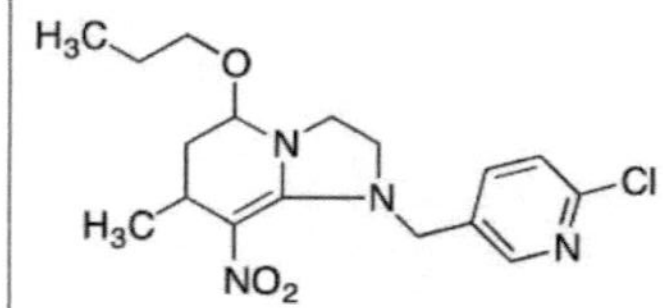	$C_{17}H_{23}ClN_4O_3$	2008	

(continued)

Table 1.14 (continued)

Parent substance	Abbreviation	CAS Number		Formula	Year introduced	Metabolite (in humans)
Neonicotinoid-like compounds						
Sulfoxaflor	SUL	946578-00-3	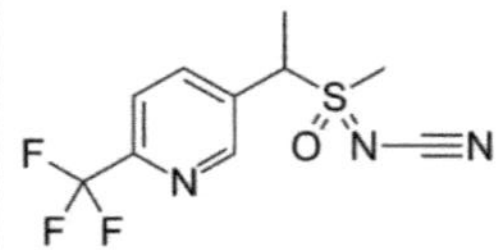	$C_{10}H_{10}F_3N_3OS$	2013	
Flonicamid	FLO	158062-67-0	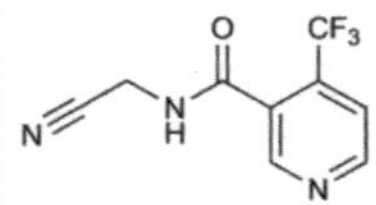	$C_9H_6F_3N_3O$	2000	

From 1999 to 2015, measurements in the US indicated that low levels of Neonics are found in commonly consumed fruits and vegetables. Trends in detection frequencies suggest an increase in the use of acetamiprid, clothianidin, and thiamethoxam as replacements for imidacloprid (Craddock et al. 2019). In parallel with application rates or restrictions, these findings underline the ongoing presence of these chemicals in the food supply.

Recent reviews by Thompson et al. (2020) and Giorio et al. (2021) provide an in-depth overview of the occurrence of Neonics in various environmental media.

In addition to their effects on pest insects, there is growing evidence of various adverse impacts on the ecosystem, particularly on nontarget organisms such as pollinators. This resulted in EU restrictions on IMI, CLO, and THX in 2018, while the approval of THI has been revoked in 2020. The discussion regarding the limitation of use is still ongoing. Pisa et al. (2021) summarized the adverse effects on pollinators and soil organisms as well as the impact on the ecosystem.

1.6.1 Health Aspects

Cimino et al. (2017) summarized the scientific literature from 2005 to 2015. They found that, regarding acute toxicity, only one workplace study, three poisoning studies, and four chronic toxicity studies had been conducted in the general population. They reported associations between chronic Neonic exposure and adverse developmental or neurological outcomes, such as congenital malformation of the heart (significant association), anencephaly (suggestive association between proximity to agricultural use), autism spectrum disorder (weak association), and a symptom cluster including memory loss and finger tremor (significant association). Furthermore, Mendy and Pinney (2022) indicated that exposure to Neonics is linked to decreased serum testosterone levels in humans. A recent Chinese study investigated 296 mother-infant pairs and reported an adverse effect on newborns' head circumference due to IMI and ACE (Pan et al. 2022).

Depending on the dose, gastrointestinal symptoms may include nausea, vomiting, abdominal pain, and corrosive lesions; central nervous system symptoms can involve headaches, agitation, confusion, fasciculations, seizures, and even coma; cardiovascular symptoms might feature tachycardia, bradycardia, hypertension, hypotension, and palpitations; and respiratory effects such as dyspnea, aspiration pneumonia, or respiratory failure can also indicate poisoning (Nistor et al. 2017).

In animal experiments involving various species, the chronic toxicity endpoint of Neonics after oral exposure primarily depends on reduced body weight and weight gain, as well as effects on the liver, nervous system, hematopoietic system, kidneys, and partially on testicular and hormone levels (e.g., diverse reports prepared by the US EPA).

1.6.2 Toxicokinetics

Information on human metabolism and the excretion of specific metabolites remains limited to date. Initially, Harada et al. (2016) orally dosed 12 healthy adults with a mixture of deuterium-labeled IMI, ACE, CLO, and DIN. They collected pooled urine samples over a 24-h period for four consecutive days following dosing. Within 96 h of ingestion, 63.7% of CLO, 92.8% of DIN, 12.7% of IMI, and 2.6% of ACE were recovered unchanged in urine. ACE was excreted primarily as DME-ACE, with 30.7% of the deuterated parent compound in the urine. The elimination half-lives were approximately 14 h for CLO, 35 h for IMI, 4.1 h for DIN, and 5.5 h for DME-ACE. Additionally, one volunteer received a single oral dose containing five Neonics (5 μg each of ACE, IMI, CLO, THI, and THX) and two Neonic-like substances (SUL and flupyradifurone) according to the currently acceptable daily intake levels set by EFSA, and urine samples were collected for up to 48 h (Wrobel et al. 2022). They found that, except for SUL (which was excreted unchanged in urine), there was extensive metabolization of the Neonics into specific metabolites that were excreted alongside the parent compounds. They confirmed previously reported key metabolites in humans, such as DME-ACE, DME-CLO, OH-IMI, IMI-olefin, CLO-urea, and THI-amide. In contrast, the significance of the previously measured metabolites 6-chloronicotinic acid (6-CNA) and 2-chloro-1,3-thiazole-5-carboxylic acid (2-CTA) was deemed somewhat questionable for humans by the authors. Moreover, two volunteers received a single oral dose of ACE, CLO, FLUP, and IMI (Wrobel et al. 2023b). The mean elimination half-lives in urine were 2.5 h for ACE, 5.3 h for THI, 11.6 h for THX, 12.2 h for IMI, 19.1 h for DME-ACE, and 49.5 h for SUL. On average, 23% of the parent compounds were excreted in the urine after 48 hours as 4-OH-THI, 27% as IMI-olefin, 50% as THX, 51% as Sul, 54% as DME-ACE, and 84% as CLO.

1.6.3 Occurrence in Urine

Table 1.15 summarizes the results of recent studies on Neonics, while Table 1.16 presents data on their metabolites in adults and in children/adolescents. Overall, concentrations are low, often near or below the limit of quantification. IMI and DME-ACE continue to be the most frequently detected. For instance, Li and Kannan (2020) analyzed spot urine samples ($n = 566$) collected from nine countries between 2010 and 2014. The median urinary concentrations of the sum of 14 Neonics varied by an order of magnitude among the nine countries, with the highest median concentration found in Vietnam (12.2 μg/l), followed by Saudi Arabia (6.1 μg/l), Greece (6.0 μg/l), China (4.6 μg/l), Japan (4.6 μg/l), the USA (3.1 μg/l), India (2.9 μg/l), Kuwait (2.5 μg/l), and Korea (1.9 μg/l). Other metabolites have rarely been determined to date. Huang et al. (2022) reported medians of IMI-guanidine and IMI-urea in 20 urine samples from Chinese children of 7.91 μg/l and 0.03 μg/l, respectively.

Table 1.15 Median levels of neonicotinoid pesticides in urine samples in μg/l

References	N	Age (years)	ACE	CLO	DIN	IMI	THX	NIT	THI	FLO	SUL	IMID	Country, year
Adults (general population)													
Jia et al. (2025)	497	6–79	0.01	1.31	0.22	0.13	0.90	–	0.03	–	–	–	China, 2023
Huang et al. (2024)	432	18–42	< LOQ	0.45	< LOQ	<	< LOQ	< LOQ	< LOQ	-	< LOQ	–	China, 2022
Li et al. (2025)	425	≥ 20	< LOQ	0.24	–	0.08	–	–	–	–	–	–	China, 2018/19
Wrobel et al. (2023a)	39	–	< 0.06	–	–	< 0.19	< 0.30	–	–	–	–	–	Germany
Wrobel et al. (2023a)	43	–	< 0.06	–	–	< 0.19	< 0.30	–	–	–	–	–	Brazil
Wang et al. (2020)	129	20 to > 60	< 0.01	0.29	0.16	0.08	0.21	–	–	–	–	–	China 2018
Li et al. (2020)	243 [b]	34[a]	< 0.011	0.193		0.091	0.057	< 0.004	< 0.003	< 0.010	0.019	0.177	USA, 2018
Taira et al. (2022)	99	6 to > 60	< 0.01	< 0.05	< 0.05	< 0.05	< 0.05	< 0.05	< 0.01	–	–	–	Philippines, 2017
Ospina et al. (2019)	3038	> 3	< 0.30	< 0.20	–	< 0.40	–	–	< 0.03	–	–	–	USA, 2015/16
Zhang et al. (2019)	324	1–97	0.014	0.240	0.14	0.21	0.15	–	0,002	–	–	–	China, 2012–17
Nishihama et al. (2023)	8538	32[a]	< LOQ	0.12[d]	0.18[d]	< LOQ	0.04[d]	< LOQ	–	< LOQ	< LOQ	–	Japan, 2011–14

(continued)

Table 1.15 (continued)

References	N	Age (years)	ACE	CLO	DIN	IMI	THX	NIT	THI	FLO	SUL	IMID	Country, year
Nimako et al. (2022)	919	< 1–57	< 0.05	< 0.10	0.26	< 0.20	< 0.10	< 0.10	< 1.33	–	–	–	Japan
Suwannarin et al. (2022)	30	–	< 0.05	0.2	0.3	< 0.5	0.07	< 0.5	–	< 1.0	< 0.05	–	Japan, 2014
Children, adolescents													
Guo et al. (2025)	88	11–12	0.02	0.71	0.21	0.12	0.30	–	0.03	–	–	–	China, 2020
Huang et al. (2022)	20	11–14	< 0.27	1.42[a]	–	0.79[a]	0.57[a]	0.29[a]	0.02[a]	–	–	–	China, 2019
Zhang et al. (2023)	103	< 1	0.031	0.019	0.25	0.088	< 0.005	–	0.017	–	–	–	China, 2018
Zhao et al. (2022)	305	8–11	0.01	0.19	1.64	0.13	0.21	–	0.002	–	–	–	China, 2017
Ikenaka et al. (2019)	46	3–6	< 0.05	< 0.1	< 0.1	< 0.2	< 0.1	< 0.1	< 0.05	–	–	–	Japan, 2016
Oya et al. (2017)	1036[c]	1.3–2	< 0.01	< 0.13	< 0.74	< 0.07	< 0.06	–	< 0.03	–	–	–	Japan, children, 2015/16
Osaka et al. 2016	233	3	< 0.03	< 1.07	0,44	< 0.31	< 0.22	< 0.13	< 0.32	–	–	–	Japan, 2012/13

[a] Mean; [b] collected from 19 volunteers; [c] collected from diapers; [d] specific gravity normalised; *LOQ* limit of quantification

Table 1.16 Median levels of neonicotinoid metabolites in urine samples in µg/l

References	N	Age (years)	DME-ACE	THI-amide	6-CNA	IMI–guanidine	IMI–urea	IMI–olefin	5-OH-IMI	DME-THX	Country, year
Adults (general population)											
Jia et al. (2025)	497	6–79	1.25	–	–	7.91	0.03	0.53	0.61	–	China, 2023
Huang et al. (2024)	432	18–42	1.31	–	–	–	–	–	< LOQ	< LOQ	China, 2022
Li et al. (2025)	425	≥ 20	0.73	–	–	–	–	0.51	1.03	–	China, 2018/19
Wrobel et al. (2023a)	39	–	0.38	–	–	–	–	< 2.1	< 1.0	–	Germany
Wrobel et al. (2023a)	43	–	0.22	–	–	–	–	< 2.1	< 1.0	–	Brazil
Wang et al. (2020)	129	20 to > 60	0.75	–	–	–	–	0.52	–	–	China 2018
Li et al. (2020)	243 [b]	34[a]	0.272	< 0.016	0.445	–	–	–	–	< 0.045	USA, 2018
Ospina et al. (2019)	3038	> 3 years	< 0.20	–	–	–	–	–	< 0.40	–	USA, 2015/16
Song et al. (2020)	324	1–91	0.45	–	–	–	–	0.48	1.02	–	China, 2012–17
Children, adolescents											
Huang et al. (2022)	20	11–14	0.72[a]	3.18[a]	0.08[a]	7.91[a]	0.03[a]	10.2[a]	–	0.21[a]	China, 2019
Zhang et al. (2023)	103	< 1	0.015	–	–	–	–	< 0.02	–	–	China, 2018

(continued)

Table 1.16 (continued)

References	N	Age (years)	DME-ACE	THI-amide	6-CNA	IMI–guanidine	IMI–urea	IMI–olefin	5-OH-IMI	DME-THX	Country, year
NRW (2020)	251	2–6	–	–	0.31	–	–	–	–	–	Germany, 2017/18
Ikenaka et al. 2019	46	3–6	0.39	–	–	–	–	–	–	–	Japan, 2016
Suwannarin et al. (2022)	30	–	0.30	< 0.1	–	–	–	–	–	–	Japan, 2014
Ichikawa et al. 2019	57	PND 1–2	0.048	–	–	–	–	–	–	–	Japan, 2009/10
Ichikawa et al. 2019	65	PND 14	0.090	–	–	–	–	–	–	–	Japan, 2009/10

[a] Mean; [b] collected from 19 volunteers; PND: post-natal day; 6-CNA: 6-chloronicotinic acid; *LOQ* limit of quantification

Additionally, Wrobel et al. (2025) found DME-THX in 227 urine samples from Ireland in no samples abo the limit of quantification of 0.18 μg/l.

Using representative samples of the U.S. general population aged three years and older from the 2015–2016 National Health and Nutrition Examination Survey (NHANES), Ospina et al. (2019) reported the 95th percentiles for DME-ACE, 5-OH-IMI, and CLO as 1.29, 1.37, and 0.40 μg/l, respectively. The weighted detection frequencies were highest for DME-ACE (35%) and 5-OH-IMI (19.7%), followed by CLO (7.7%) and IMI (4.3%), while AMI and THI were rarely detected (< 0.5%). Children aged 3–5 years had higher concentrations of DME-ACE than any other age group in this study. After accounting for various confounders, the detection of Neonics in urine was associated with age, being higher in participants over 65 years old (Akliku 2023). Additionally, Jia et al. (2025) observed the highest concentrations (except DIN) in the 18–59 age group and lower concentrations in children, adolescents, and people over 59 years. In contrast, Wang et al. (2020) reported no significant age-related differences among 129 volunteers aged 20 to over 60 years in China. Some researchers observed higher concentrations of Neonics in urine samples from women compared to men (Nimako et al. 2022; Akliku 2023). Conversely, no significant differences were found for individual Neonics between females and males in a global study of 566 urine samples (Li and Kannan 2020) and in the study conducted in China (Wang et al. 2020). No gender-related differences were found in 227 Irisah urine samples either (Wrobel et al. 2025). In a nationwide survey in China, the concentrations of IME and ACE were significantly higher in males than in females (Zhang et al. 2019). In another study in China involving 497 urine samples, higher concentrations of DIN, THI, and ACE were found in men compared to women (Jia et al. 2025).

Li et al. (2020) analyzed 13 Neonics and their metabolites in single-spot and first-morning-void urine samples collected over 44 days from 19 individuals. They concluded that overall exposure could only be weakly reproduced from a single-spot urine sample and that analysis of repeated samples is necessary.

Fruit and vegetable consumption was found to be associated with Neonics exposure in Japan (Harada et al. 2016; Ueyama et al. 2014). Additionally, in the U.S., Neonics are often present in fruits and everyday food staples in children's diets, which partly explains the higher internal exposure among children (Ospina et al. 2019).

Ueyama et al. (2015) quantified urinary concentrations of 7 Neonics in 95 females aged 45−75 in the years 1994, 2000, 2003, 2009, and 2011 (17−20 individuals each year). The results indicate that the detection rates increased significantly between 1994 and 2011, and the rise in applications can effectively explain this increase.

1.6.4 Occurrence in Breast Milk

Data on measurement results of Neonics in breast milk are extremely limited and often represent findings from a newly developed analytical method. An overview of the results is provided in Table 1.17. Only very low median concentrations were observed in the USA, India, and Switzerland. More comprehensive and recent studies with significantly higher levels come from China. For example, Chen et al. (2020) examined pooled breast milk samples from 33 urban and 64 rural areas across 23 provinces and found the highest concentrations for DME-ACE, ranging from 0.008 to 1.48 μg/l, followed by IMI at < 0.005 to 0.19 μg/l. Only the sum of ACE and its metabolite DME-ACE in breast milk is positively correlated with the corresponding dietary exposure, and no significant difference between the Neonic profiles of urban and rural areas was noted. In another study, 22 breast milk samples were analyzed and found to have comparable medians for CLO and THI when compared to the previous survey, but slightly lower for DME-ACE (< 0.01–0.21 μg/l) and IMI (< 0.01–2.06 μg/l) (Zhang et al. 2023).

Simultaneously, 22 infant formulas dissolved in Milli-Q water were analyzed in the Zhang et al. (2023) study. Here, IMI exhibited a median concentration of 0.22 μg/l, followed by THX at 0.080 μg/l, ACE at 0.018 μg/l, and THI at 0.011 μg/l, all of which were not significantly different from the concentrations found in breast milk. In contrast, among 4935 samples of dairy-based infant formula from the USA collected between 1999 and 2015, no values above the limit of quantification were found in any sample (Craddock et al. 2019).

Studies on the temporal development of Neonics in breast milk are limited. Huang et al. (2024b) analyzed Chinese samples over eight years from 2014 to 2022 and described a statistically significant decrease in the mean values of total NNIs from 0.567 to 0.140 μg/l and of DME-ACE and ACE from 0.53 to 0.049 μg/l and from 0.008 to 0.002 μg/l, respectively. However, there were upward trends in CLO and THM values (from 0.006 to 0.024 μg/l and from <BG to 0.050 μg/l) during the same period. In another Chinese study, in which colostrum samples were examined in 2007, 2013, and 2019, the medians for IMI decreased significantly from 0.029 μg/l to 0.009 μg/l and for ACE from 2.1 μg/l to <0.001 μg/l, while THM and CLO increased slightly during this period (Li et al. 2025).

Metabolites have been investigated in only a few studies to date, and, except DME-ACE, are found at low concentrations. In three Chinese studies, DME-ACE concentrations ranged from <0.01 to 0.21 μg/l (Zhang et al. 2023), 0.008 and 1.48 μg/l (Chen et al. 2020), and 0.013 and 3.29 μg/l (Sha et al. 2025). While Sha et al. (2025) were able to determine 5-OH-IMI and DME-THX with medians of 0.023 μg/l and 0.144 μg/l, the medians in Li et al. (2025) were below the limits of quantification of 0.001 μg/l and 0.002 μg/l for 5-OH-IMI and DIN urea, respectively. IMI-olefin could only be determined in 1.3% and 26% of samples with values of <0.003-0.055 μg/l and <0.08-0.327 μg/l, respectively (Li et al. 2025, Sha et al. 2025). Chen et al. (2020) observed only a positive correlation between the sum of ACE and its metabolite DME-ACE in breast milk and the corresponding exposure via food. No significant difference between concentrations in urban and rural areas was found.

Table 1.17 Median concentrations of neonicotinoid pesticides in breast milk in µg/l

References	N	Acetamiprid	Desmethyl-acetamiprid	Imidacloprid	Clothianidin	Dinotefuran	Thiamethoxam	Nitenpyram	Thiacloprid	Country, year
Sha et al. (2025)	–	0.040	0.163	0.069	0.076	0.269	0.026	–	0.037	China, 2023
Zhang et al. (2023)	22	0.015	0.041	0.27	< 0.025	< 0.050	0.093	–	0.010	China, 2022
Huang et al. (2025)	10	< 0.003	0.046	0.019	0.012	< 0.03	0.022	–	< 0.002	China,
Chen et al. (2020)	97 [b]	0.014	0.103	0.039	0.011	–	–	< 0.010	0.017	China, 2017–19
Lachat and Glauser (2018)	3	< 0.001–0.019	–	0.004–0.007	< 0.005–0.006	–	< 0.002–0.002	–	< 0.001	Switzerland, 2018
Anand et al. (2021)	81	< 0.001	–	< 0.001	–	–	–	–	< 0.002	India, 2013–16
Pedersen et al. (2021)	79	–	–	0.02[a]	–	–	–	–	–	USA

[a] Mean; [b] pooled samples from 3570 lactating women

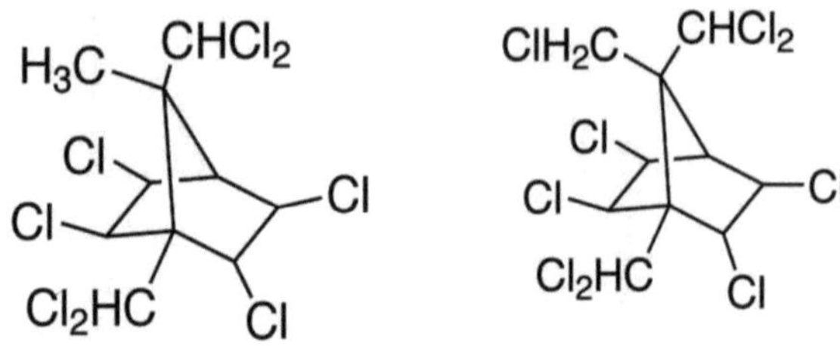

Fig. 1.14 Chemical structures of Parlar 26 (left) and Parlar 50 (right)

1.7 Toxaphenes

Technical toxaphene is a complex mixture of several hundred chlorinated bornanes, bornenes, bornadienes, camphenes, and dihydrocamphenes that appears as a yellowish-waxy substance. The chemical structures of two important congeners are shown in Fig. 1.14. The nomenclature of toxaphenes follows the systematics proposed by Coehlhan and Parlar (1996), where each congener is assigned an ascending number after the word "Parlar".

Toxaphenes are synthetic chemicals with no natural sources. They have been widely used as insecticides since around 1940, with use peaking in the early 1970s, totaling nearly 23,000 tons in the US per year (ATSDR 2014). Toxaphene, often utilized as a substitute for DDT due to its simple and cost-effective manufacturing process, became the most-produced organochlorine insecticide globally (HC 2010). Overall, toxaphene use from 1950 to 1992 exceeded 1.3 million tonnes (Oehme 2000; HC 2010). It was primarily employed as a nonsystemic stomach and contact insecticide with some acaricidal effects. In certain situations, extensive use in aquatic environments was also implemented. Toxaphenes persist in the environment, and due to their volatility, they can be transported over long distances, even to the Arctic (ATSDR 2014).

Toxaphenes are listed in Annex A of the Stockholm Convention. Therefore, parties are required to take action to eliminate the production and use of these chemicals.

1.7.1 *Health Aspects*

Extensive substance dossiers from scientific organizations are available regarding the health effects of toxaphenes (ATSDR 2014).

Here, in brief:

Animal studies demonstrate that toxaphenes can also impact the liver, kidneys, endocrine system, immune response, body weight, and nervous system. Some case–control studies of farm workers and prospective cohort studies of pesticide applicators have reported statistically significant associations between exposure to toxaphene and the risk of cancers such as leukemia, certain types of non-Hodgkin lymphoma, rectal cancer, and melanoma. Given insufficient human data but sufficient evidence of carcinogenicity in animals, the IARC has classified toxaphene as possibly carcinogenic to humans (Group 2B).

1.7.2 Toxicokinetics

Animal studies indicate that toxaphenes are rapidly absorbed following oral administration and likely after inhalation. They are quickly and extensively degraded in mammals, primarily through dechlorination, dehydrodechlorination, and oxidation. Excretion occurs predominantly (approximately 60%) via feces, with only about 10% via urine. In animal experiments, a half-life of several days was observed following oral administration.

1.7.3 Occurrence in the Blood

Measuring toxaphenes in human test samples is challenging due to the complexity of the congeners in the technical mixture. Few study results have been published, particularly for Parlar 26 (CAS No: 142534-71-2) and Parlar 50 (CAS No: 66860-80-8).

In the US, Parlar 26 and 50 were found at concentrations ranging from about 0.003–0.03 μg/l (0.7–7.0 ng/g lipid) in pooled human serum collected in Atlanta, Chicago, and Cincinnati between 1987 and 1994 (Barr et al. 2004). In Northern Canada, Butler Walker et al. (2003) analyzed 386 maternal and 407 cord blood samples collected from 1994 to 1999, with information available for 351 maternal-cord pairs. Parlar 26 and 50 were detected in 51 and 53% of the 148 maternal samples, respectively, with mean (maximum) values of 0.07 μg/l (0.57 μg/l) and 0.09 μg/l (0.66 μg/l). The total toxaphene concentration, as the sum of 21 peaks, averaged 0.70 μg/l (maximum: 6.40 μg/l). In cord blood, Parlar 26 and 50 were detected more frequently in Inuit participants than in other ethnic groups. However, levels of toxaphene were, on average, close to or at the detection limit of 0.02 μg/l. In a study conducted from 2007 to 2008, a total of 2162 Inuit in Northern Canada were recruited, revealing a geometric mean for the sum of Parlar 26 and 50 of 0.17 μg/l (0.01–8.3 μg/l), while the geometric mean in women of childbearing age was significantly lower at 0.073 μg/l (Laird et al. 2013). Overall, the plasma concentration of toxaphene was considerably higher in men than in women and increased with age. Parlar 26 and 50 were also measured in 1666 samples from volunteers aged 20–79 participating in the Canadian Health Measures Survey from 2007 to 2009 (HC 2010). The low medians (95th percentiles) of Parlar 26 and 50 were < 0.005 μg/l (0.01 μg/l) and < 0.005 μg/l (0.01 μg/l), respectively.

In a Russian peninsula known for a traditionally high intake of aquatic food, Parlar 26 and Parlar 50 were found in 82 and 86% of 50 samples, respectively, with geometric mean levels of 0.17 and 0.15 μg/l, and a sum range of the two congeners of 0.05–1.58 μg/l (Sandanger et al. 2003). As part of the Arctic Monitoring and Assessment Programme, 48 maternal and cord blood samples, each from Northeast Russia, were analyzed in 2001–2002, revealing a geometric mean of 0.06 μg/l (< LOD-0.64 μg/l) for the sum of Parlar 26 and Parlar 50 in maternal blood (Eik Anda et al. 2007).

As part of a prospective birth cohort study in Japan, 379 maternal blood samples were collected in late pregnancy from 2002 to 2005 (Araki et al. 2018). Of the six congeners measured, only Parlar 26 and 50 were detected in 97 and 96% of the samples, respectively, with medians (maximum) of 4.39 pg/g ww (20.8 pg/g ww) and 6.52 pg/g ww (29.3 pg/g ww).

1.7.4 Occurrence in Breast Milk

Newsome and Ryan (1999) compared 58 milk samples from Canada and 24 from the Great Lakes Basin, both collected in 1992, with 30 Canadian samples from 1986, and 12 samples from indigenous residents of the Keewatin region of the Arctic. The concentrations reported in the 1992 survey were significantly lower than those reported in a 1986 survey (medians of the sum Parlar 26 and 50: 4.94 vs. 7.84 ng/g lw), while the milk from donors in the Arctic region had a higher median of 56.4 ng/g lw.

Polder et al. (2008a) collected 14 and 23 breast milk samples from two cities in Northern Russia in 2000 and reported medians of 19 ng/g lw (7–45 ng/g lw) and 10 ng/g lw (2–18 ng/g lw) for the sum of five congeners. As part of the Arctic Monitoring and Assessment Programme, 48 samples from Northeast Russia were analyzed during 2001–2002, revealing a geometric mean of 16 ng/g lw (< 0.5–213 ng/g lw) for the sum of Parlar 26 and 50 (Eik Anda et al. 2007).

Parlar 26 and 50 were the most abundant congeners found in six German breast milk samples collected in 1998 and 1999, with levels ranging from 1.3 to 7.2 ng/g lw and 4.4 to 13.0 ng/g lw (Skopp et al. 2002). The total levels for the sum of four congeners varied from 7 to 24 ng/g lw. In 10 breast milk samples obtained from primipara women in Northern Norway between 2000 and 2002, a median sum of six congeners of 11 ng/g lw (ranging from 2.3 to 19 ng/g lw) was reported, with Parlar 50 being the predominant congener, contributing 55% to the total sum of toxaphenes (Polder et al. 2008b). In a single pooled milk sample collected in 2009–2010 from a rural area of Belgium, concentrations of 0.5 ng/g lw for Parlar 26 and 1.8 ng/g lw for Parlar 50 were reported, comparable to the levels found in Belgian samples analyzed in the WHO 2006 global human milk survey (Croes et al. 2012).

In 70 pooled human breast milk samples from 210 individuals across China, Korea, and Japan collected in 2007–2009, the mean concentration (the sum of Parlar 26 and Parlar 50) in Japan was 2.5 ng/g lw, significantly higher than those in Korea (0.73 ng/g lw) and China (0.36 ng/g lw) (Fujii et al. 2012). The source of toxaphenes in breast milk from Japan may be attributed to the dietary intake of imported foods or long-distance transport, as technical toxaphene was not registered in Japan. Comparable levels were reported from other East Asian countries. For instance, Hedley et al. (2010) found 0.8 ng/g lw as the sum of three congeners in 10 pools of human milk samples collected at 2–6 weeks postpartum from 238 primiparous women living in Hong Kong and South China. Additionally, 200 breast milk samples were collected between 2005 and 2010 from 10 Japanese regions, with levels ranging from < 0.08–5.6 ng/g lw for Parlar 26 and < 0.1–8.5 ng/g lw for Parlar 50 (Fujii et al. 2023).

There is no clear, unambiguous picture of the time course. Under a UNEP program, using a uniform method, toxaphenes were studied from 2000 to 2004, 2005 to 2009, and 2010–2019 in 27–51 national pools worldwide (Fiedler et al. 2023). While the medians consistently remained below the LOQ, the mean values started at 2.23 ng/g lw, steadily decreased to 0.46 ng/g lw by 2010–2014, and then increased again to 1.27 ng/g lw.

1.8 Insect Repellent DEET

N,N-diethyl-meta-toluamide (DEET; CAS No. 134-62-3) has high water solubility and low volatility, and it is available in various isomeric forms (ECHA 2010). The chemical structures of DEET and two metabolites are shown in Fig. 1.15. Since the 1950s, technical DEET has generally been formulated with carriers and solvents, and used in microencapsulated form, as sprays or aerosols, in impregnated materials such as freshening wipes and roll-ons, as well as ready-to-use liquids, creams, lotions, and sunscreens, with the proportion of active substance in the product varying greatly. DEET is highly effective as a repellent, although the time it takes to repel insects after the product is applied can range from less than an hour to 6 h (ECHA 2010). DEET is a very common repellent used by around 50–100 million people in the USA annually (Nguyen et al. 2023). Overall, repellents contribute significantly to the prevention of vector-borne diseases, which account for over 17% of all infectious diseases and are responsible for 700,000 deaths annually, according to the WHO in 2021 (WHO 2024).

1.8.1 Health Aspects

Various studies have found evidence of an increased incidence of dizziness, disorientation, concentration difficulties, and skin rashes among very high-exposure DEET users compared to nonusers. At the same time, data from poison control centers showed no clear dose dependence of the effects (ATSDR 2017; Ghali and Albers 2024). Comparable neurological effects have also been described at high doses in

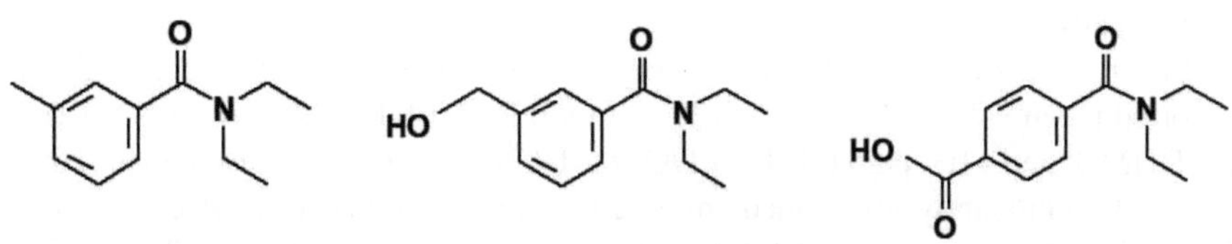

Fig. 1.15 Chemical structures of DEET (left), DHMB (middle), and DCBA (right)

animal experiments following oral administration. DEET exhibits low systemic toxicity in animal studies after oral, inhalation, and dermal exposure, showing reduced weight gain and decreases in hematological and clinical chemical parameters without morphological changes (ATSDR 2017). No reproductive or developmental toxic effects were observed in long-term studies involving rats.

1.8.2 Toxicokinetics

Based on limited data, DEET is almost completely absorbed after oral administration in experimental animals. In contrast, after dermal application in test animals and volunteers, absorption can vary between approximately 4–17%, depending on the dose, formulation, species, and sex (Moody et al. 1992; Selim et al. 1995; ATSDR 2017). In humans, rapid metabolization occurs, producing various metabolites, particularly in the liver. One primary metabolite, m-(diethylaminocarbonyl) benzoic acid (DCBA; CAS No. 34231-49-7), is formed through ring methyl oxidation via the intermediate N,N-diethyl-3-hydroxymethyl-benzamide (DHMB) (ATSDR 2017). The other primary metabolite, m-(ethylaminocarbonyl) benzoic acid (EACB), results from N-dealkylation of the intermediate N-ethyl-m-toluamide. Excretion primarily occurs through urine, with plasma elimination half-lives ranging from 2.5 to 9 h in experimental animals.

Selim et al. (1995) applied ^{14}C-labeled undiluted technical DEET and DEET dissolved in ethanol to the forearms of 6 volunteers for 8 h. The dermal absorption of DEET ranged from 3 to 8% (mean value: 5.6%) in volunteers who applied undiluted technical DEET and from 4 to 14% (mean value: 8.4%) in subjects who used the diluted substance. In a further study with 4 volunteers who used DEET for one week, maximum concentrations of 5.1–70 μg/l were observed for DEET and > 7000 μg/l for DCBA in the urine 4–14.8 h after a single topical application (Nishihara et al. 2025).

1.8.3 Occurrence in Urine

Depending on the specific application conditions, urinary concentrations can vary significantly.

In 2012, Kuklenyik et al. (2013) analyzed 75 urine samples from the USA and reported mean values of 4.4 μg/l (12% > LOQ) for DEET, 10.2 μg/l (16% > LOQ) for DHMB, and 1103 μg/l (48% > LOQ) for DCBA. In contrast, Lewis et al. (2014) reported significantly lower medians in 54 pregnant women from Puerto Rico in 2012/13, showing < 0.1 μg/l for DEET, < 0.1 μg/l for DHBM, and 3.5 μg/l for DCBA. Comparable medians of < 0.08 μg/l for DEET, < 0.09 μg/l for DHMB, and 1.26 μg/l for DCBA were also found in 171 samples from 90 men in the USA from 2007 to 2015 (Segal et al. 2017). In 60 urine samples analyzed in the USA in 2017,

DCBA was detected in 95% of the samples, with a median of 4.0 μg/l (< 0.06–433 μg/l), and ECBA was found in 83% of the samples, with a median of 0.92 μg/l (< 0.05–142 μg/l) (Baker et al. 2019). In the representative US National Health and Nutrition Examination Survey, the 95th percentiles for DEET in urine were below the LOQ of 0.083 μg/l in the 2007–2008, 2009–2010, and 2011–2012 surveys, while it was 0.103 μg/l in the 2013–2014 survey (CDC 2023). The concentrations were slightly higher for children aged 6–11 years than for adults over 20 years (0.205 vs. 0.099 μg/l). In these surveys, the medians of DCBA increased from 3.50 μg/l in 2007–2008 to 4.54 μg/l in 2009–2010, then rose again to 2.89 μg/l after a slight drop between 2011 and 2014. ECBA was only determined in the 2015–2016 survey, with a median of 1.02 and a 95th percentile of 59.2 μg/l. The intermediate DHMB was found in very low concentrations in urine.

Chronister et al. (2023) analyzed 519 urine samples from 11- to 17-year-olds living in agricultural communities in Ecuador in 2016, reporting geometric mean values of 1.25 μg/l for DCBA and 1.54 μg/l for ECBA. Similarly, in 20 urine samples from children in South Africa, Fišerová et al. (2021) could not detect DEET above the LOD of 0.03 μg/l in 2019. In 24 pools of 100 urine samples from the general population collected in Australia during 2012–2013, geometric mean values of 65.9 and 0.83 μg/l were found for DCBA and DHMB, respectively. DEET was only detected above the LOQ of 0.08 μg/l in two pools at concentrations of 3.43 and 0.55 μg/l (Heffernan et al. 2016).

In Canada, the urine of children aged 7–13 at a summer camp was repeatedly analyzed, and the frequency and amount of DEET application were documented (Gibson et al. 2023). In participants who used insect repellent, urinary DEET and DCBA levels were elevated between 2 and 8 h after application (12.4 and 12,102 μg/l) compared to those who did not use insect repellent on the study day (1.6 and 1966 μg/l). Children who used more insect repellent or higher concentrations excreted greater levels of DEET and DCBA.

References

Abdeen Z, Berman T, Azmi K, Abu Seir R, Agha H, Ein-Mor E, Göen T, Stein Y, Richter E, Calderon-Margalit R (2016) Urinary organophosphate metabolite levels in Palestinian pregnant women: results of the Middle East Regional Cooperation Project. Int J Environ Health Res 26:254–266

Aerts R, Van Overmeire I, Colles A, Andjelković M, Malarvannan G, Poma G, Den Hond E, Van de Mieroop E, Dewolf MC, Charlet F, Van Nieuwenhuyse A, Van Loco J, Covaci A (2019) Determinants of persistent organic pollutant (POP) concentrations in human breast milk of a cross-sectional sample of primiparous mothers in Belgium. Environ Int 131:104979

Agus S, Akkaya H, Dagliogl N, Eyuboglu S, Atasayan O, Mete F, Colak C, Sandal S, Yilmaz B (2022) Polychlorinated biphenyls and organochlorine pesticides in breast milk samples and their correlation with dietary and reproductive factors in lactating mothers in Istanbul. Environ Sci Pollut Res Int 29:3463–3473

Ahamad A, Kumar J (2023) Pyrethroid pesticides: An overview on classification, toxicological assessment and monitoring. J Hazard Mat Adv 10:100284

Akliku KN (2023) Association between socioeconomic and behavioral characteristics with urinary neonicotinoid levels in the United States, 2015–2016. In: Capstone experience, p 251. Online: https://digitalcommons.unmc.edu/coph_slce/251

Alengebawy A, Abdelkhalek ST, Qureshi SR, Wang MQ (2021) Heavy metals and pesticides toxicity in agricultural soil and plants: ecological risks and human health implications. Toxics 9(3):42

Anand N, Chakraborty P, Ray S (2021) Human exposure to organochlorine, pyrethroid and neonicotinoid pesticides: comparison between urban and semi-urban regions of India. Environ Pollut 270:116156

Andreotti G, Koutros S, Hofmann JN, Sandler DP, Lubin JH, Lynch CF, Lerro CC, De Roos AJ, Parks CG, Alavanja MC, Silverman DT, Beane Freeman LE (2018) Glyphosate use and cancer incidence in the Agricultural Health Study. J Natl Cancer Inst 110:509–516

Antignac JP, Main KM, Virtanen HE, Boquien CY, Marchand P, Venisseau A, Guiffard I, Bichon E, Wohlfahrt-Veje C, Legrand A, Boscher C, Skakkebæk NE, Toppari J, Le Bizec B (2016) Country-specific chemical signatures of persistent organic pollutants (POPs) in breast milk of French, Danish and Finnish women. Environ Pollut 218:728–738

Antonangeli LM, Kenzhebekova S, Colosio C (2023) Neurobehavioral effects of low-dose chronic insecticide exposure: a review. Toxics 11:192

Araki A, Miyashita C, Mitsui T, Goudarzi H, Mizutani F, Chisaki Y, Itoh S, Sasaki S, Cho K, Moriya K, Shinohara N, Nonomura K, Kishi R (2018) Prenatal organochlorine pesticide exposure and the disruption of steroids and reproductive hormones in cord blood: the Hokkaido study. Environ Int 110:1–13

Araújo MF, Castanheira EMS, Sousa SF (2023) The buzz on insecticides: a review of uses, molecular structures, targets, adverse effects, and alternatives. Molecules 28:3641

Arcury TA, Chen H, Laurienti PJ, Howard TD, Barr DB, Mora DC, Quandt SA (2018a) Farmworker and nonfarmworker Latino immigrant men in North Carolina have high levels of specific pesticide urinary metabolites. Arch Environ Occup Health 73:219–227

Arcury TA, Laurienti PJ, Talton JW, Chen H, Howard TD, Barr DB, Mora DC, Quandt SA (2018b) Pesticide urinary metabolites among Latina farmworkers and nonfarmworkers in North Carolina. J Occup Environ Med 60:e63–e71

Aris A, Leblanc S (2011) Maternal and fetal exposure to pesticides associated to genetically modified foods in Eastern Townships of Quebec. Canada. Reprod Toxicol 31:528–533

Ashley-Martin J, Huang R, MacPherson S, Brion O, Owen J, Gaudreau E, Bienvenu JF, Fisher M, Borghese MM, Bouchard MF, Lanphear B, Foster WG, Arbuckle TE (2023) Urinary concentrations and determinants of glyphosate and glufosinate in pregnant Canadian participants in the MIREC study. Environ Res 217:114842

ATSDR (1997) Toxicological profile for chlorpyrifos. https://wwwn.cdc.gov/TSP/ToxProfiles/ToxProfiles.aspx?id=495&tid=88

ATSDR (2003) Toxicological profile for pyrethrins and pyrethroids. Online: https://www.atsdr.cdc.gov/ToxProfiles/tp155-p.pdf

ATSDR (2014) Toxicological profile for toxaphene. Online: https://www.atsdr.cdc.gov/ToxProfiles/tp94.pdf

ATSDR (2015) Toxicological profile forhexachlorobenzene. Atlanta, GA, USA. Online: https://www.atsdr.cdc.gov/toxprofiles/tp90.pdf

ATSDR (2017) Toxicological Profile for DEET (N,N-Diethyl-meta-toluamide). https://www.atsdr.cdc.gov/toxprofiles/tp185.pdf

ATSDR (2019) Toxicological profile for DDT, DDE, DDD (Draft for Public Comment). Atlanta, GA, USA. Online: https://www.atsdr.cdc.gov/ToxProfiles/tp.asp?id=81&tid=20

ATSDR (2020) Toxicological profile for glyphosate. Online:https://wwwn.cdc.gov/tsp/ToxProfiles/ToxProfiles.aspx?id=1488&tid=293

ATSDR (2023a) Toxicological profile for hexachlorocyclohexane. Atlanta, GA, USA. Online: https://www.atsdr.cdc.gov/Toxprofiles/tp43.pdf

ATSDR (2023b) Interaction profile for mixtures of insecticides: Pyrethroids, organophosphate compounds, and carbamates. https://www.atsdr.cdc.gov/interactionprofiles/ip-15/ip15.pdf

Babina K, Dollard M, Pilotto L, Edwards JW (2012) Environmental exposure to organophosphorus and pyrethroid pesticides in South Australian preschool children: a cross sectional study. Environ Int 48:109–120

Baker SE, Serafim AB, Morales-Agudelo P, Vidal M, Calafat AM, Ospina M (2019) Quantification of DEET and neonicotinoid pesticide biomarkers in human urine by online solid-phase extraction high-performance liquid chromatography-tandem mass spectrometry. Anal Bioanal Chem 411:669–678

Bandow N, Conrad A, Kolossa-Gehring M, Murawski A, Sawal G (2020) Polychlorinated biphenyls (PCB) and organochlorine pesticides (OCP) in blood plasma - Results of the German environmental survey for children and adolescents 2014–2017 (GerES V). Int J Hyg Environ Health 224:113426

Barr DB (2008) Biomonitoring of exposure to pesticides. J Chem Health Saf Nov/Dec: 20–29

Barr DB, Olsson AO, Wong LY, Udunka S, Baker SE, Whitehead RD, Magsumbol MS, Williams BL, Needham LL (2010) Urinary concentrations of metabolites of pyrethroid insecticides in the general U.S. population: National Health and Nutrition Examination Survey 1999–2002. Environ Health Perspect 118:742–748

Barr JR, Woolfitt AR, Maggio VL, Patterson DG Jr (2004) Measurement of toxaphene congeners in pooled human serum collected in three U.S. cities using high-resolution mass spectrometry. Arch Environ Contam Toxicol 46:551–556

Baumert BO, Fiedler N, Prapamontol T, Naksen W, Panuwet P, Hongsibsong S, Wongkampaun A, Thongjan N, Lee G, Sittiwang S, Dokjunyam C, Promkam N, Pingwong S, Suttiwan P, Siriwong W, Barry Ryan P, Boyd Barr D; SAWASDEE birth cohort investigative team (2022) Urinary concentrations of dialkylphosphate metabolites of organophosphate pesticides in the study of Asian Women and their Offspring's Development and Environmental Exposures (SAWASDEE). Environ Int 158:106884

Bawa P, Bedi JS, Gill JPS, Aulakh RS, Kumar A, Arora K (2018) Persistent organic pollutants residues in human breast milk from Bathinda and Ludhiana districts of Punjab, India. Arch Environ Contam Toxicol 75:512–520

Bedi JS, Gill JP, Aulakh RS, Kaur P, Sharma A, Pooni PA (2013) Pesticide residues in human breast milk: risk assessment for infants from Punjab, India. Sci Total Environ 463–464:720–726

Ben Khadda Z, Bungau SG, El Balkhi S, Ezrari S, Radu AF, Houssaini TS, Achour S (2025) Urinary biomonitoring of exposure to glyphosate and its metabolite amino-methyl phosphonic acid among farmers and non-farmers in Morocco. Environ Toxicol Pharmacol 113:104620

Benbrook CM (2016) Trends in glyphosate herbicide use in the United States and globally. Environ Sci Eur 28:3

Benbrook CM (2019) How did the US EPA and IARC reach diametrically opposed conclusions on the genotoxicity of glyphosate-based herbicides? Environ Sci Eur 31:2

Bennett DH, Busgang SA, Kannan K, Parsons PJ, Takazawa M, Palmer CD, Schmidt RJ, Doucette JT, Schweitzer JB, Gennings C, Hertz-Picciotto I (2022) Environmental exposures to pesticides, phthalates, phenols and trace elements are associated with neurodevelopment in the CHARGE study. Environ Int 161:107075

Bergonzi R, Specchia C, Dinolfo M, Tomasi C, De Palma G, Frusca T, Apostoli P (2009) Distribution of persistent organochlorine pollutants in maternal and foetal tissues: data from an Italian polluted urban area. Chemosphere 76:747–754

Berman T, Göen T, Novack L, Beacher L, Grinshpan L, Segev D, Tordjman K (2016) Urinary concentrations of organophosphate and carbamate pesticides in residents of a vegetarian community. Environ Int 96:34–40

Berman T, Goldsmith R, Göen T, Spungen J, Novack L, Levine H, Amitai Y, Shohat T, Grotto I (2013) Urinary concentrations of environmental contaminants and phytoestrogens in adults in Israel. Environ Int 59:478–484

Berni I, Menouni A, Creta M, El Ghazi I, Duca RC, Godderis L, El Jaafari S (2023) Exposure of children to glyphosate in Morocco: Urinary levels and predictors of exposure. Environ Res 217:114868

Bienvenu JF, Bélanger P, Gaudreau É, Provencher G, Fleury N (2021) Determination of glyphosate, glufosinate and their major metabolites in urine by the UPLC-MS/MS method applicable to biomonitoring and epidemiological studies. Anal Bioanal Chem 413:2225–2234

Björvang RD, Vinnars MT, Papadogiannakis N, Gidlöf S, Mamsen LS, Mucs D, Kiviranta H, Rantakokko P, Ruokojärvi P, Lindh CH, Andersen CY, Damdimopoulou P (2021) Mixtures of persistent organic pollutants are found in vital organs of late gestation human fetuses. Chemosphere 283:131125

Bliznashka L, Roy A, Christiani DC, Calafat AM, Ospina M, Diao N, Mazumdar M, Jaacks LM (2023) Pregnancy pesticide exposure and child development in low- and middle-income countries: a prospective analysis of a birth cohort in rural Bangladesh and meta-analysis. PLoS ONE 18(6):e0287089

Bouwman H, Kylin H (2009) Malaria control insecticide residues in breast milk: The need to consider infant health risks. Environ Health Perspect 117:1477–1480

Bouwman H, Sereda B, Meinhardt HM (2006) Simultaneous presence of DDT and pyrethroid residues in human breast milk from a malaria endemic area in South Africa. Environ Pollut 144:902–917

Bradman A, Quirós-Alcalá L, Castorina R, Aguilar Schall R, Camacho J, Holland NT, Barr DB, Eskenazi B (2015) Effect of organic diet intervention on pesticide exposures in young children living in low-income urban and agricultural communities. Environ Health Perspect 123:1086–1093

Brahmand MB, Yunesian M, Nabizadeh R, Nasseri S, Alimohammadi M, Rastkari N (2019) Evaluation of chlorpyrifos residue in breast milk and its metabolite in urine of mothers and their infants feeding exclusively by breast milk in north of Iran. J Environ Health Sci Eng 17:817–825

Bravo N, Garí M, Grimalt JO (2022) Occupational and residential exposures to organophosphate and pyrethroid pesticides in a rural setting. Environ Res 214(Pt 4):114186

Bravo N, Grimalt JO, Bocca B, Pino A, Bin M, Brumatti LV, Rosolen V, Barbone F, Ronfani L, Alimonti A, Calamandrei G (2019) Urinary metabolites of organophosphate and pyrethroid pesticides in children from an Italian cohort (PHIME, Trieste). Environ Res 176:108508

Bravo N, Grimalt JO, Mazej D, Tratnik JS, Sarigiannis DA, Horvat M (2020) Mother/child organophosphate and pyrethroid distributions. Environ Int 134:105264

Buchholz BA, Ahn KC, Huang H, Gee SJ, Stewart BJ, Ognibene TJ, Hammock BD (2021) Pharmacokinetics, metabolite measurement, and biomarker identification of dermal exposure to permethrin using accelerator mass spectrometry. Toxicol Sci 183:49–59

Buekers J, Remy S, Bessems J, Govarts E, Rambaud L, Riou M, Halldorsson TI, Olafsdottir K, Probst-Hensch N, Ammann P, Weber T, Kolossa-Gehring M, Esteban-Lopez M, Castano A, Andersen HR, Schoeters G (2022a) Glyphosate and AMPA in human urine of HBM4EU-aligned studies: Part B adults. Toxics 10:552

Buekers J, Remy S, Bessems J, Govarts E, Rambaud L, Riou M, Tratnik JS, Stajnko A, Katsonouri A, Makris KC, De Decker A, Morrens B, Vogel N, Kolossa-Gehring M, Esteban-Lopez M, Castano A, Andersen HR, Schoeters G (2022b) Glyphosate and AMPA in human urine of HBM4EU aligned studies: Part A children. Toxics 10:470

Burns CJ, LaKind JS (2022) Elements to increase translation in pyrethroid epidemiology research: a review. Sci Total Environ 813:152568

Burns CJ, Pastoor TP (2018) Pyrethroid epidemiology: a quality-based review. Crit Rev Toxicol 48:297–311

Bustamante CM, Polanska K, Jankowska A, Jerzyńska J, Grimalt JO, Garí M (2025) In-depth screening of pesticide exposure in children from urban and agricultural settings in Poland. Sci Total Environ 964:178501

Butler Walker J, Seddon L, McMullen E, Houseman J, Tofflemire K, Corriveau A, Weber JP, Mills C, Smith S, Van Oostdam JV (2003) Organochlorine levels in maternal and umbilical cord blood plasma in Arctic Canada. Sci Total Environ 302:27–52

Camiccia M, Candiotto LZP, Gaboardi SC, Panis C, Kottiwitz LBM (2022) Determination of glyphosate in breast milk of lactating women in a rural area from Paraná state. Brazil Braz J Med Biol Res 55:e12194

Campbell G, Mannetje A, Keer S, Eaglesham G, Wang X, Lin CY, Hobson P, Toms LM, Douwes J, Thomas KV, Mueller JF, Kaserzon S (2022) Characterization of glyphosate and AMPA concentrations in the urine of Australian and New Zealand populations. Sci Total Environ 847:15758

CARC (Cancer Assessment Review Committee) (2015) Cancer Assessment Document. evaluation of the carcinogenic potential of glyphosate. Final report, October 1, 2015. Online: file:///C:/Users/herma/Downloads/EPA-HQ-OPP-2016-0385-0014_content.pdf

Casida JE, Quistad GB (2004) Organophosphate toxicology: safety aspects of nonacetylcholinesterase secondary targets. Chem Res Toxicol 17:983–998

Castiello F, Suárez B, Gómez-Vida J, Torrent M, Fernández MF, Olea N, Freire C (2023) Exposure to non-persistent pesticides and sexual maturation of Spanish adolescent males. Chemosphere 324:138350

Castorina R, Bradman A, Fenster L, Barr DB, Bravo R, Vedar MG, Harnly ME, McKone TE, Eisen EA, Eskenazi B (2010) Comparison of current-use pesticide and other toxicant urinary metabolite levels among pregnant women in the CHAMACOS cohort and NHANE. Environ Health Perspect 118:856–863

CDC (2022) Biomonitoring data tables for environmental chemicals. Pyrethroid metabolites. Online: https://www.cdc.gov/exposurereport/report/pdf/Pyrethroid%20Metabolites%20%20NHANES-p.pdf

CDC (2023) National report on human exposure to environmental chemicals. In: Bomonitoring data tables for environmental chemicals. https://www.cdc.gov/exposurereport/data_tables.html

CDPR (California Department of Pesticide Regulation) (2018) California neonicotinoid risk determination. Sacramento, California Online: https://www.cdpr.ca.gov/docs/registration/reevaluation/chemicals/neonicotinoid_risk_determination.pdf

Čechová E, Seifertová M, Kukučka P, Vojta Š, Quaak I, de Cock M, van de Bor M, Kočan A (2017) An effective clean-up technique for GC/EI-HRMS determination of developmental neurotoxicants in human breast milk. Anal Bioanal Chem 409:1311–1322

Cequier E, Sakhi AK, Haug LS, Thomsen C (2017) Exposure to organophosphorus pesticides in Norwegian mothers and their children: Diurnal variability in concentrations of their biomarkers and associations with food consumption. Sci Total Environ 590–591:655–662

Chen D, Liu Z, Barrett H, Han J, Lv B, Li Y, Li J, Zhao Y, Wu Y (2020) Nationwide biomonitoring of neonicotinoid insecticides in breast milk and health risk assessment to nursing infants in the Chinese population. J Agric Food Chem 68:13906–13915

Chen MW, Santos HM, Que DE, Gou YY, Tayo LL, Hsu YC, Chen YB, Chen FA, Chao HR, Huang KL (2018) Association between organochlorine pesticide levels in breast milk and their effects on female reproduction in a Taiwanese population. Int J Environ Res Public Health 15:931

Chen X, Panuwet P, Hunter RE, Riederer AM, Bernoudy GC, Barr DB, Ryan PB (2014) Method for the quantification of current use and persistent pesticides in cow milk, human milk and baby formula using gas chromatography tandem mass spectrometry. J Chromatogr B Analyt Technol Biomed Life Sci 970:121–130

Chronister BNC, Yang K, Yang AR, Lin T, Tu XM, Lopez-Paredes D, Checkoway H, Suarez-Torres J, Gahagan S, Martinez D, Barr D, Moore RC, Suarez-Lopez JR (2023) Urinary glyphosate, 2,4-D and DEET biomarkers in relation to neurobehavioral performance in ecuadorian adolescents in the ESPINA Cohort. Environ Health Perspect 131(10):107007

Chrustek A, Hołyńska-Iwan I, Dziembowska I, Bogusiewicz J, Wróblewski M, Cwynar A, Olszewska-Słonina D (2018) Current research on the safety of pyrethroids used as insecticides. Medicina (Kaunas) 54:61

Cimino AM, Boyles AL, Thayer KA, Perry MJ (2017) Effects of neonicotinoid pesticide exposure on human health: a systematic review. Environ Health Perspect 125:155–162

Clune AL, Ryan PB, Barr DB (2012) Have regulatory efforts to reduce organophosphorus insecticide exposures been effective? Environ Health Perspect 120:521–525

CMHS (Canadian Health Measures Survey) (2019) Fifth report on human biomonitoring of environmental chemicals in Canada. https://www.canada.ca/content/dam/hc-sc/documents/services/environmental-workplace-health/reports-publications/environmental-contaminants/fifth-report-human-biomonitoring/pub1-eng.pdf

Coelhan M, Parlar H (1996) The nomenclature of chlorinated bornanes and camphenes relevant to toxaphene. Chemosphere 32:217–228

Colosio C, Fustinoni S, Birindelli S, Bonomi I, De Paschale G, Mammone T, Tiramani M, Vercelli F, Visentin S, Maroni M (2002) Ethylenethiourea in urine as an indicator of exposure to mancozeb in vineyard workers. Toxicol Lett 134:133–140

Connolly A, Coggins MA, Koch HM (2020) Human biomonitoring of glyphosate exposures: State-of-the-art and future research challenges. Toxics 8:60

Connolly A, Koch HM, Bury D, Koslitz S, Kolossa-Gehring M, Conrad A, Murawski A, McGrath JA, Leahy M, Brüning T, Coggins MA (2022) A human biomonitoring study assessing glyphosate and aminomethylphosphonic acid (AMPA) exposures among farm and non-farm families. Toxics 10:690

Conrad A, Schroter-Kermani C, Hoppe W, Ruther M, Pieper S, Kolossa-Gehring M (2017) Glyphosate in German adults—time trend (2001 to 2015) of human exposure to a widely used herbicide. Int J Hyg Environ Health 220:8–16

Corcellas C, Feo ML, Torres JP, Malm O, Ocampo-Duque W, Eljarrat E, Barceló D (2012) Pyrethroids in human breast milk: occurrence and nursing daily intake estimation. Environ Int 47:17–22

COT (Committee on Toxicity of Chemicals in Food, Consumer Products and the Environment) (1999) Organophosphates

Craddock HA, Huang D, Turner PC, Quirós-Alcalá L, Payne-Sturges DC (2019) Trends in neonicotinoid pesticide residues in food and water in the United States, 1999–2015. Environ Health 18:7

Croes K, Colles A, Koppen G, Govarts E, Bruckers L, Van de Mieroop E, Nelen V, Covaci A, Dirtu AC, Thomsen C, Haug LS, Becher G, Mampaey M, Schoeters G, Van Larebeke N, Baeyens W (2012) Persistent organic pollutants (POPs) in human milk: a biomonitoring study in rural areas of Flanders (Belgium). Chemosphere 89:988–994

Curwin BD, Hein MJ, Sanderson WT, Striley C, Heederik D, Kromhout H, Reynolds SJ, Alavanja MC (2007) Urinary pesticide concentrations among children, mothers and fathers living in farm and non-farm households in Iowa. Ann Occup Hyg 51:53–65

Dalsager L, Fage-Larsen B, Bilenberg N, Jensen TK, Nielsen F, Kyhl HB, Grandjean P, Andersen HR (2019) Maternal urinary concentrations of pyrethroid and chlorpyrifos metabolites and attention deficit hyperactivity disorder (ADHD) symptoms in 2-4-year-old children from the Odense Child Cohort. Environ Res 176:108533

Danish EPA (Danish Environmental Protection Agency) (2016) Organophosphate metabolites in urine samples from Danish children and women Measured in the Danish DEMOCOPHES population. Pesticide research No. 164. https://www2.mst.dk/udgiv/publications/2016/08/978-87-93529-03-8.pdf

Davis MD, Wade EL, Restrepo PR, Roman-Esteva W, Bravo R, Kuklenyik P, Calafat AM (2013) Semi-automated solid phase extraction method for the mass spectrometric quantification of 12 specific metabolites of organophosphorus pesticides, synthetic pyrethroids, and select herbicides in human urine. J Chromatogr B Analyt Technol Biomed Life Sci 929:18–26

De Roos AJ, Fritschi L, Ward MH, Monnereau A, Hofmann J, Bernstein L, Bhatti P, Benavente Moreno Y, Benke G, Casabonne D, Clavel J, Cocco P, Huynh T, 't Mannetje A, Miligi L, Piro S, Rothman N, Schinasi LH, Vajdic CM, Wang SS, Zhang Y, Slager SL, Cerhan JR (2022) Herbicide use in farming and other jobs in relation to non-Hodgkin's lymphoma (NHL) risk Occup Environ Med 79:795–806

Deng F, He J, Dai Y, Peng R, Pan X, Yuan J, Tan L (2024) Biomonitoring urinary pesticide metabolites in preschool children by supported liquid extraction and ultra-high performance liquid chromatography-tandem mass spectrometry and their association with oxidative stress. J Chromatogr A 1725:464944

Denghel H, Göen T (2018) Simultaneous assessment of phenolic metabolites in human urine for a specific biomonitoring of exposure to organophosphate and carbamate pesticides. Toxicol Lett 298:33–41

Dereumeaux C, Mercier F, Soulard P, Hulin M, Oleko A, Pecheux M, Fillol C, Denys S, Quenel P (2022) Identification of pesticides exposure biomarkers for residents living close to vineyards in France. Environ Int 159:107013

Dewailly E, Forde M, Robertson L, Kaddar N, Laouan Sidi EA, Côté S, Gaudreau E, Drescher O, Ayotte P (2014) Evaluation of pyrethroid exposures in pregnant women from 10 Caribbean countries. Environ Int 63:201–206

DFG (German Research Foundation) (2008) Neuropathy target esterase (NTE) The MAK-collection Part IV: biomonitoring methods, vol 11

Du J, Gridneva Z, Gay MC, Lai CT, Trengove RD, Hartmann PE, Geddes DT (2016) Longitudinal study of pesticide residue levels in human milk from Western Australia during 12 months of lactation: exposure assessment for infants. Sci Rep 6:38355

Du J, Gridneva Z, Gay MCL, Trengove RD, Hartmann PE, Geddes DT (2017) Pesticides in human milk of Western Australian women and their influence on infant growth outcomes: a cross-sectional study. Chemosphere 167:247–254

Dwivedi N, Mahdi AA, Deo S (2023) Assessment of endocrine disrupting chemicals in breast milk: association with dietary habits and duration of lactation. Environ Res 221:115216

Eaton JL, Cathey AL, Fernandez JA, Watkins DJ, Silver MK, Milne GL, Velez-Vega C, Rosario Z, Cordero J, Alshawabkeh A, Meeker JD (2022) The association between urinary glyphosate and aminomethyl phosphonic acid with biomarkers of oxidative stress among pregnant women in the PROTECT birth cohort study. Ecotoxicol Environ Saf 233:113300

ECHA (2010) Assessment Report. N,N- diethyl-meta-toluamide (DEET) Product-type 19 (Repellents and attractants). https://echa.europa.eu/documents/10162/a9b111f6-37b7-c179-dce4-361b62174844d

ECHA (European Chemicals Agency, Committee for Risk Assessment) (2022) Opinion proposing harmonised classification and labelling at EU level of glyphosate (ISO); N-(phosphonomethyl)glycine. Online: https://echa.europa.eu/documents/10162/5702e99d-d503-f154-226f-d8ab070ac47a

EEA (European Environment Agency) (2023) How pesticides impact human health and ecosystems in Europe. https://www.eea.europa.eu/en/analysis/publications/how-pesticidesimpact-human-health

EFSA (2005a) Opinion of the Scientific Panel on contaminants in the food chain [CONTAM] related to gamma-HCH and other hexachlorocyclohexanes as undesirable substances in animal feed. EFSA J 250:1–39. Online: https://www.efsa.europa.eu/de/efsajournal/pub/250

EFSA (2005b) Conclusion regarding the peer review of the pesticide risk assessment of the active substance glufosinate. EFSA Sci Rep 27:1–81. https://www.efsa.europa.eu/en/efsajournal/pub/rn-27

EFSA (2006a) Opinion of the Scientific Panel on contaminants in the food chain [CONTAM] related to Hexachlorobenzene as undesirable substance in animal feed. EFSA J 402:1–49. Online: https://www.efsa.europa.eu/en/efsajournal/pub/402

EFSA (2006b) Opinion of the scientific panel on contaminants in the food chain [CONTAM] related to DDT as an undesirable substance in animal feed. EFSA J 433:1–69. Online: https://www.efsa.europa.eu/en/efsajournal/pub/433

EFSA (2015) Conclusion on the peer review of the pesticide risk assessment of the active substance glyphosate. EFSA J 13:4302. Online: https://efsa.onlinelibrary.wiley.com/doi/abs/https://doi.org/10.2903/j.efsa.2015.4302

EFSA (2019) Statement on the available outcomes of the human health assessment in the context of the pesticides peer review of the active substance chlorpyrifos. https://www.efsa.europa.eu/en/efsajournal/pub/5809
EFSA (2020) Conclusion on the peer review of the pesticide risk assessment of the active substance mancozeb. EFSA J 18:5755
EFSA (2022) Scientific opinion on toxicity of pyrethroid common metabolites. EFSA J 20:7582. Online: https://www.efsa.europa.eu/en/efsajournal/pub/7582
EFSA (2023) Peer review of the pesticide risk assessment of the active substance glyphosate. EFSA J 21:8164. Online: file:///C:/Users/herma/Downloads/EFSA%20Journal%20-%202023%20-%20-%20Peer%20review%20of%20the%20pesticide%20risk%20assessment%20of%20the%20active%20substance%20glyphosate.pdf
Egeghy PP, Cohen Hubal EA, Tulve NS, Melnyk LJ, Morgan MK, Fortmann RC, Sheldon LS (2011) Review of pesticide urinary biomarker measurements from selected US EPA children's observational exposure studies. Int J Environ Res Public Health 8:1727–1754
Eik Anda E, Nieboer E, Dudarev AA, Sandanger TM, Odland JØ (2007) Intra- and intercompartmental associations between levels of organochlorines in maternal plasma, cord plasma and breast milk, and lead and cadmium in whole blood, for indigenous peoples of Chukotka, Russia. J Environ Monit 9:884–893
El-Saeid MH, Hassanin AS, Bazeyad AY (2021) Levels of pesticide residues in breast milk and the associated risk assessment. Saudi J Biol Sci 28:3741–3744
EMR (Expert Market Research) (2023) Market report. Historical and forecast market analysis. Online: https://www.expertmarketresearch.com/reports/pyrethroids-market
EUROSTAT (2023) Data browser. Pesticide sales. https://ec.europa.eu/eurostat/databrowser/view/AEI_FM_SALPEST09$DEFAULTVIEW/default/table
Fagan J, Bohlen L, Patton S, Klein K (2020) Organic diet intervention significantly reduces urinary glyphosate levels in U.S. children and adults. Env Res 189:109898.
Faniband MH, Nofen E, Littorin M, Lindh CH (2021) Human experimental exposure to glyphosate and biomonitoring of young Swedish adults. Int J HygEnviron Health 231:113657
Feo ML, Eljarrat E, Manaca MN, Dobaño C, Barcelo D, Sunyer J, Alonso PL, Menendez C, Grimalt JO (2012) Pyrethroid use-malaria control and individual applications by households for other pests and home garden use. Environ Int 38:67–72
Ferreira C, Duarte SC, Costa E, Pereira AMPT, Silva LJG, Almeida A, Lino C, Pena A (2021) Urine biomonitoring of glyphosate in children: Exposure and risk assessment. Env Res 198:111294
Fiedler H, Li X, Zhang J (2023) Persistent organic pollutants in human milk from primiparae—correlations, global, regional, and national time-trends. Chemosphere 313:137484
Fišerová PS, Kohoutek J, Degrendele C, Dalvie MA, Klánová J (2021) New sample preparation method to analyse 15 specific and non-specific pesticide metabolites in human urine using LC-MS/MS. J Chromatogr B Analyt Technol Biomed Life Sci 1166:122542
Forde MS, Robertson L, Laouan Sidi EA, Côté S, Gaudreau E, Drescher O, Ayotte P (2015) Evaluation of exposure to organophosphate, carbamate, phenoxy acid, and chlorophenol pesticides in pregnant women from 10 Caribbean countries. Environ Sci Process Impacts 17:1661–1671
Fujii Y, Harada KH, Ito Y, Yoshitake M, Matsunobu C, Kato Y, Ohta C, Koga N, Kimura O, Endo T, Koizumi A, Haraguchi K (2023) Profiles and determinants of dicofol, endosulfans, mirex, and toxaphenes in breast milk samples from 10 prefectures in Japan Chemosphere 311(Pt 1):137002
Fujii Y, Ito Y, Harada KH, Hitomi T, Koizumi A, Haraguchi K (2012) Comparative survey of levels of chlorinated cyclodiene pesticides in breast milk from some cities of China, Korea and Japa. Chemosphere 89:452–457
Fustinoni S, Campo L, Liesivuori J, Pennanen S, Vergieva T, van AL, Bosetti C, Van LH, Colosio C (2008) Biological monitoring and questionnaire for assessing exposure to ethylenebisdithiocarbamates in a multicenter European field study. Hum Exp Toxicol 27:681–691
Gao B, Poma G, Govindan M, Dumitrascu C, Bastiaensen M, Wang M, Covaci A (2022) Development of an analytical method based on solid-phase extraction and LC-MS/MS for the monitoring

of current-use pesticides and their metabolites in human urine. J Environ Sci: China Eng Edn 111:153–163

García-Villarino M, Riaño-Galán I, Rodríguez-Dehli AC, Freire C, Vizcaíno E, Grimalt JO, Tardón A, Fernández-Somoano A (2020) Association between pre/perinatal exposure to POPs and children's anogenital distance at age 4 years: a study from the INMA-Asturias cohort. Int J Hyg Environ Health 229:113563

Garí M, González-Quinteiro Y, Bravo N, Grimalt JO (2018) Analysis of metabolites of organophosphate and pyrethroid pesticides in human urine from urban and agricultural populations (Catalonia and Galicia). Sci Total Environ 622–623:526–533

Ghali H, Albers SE (2024) An updated review on the safety of N, N-diethyl-meta-toluamide insect repellent use in children and the efficacy of natural alternatives. Pediatr Dermatol 41(3):403–409

Gibson JC, Marro L, Brandow D, Remedios L, Fisher M, Borghese MM, Kieliszkiewicz K, Lukina AO, Irwin K (2023) Biomonitoring of DEET and DCBA in Canadian children following typical protective insect repellent use. Int J Hyg Environ Health 248:114093

Gillezeau C, van Gerwen M, Shaffer RM, Rana I, Zhang L, Sheppard L, Taioli E (2019) The evidence of human exposure to glyphosate: a review. Environ Health 18:2

Giorio C, Safer A, Sánchez-Bayo F, Tapparo A, Lentola A, Girolami V, van Lexmond MB, Bonmatin JM (2021) An update of the Worldwide Integrated Assessment (WIA) on systemic insecticides. Part 1: new molecules, metabolism, fate, and transport. Environ Sci Pollut Res Int 28:11716–11748

Göl E, Çok İ, Battal D, Şüküroğlu AA (2023) Assessment of preschool children's exposure levels to organophosphate and pyrethroid pesticide: A human biomonitoring study in two Turkish provinces. Arch Environ Contam Toxicol 84:318–331

Goulson D (2013) An overview of the environmental risks posed by neonicotinoid insecticides. J Appl Ecol 50:977–987

Govarts E, Gilles L, Rodriguez Martin L, Santonen T, Apel P, Alvito P, Anastasi E, Andersen HR, Andersson AM, Andryskova L, Antignac JP, Appenzeller B, Barbone F, Barnett-Itzhaki Z, Barouki R, Berman T, Bil W, Borges T, et al (2023) Harmonized human biomonitoring in European children, teenagers and adults: EU-wide exposure data of 11 chemical substance groups from the HBM4EU Aligned Studies (2014–2021) Int J Hyg Environ Health 249:114119

Grau D, Grau N, Gascuel Q, Paroissin C, Stratonovitch C, Lairon D, Devault DA, Di Cristofaro J (2022) Quantifiable urine glyphosate levels detected in 99% of the French population, with higher values in men, in younger people, and in farmers. Environ Sci Pollut Res 29:32882–32893

Guo LC, Zhu P, Gui C, Deng J, Gao Y, Long C, Zhang H, Lv Z, Yu S (2025) Disrupting effects of neonicotinoids and their interaction with metals on thyroid hormone, an evidence of children in a rural area South China. Ecotoxicol Environ Saf 290:117788

Haines DA, Saravanabhavan G, Werry K, Khoury C (2017) An overview of human biomonitoring of environmental chemicals in the Canadian Health Measures Survey: 2007–2019 Int J Hyg Environ Health 220(2 Pt A):13–28

Han J, Zhou L, Luo M, Liang Y, Zhao W, Wang P, Zhou Z, Liu D (2017) Nonoccupational exposure to pyrethroids and risk of coronary heart disease in the Chinese population. Environ Sci Technol 51:664–670

Handal AJ, Hund L, Paez M, Bear S, Greenberg C, Fenske RA, Barr DB (2016) Characterization of pesticide exposure in a sample of pregnant women in Ecuador. Arch Environ Contam Toxicol 70:627–639

Harada KH, Tanaka K, Sakamoto H, Imanaka M, Niisoe T, Hitomi T, Kobayashi H, Okuda H, Inoue S, Kusakawa K (2016) Biological monitoring of human exposure to neonicotinoids using urine samples, and neonicotinoid excretion kinetics. PLoS ONE 11:e0146335

Hardt J, Angerer J (1999a) Determination of metabolites of pirimicarb in human urine by gas chromatography-mass spectrometry. J Chromatogr B Biomed Sci Appl 730:229–238

Hardt J, Angerer J (1999b) Gas chromatographic method with mass-selective detection for the determination of 2-isopropoxyphenol in human urine. J Chromatogr B Biomed Sci Appl 723:139–145

HC (2010) Report on human biomonitoring of environmental chemicals in Canada results of the Canadian health measures survey cycle 1 (2007–2009). Online: https://www.canada.ca/content/dam/hc-sc/migration/hc-sc/ewh-semt/alt_formats/hecs-sesc/pdf/pubs/contaminants/chms-ecms/report-rapport-eng.pdf

HC (2017) Re-evaluation Decision RVD2017-01 Glyphosate. Online: https://publications.gc.ca/collections/collection_2017/sc-hc/H113-28/H113-28-2017-1-eng.pdf

HC (2019) Fifth Report on Human biomonitoring of environmental chemicals in Canada. Results of the Canadian Health Measures Survey Cycle 5 (2016–2017)

HC (2021) Sixth report on human biomonitoring of environmental chemicals in Canada. Minister of Health, Ottawa. Available: www.canada.ca/en/health-canada/services/environmental-workplacehealth/reports-publications/environmental-contaminants/sixth-report-human-biomonitoring.html

Heffernan AL, English K, Toms L, Calafat AM, Valentin-Blasini L, Hobson P, Broomhall S, Ware RS, Jagals P, Sly PD, Mueller JF (2016) Cross-sectional biomonitoring study of pesticide exposures in Queensland, Australia, using pooled urine samples. Environ Sci Pollut Res Int 23:23436–23448

Hernández AF, Lozano-Paniagua D, González-Alzaga B, Kavvalakis MP, Tzatzarakis MN, López-Flores I, Aguilar-Garduño C, Caparros-Gonzalez RA, Tsatsakis AM, Lacasaña M (2019) Biomonitoring of common organophosphate metabolites in hair and urine of children from an agricultural community. Environ Int 131:104997

Hirose Y, Kobayashi M, Koyama K, Kohda Y, Tanaka T, Honda H, Hori Y, Yoshida K, Kikuchi M (1999) A toxicokinetic analysis in a patient with acute glufosinate poisoning. Hum Exp Toxicol 18:305–308

Honeycutt Z, Rowlands H (2014) Glyphosate test results. Glyphosate testing full report: findings in American mothers' breast milk, urine and water. Online: https://www.momsacrossamerica.com/glyphosate_testing_results

Hoppe HW (2013) Determination of glyphosate residues in human urine samples from 18 European countries. Online: https://www.bund.net/fileadmin/user_upload_bund/publikationen/umweltgifte/glyphosat_urin_analyse.pdf

Huang K, Yi J, Liu G, Liu Y, Jiang K, Li Z, Qu Y, Lin M, Ma S (2025) Automatic magnetic solid phase extraction for rapid and high-throughput determination of neonicotinoid insecticides and their metabolites in serum, breast milk and urine samples. Anal Methods 17(4):772–782

Huang M, Wang Y, Wang Y, Lin G, Wen X, Xu X, Hong S, Chen Y, Lin H, Yang Z, Zhao K, Liu J, Wang J, Wang H, Wang N, Chen Y, Jiang Q (2024a) Exposure of pregnant women to neonicotinoids in Wenzhou City, East China: a biomonitoring study. Environ Int 189:108811

Huang K, Lin M, Yi J, Liu G, Hua R, Liu Y, Qu Y, Chen C, Ma S (2024b) Comparison of prenatal and postnatal exposure to neonicotinoids and their temporal trends in breast milk. Sci Total Environ 950: 175386.

Huang Z, Li H, Xiong J, You J (2022) Target and suspect screening of urinary biomarkers for current-use pesticides: application of a simple extraction method. Environ Toxicol Chem 41:73–80

Hung C-C, Simaremare SRS, Hsieh C-J, Yiin L-M (2019) Simultaneous determination of pyrethroid, organophosphate and carbamate metabolites in human urine by gas chromatography–mass spectrometry (GCMS). Appl Sci 9:879

IARC (2001) Hexachlorobenzene. In Some thyrotropic agents. IARC monographs on the evaluation of carcinogenic risk of chemicals to humans, vol 79. Lyon, France. Online: http://www.inchem.org/documents/iarc/vol79/79-13.html

IARC (2015a) DDT, Lindane, and 2,4-D. IARC Working Group. Monographs Volume 113: Lyon, France. Online: https://publications.iarc.fr/Book-And-Report-Series/Iarc-Monographs-On-The-Identification-Of-Carcinogenic-Hazards-To-Humans/DDT-Lindane-And-2-4-D-2016

IARC (2015b) Some organophosphate insecticides and herbicides/IARC Working Group on the evaluation of carcinogenic risks to humans. Lyon, France) (IARC monographs on the evaluation of carcinogenic risks to humans, vol 112)

Ichikawa G, Kuribayashi R, Ikenaka Y, Ichise T, Nakayama SMM, Ishizuka M, Taira K, Fujioka K, Sairenchi T, Kobashi G, Bonmatin JM, Yoshihara S (2019) LC-ESI/MS/MS analysis of neonicotinoids in urine of very low birth weight infants at birth. PLoS ONE 14:e0219208

Ikenaka Y, Miyabara Y, Ichise T, Nakayama S, Nimako C, Ishizuka M, Tohyama C (2019) Exposures of children to neonicotinoids in pine wilt disease control areas. Environ Toxicol Chem 38:71–79

JECFA (2012) Pesticide residues in food—2012: toxicological evaluations/joint meeting of the FAO panel of experts on pesticide residues in food and the environment and the WHO core assessment group on pesticide residues, Rome, Italy, 11–20 Sept 2012. file:///C:/Users/herma/Downloads/9789241665285_eng.pdf

Jia X, Li X, Deng F, He J, Li Q, Guo C, Yuan J, Tan L (2024) Longitudinal study and predictive modelling of urinary pesticide metabolite concentrations in residents of Guangzhou, China. Chemosphere 365:143353

Jia X, Guo C, Deng F, Li X, Bi H, Yuan J, Tan L (2025) Distribution and differential associations of neonicotinoids and their metabolites in paired urine and serum with liver function. Environ Sci Technol 59(27):13647–13659

JMPR (Joint FAO/WHO Meeting on Pesticide Residues) (2016) Summary report. Geneva, 9–13 May 2016. Online: https://www.fao.org/3/i8267en/I8267EN.pdf

Jokanović M, Oleksak P, Kuca K (2023) Multiple neurological effects associated with exposure to organophosphorus pesticides in man. Toxicology 484:153407

Jones K, Everard M, Harding AH (2014) Investigation of gastrointestinal effects of organophosphate and carbamate pesticide residues on young children. Int J Hyg Environ Health 217:392–398

Junqué E, Fernández P, Filippi I, Grimalt JO (2023) Determination of glyphosate and its derivative, aminomethylphosphonic acid, in human urine by gas chromatography coupled to tandem mass spectrometry and isotope pattern deconvolution. J Chromatogr Open 4:100087

Kaul L, Süss R, Zannettino A, Richter K (2021) The revival of dithiocarbamates: from pesticides to innovative medical treatments. iScience 24:102092

Kazui Y, Seto Y, Inoue H (2014) Phosphorus-specific determination of glyphosate, glufosinate, and their hydrolysis products in biological samples by liquid chromatography–inductively coupled plasma–mass spectrometry. Forensic Toxicol 32:317–322

Khemiri R, Cote J, Fetoui H, Bouchard M (2017) Documenting the kinetic time course of lambda-cyhalothrin metabolites in orally exposed volunteers for the interpretation of biomonitoring data. Toxicol Lett 276:115–121

Khemiri R, Côté J, Fetoui H, Bouchard M (2018) Kinetic time courses of lambda-cyhalothrin metabolites after dermal application of Matador EC 120 in volunteers. Toxicol Lett 296:132–138

Kim J, Leon ME, Schinasi LH, Baldi I, Lebailly P, Freeman LEB, Nordby KC, Ferro G, Monnereau A, Brouwer M, Kjaerheim K, Hofmann JN, Straif K, Kromhout H, Schüz J, Togawa K (2023) Exposure to pesticides and risk of Hodgkin lymphoma in an international consortium of agricultural cohorts (AGRICOH). Cancer Causes Control Jul 7

Knudsen LE, Hansen PW, Mizrak S, Hansen HK, Morck TA, Nielsen F, Siersma V, Mathiesen L (2017) Biomonitoring of Danish school children and mothers including biomarkers of PBDE and glyphosate. Rev Environ Health 32:279–290

Kokkinaki A, Kokkinakis M, Kavvalakis MP, Tzatzarakis MN, Alegakis AK, Maravgakis G, Babatsikou F, Fragkiadakis GA, Tsatsakis AM (2014) Biomonitoring of dialkylphosphate metabolites (DAPs) in urine and hair samples of sprayers and rural residents of Crete, Greece. Environ Res 134:181–187

Kuang L, Hou Y, Huang F, Guo A, Deng W, Sun H, Shen L, Lin H, Hong H (2020) Pesticides in human milk collected from Jinhua, China: Levels, influencing factors and health risk assessment. Ecotoxicol Environ Saf 205:111331

Kuklenyik P, Baker SE, Bishop AM, Morales-A P, Calafat AM (2013) On-line solid phase extraction-high performance liquid chromatography-isotope dilutiontandem mass spectrometry approach to quantify N,N-diethyl-m-toluamide and oxidative metabolites in urine. Anal Chim Acta 787:267–273

Lachat L, Glauser G (2018) Development and validation of an ultra-sensitive UHPLC-MS/MS method for neonicotinoid analysis in milk. J Agric Food Chem 66:8639–8646

Laird BD, Goncharov AB, Chan HM (2013) Body burden of metals and persistent organic pollutants among Inuit in the Canadian Arctic. Environ Int 59:33–40

Lee S, Ahn RM, Kim JH, Han YD, Lee JH, Son BS, Lee K (2019) Study design, rartionale and procedures for human biomonitoring of hazardous chemicals from foods and cooking in Korea. Int J Environ Res Public Health 16:2583

Lee S, Kim S, Lee HK, Lee IS, Park J, Kim HJ, Lee JJ, Choi G, Choi S, Kim S, Kim SY, Choi K, Kim S, Moon HB (2013) Contamination of polychlorinated biphenyls and organochlorine pesticides in breast milk in Korea: Time-course variation, influencing factors, and exposure assessment. Chemosphere 93:1578–1585

Lehmler HJ, Simonsen D, Garcia AQ, Irfan NM, Dean L, Wang H, von Elsterman M, Li X (2022) A systematic review of human biomonitoring studies of 3-phenoxybenzoic acid, a urinary biomarker pyrethroid insecticide exposure, 1997 to 2019. Hyg Environ Health Adv 4:100018

Lemke N, Murawski A, Schmied-Tobies MIH, Rucic E, Hoppe HW, Conrad A, Kolossa-Gehring M (2021) Glyphosate and aminomethylphosphonic acid (AMPA) in urine of children and adolescents in Germany – Human biomonitoring results of the German Environmental Survey 2014–2017 (GerES V). Environ Int 156:106769

Lesseur C, Pathak KV, Pirrotte P, Martinez MN, Ferguson KK, Barrett ES, Nguyen RHN, Sathyanarayana S, Mandrioli D, Swan SH, Chen J (2022) Urinary glyphosate concentration in pregnant women in relation to length of gestation. Environ Res 203:111811

Lewis KA, Tzilivakis J, Warner D, Green A (2016) An international database for pesticide risk assessments and management. Hum Ecological Risk Assess 22:1050–1064

Lewis RC, Cantonwine DE, Anzalota Del Toro LV, Calafat AM, Valentin-Blasini L, Davis MD, Baker SE, Alshawabkeh AN, Cordero JF, Meeker JD (2014) Urinary biomarkers of exposure to insecticides, herbicides, and one insect repellent among pregnant women in Puerto Rico. Environ Health 19(13):97

Li AJ, Chen Z, Lin TC, Buck Louis GM, Kannan K (2020) Association of urinary metabolites of organophosphate and pyrethroid insecticides, and phenoxy herbicides with endometriosis. Environ Int 136:105456

Li AJ, Kannan K (2018) Urinary concentrations and profiles of organophosphate and pyrethroid pesticide metabolites and phenoxyacid herbicides in populations in eight countries. Environ Int 121:1148–1154

Li AJ, Kannan K (2020) Profiles of urinary neonicotinoids and dialkylphosphates in populations in nine countries. Environ Int 145:106120

Li AJ, Martinez-Moral MP, Kannan K (2020) Variability in urinary neonicotinoid concentrations in single-spot and first-morning void and its association with oxidative stress markers. Environ Int 135:105415

Li M, Wang R, Su C, Li J, Wu Z (2022) Temporal trends of exposure to organochlorine pesticides in the United States: a population study from 2005 to 2016. Int J Environ Res Public Health 19:3862

Li R, Wan Y, He Z, Wang D, Xu S, Zhao X, Xia W (2025) Exposure to organophosphates, pyrethroids, neonicotinoids, and pentachlorophenol: spatial variations in urinary biomarkers and associations with oxidative stress based on a repeated-measure study. Sci Total Environ 969:178934

Li Y, Wang X, Toms LL, He C, Hobson P, Sly PD, Aylward LL, Mueller JF (2019) Pesticide metabolite concentrations in Queensland pre-schoolers – Exposure trends related to age and sex using urinary biomarkers. Environ Res 176:108532

Li ZM, Kannan K (2022) A Method for the analysis of glyphosate, aminomethylphosphonic acid, and glufosinate in human urine using liquid chromatography-tandem mass spectrometry. Int J Environ Res Public Health 19:4966

Li K, Fu M, Lei B, Shen X, Zhang X, Xu J, Zhang X (2025) Characteristics of neonicotinoids in colostrum from Shanghai, China (2007–2019): concentration levels, temporal trends, and potential health risk. Toxics 13(5):366

Lu D, Feng C, Wang D, Lin Y, Ip HS, She J, Xu Q, Wu C, Wang G, Zhou Z (2015a) Analysis of twenty phenolic compounds in human urine: hydrochloric acid hydrolysis, solid-phase extraction based on K2CO 3-treated silica, and gas chromatography tandem mass spectrometry. Anal Bioanal Chem 407:4131–4141

Lu D, Wang D, Ni R, Lin Y, Feng C, Xu Q, Jia X, Wang G, Zhou Z (2015b) Organochlorine pesticides and their metabolites in human breast milk from Shanghai, China. Environ Sci Pollut Res Int 22:9293–9306

Lucia RM, Liao X, Huang WL, Forman D, Kim A, Ziogas A, Norden-Krichmar TM, Goodman D, Alvarez A, Masunaka I, Pathak KV, McGilvrey M, Hegde AM, Pirrotte P, Park HL (2023) Urinary glyphosate and AMPA levels in a cross-sectional study of postmenopausal women: associations with organic eating behavior and dietary intake. Int J Hyg Environ Health 252:114211

Makris KC, Efthymiou N, Konstantinou C, Anastasi E, Schoeters G, Kolossa-Gehring M, Katsonouri A (2022) Oxidative stress of glyphosate, AMPA and metabolites of pyrethroids and chlorpyrifos pesticides among primary school children in Cyprus. Environ Res 212(Pt B):113316

Mat Sutris J, Md Isa Z, Sumeri SA, Ghazi HF (2016) Predictors of detected organophosphorus pesticides among Orang Asli children living in Malaysia. Ann Glob Health 82:770–778

McGuire MK, McGuire MA, Price WJ, Shafii B, Carrothers JM, Lackey KA, Goldstein DA, Jensen PK, Vicini JL (2016) Glyphosate and aminomethylphosphonic acid are not detectable in human milk. Am J Clin Nutr 103:1285–1290

Mdeni NL, Adeniji AO, Okoh AI, Okoh OO (2022) Analytical evaluation of carbamate and organophosphate pesticides in human and environmental matrices: a review. Molecules 27:618

Meeker JD, Barr DB, Serdar B, Rappaport SM, Hauser R (2007) Utility of urinary 1-naphthol and 2-naphthol levels to assess environmental carbaryl and naphthalene exposure in an epidemiology study. J Exp Sci Environ Epidemiol 17:314–320

Mellor E, Trasande L, Albergamo V, Kannan K, Li Z, Ghassabian A, Afanasyeva Y, Liu M, Cowell W (2024) Sociodemographic and dietary determinants of glyphosate exposure in a NYC-based pregnancy cohort. Environ Pollut 363(Pt 1):125083

Mendy A, Pinney SM (2022) Exposure to neonicotinoids and serum testosterone in men, women, and children. Environ Toxicol 37:1521–1528

Mesnage R, Ferguson S, Brandsma I, Moelijker N, Zhang G, Mazzacuva F, Caldwell A, Halket J, Antoniou MN (2022) The surfactant co-formulant POEA in the glyphosate-based herbicide RangerPro but not glyphosate alone causes necrosis in Caco-2 and HepG2 human cell lines and ER stress in the ToxTracker assay. Food Chem Toxicol 168:113380

Mikó Z, Hettyey A (2023) Toxicity of POEA-containing glyphosate-based herbicides to amphibians is mainly due to the surfactant, not to the active ingredient. Ecotoxicol 32:150–159

Mills PJ, Kania-Korwel I, Fagan J, McEvoy LK, Laughlin GA, Barrett-Connor E (2017) Excretion of the herbicide glyphosate in older adults between 1993 and 2016. JAMA 318:1610–1611

Moody RP, Wester RC, Melendres JL, Maibach HI (1992) Dermal absorption of the phenoxy herbicide 2,4-D dimethylamine in humans: effect of DEET and anatomic site. J Toxicol Environ Health 36(3):241–250

Mora AM, Hoppin JA, Córdoba L, Cano JC, Soto-Martínez M, Eskenazi B, Lindh CH, van Wendel de Joode B (2020) Prenatal pesticide exposure and respiratory health outcomes in the first year of life: results from the infants' Environmental Health (ISA) study. Int J Hyg Environ Health 225:113474

Mora-Gutiérrez A, Rubio C, Romero-López ÁA, Rubio-Osornio M (2021) Neurotoxic effects of insecticides chlorpyrifos, carbaryl, imidacloprid, in different animal species. In: Sabuncuoglu S (ed) Neurotoxicity-new advances. IntechOpen: Rijeka, Croatia

Mørck TA, Erdmann SE, Long M, Mathiesen L, Nielsen F, Siersma VD, Bonefeld-Jørgensen EC, Knudsen LE (2014) PCB concentrations and dioxin-like activity in blood samples from Danish school children and their mothers living in urban and rural areas. Basic Clin Pharmacol Toxicol 115:134–144

Morgan MK (2012) Children's exposures to pyrethroid insecticides at home: a review of data collected in published exposure measurement studies conducted in the United States. Int J Environ Res Public Health 9:2964–2985

Murínová ĽP, Wimmerová S, Lancz K, Patayová H, Koštiaková V, Richterová D, Govarts E, Jusko TA, Trnovec T (2017) Partitioning of hexachlorobenzene between human milk and blood lipid. Environ Pollut 229:994–999

Myridakis A, Chalkiadaki G, Fotou M, Kogevinas M, Chatzi L, Stephanou EG (2016) Exposure of preschool-age Greek children (RHEA cohort) to bisphenol A, parabens, phthalates, and organophosphates. Environ Sci Technol 50(2):932–941

Naksen W, Prapamontol T, Mangklabruks A, Chantara S, Thavornyutikarn P, Robson MG, Ryan PB, Barr DB, Panuwet P (2016) A single method for detecting 11 organophosphate pesticides in human plasma and breastmilk using GC-FPD. J Chromatogr B Analyt Technol Biomed Life Sci 1025:92–104

Nermo KR, Bakken KS, Lyche JL, Polder A, Jansen A, Kaldenbach S, Haddad-Weiser G, Strand TA, Eggesbø MÅ (2025) Quantification of persistent organic pollutants in breastmilk and estimated infant intake, Norway. Int J Hyg Environ Health 263:114458

Newsome WH, Ryan JJ (1999) Toxaphene and other chlorinated compounds in human milk from northern and southern Canada: a comparison. Chemosphere 39:519–526

Nguyen QD, Vu MN, Hebert AA (2023) Insect repellents: An updated review for the clinician. J Am Acad Dermatol 88(1):123–130

Nimako C, Ichise T, Hasegawa H, Akoto O, Boadi NO, Taira K, Fujioka K, Isoda N, Nakayama SMM, Ishizuka M, Ikenaka Y (2022) Assessment of ameliorative effects of organic dietary interventions on neonicotinoid exposure rates in a Japanese population. Environ Int 162:107169

Nishihama Y, Nakayama SF, Isobe T, Jung CR, Iwai-Shimada M, Kobayashi Y, Michikawa T, Sekiyama M, Taniguchi Y, Yamazaki S, On behalf of the Japan Environment And Children's Study Group (2021) Urinary metabolites of organophosphate pesticides among pregnant women participating in the Japan Environment and Children's Study (JECS). Int J Environ Res Public Health 18:5929

Nishihama Y, Nakayama SF, Isobe T, Kamijima M (2023) Association between maternal urinary neonicotinoid concentrations and child development in the Japan Environment and Children's Study. Environ Int 181:108267

Nishihara N, Isobe T, Takagi M, Tajima T, Kitahara Y, Hayashi M, Saito I, Watanabe S, Iwai-Shimada M, Ueyama J (2025) Determination of skin-insect repellent icaridin and DEET in human urine using solid-phase extraction and liquid chromatography with tandem mass spectrometry and its application to a sample of Japanese adults. Environ Health Prev Med 30:18

Nistor N, Frăsinariu OE, Ştreangă V (2017) Acute poisoning with neonicotinoid insecticide. Online. https://doi.org/10.5772/intechopen.72004

Nomura H, Hamada R, Saito I, Nakane K, Sawa R, Ukai M, Shibata E, Sato M, Kamijima M, Ueyama J (2020) Optimization and validation of a highly sensitive method for determining glyphosate in human urine by solid-phase extraction and liquid chromatography with tandem mass spectrometry: a methodological study. Environ Health Prev Med 25:83

Nomura H, Hamada R, Wada K, Saito I, Nishihara N, Kitahara Y, Watanabe S, Nakane K, Nagata C, Kondo T, Kamijima M, Ueyama J (2022) Temporal trend and cross-sectional characterization of urinary concentrations of glyphosate in Japanese children from 2006 to 2015. Int J Hyg Environ Health 242:113963

Norén E, Lindh C, Rylander L, Glynn A, Axelsson J, Littorin M, Faniband M, Larsson E, Nielsen C (2020) Concentrations and temporal trends in pesticide biomarkers in urine of Swedish adolescents, 2000–2017. J Expo Sci Environ Epidemiol 30:756–767

NRW (North Rhine-Westphalia Office of Nature, Environment and Consumer Protection) (2020) Determination of pollutants and pollutant metabolites in the urine of 2- to 6-year-old children from North Rhine-Westphalia (in German). https://www.lanuk.nrw.de/fileadmin/lanuv/gesundheit/pdf/2024/2024-01_Nachuntersuchung_DnHexP.pdf

Oehme M (2000) Toxaphene, the underestimated pesticide? Chemosphere 41:459

Ojelade BS, Durowoju OS, Adesoye PO, Gibb SW, Ekosse G-I (2022) Review of glyphosate-based herbicide and aminomethylphosphonic Acid (AMPA): environmental and health impacts. Appl Sci 12:8789

Osaka A, Ueyama J, Kondo T, Nomura H, Sugiura Y, Saito I, Nakane K, Takaishi A, Ogi H, Wakusawa S (2016) Exposure characterization of three major insecticide lines in urine of young children in Japan—neonicotinoids, organophosphates, and pyrethroids. Environ Res 147:89–96

Ospina M, Schütze A, Morales-Agudelo P, Vidal M, Wong LY, Calafat AM (2022) Exposure to glyphosate in the United States: data from the 2013–2014 National Health and Nutrition Examination Survey. Environ Int 170:107620

Ospina M, Wong L-Y, Baker S, Serafim AB, Morales-Agudelo P, Calafat AM (2019) Exposure to neonicotinoid insecticides in the US general population: Data from the 2015–2016 national health and nutrition examination survey. Environ Res 108555

Oya N, Ito Y, Hioki K, Asai Y, Aoi A, Sugiura Y, Ueyama J, Oguri T, Kato S, Ebara T, Kamijima M (2017) Quantitative analysis of organophosphate insecticide metabolites in urine extracted from disposable diapers of toddlers in Japan. Int J Hyg Environ Health 220(2 Pt A):209–216

Palaniswamy S, Salah S, Přibylová P, Kukučka P, Laatio L, Sieppi E, Kummu M, Rautio A, Myllynen P, Abass K (2015) Mother-child exposure to polychlorinated biphenyls and organochlorine pesticides in the Northern Finland cohort (NUGEN). Environ Toxicol Pharmacol 115:104679

Pan C, Yu J, Yao Q, Lin N, Lu Z, Zhang Y, Zhao S, Wang Z, Lei X, Tian Y, Gao Y (2022) Prenatal neonicotinoid insecticides exposure, oxidative stress, and birth outcomes. Environ Int 163:107180

Parizek O, Gramblicka T, Parizkova D, Polachova A, Bechynska K, Dvorakova D, Stupak M, Dusek J, Pavlikova J, Topinka J, Sram RJ, Pulkrabova J (2023) Assessment of organohalogenated pollutants in breast milk from the Czech Republic. Sci Total Environ 871:161938

Parvez S, Gerona RR, Proctor C, Friesen M, Ashby JL, Reiter JL, Lui Z, Winchester PD (2018) Glyphosate exposure in pregnancy and shortened gestational length: a prospective Indiana birth cohort study. Environ Health 17:23

Pathak VM, Verma VK, Rawat BS, Kaur B, Babu N, Sharma A, Dewali S, Yadav M, Kumari R, Singh S, Mohapatra A, Pandey V, Rana N, Cunill JM (2022) Current status of pesticide effects on environment, human health and it's eco-friendly management as bioremediation: a comprehensive review. Front Microbiol 13: 962619

Pedersen TL, Smilowitz JT, Winter CK, Emami S, Schmidt RJ, Bennett DH, Hertz-Picciotto I, Taha AY (2021) Quantification of nonpersistent pesticides in small volumes of human breast milk with ultrahigh performance liquid chromatography coupled to tandem mass spectrometry. J Agric Food Chem 69:6676–6689

Petropoulou SE, Gikas E, Tsarbopoulos A, Siskos PA (2006) Gas chromatographic–tandem mass spectrometric method for the quantitation of carbofuran, carbaryl and their main metabolites in applicators' urine. J Chromatogr A 1108:99–110

Pineda S, Lignell S, Gyllenhammar I, Lampa E, Benskin JP, Lundh T, Lindh C, Kiviranta H, Glynn A (2023) Exposure of Swedish adolescents to elements, persistent organic pollutants (POPs), and rapidly excreted substances - The Riksmaten adolescents 2016–17 national survey. Int J Hyg Environ Health 251:114196

Pisa L, Goulson D, Yang EC, Gibbons D, Sánchez-Bayo F, Mitchell E, Aebi A, van der Sluijs J, MacQuarrie CJK, Giorio C, Long EY, McField M, Bijleveld van Lexmond M, Bonmatin JM (2021) An update of the Worldwide Integrated Assessment (WIA) on systemic insecticides. Part 2: impacts on organisms and ecosystems. Environ Sci Pollut Res Int 28:11749–11797

Polder A, Gabrielsen GW, Odland JØ, Savinova TN, Tkachev A, Løken KB, Skaare JU (2008a) Spatial and temporal changes of chlorinated pesticides, PCBs, dioxins (PCDDs/PCDFs) and brominated flame retardants in human breast milk from Northern Russia. Sci Total Environ 391:41–54

Polder A, Thomsen C, Lindström G, Løken KB, Skaare JU (2008b) Levels and temporal trends of chlorinated pesticides, polychlorinated biphenyls and brominated flame retardants in individual human breast milk samples from Northern and Southern Norway. Chemosphere 73:14–23

Pollock T, Karthikeyan S, Walker M, Werry K, St-Amand A (2021) Trends in environmental chemical concentrations in the Canadian population: Biomonitoring data from the Canadian Health Measures Survey 2007–2017. Environ Int 155:106678

Raab U, Albrecht M, Preiss U, Völkel W, Schwegler U, Fromme H (2013) Organochlorine compounds, nitro musks and perfluorinated substances in breast milk - results from Bavarian Monitoring of Breast Milk 2007/8. Chemosphere 93:461–467

Radwan P, Wielgomas B, Radwan M, Krasiński R, Kilanowicz-Sapota A, Banaszczyk R, Jurewicz J (2022) Synthetic pyrethroids exposure and embryological outcomes: a cohort study in women from fertility clinic. Int J Environ Res Public Health 19:5117

Ratelle M, Cote J, Bouchard M (2015a) Time profiles and toxicokinetic parameters of key biomarkers of exposure to cypermethrin in orally exposed volunteers compared with previously available kinetic data following permethrin exposure. J Appl Toxicol 35:1586–1593

Ratelle M, Cote J, Bouchard M (2015b) Toxicokinetics of permethrin biomarkers of exposure in orally exposed volunteers. Toxicol Lett 232:369–375

Reid A, Callan A, Stasinska A, Heyworth J, Phi DT, Odland JO, Hinwood A (2013) Maternal exposure to organochlorine pesticides in Western Australia. Sci Total Environ 449:208–213

Rivero A, Vézilier J, Weill M, Read AF, Gandon S (2010) Insecticide control of vector-borne diseases: when is insecticide resistance a problem? PLoS Pathog 6:e1001000

Roca M, Miralles-Marco A, Ferré J, Pérez R, Yusà V (2014) Biomonitoring exposure assessment to contemporary pesticides in a school children population of Spain. Environ Res 131:77–85

Rodríguez-Gil JL, Prosser RS, Duke SO, Solomon KR (2021) Ecotoxicology of glyphosate, its formulants, and environmental degradation products. Rev Environ Contam Toxicol 255:129–205

Ruiz P, Dualde P, Coscolla C, Fernandez SF, Carbonell E, Yusa V (2021) Biomonitoring of glyphosate and AMPA in the urine of Spanish lactating mothers. Sci Total Environ 801:149688

Runkle JD, Tovar-Aguilar JA, Economos E, Flocks J, Williams B, Muniz JF, Semple M, McCauley L (2013) Pesticide risk perception and biomarkers of exposure in Florida female farmworkers. J Occup Environ Med 55:1286–1292

Sabbioni G, Castaño A, Esteban López M, Göen T, Mol H, Riou M, Tagne-Fotso R (2022) Literature review and evaluation of biomarkers, matrices and analytical methods for chemicals selected in the research program human biomonitoring for the european union (HBM4EU). Environ Int 169:107458

Samarawickrema N, Pathmeswaran A, Wickremasinghe R, Peiris-John R, Karunaratna M, Buckley N, Dawson A, de Silva J (2008) Fetal effects of environmental exposure of pregnant women to organophosphorus compounds in a rural farming community in Sri Lanka. Clin Toxicol (Phila) 46:489–495

Sams C, Patel K, Jones K (2010) Biological monitoring for exposure to pirimicarb: method development and a human oral dosing study. Toxicol Lett 192:56–60

Sandanger TM, Brustad M, Odland JO, Doudarev AA, Miretsky GI, Chaschin V, Burkow IC, Lund E (2003) Human plasma levels of POPs, and diet among native people from Uelen, Chukotka. J Environ Monit 5:689–696

Sanguos CL, Suárez OL, Martínez-Carballo E, Couce ML (2023) Postnatal exposure to organic pollutants in maternal milk in north-western Spain. Environ Pollut 318:120903

Santos ASE, Moreira JC, Rosa ACS, Câmara VM, Azeredo A, Asmus CIRF, Meyer A (2022) Persistent organic pollutant levels in maternal and cord blood plasma and breast milk: results from the Rio Birth Cohort Pilot study of environmental exposure and childhood development (PIPA Study). Int J Environ Res Public Health 20:778

Schettgen T, Dewes P, Kraus T (2016) A method for the simultaneous quantification of eight metabolites of synthetic pyrethroids in urine of the general population using gas chromatography-tandem mass spectrometry. Anal Bioanal Chem 408:5467–5478

Schoeters G, Verheyen VJ, Colles A, Remy S, Martin LR, Govarts E, Nelen V, Den Hond E, De Decker A, Franken C, Loots I, Coertjens D, Morrens B, Bastiaensen M, Gys C, Malarvannan G, Covaci A, Nawrot T, De Henauw S, Bellemans M, Leermakers M, Van Larebeke N, Baeyens W, Jacobs G, Voorspoels S, Nielsen F, Bruckers L (2022) Internal exposure of Flemish teenagers to environmental pollutants: results of the Flemish Environment and Health Study 2016–2020 (FLEHS IV). Int J Hyg Environ Health 242:113972

Schütze A, Morales-Agudelo P, Vidal M, Calafat AM, Ospina M (2021) Quantification of glyphosate and other organophosphorus compounds in human urine via ion chromatography isotope dilution tandem mass spectrometry. Chemosphere 274:129427

Segal TR, Mínguez-Alarcón L, Chiu YH, Williams PL, Nassan FL, Dadd R, Ospina M, Calafat AM, Hauser R (2017) Urinary concentrations of 3-(diethylcarbamoyl)benzoic acid (DCBA), a major metabolite of N,N-diethyl-m-toluamide (DEET) and semen parameters among men attending a fertility center. Hum Reprod 32(12):2532–2539

Selim S, Hartnagel RE Jr, Osimitz TG, Gabriel KL, Schoenig GP (1995) Absorption, metabolism, and excretion of N,N-diethyl-m-toluamide following dermal application to human volunteers. Fundam Appl Toxicol 25(1):95–100

Sepahi S, Gerayli S, Delirrad M, Taghavizadeh Yazdi ME, Zare-Zardini H, Bushehri B, Ghorani-Azam A (2023) Biochemical responses as early and reliable biomarkers of organophosphate and carbamate pesticides intoxication: a systematic literature review. J Biochem Mol Toxicol 37:e23285

Sha Y, Zhu W, Lin K, Yu S, Jin H, Chen Y, Lu S (2025) Dietary exposure to neonicotinoids and related metabolites in early-life nutrition: contamination profiles and health implications. Ecotoxicol Environ Saf 306:119315

Shahmoradi B, Maleki A, Kohzadi S, Khoubi J, Zandi S (2019) Levels of organochlorine pesticides in human breast milk in Marivan, west of Iran. J Advanc Environ Health Res 7:32–37

Sharma A, Kumar V, Shahzad B, Tanveer M, Sidhu GPS, Handa N, Kohli SK, Yadav P, Bali AS, Parihar RD, Dar OI, Sing K, Jasrotia S, Bakshi P, Ramakrishnan M, Kumar S, Bhardwaj R, Kumar A, Thukral AK (2019) Worldwide pesticide usage and its impacts on ecosystem. SN Appl Sci 1:1446

Silver MK, Fernandez J, Tang J, McDade A, Sabino J, Rosario Z, Velez-Vega C, Alshawabkeh A, Cordero JF, Meeker JD (2021) Prenatal exposure to glyphosate and Its environmental degradate, aminomethyl phosphonic acid(AMPA), and preterm birth: a nested case–control Study in the PROTECT Cohort (PuertoRico). Environ Health Perspect 129:57011

Simon-Delso N, Amaral-Rogers V, Belzunces LP, Bonmatin JM, Chagnon M, Downs C, Furlan L, Gibbons DW, Giorio C, Girolami V, Goulson D, Kreutzweiser DP, Krupke CH, Liess M, Long E, McField M, Mineau P, Mitchell EA, Morrissey CA, Noome DA, Pisa L, Settele J, Stark JD, Tapparo A, Van Dyck H, Van Praagh J, Van der Sluijs JP, Whitehorn PR, Wiemers M (2015) Systemic insecticides (neonicotinoids and fipronil): trends, uses, mode of action and metabolites. Environ Sci Pollut Res Int 22(1):5–34

Skopp S, Oehme M, Fürst P (2002) Enantiomer ratios, patterns and levels of toxaphene congeners in human milk from Germany. J Environ Monit 4:389–394

Smadi N, Jammoul A, El Darra N (2019) Assessment of antibiotic and pesticides residues in breast milk of Syrian refugee lactating mothers. Toxics 7:39

Soderlund DM (2012) Molecular mechanisms of pyrethroid insecticide neurotoxicity. Recent Adv Arch Toxicol 86:165–181

Song M, Kang K, Song F (2021) SARM1-mediated wallerian degeneration: a possible mechanism underlying organophosphorus-induced delayed neuropath. Med Hypotheses 155:110666

Song S, Ma X, Tong L, Tian Q, Huang Y, Yin S, Sun H (2013) Residue levels of hexachlorocyclohexane and dichlorodiphenyltrichloroethane in human milk collected from Beijing. Environ Monit Assess 185:7225–7229

Song S, Zhang T, Huang Y, Zhang B, Guo Y, He Y, Huang X, Bai X, Kannan K (2020) Urinary metabolites of neonicotinoid insecticides: Levels and recommendations for future biomonitoring studies in China. Environ Sci Technol 54:8210–8220

Soukup ST, Merz B, Bub A, Hoffmann I, Watzl B, Steinberg P, Kulling SE (2020) Glyphosate and AMPA levels in human urine samples and their correlation with food consumption: results of the cross-sectional KarMeN study in Germany. Arch Toxicol 94:1575–1584

Srivastava S, Narvi SS, Prasad SC (2011) Levels of select organophosphates in human colostrum and mature milk samples in rural region of Faizabad district, Uttar Pradesh, India. Hum Exp Toxicol 30:1458–1463

Stadler K, Li X, Liu B, Bao W, Wang K, Lehmler HJ (2022) Systematic review of human biomonitoring studies of ethylenethiourea, a urinary biomarker for exposure to dithiocarbamate fungicides. Environ Pollut 292(Pt B):118419

Stajnko A, Snoj Tratnik J, Kosjek T, Mazej D, Jagodic M, Eržen I, Horvat M (2020) Seasonal glyphosate and AMPA levels in urine of children and adolescents living in rural regions of Northeastern Slovenia. Environ Int 143:105985

Steinborn A, Alder L, Michalski B, Zomer P, Bendig P, Martinez SA, Mol HG, Class TJ, Pinheiro NC (2016) Determination of Glyphosate Levels in Breast Milk Samples from Germany by LC-MS/MS and GC-MS/MS. J Agric Food Chem 64:1414–1421

Šulc L, Janoš T, Figueiredo D, Ottenbros I, Šenk P, Mikeš O, Huss A, Čupr P (2022) Pesticide exposure among Czech adults and children from the CELSPAC-SPECIMEn cohort: Urinary biomarker levels and associated health risks. Environ Res 214(Pt 3):114002

Suwannarina N, Isobe T, Nishihama Y, Ito Y, Kamijima M, Ebara T, Sugiura-Ogasawara M, Nishikawa N, Nakai K, Minamikawa Y, Nakayama SF (2022) Concentrations of neonicotinoid insecticides and their metabolites in multiple urine samples collected from pregnant women in Japan. SSRN. Online: https://papers.ssrn.com/sol3/papers.cfm?abstract_id=4496788

Svarcova A, Lankova D, Gramblicka T, Stupak M, Hajslova J, Pulkrabova J (2019) Integration of five groups of POPs into one multi-analyte method for human blood serum analysis: an innovative approach within biomonitoring studies. Sci Total Environ 667:701–709

Tagne-Fotso R, Zeghnoun A, Saoudi A, Balestier A, Pecheux M, Chaperon L, Oleko A, Marchand P, Le Bizec B, Vattier L, Bouchart V, Limon G, Le Gléau F, Denys S, Fillol C (2023) Exposure of the general French population to herbicides, pyrethroids, organophosphates, organochlorines, and carbamate pesticides in 2014–2016: results from the Esteban Study. SSRN. https://papers.ssrn.com/sol3/papers.cfm?abstract_id=4452103

Taira K, Fujioka K, Aebi A, Glauser G, MitchellEAD, Ikenaka Y, Sanchez-Bayo F, Tapparo A, Lumawig-Heitzmann L, Bonmatin J-M, Safer A, Bijleveld van Lexmond M, Claveria F, Aoyama Y, Miyabara Y, Komori M (2022) Contrasted concentrations of neonicotinoids in human hair and urine suggest long term vs short-term exposure in two observational studies, international academics and rural population of the Philippines. Research Square. Online: https://doi.org/10.21203/rs.3.rs-1487454/v1

Tang W, Wang D, Wang J, Wu Z, Li L, Huang M, Xu S, Yan D (2018) Pyrethroid pesticide residues in the global environment: an overview. Chemosphere 191:990–1007

Tarazona JV, Court-Marques D, Tiramani M, Reich H, Pfeil R, Istace F, Crivellente F (2017) Glyphosate toxicity and carcinogenicity: a review of the scientific basis of the European Union assessment and its differences with IARC. Arch Toxicol 91:2723–2743

Thomas A, Toms LL, Harden FA, Hobson P, White NM, Mengersen KL, Mueller JF (2017) Concentrations of organochlorine pesticides in pooled human serum by age and gender. Environ Res 154:10–18

Thompson DA, Lehmler HJ, Kolpin DW, Hladik ML, Vargo JD, Schilling KE, LeFevre GH, Peeples TL, Poch MC, LaDuca LE, Cwiertny DM, Field RW (2020) A critical review on the potential impacts of neonicotinoid insecticide use: current knowledge of environmental fate, toxicity and implications for human health. Environ Sci Process Impacts 22:1315–1346

Torres-Moreno AC, Mejia-Grau K, Puente-DelaCruz L, Codling G, Villa AL, Ríos-Marquez O, Patequiva-Chauta L, Cobo M, Johnson-Restrepo B (2023) Polybrominated diphenyl ethers (PBDEs), polychlorinated biphenyls (PCBs), organochlorine pesticides (OCPs) in human breast milk from Colombia: a probabilistic risk assessment approach. Chemosphere 339:139597

Trasande L, Aldana SI, Trachtman H, Kannan K, Morrison D, Christakis DA, Whitlock K, Messito MJ, Gross RS, Karthikraj R, Sathyanarayana S (2020) Glyphosate exposures and kidney injury biomarkers in infants and young children. Environ Pollut 256:11333

Tush D, Loftin KA, Meyer MT (2013) Characterization of polyoxyethylene tallow amine surfactants in technical mixtures and glyphosate formulations using ultra-high performance liquid chromatography and triple quadrupole mass spectrometry. J Chromatogr A 1319:80–87

Ueyama J, Harada KH, Koizumi A, Sugiura Y, Kondo T, Saito I, Kamijima M (2015) Temporal levels of urinary neonicotinoid and dialkylphosphate concentrations in Japanese women between 1994 and 2011. Environ Sci Technol 49:14522–14528

Ueyama J, Ito Y, Hamada R, Oya N, Kato S, Matsuki T, Tamada H, Kaneko K, Saitoh S, Sugiura-Ogasawara M, Ebara T, Kamijima M (2022) Simultaneous quantification of pyrethroid metabolites in urine of non-toilet-trained children in Japan. Environ Health Prev Med 27:25

Ueyama J, Saito I, Takaishi A, Nomura H, Inoue M, Osaka A, Sugiura Y, Hayashi Y, Wakusawa S, Ogi H, Inuzuka K, Kamijima M, Kondo T (2014) A revised method for determination of dialkylphosphate levels in human urine by solid-phase extraction and liquid chromatography with tandem mass spectrometry: Application to human urine samples from Japanese children. Environ Health Prev Med 19:405–413

Umetsu N, Shirai Y (2020) Development of novel pesticides in the 21st century. J Pestic Sci 45:54–74

UNEP (2015) Road map for the development of alternatives to DDT. UNEP/DTI Chemicals Branch, Switzerland. Online: https://wedocs.unep.org/bitstream/handle/20.500.11822/29028/DDT_Road_Map_2015.pdf?sequence=1&isAllowed=y

US EPA (2017a) Revised glyphosate issue paper: evaluation of carcinogenic potential EPA's office of pesticide programs. December 12, 2017. Online: file:///C:/Users/herma/Downloads/EPA-HQ-OPP-2009-0361-0073_content.pdf

US EPA (2017b) Pesticides industry sales and usage 2008–2012. Market estimates. https://www.epa.gov/pesticides/pesticides-industry-sales-and-usage-2008-2012-market-estimates

US EPA (2019) Memorandum. Pyrethroids: Tier II epidemiology report. Online: https://www.epa.gov/sites/default/files/2019-08/documents/tier-ii-epidemiology-report.pdf

US EPA (2023a) Pesticide poisoning handbook—Section II insecticides. Chapter 5. In: Organophosphate insecticides. https://www.epa.gov/pesticide-worker-safety/pesticide-poisoning-handbook-section-ii-insecticides

US EPA (2023b) Pesticide poisoning handbook—Section II insecticides. Chapter 6. N-Methyl carbamate insecticides. https://www.epa.gov/pesticide-worker-safety/pesticide-poisoning-handbook-section-ii-insecticides

US EPA (2023c) What is a pesticide? Online: https://www.epa.gov/minimum-risk-pesticides/what-pesticide

van den Berg H, da Silva Bezerra HS, Al-Eryani S, Chanda E, Nagpal BN, Knox TB, Velayudhan R, Yadav RS (2021) Recent trends in global insecticide use for disease vector control and potential implications for resistance management. Sci Rep 11:23867

Viera MS, Ferronato G, Abreu DF, Prestes OD, Adaimea MB, Zanella R (2022) Simultaneous GC-NCI-MS determination of persistent organic pollutants and current-use pesticides in breast milk samples. J Braz Chem Soc 33:790–799

Wang A, Mahai G, Wan Y, Yang Z, He Z, Xu S, Xia W (2020) Assessment of imidacloprid related exposure using imidacloprid-olefin and desnitro-imidacloprid: Neonicotinoid insecticides in human urine in Wuhan, China. Environ Int 141:105785

Wang Y, Zhang Y, Ji L, Hu Y, Zhang J, Wang C, Ding G, Chen L, Kamijima M, Ueyama J, Gao Y, Tian Y (2017) Prenatal and postnatal exposure to organophosphate pesticides and childhood neurodevelopment in Shandong, China. Environ Int 108:119–126

Weldon RH, Barr DB, Trujillo C, Bradman A, Holland N, Eskenazi B (2011) A pilot study of pesticides and PCBs in the breast milk of women residing in urban and agricultural communities of California. J Environ Monit 13:3136–3144

Wendel V, de Joode B, Mora AM, Lindh CH, Hernández-Bonilla D, Córdoba L, Wesseling C, Hoppin JA, Mergler D (2016) Pesticide exposure and neurodevelopment in children aged 6–9 years from Talamanca, Costa Rica. Cortex 85:137–150

WHO (1986a) Organophophorus insecticides: a general introduction—Environmental Health Criteria No. 63. Geneva, Switzerland. https://apps.who.int/iris/bitstream/handle/10665/40198/9241542632-eng.pdf

WHO (1986b) Carbamate pesticides: a general introduction. Environmental Health Criteria 64. https://apps.who.int/iris/bitstream/handle/10665/38687/9241542640-eng.pdf

WHO (1988) Dithiocarbamate pesticides, ethylenethiourea and propylenethiourea: a general introduction. Environmental Health Criteria 78. https://apps.who.int/iris/bitstream/handle/10665/39117/9241542780-eng.pdf?sequence=1

WHO (2020) Vector-borne diseases. https://www.who.int/news-room/fact-sheets/detail/vector-borne-diseases

WHO (2024) Vector-borne diseases. https://www.who.int/news-room/fact-sheets/detail/vector-borne-diseases

Witczak A, Pohoryło A, Abdel-Gawad H (2021) Endocrine-disrupting organochlorine pesticides in human breast milk: changes during lactation. Nutrients 13:229

Woollen BH, Marsh JR, Laird WJD, Lesser JE (1992) The metabolism of cypermethrin in man: differences in urinary metabolite profiles following oral and dermal administration. Xenobiotica 22:983–991

Wrobel SA, Bury D, Belov VN, Klenk JM, Hauer B, Hayen H, Martino-Andrade AJ, Koch HM, Brüning T, Käfferlein HU (2023a) Rapid quantification of seven major neonicotinoids and neonicotinoid-like compounds and their key metabolites in human urine. Anal Chim Acta 25(1239):340680

Wrobel SA, Bury D, Hayen H, Koch HM, Brüning T, Käfferlein HU (2022) Human metabolism and urinary excretion of seven neonicotinoids and neonicotinoid-like compounds after controlled oral dosages. Arch Toxicol 96:121–134

Wrobel SA, Bury D, Koslitz S, Hayen H, Koch HM, Brüning T, Käfferlein HU (2023b) Quantitative metabolism and urinary elimination kinetics of seven neonicotinoids and neonicotinoid-like compounds in humans. Environ Sci Technol 57(48):19285–19294

Wrobel SA, Doherty DM, Koch HM, Käfferlein HU, Bury D, Slattery C, Coggins MA, Connolly A (2025) Characterising neonicotinoid insecticide exposures among the Irish population using human biomonitoring. Int J Hyg Environ Health 268:114610

Xu S, Hansen S, Rautio A, Järvelin MR, Abass K, Rysä J, Palaniswamy S, Huber S, Grimalt JO, Dumas P, Odland JØ (2022) Monitoring temporal trends of dioxins, organochlorine pesticides and chlorinated paraffins in pooled serum samples collected from Northern Norwegian women: the MISA cohort study. Environ Res 204(Pt A):111980

Zhang H, Wang Y, Zhu H, Lu S, Wang Y, Xue J, Zhang T, Kannan K, Sun H (2023) Infantile internal and external exposure to neonicotinoid insecticides: a comparison of levels across various sources. Environ Sci Technol 57:5358–5367

Zhang J, Guo J, Wu C, Qi X, Jiang S, Lv S, Lu D, Liang W, Chang X, Zhang Y, Cao Y, Zhou Z (2022) Carbamate pesticides exposure and delayed physical development at the age of seven: evidence from the SMBCS study. Environ Int 160:107076

Zhang T, Song S, Bai X, He Y, Zhang B, Gui M, Kannan K, Lu S, Huang Y, Sun H (2019) A nationwide survey of urinary concentrations of neonicotinoid insecticides in China. Environ Int 132:105114

Zhao Y, Zhu Z, Xiao Q, Li Z, Jia X, Hu W, Liu K, Lu S (2022) Urinary neonicotinoid insecticides in children from South China: concentrations, profiles and influencing factors. Chemosphere 291(Pt 2):132937

Zhu Q, Yang Y, Zhong Y, Lao Z, O'Neill P, Hong D, Zhang K, Zhao S (2020) Synthesis, insecticidal activity, resistance, photodegradation and toxicity of pyrethroids—a review. Chemosphere 254:126779

Zoller O, Rhyn P, Zarn JA, Dudler V (2020) Urine glyphosate level as a quantitative biomarker of oral exposure. Int J Hyg Environ Health 228:113526

Chapter 2
Polychlorinated Compounds

Abstract The polychlorinated compounds discussed in this chapter are man-made chemicals, including polychlorinated biphenyls and polychlorinated naphthalenes, that have been used for decades in various open and closed applications, such as lubricants, gear oils, flame retardants for wood, paper, fabrics, leather, coatings, and additives in adhesives, sealants, transformers, condensers, and hydraulic fluids. Conversely, polychlorinated and polybrominated dibenzodioxins and dibenzofurans come from anthropogenic sources and are unintentionally generated as byproducts of various industrial, chemical, and incomplete thermal processes. Overall, these substances are prevalent in the environment and humans and face restrictions due to their persistence and toxicological properties, as outlined in the Stockholm Convention on Persistent Organic Pollutants. Therefore, this text provides a brief description of the chemical-physical characteristics, use patterns, primary health risks, and toxicokinetic properties of these substances, as well as references for further scientific literature. The primary objective is to detail their occurrence in human matrices, such as blood, urine, and breast milk. To this end, comprehensive information is presented on polychlorinated biphenyls (PCBs), polychlorinated dibenzodioxins/furans (PCDD/PCDFs), dioxin-like PCBs, polybrominated dibenzodioxins/furans (PBDD/PBDFs), and polychlorinated naphthalenes (PCNs).

Keywords Polychlorinated compounds · PCB · PCDD/F · PBDD/F · Polychlorinated naphthalenes

2.1 Polychlorinated Biphenyls (PCBs)

PCBs consist of two linked aromatic rings and are technically synthesized through the chlorination of biphenyl. Due to the number and positional variants of the chlorine atoms on the rings, 209 congeners are theoretically possible. According to Ballschmiter and Zell (1980), the individual mono- to decachlorobiphenyls are

© The Author(s), under exclusive license to Springer Nature Switzerland AG 2026

H. Fromme, *Human Biomonitoring II*, https://doi.org/10.1007/978-3-032-10036-8_2

designated by consecutive numbers. Two major groups of PCB congeners are distinguished based on their health effects. Firstly, there are those without a specific dioxin-like effect (non-dioxin-like PCBs, ndl-PCBs) and the 12 non-ortho and mono-ortho substituted PCB congeners, which exhibit similar biological effects to the polychlorinated dibenzodioxins and furans (PCDD/Fs) due to their spatial orientation and are therefore addressed in the next chapter.

Since quantifying all 209 PCB possible congeners requires a significant analytical effort, determining individual congeners—especially the measurement of the so-called indicator congeners—has become standard practice. These include PCB 28, PCB 52, PCB 101, PCB 138, PCB 153, and PCB 180, which are effective indicators of various PCB patterns across different sample matrices. Table 2.1 presents the IUPAC names, CAS numbers, and chemical structures of the six indicator PCBs listed above, along with PCB 118, which is also frequently measured.

PCBs were used in mixtures called Aroclor, Clophen, Phenochlor, Kanechlor, Fenclor, and Delor. For the technical products produced in the USA (Aroclor) and Germany (Clophen), the period from around 1930–1983 is summarized in Table 2.2, which details the percentages of individual homolog groups and the degree of chlorination. Since the 1930s, PCBs have been produced, especially in the USA, West Germany, the USSR, and France, and have been utilized in many countries. It is estimated that over 1.5 million tons of pure PCBs were produced between 1930 and 1993 (IARC 2015). Production ceased from the late 1970s to the mid-1980s, and use was banned in many countries (Melymuk et al. 2022). PCB mixtures are viscous liquids, with viscosity and volatility decreasing as the degree of chlorination increases. They are poorly soluble in water.

Until their use was restricted or banned, PCBs were produced on a large scale due to their physicochemical properties, including high heat stability, low flammability, relative resistance to acids, alkalis, and other chemicals, good thermal conductivity, and very low electrical conductivity. They were utilized for decades in various open and closed applications. In the past, leading open applications included lubricants, gear oils, screw greases, water-repellent impregnating agents, and flame retardants for wood, paper, fabrics, and leather, coating for transparent and carbon copy paper, additives in adhesives, sealants, and joint sealants, as well as dispersing agents in printing inks, color pigments, and waxes. They were primarily used in closed systems such as transformers, condensers, heat exchangers, and hydraulic fluids. Figure 2.1 extrapolates the total PCB consumption by country from 1930 to 2000 based on data from Breivik et al. (2002), clearly demonstrating the different uses worldwide (Melymuk et al. 2022).

PCBs remain widespread due to the large quantities used and their persistence in the environment and biosphere. Higher chlorinated congeners, in particular, have a strong tendency to bioaccumulate in the food chain because of their high lipid solubility. For humans, food is the primary source of exposure outside specific workplaces and exposure hot spots.

Table 2.1 Informations and chemical structures of polychlorinated biphenyls (PCBs)

Abbreviation	IUPAC name	CAS No	Structure
PCB 28	2,4,4′-trichlorobiphenyl	7012-37-5	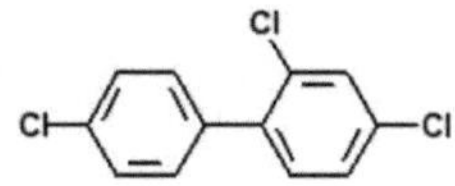
PCB 52	2,2′,5,5′-tetrachlorobipheny	35693-99-3	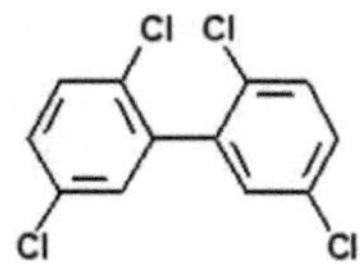
PCB 101	2,2′,4,5,5′-pentachlorobiphenyl	37680-73-2	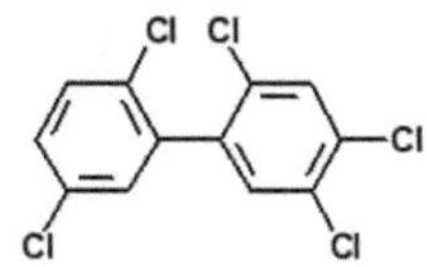
PCB 138	2,2′,3,4,4′,5′-hexachlorobiphenyl	35065-28-2	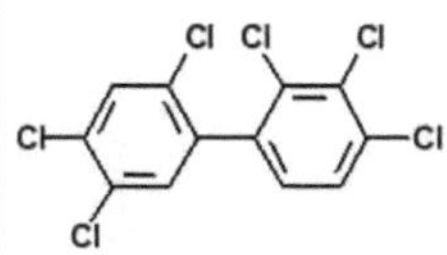
PCB 153	2,2′,4,4′,5,5′-hexachlorobiphenyl	35065-27-1	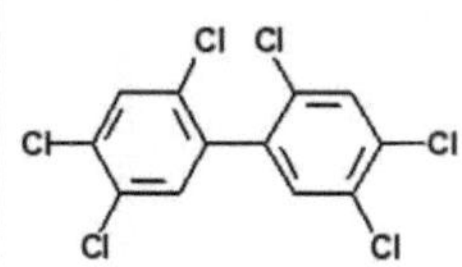
PCB 180	2,2′,3,4,4′,5,5′-heptachlorobiphenyl	35065-29-3	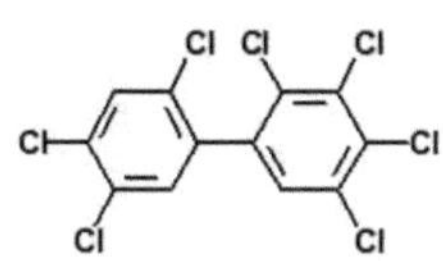
PCB 118	2,3′,4, 4′,5,-pentachlorobiphenyl	31508-00-6	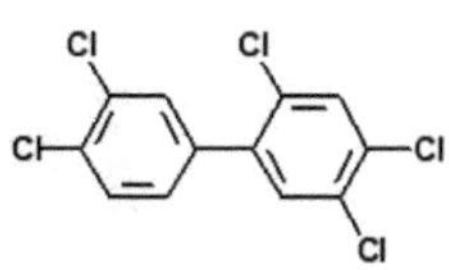

PCBs are listed in Annex A of the Stockholm POPs Convention (SC 2023). This requires each Party to phase out the use of PCBs in appliances by 2025 and to make determined efforts to achieve environmentally sound waste management of PCB-containing liquids and PCB-contaminated appliances that contain more than 0.005% (50 mg/kg) of PCBs as soon as possible, but no later than 2028.

Table 2.2 Percentage of homologs on two PCB products and range of chlorination (after VDI 2009; IARC 2015)

	Congener-number	Aroclor							Clophen			
		1221	1232	1016	1242	1248	1254	1260	A 30	A 40	A 50	A 60
Monochlorbiphenyl	1–3	51	26	1	1	–	–	–	2	< 1	–	–
Dichlorbiphenyl	4–15	32	29	20	17	1	–	–	20	2	< 1	< 1
Trichlorbiphenyl	16–39	4	24	57	40	23	–	–	56	23	9	2
Tetrachlorbiphenyl	40–81	2	15	21	32	50	16	–	20	50	28	3
Pentachlorbiphenyl	82–127	0,5	0,5	1	10	20	60	12	2	19	44	20
Hexachlorbiphenyl	128–169	–	–	< 0,01	0,5	1	23	46	< 1	4	16	43
Heptachlorbiphenyl	170–193	–	–	–	–	–	1	36		1	2	25
Octachlorbiphenyl	194–205	–	–	–	–	–	–	6		< 1	< 1	5
Nonachlorbiphenyl	206–208	–	–	–	–	–	–	–				< 1
Decachlorbiphenyl	209	–	–	–	–	–	–	–				–
Degree of chlorination		**21**	**32–33**	**40–42**	**40–42**	**48**	**52–54**	**60–62**	**30**	**40**	**50**	**60**

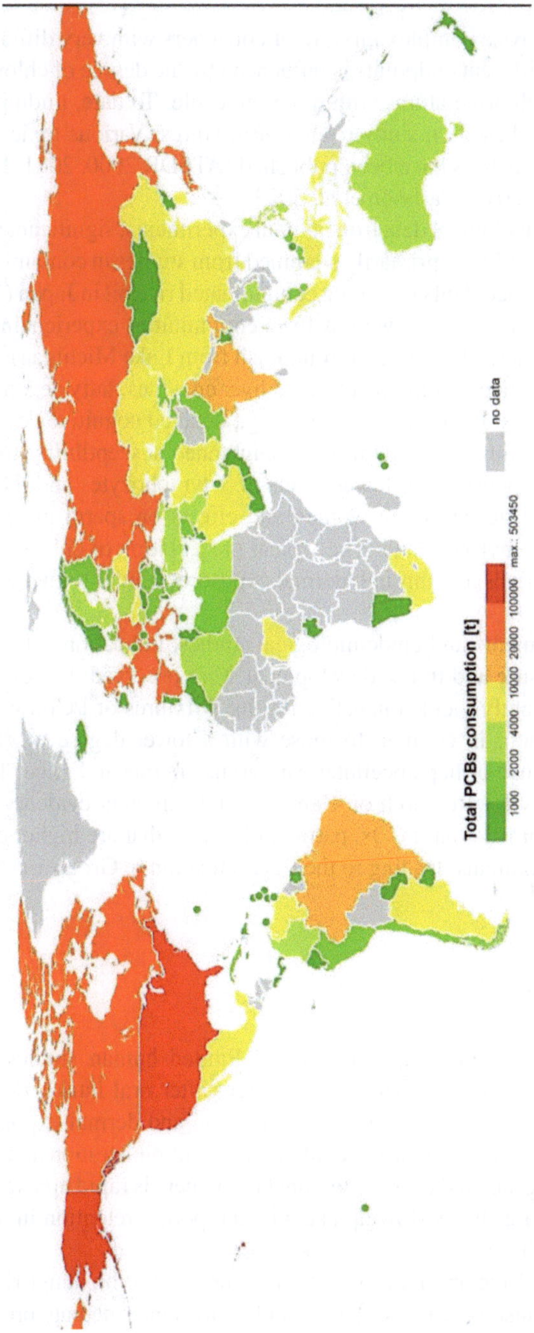

Fig. 2.1 Total PCB consumption by country throughout 1930–2000 (according to Melymuk et al. 2022)

2.1.1 Health Aspects

Since PCBs are always a complex mixture of congeners with very different compositions, toxicity at different endpoints is influenced by the degree of chlorination and the position of the chlorine atoms within the molecule. To date, findings on health effects are primarily based on study results of mixtures. Various toxicological and epidemiological summaries have been published (ATSDR 2000, 2011; EFSA 2005; JECFA 2016; IARC 2015; Carlsson et al. 2023).

In addition to toxicological data from animal experiments, significant information on the health effects of PCBs primarily stemmed from studies in contaminated workplaces, the context of accidents involving contaminated rice oil in Japan (Yusho case) and Taiwan (Yu-Cheng case), as well as from communities experiencing increased exposure to contaminated food (e.g., through fish from Lake Michigan). Key effects were observed on the liver, including elevated liver enzymes, fatty degeneration, and hepatocellular necrosis; the nervous system (e.g., delayed cognitive development in young children); the immune system (e.g., heightened susceptibility to infections, reduced antibody responses, and lower levels of T-lymphocytes and their subsets); reproduction (e.g., disturbances in women and effects on sperm morphology and production); and the thyroid gland. The typical presentation of chloracne occurs at very high concentrations and through dermal exposure, similar to many chlorinated compounds.

In animal experiments and epidemiological studies, indications of a connection between PCB exposure and tumor development were observed; however, genotoxicity tests predominantly yielded negative results. Mixtures of PCBs with a higher degree of chlorination, in contrast to those with a lower degree of chlorination, increased the incidence of hepatocellular carcinomas in rats and mice. The WHO's International Agency for Research on Cancer found sufficient evidence in humans and experimental animals that PCBs, particularly those that are higher chlorinated, are carcinogenic to humans, leading to their classification in Group 1 (IARC 2015).

2.1.2 Toxicokinetics

Based on data from animal experiments and limited human studies, PCBs are absorbed efficiently from the gastrointestinal tract after oral intake, depending on the specific congener; this is also true for inhalation and dermal exposure. Redistributive processes into fatty tissue occur after enteral administration and absorption, which varies by congener and dose. After inhalation, there is rapid uptake in the liver, while dermal exposure shows slow uptake with temporary retention in the subcutis of the skin (IARC 2015).

In the first step of metabolizm, oxidation occurs on the biphenyl rings. In this complex process, transient reactive species and various intermediate products, such as PCB arene oxides and dihydrodiol PCBs, are formed (EFSA 2005; Grimm et al.

2015). In carcinogenesis, biotransformation into more reactive electrophilic metabolites, such as arenes, is an important mechanism. The extent of metabolism of individual congeners depends on the number and spatial distribution of chlorine substituents, the species, age, and, in some cases, the sex and exposure conditions (single or continuous) (IARC 2015; Liu et al. 2020). Generally, congeners with more than four chlorine atoms and substitutions at the 3,4-positions in one or both rings tend to be metabolized more slowly than those with fewer substituents or unsubstituted congeners at that position. Higher chlorinated PCBs are metabolized very slowly. Hydroxy-PCBs (OH-PCBs) constitute the most important group of metabolites, which are bound to proteins in blood and excreted in urine or feces either in free form or conjugated to glutathione or sulfate (IARC 2015; Grimm et al. 2015).

When measuring the blood of 229 English adults without known PCB exposure, Ritter et al. (2011) determined elimination half-lives of 5.5 years for PCB 28, 2.6 years for PCB 52, 9.3 years for PCB 118, 10.8 years for PCB 138, 14.4 years for PCB 153, and 11.5 years for PCB 180 using a pharmacokinetic model. Slightly shorter half-lives were reported for 200 children aged 4.5–14 years with higher environmental exposure, showing 5.7 years for PCB 118, 3.7 years for PCB 138, 8.4 years for PCB 153, and 9.1 years for PCB 180 (Grandjean et al. 2008).

The passage of PCBs into breast milk is an important elimination step, while the transfer of OH-PCBs into breast milk is limited (Fangström et al. 2005). However, OH-PCBs can more easily cross the placental barrier than the parent substances (Soechitram et al. 2004).

2.1.3 *Occurrence in the Blood*

The blood matrix is especially suitable for determining the body burden of persistent substances with long elimination half-lives, such as PCBs. It is important to distinguish between the parent substances themselves, specifically the individual ndl-PCB congeners, and their hydroxylated metabolites, the OH-PCBs. A significant number of test results on blood measurements have been published in the scientific literature and summarized in reviews (e.g., Aylward et al. 2014b; IARC 2015).

To determine the total PCB concentration in blood (total PCB), the levels of PCB 138, PCB 153, and PCB 180 in serum are often measured, and the resulting sum is multiplied by a factor of 2 (HBM-Kom 2012). With this determination of the total levels, the combined concentrations of all PCBs in the blood and the toxicity of unmeasured congeners are considered with sufficient certainty. For instance, using in vitro tests to determine the potency of 24 PCBs (as pure substances) across 10 test systems, Hamers et al. (2011) demonstrated that the six indicator congeners accounted for 55–81% of the antiandrogenic, (anti)estrogenic, antithyroid, tumor-promoting, and neurotoxic effects. Many study results also indicated that the three higher chlorinated PCB 138, PCB 152, and PCB 180, significantly contributed to the total PCB content in the blood. This approach is further supported by studies that attempted to quantify all 209 congeners as completely as possible. For example,

Marek et al. (2013) determined all 209 congeners in 177 blood samples from women across two locations in the USA in 2008/09 and found a median of 33.5 ng/g lw for their total. The representation of PCB 138, PCB 153, and PCB 180 in the total content was approximately 74% in both regions. Among 87 Korean workers in a waste incineration plant and residents, the median proportion of these three congeners was also around 47%, with a total content of 180 ng/g lw (Park et al. 2007). A German study, which analyzed all congeners in the whole blood of 43 individuals from North Rhine-Westphalia and the plasma of 42 individuals from Bavaria, concluded that the sum of PCB 138, PCB 153, and PCB 180 multiplied by a factor of 2 accurately reflects the total content in the blood samples (Kraft et al. 2017). Figure 2.2 illustrates the percentage of indicator congeners and other congeners that contribute significantly to the total PCB content in the Bavarian plasma samples. In this study, the median total PCB level in whole blood (plasma), accounting for half the limit of quantification, was 454 ng/g lw (226 ng/g lw), with the 95th percentile at 1404 ng/g lw (642 ng/g fat). If a very high exposure to low-chlorinated PCBs is anticipated, it makes sense to include the congeners PCB 28, PCB 52, and PCB 101 in the calculation of total PCBs.

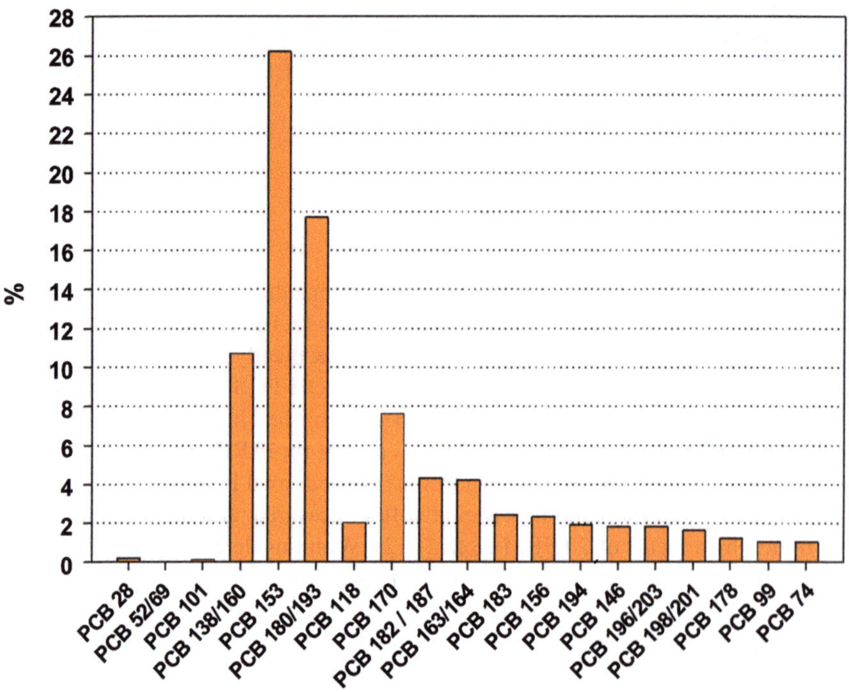

Fig. 2.2 Percentage of indicator congeners and other congeners with a high proportion of total PCBs in 42 German plasma samples (Kraft et al. 2017)

2.1.3.1 Occurrence in the Blood of Adults

Table 2.3 contains a compilation of results on ndl-PCBs from more recent studies expressed in ng/g lw and Table 2.4 in μg/l. They examined the exposure of the general, non-occupationally exposed population.

The tables mentioned clearly indicate that the body burden has decreased in recent decades, at least in Europe. Compared to an initial study in Germany by Wuthe et al. (1996) at the beginning of the 1990s, total PCB levels were approximately 80% lower in 2013–2014 (Fromme et al. 2016). In pooled serum samples from 4-, 8-, and 12-year-old children in Sweden, an increase in total PCB was observed in all age groups from 2008 to 2011, followed by a decrease to levels below those of 2008 in 2021 (Gyllenhammar et al. 2023).

However, considerable differences in exposure across individual countries and geographical areas must be taken into account. This can be attributed to variations in PCB usage and subsequent dietary exposure, with high-fat animal foods linked to higher intake.

In Europe, for example, unusually low concentrations were reported in the blood of 58 women aged 20–41 years from Istanbul, Turkey, between 2010 and 2012 (total PCBs: 2 ng/g lw) (Ulutas et al. 2015). This contrasts sharply with very high concentrations found in the blood of 38 adults from Prague, Czech Republic, in 2015 (total PCB: 732 ng/g lw) (Svarcova et al. 2019). In North America, the concentrations of ndl-PCBs in the blood are lower than those in Europe. For instance, the median level in the Canadian MIREC study (Maternal-Infant Research on Environmental Chemicals) (Fisher et al. 2016), which examined a total of 1935 mothers from 2008 to 2011, was significantly lower than in studies from the 1990s, with total PCBs at 32 ng/g lw. During that time, total PCB levels of 414 ng/g lw were recorded in Canada (Muckle et al. 2001). In a study involving 920 test subjects as part of the statewide New Jersey Biomonitoring Survey from 2016 to 2018, only 50 ng/g lw of total PCBs was detected in individuals aged 20–74 years (Du et al. 2020), compared to 127 ng/g lw in a nationwide, population-representative study in the USA in 2003/04 (CDC 2023). Additionally, lower total PCB concentrations were observed in various Asian studies compared to Europe, and they were comparable to those in the USA.

Overall, PCB 153 accounts for the largest proportion of total PCB levels of indicator PCBs in most studies. According to data from Table 2.3, between 32 and 50% of the total content in Europe is attributed to this congener, 31–42% in North America, and 43–46% in Asia. Congener 180 accounts for 23–46% of the total content in Europe, 21–66% in North America, and 27–30% in Asia.

2.1.3.2 Occurrence in Umbilical Cord Blood

The results of studies on ndl-PCB concentrations in umbilical cord blood are summarized in Table 2.5. Compared to earlier studies from the early 2000s, there is evidence of decreased cord blood levels. For example, in the third round of the Flemish Environment and Health Studies 2013–2014, total PCB concentrations of only 73 ng/g lw

Table 2.3 Median ndl-PCB levels in the blood of the adult general population in ng/g lw

References	N	28	52	101	118	138	153	180	Total-PCBs	Age (years)	Country, year
Europe											
Oltramare et al. (2025)	80	1.07[a]	0.84[a]	1.08[a]	6.17[a]	21.0[a]	47.4[a]	41.5[a]	220	>19	Switzerland, 2023
Frederiksen et al. (2020)	23	< 0.03	< 0.005	< 0.1	5.8	25.2	72.7	48.9	294	33–87	Denmark, 2017
Svarcova et al. (2019)	38	1.88	< 0.1	< 0.1	3.18	71.0	137	158	732	–	Czech Republic, 2015
Fromme et al. (2016)	42	0.42	< LOQ	0.19	5.9	25.8	55.8	45.6	254	20–68	Germany, 2013/14
Fromme et al. (2015)	70	–	–	–	3.6	20	34	33	174	4–76	Germany, 2013
Esposito et al. (2014)	58	35.4	0.7	2.0	–	28.5	42.2	43.1	228	18–64	Italy, 2013
Magoni et al. (2016)	488	–	–	–	–	62	105	132	598	20–79	Italy, 2013
Ulutas et al. (2015)	57	< LOQ	< LOQ	< LOQ	< LOQ	0.39	0.32	0.30	2	20–41	Turkey, 2010–12
Amodio et al. (2012)	101	4.4	4.2	4.2	4.1	22	33	23	156	< 30–> 69	Italy, 2009
Stubleski et al. (2018)	788	–	–	–	19	89	153	125	734	75	Sweden, 2006–09
Porta et al. (2012)	231	–	–	–	13.5	48.3	68.6	63.4	361	48[b]	Spain, 2006
Cerna et al. (2008)	202	14	5	–	15	188	438	397	2046	33[b]	Czech Republic, 2006

(continued)

Table 2.3 (continued)

References	N	28	52	101	118	138	153	180	Total-PCBs	Age (years)	Country, year
Agay-Shay et al. (2015)	457	–	–	–	–	16.9	31.5	20.7	138	> 16	Spain, 2004–06
Jaraczewska et al. (2006)	22	–	–	–	4.0	12.2	20.9	15.1	96	22–38	Poland, 2004
Turrio-Baldassarri et al. (2008)	42	< 0.3	< 0.4	< 0.4	31	108	242	303	1306	51^{b}	Italy, 2004
Montazeri et al. (2023)	1694	–	–	–	–	26.8	44.6	31.9	207	32^{b}	Spain, 2003–08
Thomas et al. (2006)	154	2.1	< 0.25	< 1.3	6.1	27.0	41.0	33.0	202	22–80	GB, 2003
Caspersen et al. (2016)	96	–	–	–	8	20	38	21	158	30^{b}	Norway, 2002–08
Costopoulou et al. (2006)	105	1.35^{b}	0.04^{b}	0.27^{b}	–	33.9^{b}	59.8^{b}	61.3^{b}	310	27–55	Greece, urban, 2002–04
Costopoulou et al. (2006)	22	0.39^{b}	0.31^{b}	1.51^{b}	–	11.8^{b}	22.6^{b}	20.4^{b}	110	28–56	Greece, rural, 2002–04
Lancz et al. (2015)	1101^{a}	9.7	5.2	6.1	13.0	94	146	134	748		Slovakia, 2002/03
America											
Du et al. (2020)	920	1.31	0.92	0.85	2.86	6.76	10.2	7.93	50	20–74	USA, 2016–18
Zota et al. (2013)	36	–	–	–	1.52	1.82	3.12	9.75	29	25^{b}	USA, 2011/12

(continued)

Table 2.3 (continued)

References	N	28	52	101	118	138	153	180	Total-PCBs	Age (years)	Country, year
Fisher et al. (2016)	1935	< 0.05	–	< 0.03	2.33	4.07	7.0	7.2	37	32[b]	Canada, 2008–11
Marek et al. (2014)	50	2.18	< LOQ	< LOQ	6.59	8.83	11.0	6.49	53		USA, 2009/10, rural
Marek et al. (2014)	50	2.76	< LOQ	< LOQ	5.35	8.89	8.28	4.44	43		USA, 2009/10, urban
HC (2010)	1666	< LOQ	< LOQ	< LOQ	4.23	10.0	19.5	16.5	92	20–79	Canada, 2007–09
CDC (2023)	~ 1280	4.98	2.70	1.67	–	17.6	24.2	21.5	127	≥ 20	USA, 2003/04
Asia											
Wang et al. (2018)	50	–	–	2.1–6.6[c]	0.5–1.2[c]	–	–	0.6–5.1[c]	–	≥ 20	China, 2014
Wang et al. (2018)	50	–	–	2.1–6.6[c]	0.5–1.2[c]	–	–	0.6–5.1[c]	–	≥ 20	China, 2014
Berlin et al. (2021)	157	–	–	–	2.23	3.31	5.44	3.69	25	32[b]	Israel, 2013–15
Kim et al. (2015)	20	2.4	1.0	0.9	2.7	5.3	9.6	–	30	31–40	Korea, 2010
Kang et al. (2008)	40	4.5	3.2	3.2	6.4	14.6	20.7	13.1	97	27–58	Korea, 2006
Todaka et al. (2011)	67	1.10	0.59	0.65	5.85	12.0	21.4	13.4	94	21–41	Japan, 2002–05
Baba et al. (2018)	379	–	–	–	5.78	–	20.5	–	–	20–48	Japan, 2002–05

[a] Geometric mean; [b] arithmetic mean; [c] range of means of pooled samples from five cities; Total-PCBs: approximately calculated as the sum of PCBs 138, 153, and 180 multiplied by 2; *LOQ* limit of quantification

Table 2.4 Median ndl-PCB levels in the blood of the adult general population in µg/l

References	N	28	52	101	118	138	153	180	Total-PCBs	Age (years)	Country, year
MacDonald et al. (2022)	62[d]	0.001	0.0001	0.009	0.003	–	0.009	0.005	0.03	≥ 18	Canada, 2013–16
Fromme et al. (2016)	42	0.002	< LOQ	0.001	–	0.15	0.32	0.29	1.52	20–68	Germany, 2013/14
Fromme et al. (2015)	70	0.003	0.001	0.001	–	0.15	0.28	0.26	1.38	4–76	Germany, 2013
Schettgen et al. (2015)	525	–	–	–	–	0.39	0.63	0.57	3.18	46–55	Germany, 2010–14
Magoni et al. (2016)	488	–	–	–	0.07	0.35	0.62	0.75	3.44	20–79	Italy, 2013
Rosofsky et al. (2017)	73	–	–	–	–	0.11	0.16	0.10	0.74	18–40	Denmark, 2011–14
Polevoy et al. (2020)		–	–	–	0.01	0.03	0.04	0.03	0.20	32[b]	Canada, 2008–11
Björvang et al. (2020)	818	–	–	–	0.81	0.07	0.11	0.08	0.52	18–43	Sweden, 2007–10
Boutot et al. (2021)	147	–	–	–	–	0,02	0,05	0,03	0.20	30[b]	Canada, 2007–09
Vafeiadi et al. (2014)	1117	–	–	–	0.16	0.63	1.17	0.06	3.72	30[b]	Greece, 2007
Porta et al. (2012)	231	–	–	–	0.09	0.30	0.45	0.39	2.28	48[b]	Spain, 2006

(continued)

Table 2.4 (continued)

References	N	28	52	101	118	138	153	180	Total-PCBs	Age (years)	Country, year
Ramos et al. (2007)	30	0.005	0.07	0.32	0.41	0.36	0.57	0.31	2.48		Spain, 2006
Schettgen et al. (2011)	15	0.02	< 0.01	–	0.05	0.35	0.53	0.53	2.82	41–50	Germany, 2003/04
Turci et al. (2010)	59	< 0.05	0.1	< 0.05	< 0.05	0.40	0.56	0.47	2.86	42	Italy, 2003/04
Lancz et al. (2015)	1101[a]	0.10	0.06	0.06	0.13	0.93	1.46	1.33	7.44		Slovakia, 2002/03
Porta et al. (2012)	147	–	–	–	0.17	0.55	0.71	0.54	3.60	48[b]	Spain, 2002
Park et al. (2008)	92	–	–	–	0.12	1.0	1.5	1.3	7.60		Slovakia, 2000–04

[a] Geometric mean; [b] arithmetic mean; [c] range of means of pooled samples from five cities; [d] 62 pools from 31,529 individuals in µg/kg ww; Total-PCBs: approximately calculated as the sum of PCBs 138, 153, and 180 multiplied by 2; *LOQ* limit of quantification

were detected in 273 samples (Govarts et al. 2020), whereas they had been 144 ng/g lw in the same country from 2002 to 2004. In this case, higher concentrations were observed in umbilical cord blood samples from Europe compared to North America and Asia.

In a review, Aylward et al. (2014b) summarize a total of six studies that found a ratio of the levels in fetal cord blood to those in maternal blood (F/M) between 0.1 and 0.4, considering the volume, and less than or close to one on a lipid basis. This is supported by a Slovakian study conducted in 2002/03 with 1134 pregnant women (Lancz et al. 2015). In this case, the F/M ratio based on the lipid of PCB 138, PCB 153, and PCB 180 was 0.8, 0.8, and 0.7, respectively. In Spain, lipid-related F/M ratios of 1.0 for PCB 118, 0.8 for PCB 138, 0.8 for PCB 153, and 0.6 for PCB 180 were determined in 101–302 pairs (Vizcaino et al. 2014). Kim et al. (2021) observed relatively high F/M ratios for the tri- to hexa-chlorinated congeners and lower ratios for the hepta-chlorinated congeners in their study of 20 Korean mother–child pairs. Specifically, the F/M ratios based on lipid and fresh weight were 0.7 and 1.6 for PCB 118, 0.4 and 0.7 for PCB 138, and 0.8 and 1.8 for PCB 153, respectively. Moreover, Takahashi et al. (2022) examined 144 mother–child pairs in Japan in 2014/15 and found a significant correlation between maternal and umbilical cord blood concentrations for the three higher chlorinated PCBs.

2.1.3.3 Occurrence in the Blood of Children and Adolescents

To date, only limited data are available for this age range, which, when considering the timing of sampling, are, as expected, lower than those for adults. A summary of the study results for ndl-PCB is presented in Table 2.6.

In Europe, current results are available from the representative German Environmental Health Study (GerES V), which examined a total of 1134 participants aged 3–17 years from 2014 to 2017 (Bandow et al. 2020). The total PCB concentration was 58 ng/g lw (0.30 μg/l). The 95th percentiles were 30.3 ng/g lw for PCB 138, 47.5 ng/g lw for PCB 153, and 28.6 ng/g lw for PCB 180. A higher mean level of the three higher chlorinated PCBs was found in male participants compared to females, in the blood of 3–10-year-olds compared to older participants, and in those of higher socioeconomic status. The concentrations are declining compared to the German Children's Environmental Survey from 2003 to 2006. In a study of 116 children (ages 6–11) and 143 mothers in Denmark conducted in 2011, a median total PCB concentration of 28 ng/g lw was found in the blood (Mørck et al. 2014). The concentrations in the blood of the children and mothers correlated significantly. The concentrations were higher in children from cities than in those from rural areas for the higher chlorinated PCBs. In a subgroup of the MoBa study (Norwegian Mother and Child Cohort), which involved 96 mothers and their children examined three years after birth, there was no significant difference in blood levels between mothers and children (Caspersen et al. 2016). However, higher concentrations were found in children who had been breastfed longer than in those who had been breastfed for a shorter time.

Table 2.5 Median ndl-PCB levels in umbilical cord blood

References	N	28	52	101	118	138	153	180	Total-PCBs	Country, year
In ng/g lw										
Govarts et al. (2020)	272	–	–	–		11	17	8.6	73	Belgium, 2013/14
Kim et al. (2015)	20	1.0	0.3	0.4	2.5	1.6	8.3	–	20	Korea, 2010
Vizcaino et al. (2014)	308	< LOQ			< LOQ	32	47	28	214	Spain, 2004–08
García-Villarino et al. (2020)	229	0.93		33	6.8	33	48	28	218	Spain, 2004–07
Lopez-Espinosa et al. (2010)	453	–	–	–	23	34	46	33	226	Spain, 2004–06
Herbstman et al. (2007)	297	–	–	–	3.0	4.9	6.4	2.6	28	USA, 2004/05
Jaraczewska et al. (2006)	17	–	–	–	3.2	11	15	12	76	Poland, 2004
Maervoet et al. (2007)	198	–	–	–	11	17	32	23	144	Belgium, 2002–04
Nakamura et al. (2008)	49	0.6	–	–	0.6	6.1	8.4	3.8	37	Japan, 2002/03
Lancz et al. (2015)	1063[a]	9.9	7.3	8.7	8.7	77	110	90	554	Slovakia, 2002–03
Guvenius et al. (2003)	15	1.0	< 0.5	< 0.5	4.0	34	44	17	190	Sweden, 2000–01
Soechitram et al. (2004)	51[b]	–	–	–	56	77	115	37	458	Netherlands, 1998–2000
in µg/l										
Hjazi et al. (2024)	450	0.09	0.27	0.27	0.93	0.28	0.34	0.90	3.0	China, 2022
Cabrera-Rodríguez et al. (2019)	447	0.11	0.06	0.04	0.04	0.042	0.044	0.032	0.24	Spain, 2015/16
Takahashi et al. (2022)	144	–	–	–	–	0.019[c]	0.021[c]	0.011[c]	0.10	Japan, 2014–15
Lopez-Espinosa et al. (2010)	1140	–	–	–	–	0.08	0.12	0.08	0.56	Spain, 2003–08

(continued)

Table 2.5 (continued)

References	N	28	52	101	118	138	153	180	Total-PCBs	Country, year
Boyer et al. (2023)	~ 450	–	–	–	0.03	0.07	0.12	0.08	0.54	France, 2003–06
Brucker-Davis et al. (2011)	84	–	–	< LOQ	0.10	0.20	0.20	0.20	1.20	France, 2002–05
Maervoet et al. (2007)	198	–	–	–	0.02	0.03	0.06	0.04	0.26	Belgium, 2002–04
Lancz et al. (2015)	1063[a]	0.02	0.02	0.02	0.02	0.19	0.27	0.22	1.36	Slovakia, 2002/03
Park et al. (2008)	92	–	–	–	0.01	0.22	0.30	0.25	1.54	Slovakia, 2000–04
Takser et al. (2005)	92	< LOQ	< LOQ	< LOQ	< LOQ	0.02	0.02	0.01	0.10	Canada
Soechitram et al. (2004)	51[b]	–	–	–	0.09	0.13	0.19	0.06	0.76	Netherlands, 1998–2000

[a] Geometric mean; [b] arithmetic mean; [c] in μg/kg ww; *LOQ* limit of quantification; Total-PCBs: approximately calculated as the sum of PCBs 138, 153, and 180 multiplied by 2

Table 2.6 Median ndl-PCB levels in the blood of children and adolescents

References	N	28	52	101	118	138	153	180	Total-PCBs	Age (years)	Country, year
In ng/g lw											
Bandow et al. (2020)	1134	–	–	2.86	1.99	9.41	13.1	6.44	58	3–17	Germany, 2014–17
Mørck et al. (2014)	116	3.0	< LOQ	< LOQ	< LOQ	16	23	14	106	6–11	Denmark, 2011
Aylward et al. (2014a)	ps	–	–	–	–	3.0	3.8	2.5	19	0–4	Australia, 2011
Aylward et al. (2014a)	ps	–	–	–	–	2.1	2.9	2.3	15	5–15	Australia, 2011
Caspersen et al. (2016)	99	–	–	–	14	29	57	28	228	3	Norway, 2010–11
Marek et al. (2013)	44	< LOQ	< LOQ	< LOQ	< LOQ	3.1	2.7	< LOQ	12		USA, 2008–09
Plaku-Alakbarova et al. (2021)	516	6.2	2.0	2.4	33	40	42	13	190	8–9	Russia, 2003–05
CDC 2023	590–598	4.88	3.22	1.67	–	4.57	5.40	2.97	26	12–19	USA, 2003/04
Sexton et al. (2011)	386	–	–	1.70	3.70	4.60	6.50	3.3	29	7–12	USA, 2000/01
Orloff et al. (2003)	37	0.62[b]	0.58[b]	0.79[b]	4.90[b]	6.90[b]	14.5[b]	7.0[b]	57	< 16	USA
Barr et al. (2006)		1.2	–	–	53.5	220	311	220	1502	7	Faroe Islands, 1986–87
Wuthe et al. (1996)	gp[a]	< 3.0–5.9	< 3	< 3.0–5.3	7.9–21	50–74	63–91	34–43	–	9–12	Germany
In µg/l									0		
Pineda et al. (2023)	1082	–	–	–	0.0067[d]	0.026[d]	0.042[d]	0.023[d]	0.18	12–19	Sweden, 2016/17

(continued)

Table 2.6 (continued)

References	N	28	52	101	118	138	153	180	Total-PCBs	Age (years)	Country, year
Rantakokko et al. (2020)	184	–	<	<	0.005	0.02	0.03	0.02	0.14	7–10	Finland, 2014–18
Bandow et al. (2020)	1135	< 0.3	< 0.2	< 0.03	< 0.02	0.02	0.07	0.03	0.24	3–17	Germany, 2014–17
Schettgen et al. (2015)	102	–	–	–	–	0.05	0.06	0.03	0.28	6–10	Germany, 2010–14
Sprowles et al. (2022)	115	–	–	–	–	0.02[c]	0.02[c]	0.01[c]	0.10	12–19	USA, 2007–12
Schettgen et al. (2011)	15	0.01	< 0.01	–	0.02	0.17	0.26	0.17	1.20	< 10	Germany, 2003–04
Lackmann et al. (2004)	25	< 0.01	< 0.01	< 0.01	–	0.38	0.49	0.31	2.36	6 weeks	Germany, 2003–04
Orloff et al. (2003)	37	0.005[b]	0.005[b]	0.004[b]	0.03[b]	0.04[b]	0.08[b]	0.04[b]	0.32	< 16	USA
Liebl et al. (2004)	218	< 0.001	< 0.001	< 0.001	–	0.12	0.16	0.06	0.68	6–16	Germany, 2001

[a] From 6 different sampling locations; [b] arithmetic mean; [c] μg/kg; [d] geometric mean; ps: pooled samples; Total-PCB: approximately calculated as the sum of PCBs 138, 153, and 180 multiplied by 2

In the fifth round of the Canadian Health Measures Survey Cycles (HC 2020), mean concentrations decreased to 6.0 and 5.2 ng/g lw for PCB 138, 9.1 and 8.9 ng/g lw for PCB 153, and 5.2 and 5.0 ng/g lw for PCB 180 in 6 pooled blood samples from 479 children aged 3–5 years during 2012–2013 and 2016–2017. In Australia in 2011/12, the mean levels in pooled blood samples from newborns and 5–10-year-old children were 19 and 15 ng/g lw of total PCBs (Aylward et al. 2014a), respectively, which were lower than those noted in the more recent studies from Germany. The 95th percentiles were also significantly lower, at 4.6 ng/g lw for PCB 138, 6.4 ng/g lw for PCB 153, and 4.8 ng/g lw for PCB 180. Differences in usage are the most likely explanation.

2.1.4 Occurrence in the Placenta

In principle, lipophilic and non-ionized compounds with low molecular weight can easily cross the placenta to the fetus via passive diffusion. There are also indications for PCBs that higher-chlorinated PCBs are less able to cross the placenta due to their molecular size (Vizcaino et al. 2014). Additionally, evidence suggests that persistent organic substances accumulate in higher concentrations in male fetuses and their placentas than in female fetuses (Björvang and Mamsen 2022). This could indicate sexually dimorphic genetic and epigenetic regulation, placental resistance, and protective strategies against external stress.

The concentrations of ndl-PCBs from recent investigations in placental tissue are summarized in Table 2.7. The studies conducted so far have shown considerable differences in concentration, particularly after adjustment for lipid content. One possible explanation could be that, within the heterogeneous placental tissue, the sampling location significantly influences the results.

Hydroxy metabolites were also measured in a Spanish study conducted in 2003/04 (Gomara et al. 2012). In 17 placentas, medians of 50 pg/g ww for 4-OH-CB187, 24 pg/g ww for 4-OH-CB146, 12 pg/g ww for 3-OH-CB153, and 6.1 pg/g ww for 4-OH-CB107 were reported in this study.

2.1.5 Occurrence of OH-PCBs in Blood and Urine

Only limited test results are available on the occurrence of OH-PCBs (see Table 2.8). Overall, the concentrations in this medium are significantly lower than those of the parent substances (approximately a factor of 10–20 or more) (Guvenius et al. 2003; Soechitram et al. 2004; Park et al. 2009; Koh et al. 2016; Haraguchi et al. 2016). There is a strong correlation between the concentrations of individual PCB congeners and their hydroxy metabolites in the blood (Sandau et al. 2002; Marek et al. 2013; Hisada et al. 2014). OH-PCBs can cross the placental barrier more effectively than parent PCBs due to the lower lipid content in fetal blood compared to maternal

Table 2.7 Median ndl-PCB levels in placenta samples

References	N	28	52	101	118	138	153	180	Country, year
In ng/g lw									
Alvarez-Silvares et al. (2021)	70	0.11	0.08	0.05	0.10	0.24	0.07	0.0006	Spain, 2017
Kim et al. (2015)	20	< LOQ	< LOQ	0.3	1.1	2.3	4.4	–	Korea, 2010
Bergonzi et al. (2009)	70	< LOQ	< LOQ	< LOQ	5.0	18	26	18	Italy, 2006
Vizcaino et al. (2014)	49	< 0.007	< LOQ	< LOQ	0.03	0.11	0.15	0.12	Spain, 2004–08
Naqvi et al. (2018)	43	–	4.1–7.2[a]	1.7–6.6[a]	1.0–2.5[a]	1.1–4.3[a]	1.6–5.2[a]	0.3–1.7[a]	Pakistan
In pg/g ww									
Iribarne-Durán et al. (2023)	60	–	–	–	–	17	16	< 3.0	Panama, 2016/17
Björvang et al. (2021)	20	–	–	–	12	35	52	32	Sweden, 2015/16
Li et al. (2020)	5	8.9	7.8	–	138	87	52	23	Germany
Želježić et al. (2018)	51	11.0	5.0	2.1	2.7	4.8	3.9	6.8	Croatia, 2012/13
Gomara et al. (2012)	17	169	658	409	100	198	258	150	Spain, 2003/04

[a] Range of means of five districts; *LOQ* limit of quantification

blood (Guvenius et al. 2003; Soechitram et al. 2004). A significant correlation exists between the levels of OH-PCBs in umbilical cord blood and maternal blood during the last trimester of pregnancy. Lower OH-PCB levels were observed in younger population groups, which is a general trend observed with parent PCB congeners (Sjödin et al. 2000; Koh et al. 2016). Sjödin et al. (2000) reported a significant correlation between OH-PCB concentrations in blood and fish consumption among 43 Swedish and 67 Lithuanian individuals with varying levels of fish intake, particularly salmon and herring from the Baltic Sea.

In most studies, the dominant OH-PCB in the blood is, in descending order, 4-OH-CB187, 4-OH-CB146, and 4-OH-CB107. In the Netherlands, medians for 4-OH-CB187 are 105 pg/g ww, for 4-OH-CB146 are 80 pg/g ww, and for 4-OH-CB107 are 43 pg/g ww, observed in 99 samples from pregnant women (Berghuis et al. 2018). In a study conducted in the USA in 2011/12, 4′-OH-CB107, 4′-OH-CB146, and 4′-OH-CB187 could not be found above the limit of quantification of 5 pg/g ww (Zota et al. 2013).

The OH-PCB content in urine can also indicate exposure to PCBs, although levels in the general population not specifically exposed are often below the limit of quantification (Quinete et al. 2016). In this study, 3-OH-CB28, 3-OH-CB153, and 4′-OH-CB28 were found at mean concentrations of 0.02, 0.01, and 0.01 μg/l among 21 individuals aged 23–64 years in the control group. In the occupationally highly exposed group, however, the mean values were 15.8, 0.16, and 0.37 μg/l. There was a statistically significant correlation between individual OH-PCBs in plasma and urine, as well as between the parent compounds in plasma and their possible metabolites in urine. Additionally, 3-OH-CB28 was the dominant congener in the control group.

Since high concentrations of sulfated PCBs were detected in animal experiments following the administration of low-chlorinated PCBs, these metabolites were also analyzed in human blood samples. For example, Zhang et al. (2022) found significantly higher serum concentrations of PCB sulfates in the urban group (mean: 3400 ng/g ww, 110–8900 ng/g ww) compared to the rural cohort (mean: 1800 ng/g ww, 530–6700 ng/g ww) among 24 subjects from a rural region and 24 from an urban area in the USA. The eight sulfate congeners 4-PCB 2 sulfate, 4′-PCB 2 sulfate, 2′-PCB 3 sulfate, 4′-PCB 3 sulfate, 4-PCB 11 sulfate, 4′-PCB 18 sulfate, 4′-PCB 25 sulfate, and 4-PCB 52 sulfate accounted for over 90% of the total concentrations in most individuals. An explanation for the differences between the two groups is not yet available.

2.1.6 Occurrence in Breast Milk

Like blood, only a limited number of congeners, primarily the six indicator congeners, are found in breast milk, with the higher chlorinated PCB congeners predominantly present in this matrix. Based on an assessment of the distribution patterns of all 209 congeners in 8 breast milk samples, Schulte and Malisch (1984) suggested calculating the sum of the congeners PCB 138, PCB 153, and PCB 180 then multiplying it by

Table 2.8 Median OH-PCB levels in blood in pg/g fresh weight

Parent congener	N	105, 107, 118	118, 120	128, 130, 138	138	138, 146, 153	146, 153, 167	170, 172, 180	172, 180, 189	183, 187	191, 193	Country, year
Metabolite		4-OH-CB107	4´-OH-CB120	4´-OH-CB130	3´-OH-CB138	4-OH-CB146	3-OH-CB153	4´-OH-CB172	3´-OH-CB180	4-OH-CB187	4-OH-CB193	
Adults												
Koh et al. (2016)	159	9..1	1	< LOQ	0.8	3	–	0.5	< LOQ	9.8*	< LOQ	USA, 2010/11
Hisada et al. (2014)	79	24	1.6	0.3	6.1	21	–	3.5	–	29	–	Japan, 2009–11
Marek et al. (2013)	41	50	–	–	6	< LOQ	–	–	–	24	–	USA, 2008/09
Haraguchi et al. (2016)	20	71			13	74	15	8.5	–	100	–	Japan, 2007–09
Dirtu et al. (2010)	53	12	–	< 3	14	37	8.5	12	7	52	2	Romania, 2006/07
Itoh et al. (2018)	222	< 1.1	–	–	–	10.53		< 1.1	–	11.1	–	Japan, 2002–05
Park et al. (2009)	147	23	–	–	56	89	63	32	–	157	–	Slovakia, 2002–04
Guvenius (2003)	15	10	2	4	9	29	7	5	2	49	2	Sweden, 2000/01
Dirtu et al. (2010)	29	78	–	16	22	92	15	11	4.5	57	< 2	Belgium, 2000
Soechitram et al. (2004)	51	70	–	–	45	63	35	15	–	22	–	Netherlands, 1998–2000
Berghuis et al. (2018)	~ 98	43	–	–	46	–	37	15	–	136	–	Netherlands, 1998–2000

(continued)

Table 2.8 (continued)

Parent congener	N	105, 107, 118	118, 120	128, 130, 138	138	138, 146, 153	146, 153, 167	170, 172, 180	172, 180, 189	183, 187	191, 193	Country, year
Metabolite		4-OH-CB107	4´-OH-CB120	4´-OH-CB130	3´-OH-CB138	4-OH-CB146	3-OH-CB153	4´-OH-CB172	3´-OH-CB180	4-OH-CB187	4-OH-CB193	
Sjödin et al. 2000	19	82[a]	–	–	18[a]	31[a]	12[a]	–	–	34[a]	–	Latvia, 1993
Children												
Marek et al. (2013)	44	44	–	–	< LOQ	< LOQ	–	–	–	16	–	USA, 2008/2009
Cord blood												
Guvenius 2003	15	5	2	3	9	21	5	4	1	24	2	Sweden, 2000/2001
Soechitram et al. (2004)	51	22	–	–	28	36	21	11	–	61	–	Netherlands, 1998–2000
Park et al. (2009)	80	12	–	–	58	73	40	58	–	115	–	Slovakia, 2002–2004

[a] Values in ng/g lipid; * and 5-OH-CB183; + arithmetic mean; *LOQ* limit of quantification

1.64 to estimate the total PCB content. This was based on findings that the proportion of the three congeners mentioned above in the samples ranged from 55 to 70%, with an average of 61%. A factor of 1.64 (100/61 = 1.64) is needed to reach 100%, representing the total PCB content.

In a study of 58 milk pools from 18 European countries, median total PCB levels of 280 ng/g fat (48–1655 ng/g lw) were still observed during 2001/2002 (EFSA 2005). In contrast, the 4th/5th global survey by the WHO, conducted with pooled milk samples from Sweden, Belgium, Switzerland, and Norway from 2005 to 2010, revealed concentrations of approximately 65–80 ng/g lw for the sum of the six indicator congeners, while in the Czech Republic it was around 380 ng/g lw (van den Berg et al. 2017). A summary of data from the WHO/UNEP program involving 90 countries worldwide showed that the pollution situation as measured by the sum of the six indicator congeners decreased by a median of 97% from the collection period of 1985–1989 to 2015–2019 (Fiedler et al. 2023).

Many studies are available on the contamination of breast milk with ndl-PCBs. A summary of more recent study results is provided in Table 2.9. A recent review has been published by Brajenović et al. (2018). Here, too, considerable differences in exposure exist between countries. In Europe, the current median concentrations for total PCB range from 11 ng/g lw in Turkey in 2009 (Cok et al. 2012) to 249 ng/g lw in Slovakia between 2010 and 2013 (Komprdová et al. 2022). Recent studies from Bulgaria and Sweden also reported very low total PCB concentrations of 45 and 52 ng/g lw, respectively. Overall, there has been a significant decrease in breast milk levels in recent decades. In Germany, for example, the median total PCB concentration in 100 Bavarian breast milk samples in 2016 was only 82 ng/g lw (Fromme et al. 2022), compared to 230 ng/g lw in the same region in 2005 (Raab et al. 2008). Gyllenhammar et al. (2021) also noted an annual decrease in the concentrations of the six indicator congeners of 4.8–6.6% for Swedish breast milk samples between 1996 and 2017.

Park et al. (2011) reported median concentrations for $\sum$PCB13 of 73 ng/g lw (22–433 ng/g lw) for 2003–2004 in the northwest USA and 2003–2005 in California. Furthermore, Ryan & Rawn (2014) stated a median of 34 ng/g lw (5.3–122 ng/g lw) for the sum of the six indicator congeners in 34 Canadian samples collected in 2005. Torres-Moreno et al. (2023) noted significantly lower concentrations in a more recent study conducted in 2014/15 in Colombia.

The limited study results from Asia generally show significantly lower levels than those from North America and Europe. Only Zhao et al. (2007) observed unusually high PCB concentrations in 16 samples from women in rural China without a known PCB source.

Very low results are reported from Ghana, although this case involved 105 breast milk samples from the region with the largest landfill for electrical and electronic waste, as well as an extensive recycling site (Asamoah et al. 2018). In an area with no known exposure, PCB 28 was found in only one of 23 samples; none of the other PCB congeners were detected above the limit of quantification.

Using data from Table 2.9, the mean percentage of total content in Europe is 27% for PCB 138, 45% for PCB 153, and 27% for PCB 180. For the three studies from

Table 2.9 Median ndl-PCB levels in breast milk in ng/g lw

References	N	28	52	101	118	138	153	180	Total-PCB	Country, year
Europe										
Hedvall Kallerman et al. (2023)	30	0.65	0.07	0.15	2.8	8.6	16	7.0	52	Sweden, 2020/21
Georgieva et al. (2023)	47/25[c]	0.43/ < 0.2[a]	0.22/ < 0.4[a]	0.34/1.9[a]	3.56/1.9[a]	7.2/4.3[a]	16.4/10.4[a]	4.1/3.2[a]	45	Bulgaria, 2019–21
Parizek et al. (2023)	231	0.49	0.05	0.24	–	19	34	27	131	Czech Republic, 2019–21
Mendas et al. (2024)	49	0.16	0.13	0.10	0.18	0.83	2.52	0.65		Croatia, 2018/19
Rovira et al. (2022)	60	–	–	–	–	14	33	19	108	Spain, 2016–19
Houlihan et al. (2021)	16	0.43	0.12	0.2	2.2	4.9	8.9	4.7	30	Ireland, 2016–18
Fromme et al. (2022)	100	0.81	0.14	0.25	4.3	14	23	13	82	Germany, 2016
Herceg Romanic et al. (2023)	150	1.15	1.60	2.13	0.82	3.04	4.73	2.35	17	Croatia, 2011–14
Čechová et al. (2016)	120	0.79	0.17	0.19	3.59	11	16	11	62	Netherlands, 2011–14
Antignac et al. (2016)	96	1.29	0.24	0.39	1.69	20	39	24	136	France, 2011–14
Čechová et al. (2016)	120	0.79	0.17	0.19	3.59	11	16	11	62	Netherlands, 2011–14
Antignac et al. (2016)	96	1.29	0.24	0.39	1.69	20	39	24	136	France, 2011–14
Komprdová et al. (2022)	298	1.3	0.1	0.4	4.1	33	71	48	249	Slovakia, 2010–2013
Komprdová et al. (2022)	396	2.2	0.2	0.5	8.9	81	150	123	581	Czech Republic, 2009–11
Čechová et al. (2016)	37	1.11	0.13	0.22	3.51	31	59	44	220	Slovakia, 2010–12
Pratt et al. (2012)	11[b]	0.77	0.08	0.26	4.1	6.9	13	7.8	45	Ireland, 2010

(continued)

Table 2.9 (continued)

References	N	28	52	101	118	138	153	180	Total-PCB	Country, year
Klinčić et al. (2014)	20	3.4	3.0	0.2	0.9	7.4	13	6.4	44	Croatia, 2009/10
Cok et al. (2012)	47	0.57[a]	0.33[a]	0.55[a]	0.97[a]	1.63[a]	3.36[a]	1.62[a]	11	Turkey, 2009
Raab et al. (2013)	516	< 2.0	< 2.0	< 2.0	–	28	41	20	146	Germany, 2007/08
Zietz et al. (2008)	523	–	–	–	–	38	46	24	177	Germany, 2006
Gascon et al. (2012)	290	1.71	2.13	1.26	5.76	26	37	33	157	Spain, 2004–08
Raab et al. (2008)	39	–	–	–	–	40	57	26	202	Germany, 2005
Jaraczewska et al. (2006)	22	–	–	0.5	6.3	23	35	27	139	Poland, 2004
Brucker-Davis et al. (2011)	43–65	–	–	10	25	54	59	36	244	France, 2002–05
Thomsen et al. (2010)	70	–	–	6.4	9.3	16	38	17	116	Norway, 2001–09
Čechová et al. (2016)	388	1.56	0.51	0.58	5.22	17	27	14	95	Norway, 2001–06
Kalantzi et al. (2004)	54	2.0	–	–	10	37	49	25	182	UK, 2001–03
Fürst et al. (1994)	113	–	–	–	–	202	254	125	953	Germany, 1991
Fängström et al. (2005)	9	–	–	–	96	350	380	210	1542	Faroe Islands, 1999
Americas										
Torres-Moreno et al. (2023)	60	0.33	0.07	–	0.97	0.58	3.63	2.04	10	Colombia, 2014/15
Ryan and Rawn (2014)	34	1.23	0.25	0.24	5.02	9.75	14.6	7.62	52	Canada, 2005
Park et al. (2011)	82	2.21	–	–	6.10	12.3	20.6	10.6	71	USA, 2003–05

(continued)

Table 2.9 (continued)

References	N	28	52	101	118	138	153	180	Total-PCB	Country, year
She et al. (2007)	122	2.78	–	–	6.61	11.2	19.8	9.15	66	USA, 2003/04
Asia, Africa										
Varghese et al. (2025)	32	3.29	1.02	0.43	0.72	1.55	1.57	0.87	6	India, 2022/23
Asamoah et al. (2018)	105[d]	0.79	0.11	0.04	–	0.62	0.11	0.36	2	Ghana, 2014–16
Lee et al. (2013)	206	0.47	0.15	–	1.11	2.35	4.13	1.80	14	Korea, 2011
Zhang et al. (2011)	24[b]	0.68	0.15	0.18	–	3.64	5.17	0.49	15	China, 2007
Zhao et al. (2007)	16	0.69	< LOQ	2.17	5.40	44.7	13.3	28.4	142	China, 2003–05
Todaka et al. (2011)	97	0.64	0.20	0.51	5.22	10.5	17.7	6.89	58	Japan, 2002–05

[a] Mean; [b] pooled samples; [c] two regions; [d] samples from the most prominent electronic waste site in Ghana; Total-PCB: approximately calculated as the sum of PCB 138, 153, and 180 multiplied by 1.64; *LOQ* limit of quantification

North America, the comparable percentages are 29% for PCB 138, 47% for PCB 150, and 24% for PCB 180. In the four Asian studies, the wide range of variation is striking. The proportions of PCB 138, PCB 153, and PCB 180 in the total content range from 28–52%, 15–56%, and 5–33%, respectively. In the study from Colombia, PCB 153 dominated at 58%, while in Ghana, PCB 138 was dominant at 57% of the total PCBs.

Various studies have shown that concentrations in breast milk increase with the woman's age at birth (Wittsiepe et al. 2007; Runkel et al. 2021; Rovira et al. 2022; Fromme et al. 2022; Georgieva et al. 2023). The concentration of breast milk in multipara is also lower than in primipara (Klinčić et al. 2016; Agus et al. 2022; Rovira et al. 2022; Georgieva et al. 2023). Rovira et al. (2022) found no correlation between the frequency of using organically produced products and breast milk PCB levels. Like other studies, Witczak et al. (2022) also described a significant decrease in the concentration of 96 mothers throughout breastfeeding in Poland. The level of 19.8 ng/g lw ($\sum PCB_6$) seven days after birth fell to 7.8 ng/g lw by the twelfth month.

Hydroxy metabolites (OH-PCBs) can also be found in breast milk. However, studies indicate that their levels are significantly lower than those of the parent PCB congeners. For example, Guvenius et al. (2003) reported a median of 3 pg/g ww while examining 12 congeners in 15 breast milk samples, with a range of < 0.1–5 pg/g ww, and a concentration ratio of maternal blood to breast milk medians of 41:1. The primary congener identified in this study was 4-OH-CB107. Fangström et al. (2005) also reported low levels of 0.35 ng/g lw for 4-OH-CB146 and 0.50 ng/g lw for 4-OH-CB187 in a pool of 10 breast milk samples collected in the Faroe Islands in 1999.

2.2 Polychlorinated Dibenzodioxins/Furans and Dioxin-Like PCBs

Polychlorinated dibenzo-p-dioxins (PCDDs) and polychlorinated dibenzofurans (PCDFs) are tricyclic aromatic hydrocarbons in which two benzene rings are connected by one or two oxygen bridges (see Fig. 2.3). Due to the bonding possibilities of chlorine atoms, there are theoretically 75 possible PCDD congeners and 135 PCDF congeners. PCDD/Fs originate from anthropogenic sources and occur in trace amounts in chemical products. They are generated as by-products in various industrial, chemical, and incomplete combustion processes. In particular, the 17 PCDD/Fs with chlorine substituents at positions 2,3,7, and 8 are lipophilic substances that are very persistent in the environment.

PCDD/Fs and dl-PCBs exposure extends to the general population outside of workplaces, such as in pesticide production and other specific exposure scenarios, primarily through high-fat animal foods (over 90%). Various exceptional exposure situations have been thoroughly documented in the scientific literature. For instance, in Missouri, USA, between 1971 and 1972, oils contaminated with TCDD from waste

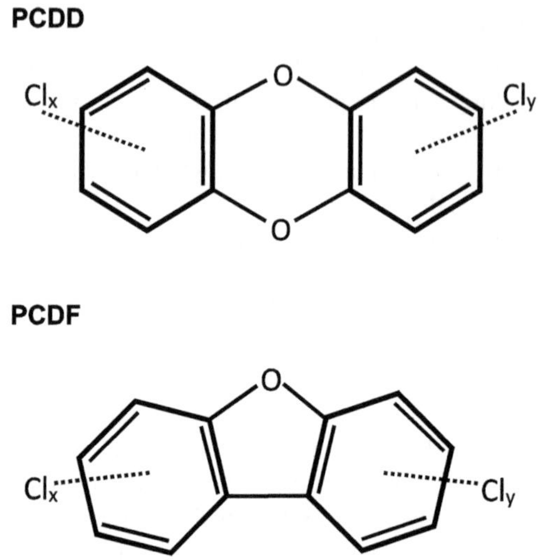

Fig. 2.3 Chemical structure of polychlorinated dibenzo-p-dioxines (PCDDs) and dibenzofuranes (PCDFs)

recycling polluted populated areas extensively, and in Oregon, decades of production led to significant environmental pollution. In Europe, a severe industrial accident at a chemical factory in Seveso, Italy, in 1976, during trichlorophenol production, resulted in the release of substantial quantities of TCDD into the environment and the contamination of a large land area with around 560,000 inhabitants. The extensive use of herbicides like 2,4,5-trichlorophenoxy acetic acid (2,4,5-T) during the US military operation Ranch Hand in 1971–1972 also led to widespread TCDD contamination in large areas of South Vietnam. Additionally, PCDD/Fs have been identified as environmental pollutants, later appearing in food and animal feed during the production of chlorine-containing substances, in the operation of waste incineration plants, and in the incineration of waste, as well as in metal recycling and paper production (EFSA 2018).

Due to their similar biological effects to PCDD/Fs, the 12 non-ortho- and mono-ortho-substituted PCBs are often regarded as dioxin-like PCBs (dl-PCBs). In contrast to PCDD/Fs, however, PCBs are a class of substances produced on a large scale and utilized in various open and closed applications until their ban in the 1980s. The PCB mixtures employed in technical applications always contain, to varying degrees, dl-PCBs (see previous chapter). Additionally, various incidents involving contaminated oil in Belgium in 1999 and bread in Ireland in 2008 have been reported, where dl-PCBs entered animal feed (EFSA 2018). Products containing PCBs can serve as a source of PCDD/Fs in thermal processes.

In the context of analysis, assessment, and regulation, concentrations of PCDDs/PCDFs and dl-PCBs are typically expressed in toxic equivalents (TEQs). For this purpose, the content in the analyzed medium is multiplied by a toxic equivalency factor (TEF) defined for each respective congener. These TEFs were established in 1998 and revised in 2005 by an expert committee of the WHO based on toxicological data (Van den Berg et al. 1998; Van den Berg et al. 2006). In 2022, the

WHO re-evaluated the 2005 TEFs using dose–response modeling and meta-analysis, proposing them for future applications (DeVito et al. 2024). The TEF value corresponds to the ratio of this substance to the most toxicologically significant substance, 2,3,7,8-tetrachlorodibenzo-p-dioxin (2,3,7,8-TCDD), with the TEF value for 2,3,7,8-TCDD set at 1. Other TEFs have previously been derived, such as by NATO/CCMS (the so-called I-TEFs, International Toxicity Equivalency Factors) (Kutz et al. 1990). Therefore, for assessing measured values, it is essential to know which TEFs were used in each case. The various TEFs are listed in Table 2.10 for comparison.

2.2.1 Health Aspects

Various reviews have been published on the toxicological and epidemiological findings (Schecter et al. 2006; Chopra and Schrenk 2011; Boffetta et al. 2011; EFSA 2015, 2018; IARC 2015). These literature references were considered for further explanations.

In the case of PCDDs/Fs, the number and position of chlorine atoms in the molecule significantly influence the biological activity and toxic effects of the congeners. In recent decades, numerous studies have been conducted to characterize the harmful effects on laboratory animals, after workplace exposures, or following accidents. Depending on the dose and degree of chlorination, reproductive and developmental toxic effects, as well as impacts on the immune system, have been observed in animal studies. Additionally, the liver, lipid and carbohydrate metabolism, and the nervous system are known target systems. After dermal exposure, typical instances of chloracne have been reported at high concentrations. Epidemiological studies involving workers or related to accidents indicate an increased incidence of tumors. However, there was often simultaneous exposure to other chlorinated compounds. For instance, follow-up studies of the Seveso accident indicated an increased risk of lymphohematopoietic neoplasia, cancer of the digestive and respiratory systems, and a higher incidence of thyroid, pleural cancers, and soft tissue sarcomas (Pesatori et al. 2003). Furthermore, effects on the cardiovascular system, endocrine effects such as diabetes, and reproductive impacts were documented in this context. Epidemiological studies in Vietnam also revealed links between dioxin exposure during the Vietnam War and the development of soft tissue sarcoma, Hodgkin's and non-Hodgkin's lymphoma, as well as lung, prostate, and liver cancer in contaminated areas (Vuong 2022).

The toxic potency of individual congeners correlates with their ability to bind to and activate the aryl hydrocarbon receptor (AhR), a cytosolic receptor (Chopra and Schrenk 2011). TCDD has a very high affinity for the AhR. For this congener, studies show that after the reaction, it is transferred to the cell nucleus, where an active heterodimer with the aryl hydrocarbon nuclear translocator (ARNT) is formed. This complex subsequently leads to increased gene expression after binding to a xenobiotic-responsive element (XRE).

Table 2.10 Names, abbreviations, and TEF (toxicity equivalency factors) values assessed from different institutions

Substances	Abbreviations	NATO/CCMS I-TEF	WHO 1998 TEF	WHO 2005 TEF	WHO 2022 TEF
Chlorinated dibenzo-p-dioxins					
2,3,7,8-Tetrachlorodibenzo-p-dioxin	2,3,7,8-TCDD	1	1	1	1
1,2,3,7,8-Pentachlorodibenzo-p-dioxin	1,2,3,7,8-PeCDD	0.5	1	1	0.4
1,2,3,4,7,8-Hexachlorodibenzo-p-dioxin	1,2,3,4,7,8-HxCDD	0.1	0.1	0.1	0.09
1,2,3,6,7,8-Hexachlorodibenzo-p-dioxin	1,2,3,6,7,8-HxCDD	0.1	0.1	0.1	0.07
1,2,3,7,8,9-Hexachlorodibenzo-p-dioxin	1,2,3,7,8,9-HxCDD	9.1	0.1	0.1	0.05
1,2,3,4,6,7,8-Heptachlorodibenzo-p-dioxin	1,2,3,4,6,7,8-HpCDD	0.01	0.01	0.01	0.05
Octachlorodibenzo-p-dioxin	OCDD	0.001	0.0001	0.0003	0.001
Chlorinated dibenzofurans					
2,3,7,8-Tetrachlorodibenzofuran	2,3,7,8-TCDF	0.1	0.1	0.1	0.07
1,2,3,7,8-Pentachlorodibenzofuran	1,2,3,7,8-PeCDF	0.05	0.05	0.03	0.01
2,3,4,7,8-Pentachlorodibenzofuran	2,3,4,7,8-PeCDF	0.5	0.5	0.3	0.1
1,2,3,4,7,8-Hexachlorodibenzofuran	1,2,3,4,7,8-HxCDF	0.1	0.1	0.1	0.3
1,2,3,6,7,8-Hexachlorodibenzofuran	1,2,3,6,7,8-HxCDF	0.1	0.1	0.1	0.09
1,2,3,7,8,9-Hexachlorodibenzofuran	1,2,3,7,8,9-HxCDF	0.1	0.1	0.1	0.2
2,3,4,6,7,8-Hexachlorodibenzofuran	2,3,4,6,7,8-HxCDF	0.1	0.1	0.1	0.1
1,2,3,4,6,7,8-Heptachlorodibenzofuran	1,2,3,4,6,7,8-HpCDF	0.01	0.01	0.01	0.02
1,2,3,4,7,8,9-Heptachlorodibenzofuran	1,2,3,4,7,8,9-HpCDF	0.01	0.01	0.01	0.1
Octachlorodibenzofuran	OCDF	0.001	0.0001	0.0003	0.002
Non-ortho substituted PCB					
3,3′,4,4′-Tetrachlorobiphenyl	PCB 77		0.0001	0.0001	0.0003
3,4,4′,5-Tetrachlorobiphenyl	PCB 81		0.0001	0.0003	0.006

(continued)

Table 2.10 (continued)

Substances	Abbreviations	NATO/CCMS I-TEF	WHO 1998 TEF	WHO 2005 TEF	WHO 2022 TEF
3,3′,4,4′,5-Pentachlorobiphenyl	PCB 126		0.1	0.1	0.05
3,3′,4,4′,5,5′-Hexachlorobiphenyl	PCB 169		0.01	0.03	0.005
Mono-ortho substituted PCB					
2,3,3′,4,4′-Pentachlorobiphenyl	PCB 105		0.0001	0.00003	0.00003
2,3,4,4′,5-Pentachlorobiphenyl	PCB 114		0.0005	0.00003	0.00003
2,3′,4,4′,5-Pentachlorobiphenyl	PCB 118		0.0001	0.00003	0.00003
2′,3,4,4′,5-Pentachlorobiphenyl	PCB 123		0.0001	0.00003	0.00003
2,3,3′,4,4′,5-Hexachlorobiphenyl	PCB 156		0.0005	0.00003	0.00003
2,3,3′,4,4′,5′-Hexachlorobiphenyl	PCB 157		0.0005	0.00003	0.00003
2,3′,4,4′,5,5′-Hexachlorobiphenyl	PCB 167		0.00001	0.00003	0.00003
2,3,3′,4,4′,5,5′-Heptachlorobiphenyl	PCB 189		0.0001	0.00003	0.00003

The WHO's International Agency for Research on Cancer has classified 2,3,7,8-TCDD as Group 1 (carcinogenic to humans). Other polychlorinated dibenzo-p-dioxins and dibenzo-p-furans could not be classified regarding their human carcinogenicity. The 12 dl-PCBs that received a TEF determination by the WHO are also categorized as carcinogenic to humans in Group 1 (IARC 2015).

2.2.2 *Toxicokinetics*

In principle, PCDD/Fs and dl-PCBs, excluding the OCDD/F congeners, are well absorbed through dermal, oral, or inhalation exposure and are distributed throughout the body, bound to serum lipids and proteins. For TCDD, the distribution between serum lipids and proteins is approximately 80–20%. Congener-specific uptake rates ranging from 70% to over 90% have been observed in animal experiments and human volunteers following oral administration. Due to their lipophilicity, these compounds are stored in fatty tissue and fat-rich organs, such as the nervous system, and they reach the highest concentrations (relative to body weight) in the liver. Rodents excrete PCDD/Fs almost exclusively via bile, while in humans, most congeners are primarily excreted unmetabolized in feces. Based on epidemiological data, the half-lives for TCDD are estimated at 4–5 years for young adults, 6.3–8 years for middle-aged adults, and 8.5–11 years for older adults. For the other PCDD/F congeners, half-lives described in the literature range from 1 to 19 years. A broad range of half-lives is also reported for the dl-PCBs.

In principle, PCDD/Fs and dl-PCBs can cross the placental barrier and transfer into breast milk. Otherwise, they are primarily excreted through feces, partly as metabolic products via bile, and partly through body fat loss.

2.2.3 *Occurrence in the Blood*

Table 2.11 summarizes more recent study results on the general population's internal exposure to PCDDs/Fs and dl-PCBs.

In Europe, current studies from Spain and Germany report median concentrations for PCDD/Fs and dl-PCBs of only 6.6 and 4.5 pgWHO$_{2005}$TEQ/g lw and 3.2 and 2.6 pgWHO$_{2005}$TEQ/g lw, respectively (Santa-Marina et al. (2023), Fromme et al. 2016). These figures are significantly lower than those observed in older studies, where median concentrations of 13.1 and 5.7 pgWHO$_{2005}$TEQ/g lw for PCDD/Fs and dl-PCBs were recorded in Germany from 2000–2002 (Wilhelm et al. 2008). Similar concentrations are reported in a more recent representative Canadian study from 2016/17 and a Chinese study from 2013–2015. Slightly lower mean concentrations were found in Ghana and South Africa. The analysis of PCDD/Fs and dl-PCBs proportions in the total TEQ indicates apparent spatial and temporal differences. For example, in Germany, the proportion of dl-PCBs in total levels has steadily risen,

Table 2.11 Median exposure to PCDD/F and dl-PCB in the blood of the general population in pg $WHO_{2005}TEQ/g$ lw (pg $WHO_{1998}TEQ/g$ lw)

References	N	PCDD/Fs	dl-PCBs	Age (years)	Country, year
Europe					
Oltramare et al. (2025)	80	6.2[c]	4.2[c]	>19	Switzerland, 2023
Santa-Marina et al. (2023)	105	6.6	3.2	18–70	Spain, 2018
Nadal et al. (2019)	40	6.8	–	17–61	Spain, 2018
Fromme et al. (2016)	70	4.5 (5.5)	2.6 (4.9)	4–76	Germany, 2013
Iamiceli et al. (2021)	44	7.9	5.9	36–50	Italy, 2013
Ingelido et al. (2017)	476	6.0 (7.1)	3.5 (5.7)	20–41	Italy, 2011/12
Ulutas et al. (2015)	57	–	0.04	20–41	Turkey, 2010–12
Esposito et al. (2014)	33	(6.7)	(6.1)	18–64	Italy, 2007
Nadal et al. (2008)	20	9.5[b]	–	17–61	Spain, 2007
Chovancová et al. (2012)	44	– (9.4)	– (13.8)	24–62	Slovakia, 2006/07
Fromme et al. (2009)	49	7.7 (10.1)	5.8 (9.5)	18–65	Germany, 2005
Fréry et al. (2007)	1030	(13.7[d])	(13.6[d])	30–65	France, 2005
Turrio-Baldassarri et al. (2008)	94	(22)	(32)	51[b]	Italy, 2004
Cerná et al. (2007)	20	(8.9)	(13.4)	43[b]	Czech Republic, 2003
Costopoulou et al. (2006)	10[d]	(6.8)	(3.2)	27–55	Greece, 2002–04
Wittsiepe et al. (2007)	169	(15.3)	(10.8)	19–41	Germany, 2000–03
Wilhelm et al. (2008)	182	13.1	5.7	18–42	Germany, 2000–02
North America					
HC 2020	67[e]	7.5[b]	1.2[b]	3–79	Canada, 2016/17
Rawn et al. (2012)	50[a]	10 (12)[c]	2.5 (5.2)[c,f]	40–59	Canada, 2007–09
Hedgeman et al. (2009)	251	15 (16)	2.9 (8.9)	50[b]	USA, 2004/05
LaKind et al. (2009)	1290	11.4	–	≥ 12	USA, 2003/04
Wong et al. (2008)	113	(20.2[c])	–	≥ 15	USA, 2002
Asia, Africa					
Bruce-Vanderpuije et al. (2021)	34	3.1[b]	1.2[b]	18–38	Ghana, 2017

(continued)

Table 2.11 (continued)

References	N	PCDD/Fs	dl-PCBs	Age (years)	Country, year
Bao et al. (2020)	55	7.3	0.6		China, 2013–15
Pieters and Focant (2014)	21[a]	4.0	1.4	37–84	South Africa, 2010
Park et al. (2013)	11	(11.4)	–	42[b]	Korea, 2006
Todaka et al. (2011)	97	9.0	3.5	22–41	Japan, 2004–08
Arisawa et al. (2011)	1656	10	5.7	15–73	Japan, 2002–07
Lampa et al. (2018)	41	10	4.6	≥ 20	Japan, 2003/04
Baba et al. (2018)	379	9.8[b]	4.9[b]	20–48	Japan, 2002–05
Park et al. (2014)	112/48	8.6	4.0		Korea, 2001–11
Nakamura et al. (2008)	49	(9.5)	(4.8)	32[b]	Japan, 2001–03
Hsu et al. (2009)	251	10.1 (11.5)	4.3 (6.1)	18–60	Taiwan, 2001–06

[a] Pooled samples; [b] mean; [c] geometric mean; [d] samples from urban sites; [e] each pool was composed of serum from 71 to 133 individuals, [f] sum of medians of non-ortho and mono-ortho PCB; + I-TEF; * Nordic-TEF

now reaching medians of 47% (using WHO 1998-TEFs) and 40% (using WHO 2005-TEFs). In a German study from 2000 to 2002, this figure was only 41% (with WHO_{1998}TEFs) and 30% (with WHO_{2005}TEFs) (Wilhelm et al. 2008). In a Spanish study, the proportion of dl-PCBs in total TEQ currently stands at only 33% (Santa-Marina et al. 2023). Conversely, North America and China report a dominance of PCDD/Fs, where the dl-PCBs account for only 14% in Canada and 13% in China (HC 2020; Bao et al. 2020).

Various studies have examined the time trend of PCDD/F concentrations in serum. For instance, LaKind et al. (2009) found a 56% decrease in the group of 12–19-year-olds and a 38% decrease in the group of 20–39-year-olds during three collection periods between 1999 and 2004 from the representative US National Health and Nutrition Examination Survey. Meanwhile, only a slight, nonsignificant decrease was noted among 40 to 59-year-olds, along with a slight, significant increase among those aged 60 and older. Park et al. (2014) studied the time course from 2001 to 2011 in Seoul, Korea, and observed a slight but continuous decrease in PCDD/F and dl-PCB TEQ levels. In Italy, data from various studies were aggregated, showing annual reductions of -2.1 pg WHO_{2005}TEQ/g lw for PCDD/F between 2009 and 2018, and for dl-PCB between 2000 and 2018 (Miniero et al. 2022). The decline was fastest for PCDDs and slowest for dl-PCBs. Seo et al. (2021) analyzed 1802 serum samples over 20 years from areas in Seoul, Korea, where municipal solid waste incinerators were located. PCDD/F concentrations decreased from 9.5 pg WHO_{2005}TEQ/g lw in 2000 to 7.9 pg WHO_{2005}TEQ/g lw in 2019, while emissions from the plant were reduced by 96%.

In the 5th cycle of the nationwide Canadian Health Measures Survey (CHMS), 6, 6, and 12 pools of 71–133 individuals were examined across three age groups of children and adolescents (HC 2020). For PCDD/Fs and dl-PCBs, the mean values in pg $WHO_{2005}TEQ$/g lw were 6.5 and 1.0 for the 3–5-year-olds, 6.1 and 1.0 for the 6–11-year-olds, and 5.2 and 0.82 for the 12–19-year-olds. The concentrations in the blood of older individuals were higher in this study. In a representative survey in Sweden conducted in 2016–2017, 2377 adolescents aged 12 years were recruited, and mean values of 3.9 pgWHO2005TEQ/g lw for PCDD/F and 1.6 pg$WHO_{2005}TEQ$/g lw for dl-PCB were found in pooled samples (Gyllenhammar et al. 2023). The same working group examined children from Uppsala, Sweden, aged 4, 8, and 12 years and found mean values in pg $WHO_{2005}TEQ$/g lw of 7.7 and 4.0, 4.9 and 2.4, as well as 5.1 and 2.7 for PCDD/Fs and dl-PCBs.

From 2001 to 2003, concentrations of 6.4 pg $WHO_{1998}TEQ$/g lw for PCDD/F and 2.8 pg WHO1998TEQ/g lw for dl-PCB were reported in 49 umbilical cord blood samples in Japan (Nakamura et al. 2008). In another Japanese study, 4.8 pg $WHO_{2005}TEQ$/g lw for PCDD/Fs and 1.8 pg $WHO_{2005}TEQ$/g lw for dl-PCBs were found in the umbilical cord blood of 41 samples (Lampa et al. 2018). In 20 cord blood samples, significantly lower medians of 2.3 pg $WHO_{2005}TEQ$/g lw and 0.11 pg $WHO_{2005}TEQ$/g lw were found for PCDD/Fs and dl-PCBs in Korea (Kim et al. 2015). The two Japanese studies showed a statistically significant correlation between maternal and umbilical cord blood. In the three studies mentioned above, the percentage of TEQ in the umbilical cord samples was 67, 48, and 49% for PCDD/Fs, and 58, 40, and 79% for dl-PCB. The analysis of 41 sample sets of maternal and umbilical cord blood also revealed a placental transfer rate of 31% for PCDD/Fs and 43% for dl-PCBs on a lipid basis in Japan (Mori et al. 2014).

2.2.4 *Occurrence in Breast Milk*

Many studies worldwide have reported the concentrations of PCDDs/Fs and dl-PCBs in breast milk, and Ulaszewska et al. (2011) and Fång et al. (2015) have summarized these findings.

Between 1987 and 2019, the WHO and UNEP conducted six rounds of a global breast milk survey in various countries, with the sharpest decline in TEQs observed for PCDDs/Fs in Western Europe and Asian countries and dl-PCBs in Western Europe and Central and Eastern European countries (Fiedler et al. 2023; Malisch et al. 2023). Overall, the decrease in median TEQs was 89% for PCDDs to 1.6 pg $WHO_{2005}TEQ$/g lw, 84% for PCDFs to 1.0 pg $WHO_{2005}TEQ$/g lw, and 84% for dl-PCBs to 1.1 pg $WHO_{2005}TEQ$/g lw in 2015–2019 (Fiedler et al. 2023). The relative contribution of PCDD/Fs TEQ to $\sum$PCDD/Fs and dl-PCBs TEQ varies between countries from 44 to 74%, while within a country, the ratio remains relatively constant over time.

Table 2.12 summarizes the more recent results of measurements published in the scientific literature. Exposure can vary widely and, along with other influencing factors such as dietary habits, is generally higher in more industrialized areas than in less industrialized regions due to potential sources. PCDDs/Fs and dl-PCBs concentrations in Europe are currently around 2.3–3.1 and 1.5–2.7 $pgWHO_{2005}TEQ/g$ lw, respectively. Higher levels were reported in studies from Denmark, Finland, and France at the end of the 2000s (Venisseau et al. 2011; Vandentorre et al. 2011). In Canada, the nationwide Maternal-Infant Research on Environmental Chemicals (MIREC) study shows slightly higher mean levels of 1.6–19 $pgWHO_{2005}TEQ/g$ lw for PCDDs/Fs compared to Europe and comparable levels of 0.4–7.8 $pgWHO_{2005}TEQ/g$ lw for dl-PCBs (Rawn et al. 2017). Similar concentrations were reported from Chile. The Asian and African studies also revealed significant differences between the countries and a current decrease in concentration to levels comparable to Europe and North America. Significantly lower TEQs were observed in 30 breast milk samples from Uganda, with ranges of only 0–2.6 for PCDD/Fs and 0.03–5.3 $pgWHO_{2005}TEQ/g$ lw for dl-PCBs (Matovu et al. 2021). Regarding the use of Agent Orange during the Vietnam War, high levels of PCDDs/Fs must still be anticipated. For example, analysis of eight breast milk samples from contaminated areas at the time and eight from a reference region still showed significantly different mean concentrations for $\sum$PCDDs/Fs and dl-PCBs of 15.7–6.0 $pgWHO_{2005}TEQ/g$ lw (Nhu et al. 2011).

Depending on the country of sampling, breast milk samples also reveal differences in the proportion of PCDD/Fs and dl-PCBs within the total TEQ content. For instance, the proportion of dl-PCBs in the total TEQs in two European studies was approximately 43% and 47% (Hedvall Kallerman et al. 2016; Fromme et al. 2022), while in Canada, it was only 28% (Rawn et al. 2017), and in two Chinese studies, it was about 28% and 29% (Zhang et al. 2023; Sun et al. 2022). Kongenere PCB 126 accounts the largest share of all PCBs in the total TEQs.

Numerous studies have shown that the overall contamination of breast milk is significantly declining, although a concentration plateau has been reached in recent years (Lignell et al. 2009; Schuhmacher et al. 2009; Vieth et al. 2011; Raab et al. 2013; Antignac et al. 2016; van den Berg et al. 2017; Fromme et al. 2022; Zhang et al. 2023). Figure 2.4 illustrates the time trend of PCDDs/Fs and dl-PCBs levels from 1986 to 2019 in German breast milk samples. A similar decrease was found in Sweden between 1996 and 2017 (Gyllenhammar et al. 2021). During this period, the total TEQ decreased by about 6% annually, from about 15 $pgWHO_{2005}TEQ/g$ lw to 4 $pgWHO_{2005}TEQ/g$ lw. The China National Human Milk Biomonitoring Program also reported a decline in the mean values of $\sum$PCDD/Fs and PCBs, from 5.6 $pgWHO_{2005}TEQ/g$ lw in 2011 to 4.7 $pgWHO_{2005}TEQ/g$ lw in 2016–2019, primarily due to the decrease in PCDD/Fs levels (Zhang et al. 2023). Nevertheless, an increase in TEQ levels was observed in this study from 2007 to 2011. In Canada, a decline in median $\sum$PCDD/Fs and PCBs TEQ was reported, dropping from 18 $pgWHO_{2005}TEQ/g$ lw in 1992 to 6.3 $pgWHO_{2005}TEQ/g$ lw in 2008–2011 (Rawn et al. 2017). This clearly demonstrates that the measures implemented to reduce pollutant emissions have positively impacted breast milk contamination.

Table 2.12 Mean concentrations of PCDDs/Fs and dl-PCBs in breast milk in pg $WHO_{2005}TEQ/g$ lw (pg $WHO_{1998}TEQ/g$ lw)

References	N	PCDDs/Fs	dl-PCBs	Age (years)	Country, year
Europe					
Hedvall Kallerman et al. (2023)	30	2.8	2.1	26–43	Sweden, 2020/21
Dellatte et al. (2025)	41	3.6	2.7	25–35	Italy, 2016–18
Houlihan et al. (2021)	16[a]	3.1	1.5	23–40	Ireland, 2016–18
Schumacher et al. (2019)	20	2.26	–	27–40	Spain, 2017
Fromme et al. (2022)	100	3.0 (3.8)	2.7 (4.3)	20–45	Germany, 2016
Hernández et al. (2020)	75	2.72	1.55	20–45	Spanian, 2015
Parsons et al. (2025)	150	3.8[b]	2.1[b]	32[d]	UK, 2013–15
Antignac et al. (2016)	96	6.1	4.3	30[d]	France, 2011–14
Antignac et al. (2016)	438	13.2	6.6	30[d]	Denmark, 2011–14
Antignac et al. (2016)	22	9.1	4.6	30[d]	Finland, 2011–14
Fång et al. (2013)	11	3.1 (3.7)	2.4 (3.8)	30[d]	Sweden, 2011
Miniero et al. (2018)	10	(8.3)	(7.9)	20–40	Italy, 2010
Runkel et al. (2021)	55[a]	1.6	1.5	18–49	Slovakia, 2008–14
Grešner et al. (2021)	110	7.9	2.67	16–38	Poland, 2007–11
Raab et al. (2013)	516	5.1 (6.0)	4.4 (7.0)	15–43	Germany, 2007–08
Pratt et al. (2012)	11[a]	6.3	3.3	20–41	Ireland, 2010
Venisseau et al. (2011)	261	(16.0)	(12.7)		Denmark/Finland
Vandentorren et al. (2011)	44	9.7 (13.4)	8.8 (14.2)	24–41	France, 2007
Schuhmacher et al. (2009)	15	(7.6)	(9.0)	25–35	Spain, 2007
Cok et al. (2009)	51	(7.5)	(3.1)	20–40	Turkey, 2007
Chovancová et al. (2012)	33	(6.1)	(6.4)	17–36	Slovakia, 2006–2007
Raab et al. (2008)	42	8.2 (9.9)	6.3 (9.9)	22–46	Germany, 2005
Fürst (2006)	175	(13.8)	13.2		Germany, 2001–03
Americas					
Rawn et al. (2017)	298	5.4	1.5	21–46	Canada, 2008–11
Foerster et al. (2021)	3[a]	3.6–5.2[e]	0.7–1.1[e]	25–29[d]	Chile
Ryan and Rawn (2014)	34	7.4 (8.3)	2.6 (4.5)		Canada, 2005
Asia					
Varghese et al. (2025)	32	4.09[b]	1.60[b]	20–37	India, 2022/23
Sun et al. (2022)	51	4.0[b]	1.6[b]	26–39	China, 2019
Xu et al. (2022)	29	2.7 (4.1)	1.4 (1.4)		China, 2019
Matovu et al. (2021)	30	0.4 (0.6)[b]	1.2 (1.1)[b]	17–38	Uganda, 2018
Leong et al. (2021)	21	1.4[c]	0.6[c]	25–45	Malaysia

(continued)

Table 2.12 (continued)

References	N	PCDDs/Fs	dl-PCBs	Age (years)	Country, year
Huang et al. (2019)	6[a]	7.0 (8.2)	2.1 (2.2)	16–35	China, 2017/18
Zhang et al. (2023)	100[a]	3.6	1.4		China, 2016–19
Bruce-Vanderpuije et al. (2021)	21	(1.7)	–		Ghana, 2017
Dong et al. (2020)	42	9.7[b]	–	25–30	China, 2015
Chen et al. (2018)	25	2.9	–	25–42	Taiwan, 2013–16
Bao et al. (2020)	55	4.5	2.0		China, 2013–15
Lu et al. (2015)	150	5.4 (6.0)	2.9 (3.1)	18–35	China, 2011–12
Zhang et al. (2016)	32[a]	(4.9)	–	25[d]	China, 2011
Nhu et al. (2011)	8	6.0	–	19–25	Vietnam, 2008
't Mannetje et al. (2013)	39	3.5	1.3	20–30	New Zealand, 2007–10
Todaka et al. (2011)	97	5.2	3.4	22–41	Japan, 2002–05
Nakamura et al. (2008)	49	9.8[b]	6.9[b]	32[d]	Japan, 2001–03

[a] Pooled samples; [b] median; [c] geometric mean; [d] mean; [e] range of three sampling sites

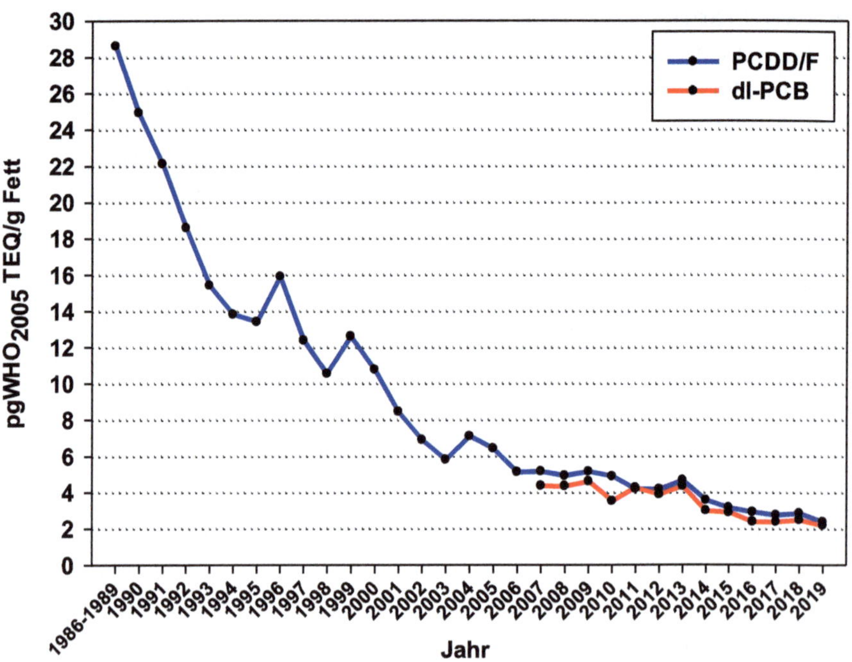

Fig. 2.4 Time course of PCDD/Fs and dl-PCBs concentrations in breast milk in Germany (after Bruns-Weller 2022)

Important factors influencing the concentration levels in breast milk include the mother's dietary habits and age, the timing during the breastfeeding period, and whether she is a primipara or multipara (EFSA 2018). For instance, a preference for high-fat animal foods is linked to higher exposure compared to a strictly vegetarian diet (Schecter et al. 1994; Lignell et al. 2011). Additionally, the concentrations in breast milk tend to increase with the age of the mother at the time of birth (Wittsiepe et al. 2007; Antignac et al. 2016; Rawn et al. 2017; Grešner et al. 2021; Fromme et al. 2022). Conversely, the concentrations in breast milk decrease with the number of children breastfed and during the breastfeeding period (Fürst et al. 1992; Uehara et al. 2007; Vigh et al. 2013). For example, Vigh et al. (2013) identified PCDD/F and dl-PCB concentrations of 2.13 and 1.04 pg $WHO_{2005}TEQ$/g lw, 1.85 and 0.85 pg $WHO_{2005}TEQ$/g lw, and 1.65 and 0.75 pg $WHO_{2005}TEQ$/g lw, respectively, in breast milk samples from 22 first-time mothers on days 5, 12, and 84 after birth.

2.3 Polybrominated Dibenzodioxins/Furans

Theoretically, similar to chlorinated dioxins, 75 polybrominated dibenzodioxins (PBDDs) and 135 dibenzofurans (PBDFs) can form, depending on the number and position of the bromine atoms in the molecule. The chemical structures are illustrated in Fig. 2.5. Compared to their chlorinated analogs, they exhibit a higher molecular weight, lower volatility, and reduced water solubility, as well as decreased environmental persistence (Birnbaum et al. 2003; Piskorska-Pliszczyńska and Maszewski 2014; Yang et al. 2021).

PBDD/Fs occur as impurities in polybrominated diphenyl ethers (PBDEs) and other brominated flame retardants and can be released during the manufacture, use,

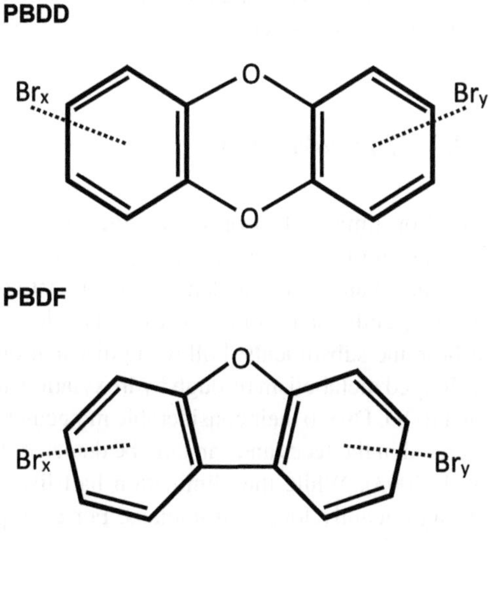

Fig. 2.5 Chemical structure of polybrominated dibenzo-p-dioxines (PBDDs) and dibenzofuranes (PBDFs)

and recycling of products like plastics containing these substances (Ebert and Bahadir 2003; Altarawneh and Dlugogorski 2013; Yang et al. 2021). It can be assumed that PBDD/Fs may also form from brominated compounds during thermal processes (UNEP 2010). In terms of quantity, this seems to be possible in the lower percentage range. PBDD/Fs have been detected in numerous environmental media and foods (Venkatesan and Halden 2014; Rose et al. 2015; Fernandes and Falandysz 2021).

2.3.1 Health Aspects

Overall, the data available on the health effects of this class of substances is minimal, and very few findings regarding their impact on humans have been published to date (Birnbaum et al. 2003; Van den Berg et al. 2013; Piskorska-Pliszczyńska and Maszewski 2014). Only Zober et al. (1992) found no adverse effects related to the levels of workplace air or blood of the workers in a study involving 42 workers in the plastics processing industry (extrusion blending of polybutylene terephthalate (PBTP) with decabromodiphenyl ethers). Studies in rats and mice showed effects for 2,3,7,8-TBDD in the same dose range as for 2,3,7,8-TCDD, indicating comparable toxic potency, and for some congeners, even higher toxicity (Van den Berg et al. 2013). Furthermore, in animal experiments, effects on the thymus and liver were particularly observed, along with effects on the immune system and the thyroid gland.

A joint expert group of the World Health Organization (WHO) and the United Nations Environment Programme (UNEP) concluded that PBDD/Fs, although based on limited data, primarily affect the liver, thymus, and weight development in a manner that is at least qualitatively comparable to the effects of chlorinated analogs (van den Berg et al. 2013). Therefore, the working group provisionally recommends using the TEFs established for chlorinated compounds for the individual PBDD/Fs as part of an assessment.

2.3.2 Toxicokinetics

Based on limited data on toxicokinetics, often referring to TBDD, it is assumed that inhalation, oral, and dermal uptake are not significantly different from those of chlorinated analogs (Van den Berg et al. 2013). At higher doses, TBDD is absorbed less efficiently than at lower doses, and the level of absorption depends on the number of bromine substituents. Following distribution mainly in the liver and fatty tissue, prolonged metabolism through hydroxylation and debromination has been described for TBDD. Due to their considerable molecular weight, the metabolites are primarily excreted in the feces and can only be detected in small quantities in urine (Birnbaum et al. 2003). While the elimination half-lives in rats are only a few weeks, they are significantly longer in humans. For example, Zober et al. (1992) found mean half-lives of 5.9 years (range: 2.9–10 years) for 2,3,7,8-TBDD and 1.5 years (range: 1.1–1.9 years) for 2,3,7,8-TBDF in the aforementioned study of blood samples from exposed workers.

2.3.3 *Occurrence in the Blood*

Only very limited results on PBDD/F concentrations in human blood samples are available. Hagberg et al. (2010) collected 17 whole blood samples in Norway and serum samples from unexposed individuals in Sweden in 2004/05. PBDD/Fs could not be detected above the respective quantification limits in either population. In a study conducted in Germany in 2013, a total of 42 plasma samples from individuals aged 20–68 years were analyzed, revealing mean values of 2.9 and 0.5 pg $WHO_{2005}TEQ$/g lw for PBDDs and PBDFs (maxima: 7.8 and 1.6 pg $WHO_{2005}TEQ$/g lw). For PCDD/Fs, the mean observed was 3.4 pg $WHO_{2005}TEQ$/g lw (maximum: 9.3 pg $WHO_{2005}TEQ$/g lw) (Fromme et al. 2016). In a rural region of Vietnam, a significantly higher exposure was found in the serum of 20 women with no known exposure, with a geometric mean of 30.6 pg $WHO_{2005}TEQ$/g lw (Peecher et al. 2022).

As expected, significantly higher concentrations of PBDD/Fs were found in serum samples from 12 Californian firefighters in 2009 (Shaw et al. 2013). The mean values of the six PBDD and six PBDF congeners analyzed were 103 pg $WHO_{2005}TEQ$/g lw (range: 0–732 pg $WHO_{2005}TEQ$/g lw).

In the aforementioned German study, the mean values of PBDD/Fs' relative contribution to the total WHO-TEQ (sum of PCDD/Fs, PBDD/Fs, and dl-PCB) amounted to 23% (Fromme et al. 2016). In contrast, the general population in Vietnam accounted for 77% of the total TEQs (Peecher et al. 2022).

2.3.4 *Occurrence in Breast Milk*

Overall, limited information on PBDDs/Fs in breast milk is often restricted to individual congeners.

Kotz et al. (2005) reported results from pooled samples of the third round of an exposure study coordinated by the WHO and UNEP, collected in 17 countries between 2000 and 2003 (Kotz et al. 2005). The mean values of the upper and lower bound estimates were 0.32 and 1.08 pg $WHO_{1998}TEQ$ /g lw, with PBDDs/Fs accounting for approximately 12% of the total TEQs (the sum of PCDD/Fs, PBDD/Fs, and dl-PCBs). The dominant PBDD/Fs congeners were 2,3,7,8-TBDF, with a mean value of 0.70 pg/g lw, and 2,3,4,7,8-PeBDF, with 0.23 pg/g lw. For 38 pooled breast milk samples collected between 2001 and 2009 in 28 countries, Schächtele et al. (2023) also reported upper bound mean values of 0.77 pg $WHO_{2005}TEQ$/g lw (range: 0.12–1.58 pg $WHO_{2005}TEQ$/g lw).

Wiberg et al. (1992) first examined a composite of breast milk in Sweden and found none of the seven congeners exceeded the detection limits of 0.3–7 pg/g lw In Belgium, 15 congeners were analyzed in a pooled sample from a rural area, revealing mean concentrations of 0.42 pg $WHO_{1998}TEQ$/g lw and 0.34 pg $WHO_{2005}TEQ$/g lw were found using medium bound approach (Croes et al. 2013). In particular, the congeners 2,3,7,8-TBDF and 2,3,4,7,8-PeBDF were detected at concentrations of 0.55 pg/g lw and 0.33 pg/g lw, respectively. Comparable mean TEQs using the upper bound were observed for PBDD/Fs in 11 pooled breast milk samples from 109 mothers from Ireland, found at 0.72 pg $WHO_{2005}TEQ$/g lw (range: 0.58–1.23 pg $WHO_{2005}TEQ$/g lw) (Pratt et al. 2012). In 22 breast milk samples from 10 Swedish women, there was a slightly higher mean value of 1.8 pg$WHO_{2005}TEQ$/g lw (range: 0.2–8.4 pg$WHO_{2005}TEQ$/g lw) (Bjurlid 2018). The analysis of 100 breast milk samples from 20–45-year-old women from Germany from 2015–2017 also showed a similar exposure level with a mean value of 0.91 $WHO_{2005}TEQ$/g lw (0.94 pg $WHO_{1998}TEQ$/g lw) and a maximum of 8.4 pg $WHO_{2005}TEQ$/g lw (21.3 pg $WHO_{1998}TEQ$/g lw) (Fromme et al. 2022).

Ohta et al. (2004) reported significantly higher levels for the sum of PBDDs/Fs, averaging 269 pg/g l.w, in five samples of multipara from Japan in 2002 compared to the 41.4 pg/g l.w found in multipara. In Taiwan, Chen et al. (2018) did not detect all 13 PBDD/F congeners above the detection limit of 0.05–25.7 pg/g l.w in 25 breast milk samples collected in 2016. In contrast, while no PBDDs were found in 40 milk samples collected in China during 2016–2017, three PBDF congeners were detected, with mean concentrations of 3.2 pg/g l.w for 2,3,7,8-TBDF, 9.5 pg/g l.w for 1,2,3,4,6,7,8-HpBDF, and 28.0 pg/g l.w for OBDF (Lin et al. 2021). In this study, the mean values using lower and upper bound estimates were 0.42 and 1.36 pg $WHO_{2005}TEQ$/g l.w, respectively, and significant correlations were observed between PBDF levels and higher brominated PBDEs, particularly BDE 183 and BDE 209.

A recent study from Ghana, which included 24 human milk samples collected in 2017, reported slightly lower average levels of 0.29 pg$WHO_{2005}TEQ$/g lw for seven congeners analyzed (Bruce-Vanderpuije et al. 2021).

Compared to PCDD/Fs and dl-PCBs, the concentrations of PBDD/Fs in breast milk are even lower. In a German study of 100 samples, the mean values of PBDDs/Fs contributed 14% to the total TEQ of PCDD/Fs, dl-PCBs, and PBDD/Fs. Schächtele et al. (2023) also reported an average contribution of PBDD/Fs of 10% to the total concentrations of PCDD/Fs and PBDD/Fs in their analysis of 38 pooled samples from 28 countries.

2.4 Polychlorinated Naphthalenes (PCN)

PCN consists of two fused aromatic rings, on which one to eight of their hydrogens are substituted by chlorine atoms. Theoretically, 75 congeners with different degrees of chlorination are possible (Table 2.13). The general chemical structure is shown in Fig. 2.6. The physicochemical properties can vary significantly. For example, the higher chlorinated tetra- to octa-CN are lipophilic, while volatility decreases with increasing chlorination (UNEP 2017). Commercially available products, ranging from liquids to hard waxes, always contain mixtures of congeners in varying proportions. PCNs were used industrially on a large scale from around 1930 until the 1980s, particularly in capacitors and transformers, as hydraulic oil, plasticizers, and wood preservatives (Klimczak et al. 2023). Various mixtures have been marketed under the trade name Halowax, which can contain predominantly mono-CN (Halowax 1031) to predominantly octa-CN (Halowax 1051). It is estimated that between 150,000 and 400,000 tons of PCN were produced worldwide (UNEP 2017). In addition to the known direct sources, unintentional formation occurs during the industrial production of other chlorine-containing substances (Klimczak et al. 2023). Although they have not been used for decades, exposure via the environment is still possible today, e.g., through thermal processes in waste incineration and old landfills (Domingo 2004; Liu et al. 2014).

PCNs are listed under Annex A (Parties must take measures to eliminate the production and use of the chemicals) and C (Parties must take measures to reduce the unintentional releases of chemicals), with specific exemptions for their use in the production of polyfluorinated naphthalenes, including octafluoronaphthalenes (SC 2023).

Table 2.13 List of polychlorinated naphthalene (PCN) homolog groups

Congener	Acronym	Formula	No. of single congeners
Monochloronaphthalenes	MonoCN	$C_{10}H_7Cl$	2
Dichloronaphthalenes	DiCN	$C_{10}H_6Cl_2$	10
Trichloronaphthalenes	TriCN	$C_{10}H_5Cl_3$	14
Tetrachloronaphthalenes	TetraCN	$C_{10}H_4Cl_4$	22
Pentachloronaphthalenes	PentaCN	$C_{10}H_3Cl_5$	14
Hexachloronaphthalenes	HexaCN	$C_{10}H_2Cl_6$	10
Heptachloronaphthalenes	HeptaCN	$C_{10}HCl_7$	2
Octachloronaphthalene	OctaCN	$C_{10}Cl_8$	1

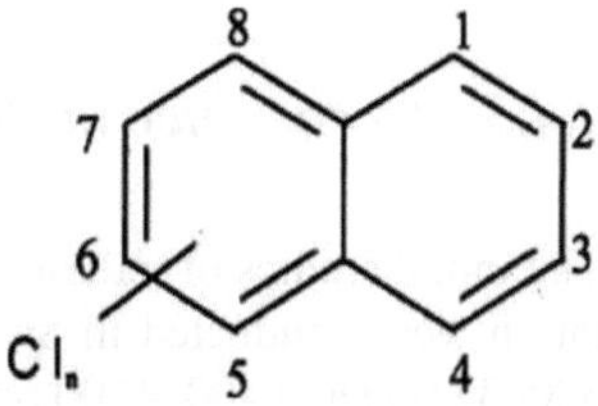

Fig. 2.6 Chemical structure of polychlorinated naphthalenes (PCNs) (n = 1–8)

2.4.1 Health Aspects

Various reviews of the health effects, including toxicokinetics, are available (WHO 2001; UNEP 2012; Falandysz et al. 2014; Fernandes et al. 2022). However, the available data is limited and often pertains to individual congeners, specifically hexa-CN or technical mixtures. Studies in experimental animals have demonstrated liver effects, including hepatomegaly and fatty liver, as well as fetotoxic and teratogenic effects that are qualitatively similar to those of other dioxin-like compounds. Additionally, neurotoxic effects such as impaired long-term memory and reduced pain sensitivity were observed in rats, and endocrine effects are being discussed. In workplace studies, long-term exposure was found to particularly affect the liver, while dermal exposure resulted in severe skin rashes and chloracne. Moreover, an association between PCN exposure and tumors of the connective tissue, as well as neoplasms of the respiratory tract, was noted, although there was co-exposure to asbestos.

As some congeners like PCDD/Fs and dl-PCBs interact with the cytosolic Ah receptor, several authors have also proposed toxicity equivalency factors (TEFs) for the group of PCNs (Falandysz et al. 2014; Suzuki et al. 2020; Fernandes et al. 2022). However, unlike dioxins, there is no international scientific consensus since the available data is still considered too limited.

2.4.2 Toxicokinetics

Data on the toxicokinetics of PCNs are limited. In animal studies, they can be well absorbed after oral administration, depending on the degree of chlorination. After distribution in the body, retention occurs in the liver and fatty or fat-rich tissues. Higher chlorinated compounds, such as penta- and octa-CNs, are hardly metabolized and are mainly excreted via feces, while lower chlorinated compounds undergo hydroxylation and can be eliminated via urine. Ryan and Masuda (1994) calculated half-lives of 1.5 and 2.4 years for three test subjects exposed to contaminated rice oil in the Yu-Cheng accident based on continuous measurements of hexa-CN66/67 in whole blood.

2.4.3 Occurrence in the Blood

The initial studies of individual congeners in the blood plasma of the general population were conducted in Sweden ($N = 37$) in the early 1990s. Concentrations of 300–32,000 and 400–4100 pg/g lw were identified for penta-CN52 and hexa-CN66/67 (Asplund et al. 1994, cited in WHO 2001). Weistrand et al. (1997) also investigated the blood plasma of six potentially exposed workers and five controls in Sweden, finding no differences between the two groups. The concentrations of penta-CN52 and hexa-CN66/67 ranged from 54 to 423 and 98 to 361 pg/g lw, respectively, while hexa-CN71, hepta-CN73, and octa-CN75 could not be quantified above the LOQ.

In the USA, a total of 43 people working near the World Trade Center were examined one week after the collapse in 2002/2003 (Horii et al., 2010; Kannan, 2015). Divided into four exposure groups, mean values of $\sum Cl_{3\text{-}8}CN$ were found in the plasma, ranging from 318 to 5360 pg/g lw. Park et al. (2010) reported results from 61 samples collected from workers and residents of a municipal waste incinerator and a control group in Korea in 2007, with no significant differences in PCN concentrations. The mean value of $\sum Cl_{4\text{-}8}CN$ was 2170 pg/g lw, with the highest mean levels of hepta-CN73 at 379 pg/g lw and tetra-CN38/40 at 248 pg/g lw. Lower concentrations, with a mean value of $\sum Cl_{4\text{-}8}CN$ at 613 pg/g lw (ranging from 101 to 1405 pg/g lw), were found in a German study involving 42 single plasma samples in 2013–2014 (Fromme et al. 2015). The dominant congeners were hepta-CN73, hexa-CN66/67, and penta-CN51, with mean values of 211, 114, and 108 pg/g lw, respectively. Table 2.14 provides an overview of the study results. No difference was found between men and women. In China, 16 pools of 480 adults living in a city with various metallurgical, machinery, and chemical industry factories were analyzed in 2016 (Jin et al., 2019). The sum of the tetra- to octa-congeners in serum was 267 pg/g lw in women, while it was 1390 pg/g lw in men, probably due to their occupation. The very high concentrations of di-PCNs, averaging 12,737 and 5046 pg/g lw for di-5/7CN and di-CN4 in men and 9413 and 3137 pg/g lw in women, are striking. The authors attribute this to specific environmental exposure. In another Chinese study, 95 serum samples from two districts with a large number of e-waste dismantling plants were examined, reporting median values of $\sum Cl_{4\text{-}8}CN$ at 1060 and 989 pg/g lw (Wang et al., 2022). In this case, the median levels of mono- to tri-congeners in both regions were also very high, at approximately 48,000 and 61,000 pg/g lw, respectively. The $\sum$di-CN accounted for about 88% of the total concentrations, with di-5/7CN alone accounting for 46 and 54%, respectively. In another Chinese study, Chen et al. (2023) analyzed the serum of 32 adults annually between 2012 and 2016, reporting mean values of $\sum$PCN at 1100 pg/g lw, with tri-CN20 alone showing values of 476 and 355 pg/g lw in men and women, respectively. No changes in exposure were observed over the years, although tri-CN20 gradually increased over time.

Figure 2.7 illustrates the average levels of the most significant congeners from studies conducted in Germany, the USA, Korea, and China. Along with substantial variations in the concentration levels of individual congeners, the distinct congener patterns are also notable, reflecting the historical use and exposure in the region.

2.4.4 Occurrence in Breast Milk

In a comprehensive first study, between 1 and 3 pooled samples from 40 to 160 individuals were analyzed in Sweden from 1972 to 1992, with results presented for seven congeners, of which two tetra-CN congeners could not be further specified

Table 2.14 Polychlorinated naphthalenes in plasma samples in pg/g lw (after Fromme et al. 2015)

Substance		Detection rate (%)	Mean	Median	95th percentile	Maximum
Polychlorinated naphthalenes (pg/g lw)						
T4CN	PCN 42	24	27.6		131	322
	PCN 33/34/37	43	48.9		245	278
	PCN 44/47	12	6.7		60.2	77.7
	PCN 36/45	5	1.0			23.2
	PCN 28/43	10	3.7		37.1	57.2
	PCN 29	0				
	PCN 30/27	0				
	PCN 39	0				
	PCN 32	0				
	PCN 48/35	0				
	PCN 38/40	12	7.1		58.5	106
P5CN	PCN 52/60	45	65.6		250	303
	PCN 58	0				
	PCN 61	0				
	PCN 57	0				
	PCN 50	14	13.9		90.4	174
	PCN 51	48	108		325	401
	PCN 54	0				
	PCN 62	0				
	PCN 53/55	0				
	PCN 59	0				
	PCN 49	0				
	PCN 56	0				
H6CN	PCN 66/67	93	114	108	209	275
	PCN 64/68	0				
	PCN 69	0				
	PCN 71/72	0				
	PCN 63	0				
	PCN 65	0				
	PCN 70	0				
H7CN	PCN 73	81	211	206	395	559
	PCN 74	0				
O8CN	PCN 75	5	5.4			131
Total PCNs			613	575	1248	1406

Values below the LOD were assigned half of the LOD

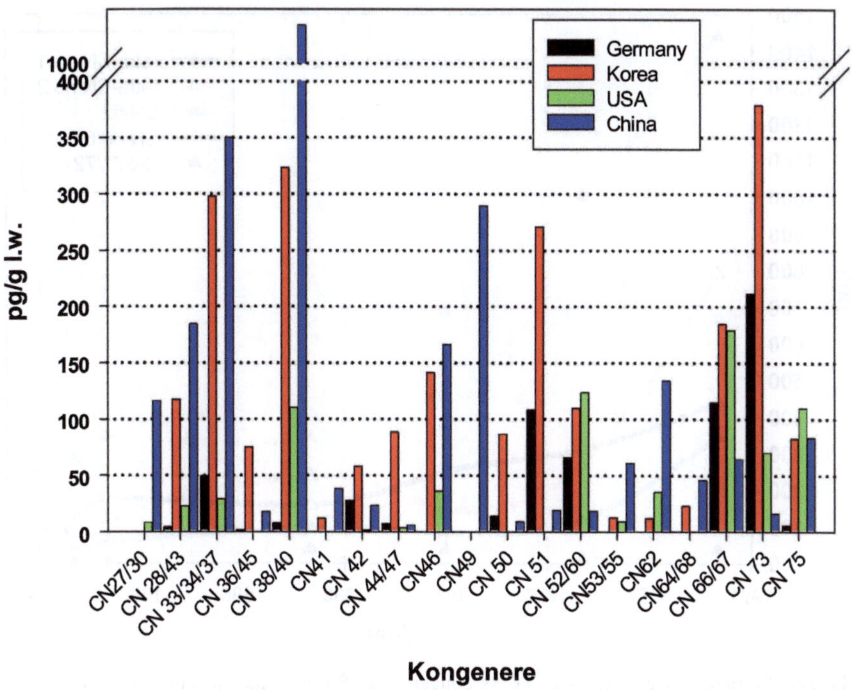

Fig. 2.7 Mean PCN serum/plasma levels of the general population in pg/g lw [Germany: Fromme et al. 2015; USA: Horii et al. 2010; Kannan 2015 (less exposure group); Korea: Park et al. 2010 (group residing away from the municipal solid waste incinerator); China: Wang et al. 2022 (Huangyan study group)]. In some cases, there are different analytical delimitation problems for individual congeners

(Lundén and Norén 1998). The temporal development is shown in Fig. 2.8, illustrating an apparent decrease of approximately 84% in the sum of the PCNs analyzed, particularly with penta-CN52, demonstrating a noticeable decline. Ålander et al. (2019) present additional results from Sweden, but these are not directly comparable to the aforementioned study due to methodological differences and the scope of the congeners analyzed. In each of the six years from 2012 to 2017, a pooled sample consisting of 30 individual samples was examined, with the $\sum$PCN ranging from 518 pg/g lw in 2012 to 375 pg/g lw in 2017. Tetra-CN and penta-CN accounted for the largest proportion of the total concentrations, registering at 165 and 111 pg/g lw in 2017. Twelve PCNs were measured, and their summed levels of 59–168 pg/g lw yielded a mean value of 93 pg/g lw (Pratt et al. 2012). Penta-CN52/60 and hexa-CN66/67 had the highest concentrations among the congeners. In the Czech Republic, 231 samples were analyzed for 15 PCNs between 2019 and 2021, but they were only found in less than 1% of the samples (Parizek et al. 2023). Tri-CN20, penta-CN52, and hexa-CN66 concentrations ranged from 250 to 832 pg/g lw.

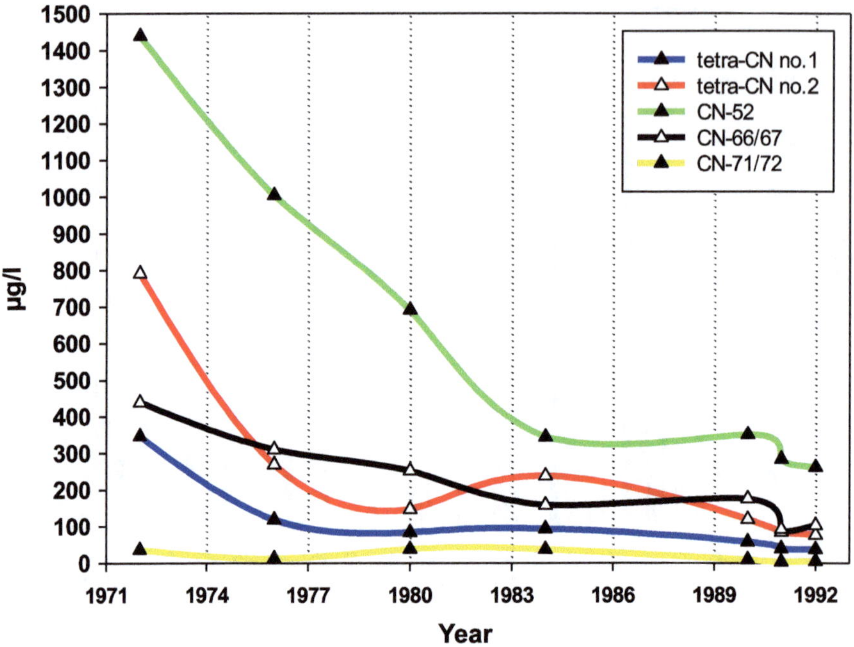

Fig. 2.8 Average concentrations of PCNs congeners in Swedish human milk in pg/g lw (after Lundén and Norén 1998)

Li et al. (2021) identified 75 PCNs nationwide from 100 pooled samples collected from 4480 individual samples in China between 2017 and 2020. The total $\sum$PCN levels ranged from 64.4 to 2497.4 pg/g lw (mean: 514.3 pg/g lw), with the highest concentrations observed for hexaCN66/67 at 44 pg/g lw and penta-CN52/60 at 20 pg/g lw. Tri-CN constituted the majority of the total content at 51%, followed by di-CN at 20%, penta-CN at 13%, and hepta-CN at 11%. The authors observed distribution patterns of congeners that are relatively characteristic of PCN exposure in areas with unintentional industrial releases acting as significant sources.

References

Agay-Shay K, Martinez D, Valvi D, Garcia-Esteban R, Basagaña X, Robinson O, Casas M, Sunyer J, Vrijheid M (2015) Exposure to endocrine-disrupting chemicals during pregnancy and weight at 7 years of age: a multi-pollutant approach. Environ Health Perspect 123:1030–1037

Agus S, Akkaya H, Daglioglu N, Eyuboglu S, Atasayan O, Mete F, Colak C, Sandal S, Yilmaz B (2022) Polychlorinated biphenyls and organochlorine pesticides in breast milk samples and their correlation with dietary and reproductive factors in lactating mothers in Istanbul. Environ Sci Pollut Res Int 29:3463–3473

Ålander J, Gyllenhammar I, Halldin Ankarberg E, Fridén U, Cantillana T, Aune M (2019) Levels of chlorinated naphthalenes in mother's milk from first-time mothers in Uppsala, Sweden:

results from the year 2012 to 2017. In: Report to the Swedish EPA (the health-related environmental monitoring program). http://www.imm.ki.se/Datavard/Rapporter/Sakrapport%20PCN%20trend%202012_2017.pdf

Altarawneh M, Dlugogorski BZ (2013) A mechanistic and kinetic study on the formation of PBDD/Fs from PBDEs. Environ Sci Technol 47:5118–5127

Alvarez-Silvares E, Fernández-Cruz T, Domínguez-Vigo P, Rubio-Cid P, Seoane-Pillado T, Martínez-Carballo E (2021) Association between placenta concentrations polybrominated and polychlorinated biphenyls and gestational diabetes mellitus: a case-control study in northwestern Spain. Environ Sci Pollut Res Int 28:10292–10301

Amodio E, Turci R, Massenti MF, Di Gaudio F, Minoia C, Vitale F, Firenze A, Calamusa G (2012) Serum concentrations of persistent organic pollutants (POPs) in the inhabitants of a Sicilian city. Chemosphere 89:970–974

Antignac JP, Main KM, Virtanen HE, Boquien CY, Marchand P, Venisseau A, Guiffard I, Bichon E, Wohlfahrt-Veje C, Legrand A, Boscher C, Skakkebæk NE, Toppari J, Le Bizec B (2016) Country-specific chemical signatures of persistent organic pollutants (POPs) in breast milk of French, Danish and Finnish women. Environ Pollut 218:728–738

Arisawa K, Uemura H, Hiyoshi M, Kitayama A, Takami H, Sawachika F, Nishioka Y, Hasegawa M, Tanto M, Satoh H, Shima M, Sumiyoshi Y, Morinaga K, Kodama K, Suzuki TI, Nagai M (2011) Dietary patterns and blood levels of PCDDs, PCDFs, and dioxin-like PCBs in 1656 Japanese individuals. Chemosphere 82:656–662

Asamoah A, Essumang DK, Muff J, Kucheryavskiy SV, Søgaard EG (2018) Assessment of PCBs and exposure risk to infants in breast milk of primiparae and multiparae mothers in an electronic waste hot spot and non-hot spot areas in Ghana. Sci Total Environ 612:1473–1479

ATSDR (2000) Toxicological profile for polychlorinated biphenyls (PCBs). https://www.atsdr.cdc.gov/toxprofiles/tp17.pdf

ATSDR (2011) Addendum for the Toxicological profile for polychlorinated biphenyls (PCBs). https://www.atsdr.cdc.gov/toxprofiles/pcbs_addendum.pdf

Aylward LL, Green E, Porta M, Toms LM, Den Hond E, Schulz C, Gasull M, Pumarega J, Conrad A, Kolossa-Gehring M, Schoeters G, Mueller JF (2014a) Population variation in biomonitoring data for persistent organic pollutants (POPs): an examination of multiple population-based datasets for application to Australian pooled biomonitoring data. Environ Int 68:127–138

Aylward LL, Hays SM, Kirman CR, Marchitti SA, Kenneke JF, English C, Mattison DR, Becker RA (2014b) Relationships of chemical concentrations in maternal and cord blood: a review of available data. J Toxicol Environ Health B Crit Rev 17:175–203

Baba T, Ito S, Yuasa M, Yoshioka E, Miyashita C, Araki A, Sasaki S, Kobayashi S, Kajiwara J, Hori T, Kato S, Kishi R (2018) Association of prenatal exposure to PCDD/Fs and PCBs with maternal and infant thyroid hormones: The Hokkaido Study on Environment and Children's Health. Sci Total Environ 615:1239–1246

Ballschmiter K, Zell M (1980) Analysis of polychlorinated biphenyls (PCB) by glass capillary gas chromatography. Fresenius' Zeitschrift Für Analytische Chemie 302:20–31

Bandow N, Conrad A, Kolossa-Gehring M, Murawski A, Sawal G (2020) Polychlorinated biphenyls (PCB) and organochlorine pesticides (OCP) in blood plasma—results of the German environmental survey for children and adolescents 2014–2017 (GerES V). Int J Hyg Environ Health 224:113426

Bao Y, Zhang L, Liu X, Shi L, Li J, Meng G, Zhao Y, Wu Y (2020) Dioxin-like compounds in paired maternal serum and breast milk under long sampling intervals. Ecotoxicol Environ Saf 194:110339

Barr DB, Weihe P, Davis MD, Needham LL, Grandjean P (2006) Serum polychlorinated biphenyl and organochlorine insecticide concentrations in a Faroese birth cohort. Chemosphere 62(7):1167–1182

Berghuis SA, Van Braeckel KNJA, Sauer PJJ, Bos AF (2018) Prenatal exposure to persistent organic pollutants and cognition and motor performance in adolescence. Environ Int 121(Pt 1):13–22

Bergonzi R, Specchia C, Dinolfo M, Tomasi C, De Palma G, Frusca T, Apostoli P (2009) Distribution of persistent organochlorine pollutants in maternal and foetal tissues: data from an Italian polluted urban area. Chemosphere 76:747–754

Berlin M, Barchel D, Brik A, Kohn E, Livne A, Keidar R, Tovbin J, Betser M, Moskovich M, Mandel D, Lubetzky R, Ovental A, Factor-Litvak P, Britzi M, Ziv-Baran T, Koren R, Klieger C, Berkovitch M, Matok I, Marom R (2021) Maternal and newborn thyroid hormone, and the association with polychlorinated biphenyls (PCBs) burden: The EHF (Environmental Health Fund) Birth Cohort. Front Pediatr 9:705395

Birnbaum LS, Staskal DF, Diliberto JJ (2003) Health effects of polybrominated dibenzo-p-dioxins (PBDDs) and dibenzofurans (PBDFs). Environ Int 29:855–860

Björvang RD, Gennings C, Lin PI, Hussein G, Kiviranta H, Rantakokko P, Ruokojärvi P, Lindh CH, Damdimopoulou P, Bornehag CG (2020) Persistent organic pollutants, pre-pregnancy use of combined oral contraceptives, age, and time-to-pregnancy in the SELMA cohort. Environ Health 19:67

Björvang RD, Mamsen LS (2022) Sexually dimorphic accumulation of persistent organic pollutants in fetuses. Front Toxicol 17:909307

Björvang RD, Vinnars MT, Papadogiannakis N, Gidlöf S, Mamsen LS, Mucs D, Kiviranta H, Rantakokko P, Ruokojärvi P, Lindh CH, Andersen CY, Damdimopoulou P (2021) Mixtures of persistent organic pollutants are found in vital organs of late gestation human fetuses. Chemosphere 283:131125

Bjurlid F (2018) Polybrominated dibenzo-p-dioxins and furans: from source of emission to human exposure. Örebro University studies in chemistry 20

Boffetta P, Mundt KA, Adami HO, Cole P, Mandel JS (2011) TCDD and cancer: a critical review of epidemiologic studies. Crit Rev Toxicol 41:622–636

Boutot ME, Whitcomb BW, Abdelouahab N, Baccarelli AA, Boivin A, Caku A, Gillet V, Martinez G, Pasquier JC, Zhu J, Takser L, St-Cyr L, Suvorov A (2021) In utero exposure to persistent organic pollutants and childhood lipid levels. Metabolites 11:657

Boyer E, Monfort C, Lainé F, Gaudreau É, Tillaut H, Bonnaure-Mallet M, Cordier S, Meuric V, Chevrier C (2023) Prenatal exposure to persistent organic pollutants and molar-incisor hypomineralization among 12-year-old children in the French mother-child cohort PELAGIE. Environ Res 231(Pt 3):116230

Brajenović N, Karačonji IB, Jurič A (2018) Levels of polychlorinated biphenyls in human milk samples in European countries. Arh Hig Rada Toksikol 69:135–153

Breivik K, Sweetman AJ, Pacyna JM, Jones KC (2002) Towards a global historical emission inventory for selected PCB congeners–a mass balance approach. 1. Global production and consumption. Sci Total Environ 290:181–198

Bruce-Vanderpuije P, Megson D, Jones GR, Jobst K, Reiner E, Clarke E, Adu-Kumi S, Gardella JA (2021) Infant dietary exposure to dioxin-like polychlorinated biphenyls (dlPCBs), polybrominated and mixed halogenated dibenzo-p-dioxins and furans (PBDD/Fs and PXDD/Fs) in milk samples of lactating mothers in Accra, Ghana. Chemosphere 263:128156

Brucker-Davis F, Ferrari P, Boda-Buccino M, Wagner-Mahler K, Pacini P, Gal J, Azuar P, Fenichel P (2011) Cord blood thyroid tests in boys born with and without cryptorchidism: correlations with birth parameters and in utero xenobiotics exposure. Thyroid 21:1133–1141

Bruns-Weller E (2022) Personal communication from the Lower Saxony State Office for Consumer Protection and Food Safety (LAVES)

Cabrera-Rodríguez R, Luzardo OP, Almeida-González M, Boada LD, Zumbado M, Acosta-Dacal A, Rial-Berriel C, Henríquez-Hernández LA (2019) Association between prenatal exposure to multiple persistent organic pollutants (POPs) and growth indicators in newborns. Environ Res 171:285–292

Carlsson H, Sreenivasan AP, Erngren I, Larsson A, Kultima K (2023) Combining the targeted and untargeted screening of environmental contaminants reveals associations between PFAS exposure and vitamin D metabolism in human plasma. Environ Sci Process Impacts 25:1116–1130

Caspersen IH, Kvalem HE, Haugen M, Brantsæter AL, Meltzer HM, Alexander J, Thomsen C, Frøshaug M, Bremnes NM, Broadwell SL, Granum B, Kogevinas M, Knutsen HK (2016) Determinants of plasma PCB, brominated flame retardants, and organochlorine pesticides in pregnant women and 3 year old children in The Norwegian Mother and Child Cohort Study. Environ Res 146:136–144

CDC (2023) National report on human exposure to environmental chemicals. In: Bomonitoring data tables for environmental chemicals. https://www.cdc.gov/exposurereport/data_tables.html

Cechová E, Scheringer M, Seifertová M, Mikeš O, Kroupová K, Kuta J, Forns J, Eggesbø M, Quaak I, de Cock M, van de Bor M, Patayová H, Palkovicová Murínová L, Kocan A (2016) Developmental neurotoxicants in human milk: Comparison of levels and intakes in three European countries. Sci Total Environ S0048-9697:32477–32479

Cerná M, Kratenová J, Zejglicová K, Brabec M, Malý M, Smíd J, Crhová S, Grabic R, Volf J (2007) Levels of PCDDs, PCDFs, and PCBs in the blood of the non-occupationally exposed residents living in the vicinity of a chemical plant in the Czech Republic. Chemosphere 67:S238–S246

Cerná M, Malý M, Grabic R, Batáriová A, Smíd J, Benes B (2008) Serum concentrations of indicator PCB congeners in the Czech adult population. Chemosphere 72:1124–1131

Chen D, Lou X, Bao J, Wang Y, Fan Y, Chen Q, Jin J, Chen L (2023) Polychlorinated naphthalene concentrations and temporal trends in serum from the general Chinese adult population and effects of polychlorinated naphthalenes on thyroid function. Environ Res 232:116309

Chen MW, Castillo BAA, Lin DY, Chao HR, Tayo LL, Gou YY, Chen FA, Huang KL (2018) Levels of PCDD/Fs, PBDEs, and PBDD/Fs in breast milk from Southern Taiwan. Bull Environ Contam Toxicol 100:369–375

Chopra M, Schrenk D (2011) Dioxin toxicity, aryl hydrocarbon receptor signaling, and apoptosis—persistent pollutants affect programmed cell death. Crit Rev Toxicol 41:292–320

Chovancová J, Čonka K, Kočan A, Sejáková ZS (2012) PCDD, PCDF, PCB and PBDE concentrations in breast milk of mothers residing in selected areas of Slovakia. Chemosphere 83:1383–1390

Cok I, Donmez MK, Uner M, Demirkaya E, Henkelmann B, Shen H, Kotalik J, Schramm KW (2009) Polychlorinated dibenzo-p-dioxins, dibenzofurans and polychlorinated biphenyls levels in human breast milk from different regions of Turkey. Chemosphere 76:1563–1571

Cok I, Mazmanci B, Mazmanci MA, Turgut C, Henkelmann B, Schramm KW (2012) Analysis of human milk to assess exposure to PAHs, PCBs and organochlorine pesticides in the vicinity Mediterranean city Mersin, Turkey. Environ Int 40:63–69

Costopoulou D, Vassiliadou I, Papadopoulos A, Makropoulos V, Leondiadis L (2006) Levels of dioxins, furans and PCBs in human serum and milk of people living in Greece. Chemosphere 65:1462–1469

Croes K, Colles A, Koppen G, De Galan S, Vandermarken T, Govarts E, Bruckers L, Nelen V, Schoeters G, Van Larebeke N, Denison MS, Mampaey M, Baeyens W (2013) Determination of PCDD/Fs, PBDD/Fs and dioxin-like PCBs in human milk from mothers residing in the rural areas in Flanders, using the CALUX bioassay and GC-HRMS. Talanta 113:99–105

Dellatte E, Abate V, Abballe A, De Filippis SP, De Luca S, Ferri F, Fulgenzi AR, Iacovella N, Iamiceli AL, Ingelido AM, Marra V, Miniero R, Valentini S, Bressanelli M, Schivardi MR, De Felip E (2025) Human biomonitoring of PCDDs, PCDFs, and PCBs in women living in a Northern Italy industrial area. Environ Sci Pollut Res Int 32(16):10392–10405

DeVito M, Bokkers B, van Duursen MBM, van Ede K, Feeley M, Antunes Fernandes Gáspár E, Haws L, Kennedy S, Peterson RE, Hoogenboom R, Nohara K, Petersen K, Rider C, Rose M, Safe S, Schrenk D, Wheeler MW, Wikoff DS, Zhao B, van den Berg M (2024) The 2022 world health organization reevaluation of human and mammalian toxic equivalency factors for polychlorinated dioxins, dibenzofurans and biphenyls. Regul Toxicol Pharmacol 146:105525

Dirtu AC, Jaspers VL, Cernat R, Neels H, Covaci A (2010) Distribution of PCBs, their hydroxylated metabolites, and other phenolic contaminants in human serum from two European countries. Environ Sci Technol 44:2876–2883

Domingo JL (2004) Polychlorinated naphthalenes in animal aquatic species and human exposure through the diet: a review. J Chromatogr A 1054:327–334

Dong JJ, Ruan MC, Hang JG, Nakayama SF, Jung CR, Kido T, Wang Z, Ma CC, Sun XL (2020) The relationship between perinatal exposure to dioxins and serum steroid hormone levels in preschool-aged children at an e-waste region in China. Int J Hyg Environ Health 229:113580

Du S, Rodenburg L, Patterson N, Chu C, Riker CD, Yu CH, Fan ZT (2020) Concentration of polychlorinated biphenyls in serum from New Jersey biomonitoring study: 2016–2018. Chemosphere 261:127730

Ebert J, Bahadir M (2003) Formation of PBDD/F from flame-retarded plastic materials under thermal stress. Environ Int 29:711–716

EFSA (2015) Scientific statement on the health-based guidance value for dioxins and dioxin-like PCBs. EFSA J 13(5):4124

EFSA (2018) Scientific Opinion on the risk for animal and human health related to the presence of dioxins and dioxin-like PCBs in feed and food. EFSA J 16:5333. https://efsa.onlinelibrary.wiley.com/https://doi.org/10.2903/j.efsa.2018.5333

EFSA (European Food Safety Authority) (2005) Opinion of the scientific panel on contaminants in the food chain on a request from the Commission related to the presence of non dioxin-like polychlorinated biphenyls (PCB) in feed and food. EFSA J 284:1–137

Esposito M, Serpe FP, Diletti G, Messina G, Scortichini G, La Rocca C, Baldi L, Amorena M, Monda M (2014) Serum levels of polychlorinated dibenzo-p-dioxins, polychlorinated dibenzofurans and polychlorinated biphenyls in a population living in the Naples area, southern Italy. Chemosphere 94:62–69

Falandysz J, Fernandes A, Gregoraszczuk E, Rose M (2014) The toxicological effects of halogenated naphthalenes: a review of aryl hydrocarbon receptor-mediated (dioxin-like) relative potency factors. J Environ Sci Health Part C 32:239–272

Fång J, Nyberg E, Bignert A, Bergman A (2013) Temporal trends of polychlorinated dibenzo-p-dioxins and dibenzofurans and dioxin-like polychlorinated biphenyls in mothers' milk from Sweden, 1972–2011. Environ Int 60:224–231

Fång J, Nyberg E, Winnberg U, Bignert A, Bergman Å (2015) Spatial and temporal trends of the Stockholm convention POPs in mothers' milk—a global review. Environ Sci Pollut Res 22:8989–9041

Fängström B, Strid A, Grandjean P, Weihe P, Bergman A (2005) A retrospective study of PBDEs and PCBs in human milk from the Faroe Islands. Environ Health 14(4):12

Fernandes AR, Falandysz J (2021) Polybrominated dibenzo-p-dioxins and furans (PBDD/Fs): contamination in food, humans and dietary exposure. Sci Total Environ 761:143191

Fernandes AR, Kilanowicz A, Stragierowicz J, Klimczak M, Falandysz J (2022) The toxicological profile of polychlorinated naphthalenes (PCNs). Sci Total Environ 837:155764

Fiedler H, Li X, Zhang J (2023) Persistent organic pollutants in human milk from primiparae—correlations, global, regional, and national time-trends. Chemosphere 313:137484

Fisher M, Arbuckle TE, Liang CL, LeBlanc A, Gaudreau E, Foster WG, Haines D, Davis K, Fraser WD (2016) Concentrations of persistent organic pollutants in maternal and cord blood from the maternal-infant research on environmental chemicals (MIREC) cohort study. Environ Health 15:59

Foerster C, Zúñiga-Venegas L, Enríquez P, Rojas J, Zamora C, Muñoz X, Pancetti F, Muñoz-Quezada MT, Lucero B, Saracini C, Salas C, Cortés S (2021) Levels of polychlorinated dibenzo-p-dioxins/furans (PCDD/Fs) and dioxin-like polychlorinated biphenyls (DL-PCBs) in human breast milk in Chile: a pilot study. Int J Environ Res Public Health 18:4825

Frederiksen M, Andersen HV, Haug LS, Thomsen C, Broadwell SL, Egsmose EL, Kolarik B, Gunnarsen L, Knudsen LE (2020) PCB in serum and hand wipes from exposed residents living in contaminated high-rise apartment buildings and a reference group. Int J Hyg Environ Health 224:113430

Fréry N, Zeghnoun A, Sarter H, Volatier JL, Falq G, Pascal M, Grange D, Schmitt M, Bérat B, Fabre P, Guillois-Becel Y, Noury U, Pouey J, Mathieu A, Heymann C, Lucas N, Thébault A, Eppe G,

Focant JF, Le Strat Y, Pelletier B, Salines G (2007) Exposure factors influencing serum dioxin concentration in the French dioxin and incinerators study. Organohalogen Comp 69:1017–1020

Fromme H, Albrecht M, Appel M, Hilger B, Völkel W, Liebl B, Roscher E (2015a) PCBs, PCDD/Fs, and PBDEs in blood samples of a rural population in South Germany. Int J Hyg Environ Health 218:41–46

Fromme H, Albrecht M, Boehmer S, Büchner K, Mayer R, Liebl B, Wittsiepe J, Bolte G (2009) Intake and body burden of dioxin-like compounds in Germany: the INES study. Chemosphere 76:1457–1463

Fromme H, Cequier E, Kim JT, Hanssen L, Hilger B, Thomsen C, Chang YS, Völkel W (2015b) Persistent and emerging pollutants in the blood of German adults: occurrence of dechloranes, polychlorinated naphthalenes, and siloxanes. Environ Int 85:292–298

Fromme H, Fuchs V, Albrecht M, Aschenbrenner B, Röhl C, Janitzki N, Herber-Jonat S, Wöckner M, Völkel W, Flemmer AW, Schober W (2022) Polychlorinated dioxins and dibenzofurans (PCDD/F), polybrominated dioxins and dibenzofurans (PBDD/F), polychlorinated biphenyls (PCB), polybrominated diphenyl ethers (PBDE), and per- and polyfluoroalkyl substances (PFAS) in German breast milk samples (LUPE 8). Sci Total Environ 825:154066

Fromme H, Hilger B, Albrecht M, Gries W, Leng G, Völkel W (2016) Occurrence of chlorinated and brominated dioxins/furans, PCBs, and brominated flame retardants in blood of German adults. Int J Hyg Environ Health 219:380–388

Fürst P (2006) Dioxins, polychlorinated biphenyls and other organohalogen compounds in human milk. Levels, correlations, trends and exposure through breastfeeding. Mol Nutr Food Res 50:922–933

Fürst P, Fürst C, Wilmers K (1992) PCDDs and PCDFs in human milk-statistical evaluation of a 6-years survey. Chemosphere 25:1029–1038

Fürst P, Fürst C, Wilmers K (1994) Human milk as a bioindicator for body burden of PCDDs, PCDFs, organochlorine pesticides, and PCBs. Environ Health Perspect 102(Suppl. 1):187–193

García-Villarino M, Riaño-Galán I, Rodríguez-Dehli AC, Freire C, Vizcaíno E, Grimalt JO, Tardón A, Fernández-Somoano A (2020) Association between pre/perinatal exposure to POPs and children's anogenital distance at age 4 years: a study from the INMA-Asturias cohort. Int J Hyg Environ Health 229:113563

Gascon M, Fort M, Martínez D, Carsin AE, Forns J, Grimalt JO, Santa Marina L, Lertxundi N, Sunyer J, Vrijheid M (2012) Polybrominated diphenyl ethers (PBDEs) in breast milk and neuropsychological development in infants. Environ Health Perspect 120:1760–1765

Georgieva SK, Trifonova T, Peteva Z (2023) Investigation of polychlorinated biphenyls in breast milk from two regions in Bulgaria. Int J Hyg Environ Health 251:114184

Gómara B, Athanasiadou M, Quintanilla-López JE, González MJ, Bergman A (2012) Polychlorinated biphenyls and their hydroxylated metabolites in placenta from Madrid mothers. Environ Sci Pollut Res Int 19:139–147

Govarts E, Portengen L, Lambrechts N, Bruckers L, Den Hond E, Covaci A, Nelen V, Nawrot TS, Loots I, Sioen I, Baeyens W, Morrens B, Schoeters G, Vermeulen R (2020) Early-life exposure to multiple persistent organic pollutants and metals and birth weight: pooled analysis in four Flemish birth cohorts. Environ Int 145:106149

Grandjean P, Budtz-Jørgensen E, Barr DB, Needham LL, Weihe P, Heinzow B (2008) Elimination half-lives of polychlorinated biphenyl congeners in children. Environ Sci Technol 42:6991–6996

Grešner P, Zieliński M, Ligocka D, Polańska K, Wąsowicz W, Gromadzińska J (2021) Environmental exposure to persistent organic pollutants measured in breast milk of lactating women from an urban area in central Poland. Environ Sci Pollut Res Int 28:4549–4557

Grimm FA, Hu D, Kania-Korwel I, Lehmler HJ, Ludewig G, Hornbuckle KC, Duffel MW, Bergman Å, Robertson LW (2015) Metabolism and metabolites of polychlorinated biphenyls. Crit Rev Toxicol 45:245–272

Guvenius DM, Aronsson A, Ekman-Ordeberg G, Bergman A, Norén K (2003) Human prenatal and postnatal exposure to polybrominated diphenyl ethers, polychlorinated biphenyls, polychlorobiphenylols, and pentachlorophenol. Environ Health Perspect 111:1235–1241

Gyllenhammar I, Aune M, Fridén U, Cantillana T, Bignert A, Lignell S, Glynn A (2021) Are temporal trends of some persistent organochlorine and organobromine compounds in Swedish breast milk slowing down? Environ Res 197:111117

Gyllenhammar I, Näslund M, Cantillana T, Fridén U, Kallerman PH, Lignell S, Ankarberg EH, Aune M (2023) Levels of PCB, PCDD/F and BFR in pooled serum samples from 4-, 8- and 12-year-old children in Uppsala, sampled 2008–2021 (POPUP), and PCB and PCDD/F in 12-year-old children from Sweden, sampled 2016–2017 (Riksmaten Adolescents). In: Report to the Swedish EPA (the health-related environmental monitoring program). https://www.diva-portal.org/smash/get/diva2:1737451/FULLTEXT01.pdf

Hagberg J, Lindström G, van Bavel B (2010) Screening of polybrominated dibenzo-p dioxins and furans (PBDD/Fs) in blood from the Nordic population. Report. Örebro University, Sweden

Hamers T, Kamstra JH, Cenijn PH, Pencikova K, Palkova L, Simeckova P, Vondracek J, Andersson PL, Stenberg M, Machala M (2011) In vitro toxicity profiling of ultrapure non-dioxin-like polychlorinated biphenyl congeners and their relative toxic contribution to PCB mixtures in humans. Toxicol Sci 121:88–100

Haraguchi K, Ito Y, Takagi M, Fujii Y, Harada KH, Koizumi A (2016) Levels, profiles and dietary sources of hydroxylated PCBs and hydroxylated and methoxylated PBDEs in Japanese women serum samples. Environ Int 97:155–162

HBM Komm (Human Biomonitoring Kommission) (2012) Human Biomonitoring (HBM) values for polychlorinated biphenyls (PCB) im Blut. Bundesgesundheitsbl Gesundheitsforsch Gesundheitsschutz 55:1069–1070 (in German)

HC (2010) Report on human biomonitoring of environmental chemicals in Canada, results of the Canadian health measures survey cycle 1 (2007–2009). Online: https://www.canada.ca/content/dam/hc-sc/migration/hc-sc/ewh-semt/alt_formats/hecs-sesc/pdf/pubs/contaminants/chms-ecms/report-rapport-eng.pdf

HC (2020) Report on human biomonitoring of environmental chemicals in pooled samples. In: Results of the Canadian health measures survey cycles 1 (2007–2009), 3 (2012–2013), 4 (2014–2015) and 5 (2016–2017). Online: https://www.canada.ca/en/health-canada/services/environmental-workplace-health/environmental-contaminants/human-biomonitoring-environmental-chemicals/report-pooled-samples.html

Hedvall Kallerman P, Fridén U, Näslund M, Cantillana T, Aune M, Gyllenhammar I (2023) Levels of persistent halogenated organic pollutants (POP's) in mother's milk from first-time mothers in Uppsala, Sweden: results from year 2020 and 2021, and temporal trends for the time period 1996–2021. In: Report to the Swedish EPA (the health-related environmental monitoring program). Online: https://www.diva-portal.org/smash/get/diva2:1741257/FULLTEXT02.pdf

Hedgeman E, Chen Q, Hong B, Chang CW, Olson K, Ladronka K, Ward B, Adriaens P, Demond A, Gillespie BW, Lepkowski J, Franzblau A, Garabrant DH (2009) The University of Michigan dioxin exposure study: population survey results and serum concentrations for polychlorinated dioxins, furans, and biphenyls. Environ Health Perspect 117(5):811–817

Herbstman JB, Sjödin A, Apelberg BJ, Witter FR, Patterson DG, Halden RU, Jones RS, Park A, Zhang Y, Heidler J, Needham LL, Goldman LR (2007) Determinants of prenatal exposure to polychlorinated biphenyls (PCBs) and polybrominated diphenyl ethers (PBDEs) in an urban population. Environ Health Perspect 115:1794–1800

Herceg Romanić S, Milićević T, Jovanović G, Matek Sarić M, Mendaš G, Fingler S, Jakšić G, Popović A, Relić D (2023) Persistent organic pollutants in Croatian breast milk: an overview of pollutant levels and infant health risk assessment from 1976 to the present. Food Chem Toxicol 179:113990

Hernández CS, Pardo O, Corpas-Burgos F, Fernández SF, López A, Coscollà C, Vento M, Yusà V; BETTERMILK (2020) Biomonitoring of polychlorinated dibenzo-p-dioxins (PCDDs), polychlorinated dibenzofurans (PCDFs) and dioxin-like polychlorinated biphenyls (dl-PCBs) in human milk: exposure and risk assessment for lactating mothers and breastfed children from Spain. Sci Total Environ 744:140710

Hisada A, Shimodaira K, Okai T, Watanabe K, Takemori H, Takasuga T, Koyama M, Watanabe N, Suzuki E, Shirakawa M, Noda Y, Komine Y, Ariki N, Kato N, Yoshinaga J (2014) Associations between levels of hydroxylated PCBs and PCBs in serum of pregnant women and blood thyroid hormone levels and body size of neonates. Int J Hyg Environ Health 217:546–553

Hjazi A, Hsu CY, Al-Attar WM, Almajidi YQ, Hussien BM, Alzahrani AA, Kareem AK, Abdulhussien Alazbjee AA, Meng X (2024) The association of exposure to polychlorinated biphenyls with lipid profile and liver enzymes in umbilical cord blood samples. Chemosphere 350:141096

Horii Y, Jiang Q, Hanari N, Lam PK, Yamashita N, Jansing R, Aldous KM, Mauer MP, Eadon GA, Kannan K (2010) Polychlorinated dibenzo-p-dioxins, dibenzofurans, biphenyls, and naphthalenes in plasma of workers deployed at the World Trade Center after the collapse. Environ Sci Technol 44:5188–5194

Houlihan M, Wemken N, Keogh M, O'Riordain C, Noone C, Tierney J, Cellarius C, Cleere K, Morrison JJ, Daly S, Harrad S, Tlustos C, Coggins MA (2021) Concentrations of polychlorinated dibenzo-p-dioxins (PCDDs), polychlorinated dibenzofurans (PCDFs) and polychlorinated biphenyls (PCBs) in human milk from Ireland: temporal trends and implications for nursing infant exposure. J Environ Expo Assess 1:2

Hsu JF, Lee CC, Su HJ, Chen HL, Yang SY, Liao PC (2009) Evaluation of background persistent organic pollutant levels in human from Taiwan: polychlorinated dibenzo-p-dioxins, dibenzofurans, and biphenyls. Environ Int 35:33–42

Huang R, Wang P, Zhang J, Chen S, Zhu P, Huo W, Jiang Y, Chen Z, Peng J (2019) The human body burden of polychlorinated dibenzo- p-dioxins/furans (PCDD/Fs) and dioxin-like polychlorinated biphenyls (DL-PCBs) in residents' human milk from Guangdong Province, China. Toxicol Res (Camb) 8:552–559

Iamiceli AL, Abate V, Abballe A, Bena A, De Filippis SP, Dellatte E, De Luca S, Fulgenzi AR, Iacovella N, Ingelido AM, Ivaldi C, Marra V, Miniero R, Valentini S, Farina E, Gandini M, Orengia M, Procopio E, Salamina G, De Felip E (2021) Biomonitoring of the adult population living near the waste incinerator of Turin: serum concentrations of PCDDs, PCDFs, and PCBs after three years from the plant start-up. Chemosphere 272:129882

IARC (International Agency for Research on Cancer) (2015) IARC Working Group on the evaluation of carcinogenic risks to humans polychlorinated biphenyls and polybrominated biphenyls, vol 107, Lyon, France

Ingelido AM, Abate V, Abballe A, Albano FL, Battista T, Carraro V, Conversano M, Corvetti R, De Luca S, Franchini S, Fulgenzi AR, Giambanco L, Iacovella N, Iamiceli AL, Maiorana A, Maneschi F, Marra V, Pirola F, Porpora MG, Procopio E, Suma N, Valentini S, Valsenti L, Vecchiè V, De Felip E (2017) Concentrations of polychlorinated dibenzodioxins, polychlorodibenzofurans, and polychlorobiphenyls in women of reproductive age in Italy: a human biomonitoring study. Int J Hyg Environ Health 220(2 Pt B): 378–386

Iribarne-Durán LM, Castillero-Rosales I, Peinado FM, Artacho-Cordón F, Molina-Molina JM, Medianero E, Nicolás-Delgado SI, Sánchez-Pinzón L, Núñez-Samudio V, Vela-Soria F, Olea N, Alvarado-González NE (2023) Placental concentrations of xenoestrogenic organochlorine pesticides and polychlorinated biphenyls and assessment of their xenoestrogenicity in the PA-MAMI mother-child cohort. Environ Res 241:117622

Itoh S, Baba T, Yuasa M, Miyashita C, Kobayashi S, Araki A, Sasaki S, Kajiwara J, Hori T, Todaka T, Fujikura K, Nakajima S, Kato S, Kishi R (2018) Association of maternal serum concentration of hydroxylated polychlorinated biphenyls with maternal and neonatal thyroid hormones: the Hokkaido birth cohort study. Environ Res 167:583–590

Jaraczewska K, Lulek J, Covaci A, Voorspoels S, Kaluba-Skotarczak A, Drews K, Schepens P (2006) Distribution of polychlorinated biphenyls, organochlorine pesticides and polybrominated diphenyl ethers in human umbilical cord serum, maternal serum and milk from Wielkopolska region, Poland. Sci Total Environ 372:20–31

JECFA (2016) Evaluation of certain food additives and contaminants Eightieth report of the Joint FAO/WHO Expert Committee on Food Additives. https://www.who.int/publications/i/item/9789241209953

Jin J, Wang S, Hu J, Wu J, Li M, Wang Y, Jin J (2019) Polychlorinated naphthalenes in human serum samples from an industrial city in Eastern China: levels, sources, and sex differences. Ecotoxicol Environ Saf 177:86–92

Kalantzi OI, Martin FL, Thomas GO, Alcock RE, Tang HR, Drury SC, Carmichael PL, Nicholson JK, Jones KC (2004) Different levels of polybrominated diphenyl ethers (PBDEs) and chlorinated compounds in breast milk from two U.K. Regions. Environ Health Perspect 112:1085–1091

Kang JH, Park H, Chang YS, Choi JW (2008) Distribution of organochlorine pesticides (OCPs) and polychlorinated biphenyls (PCBs) in human serum from urban areas in Korea. Chemosphere 73:1625–1631

Kannan K (2015) Polychlorinated dibenzo-p-dioxins, dibenzofurans, biphenyls, and naphthalenes in plasma of workers deployed at the World Trade Center after the collapse. Personal communication

Kim JT, Oh D, Choi SD, Chang YS (2021) Factors associated with partitioning behavior of persistent organic pollutants in a feto-maternal system: a multiple linear regression approach. Chemosphere 263:128247

Kim JT, Son MH, Lee DH, Seong WJ, Han S, Chang YS (2015) Partitioning behavior of heavy metals and persistent organic pollutants among feto-maternal bloods and tissues. Environ Sci Technol 49:7411–7422

Klimczak M, Liu G, Fernandes AR, Kilanowicz A, Falandysz J (2023) An updated global overview of the manufacture and unintentional formation of polychlorinated naphthalenes (PCNs). J Hazard Mater 457:131786

Klinčić D, Herceg Romanić S, Brčić Karačonji I, Matek Sarić M, Grzunov Letinić J, Brajenović N (2016) Organochlorine pesticides and PCBs (including dl-PCBs) in human milk samples collected from multiparae from Croatia and comparison with primiparae. Environ Toxicol Pharmacol 45:74–79

Klinčić D, Herceg Romanić S, Matek Sarić M, Grzunov J, Dukić B (2014) Polychlorinated biphenyls and organochlorine pesticides in human milk samples from two regions in Croatia. Environ Toxicol Pharmacol 37:543–552

Koh WX, Hornbuckle KC, Marek RF, Wang K, Thorne PS (2016) Hydroxylated polychlorinated biphenyls in human sera from adolescents and their mothers living in two U.S. Midwestern Communities. Chemosphere 147:389–395

Komprdová K, Domínguez-Romero E, Sharma BM, Komprda J, Melymuk L, Murínová ĽP, Čonka K, Trnovec T, Černá M, Drobná B, Fabišiková A, Sejáková ZS, Scheringer M (2022) Application of a pharmacokinetic model in characterizing sources of polychlorinated biphenyl exposure and determining threshold daily intakes for adverse health effects in infants and toddlers. Sci Total Environ 830:154734

Kotz A, Malisch R, Kypke K, Oehme M (2005) PBDE, PBDD/F and mixed chlorinated-brominated PXDD/F in pooled human milk samples from different countries. Organohal Comp 67:1540–1544

Kraft M, Rauchfuss K, Sievering S, Wöckner M, Neugebauer F, Fromme (2017) Quantification of all 209 PCB congeners in blood—can indicators be used to calculate the total PCB blood load? Int J Hyg Environ Health 220:201–208

Kutz FW, Barnes DG, Bottinore DP, Greim H, Bretthauer EW (1990) The international toxicity equivalency factor (I-TEF) method of risk assessment for complex mixtures of dioxins and related compounds. Chemosphere 7–9:751–757

Lackmann GM, Schaller KH, Angerer J (2004) Organochlorine compounds in breast-fed vs. bottle-fed infants: preliminary results at six weeks of age. Sci Total Environ 329:289–293

LaKind JS, Hays SM, Aylward LL, Naiman DQ (2009) Perspective on serum dioxin levels in the United States: an evaluation of the NHANES data. J Expo Sci Environ Epidemiol 19:435–441

Lampa E, Eguchi A, Todaka E, Mori C (2018) Fetal exposure markers of dioxins and dioxin-like PCBs. Environ Sci Pollut Res Int 25:11940–11947

Lancz K, Murínová L, Patayová H, Drobná B, Wimmerová S, Sovčíková E, Kováč J, Farkašová D, Hertz-Picciotto I, Jusko TA, Trnovec T (2015) Ratio of cord to maternal serum PCB concentrations in relation to their congener-specific physicochemical properties. Int J Hyg Environ Health 218:91–98

Lee S, Kim S, Lee HK, Lee IS, Park J, Kim HJ, Lee JJ, Choi G, Choi S, Kim S, Kim SY, Choi K, Kim S, Moon HB (2013) Contamination of polychlorinated biphenyls and organochlorine pesticides in breast milk in Korea: time-course variation, influencing factors, and exposure assessment. Chemosphere 93:1578–1585

Leong YH, Azmi NI, Majid MIA, Wen S (2021) Exposure and risk assessment of polychlorinated dibenzo-p-dioxins (PCDDs), polychlorinated dibenzofurans (PCDFs) and dioxin-like polychlorinated biphenyls (dl-PCBs) for primiparous mothers and breastfed infants in Penang, Malaysia. Food Addit Contam Part A 38:1416–1426

Li C, Li J, Lyu B, Wu Y, Yang L, Zheng M, Min Y, Zhang L, Liu G (2021) Burden and Risk of Polychlorinated Naphthalenes in Chinese Human Milk and a Global Comparison of Human Exposure. Environ Sci Technol 55:6804–6813

Li ZM, Benker B, Bao Q, Henkelmann B, Corsten C, Michalke B, Pauluschke-Fröhlich J, Flisikowski K, Schramm KW, De Angelis M (2020) Placental distribution of endogenous and exogenous substances: a pilot study utilizing cryo-sampled specimen off delivery room. Placenta 100:45–53

Liebl B, Schettgen T, Kerscher G, Broding HC, Otto A, Angerer J, Drexler H (2004) Evidence for increased internal exposure to lower chlorinated polychlorinated biphenyls (PCB) in pupils attending a contaminated school. Int J Hyg Environ Health 207(4): 315–324

Lignell S, Aune M, Darnerud PO, Cnattingius S, Glynn A (2009) Persistent organochlorine and organobromine compounds in mother's milk from Sweden 1996–2006: compound-specific temporal trends. Environ Res 109:760–767

Lignell S, Aune M, Darnerud PO, Soeria-Atmadja D, Hanberg A, Larsson S, Glynn A (2011) Large variation in breast milk levels of organohalogenated compounds is dependent on mother's age, changes in body composition and exposures early in life. J Environ Monit 13:1607–1616

Lin Y, Le S, Feng C, Qiu X, Xu Q, Jin S, Zhang H, Jin Y, Wen Y, Xu H, Liu P, Rao Q, She J, Lu D (2021) Exposure and health risk assessment of secondary contaminants closely related to brominated flame retardants (BFRs): polybrominated dibenzo-p-dioxins and dibenzofurans (PBDD/Fs) in human milk in Shanghai. Environ Pollut 268(Pt A):115121

Liu G, Cai Z, Zheng M (2014) Sources of unintentionally produced polychlorinated naphthalene. Chemosphere 94:1–12

Liu J, Tan Y, Song E, Song Y (2020) A critical review of polychlorinated biphenyls metabolism, metabolites, and their correlation with oxidative stress. Chem Res Toxicol 33:2022–2042

Lopez-Espinosa MJ, Vizcaino E, Murcia M, Fuentes V, Garcia AM, Rebagliato M, Grimalt JO, Ballester F (2010) Prenatal exposure to organochlorine compounds and neonatal thyroid stimulating hormone levels. J Expo Sci Environ Epidemiol 20:579–588

Lu D, Lin Y, Feng C, Wang D, She J, Shen H, Wang G, Zhou Z (2015) Levels of polychlorinated dibenzo-p-dioxins/furans (PCDD/Fs) and dioxin-like polychlorinated biphenyls (DL-PCBs) in breast milk in Shanghai, China: a temporal upward trend. Chemosphere 137:14–24

Lundén A, Norén K (1998) Polychlorinated naphthalenes and other organochlorine contaminants in Swedish human milk, 1972–1992. Arch Environ Contam Toxicol 34:414–423

MacDonald AM, Gabos S, Braakman S, Cheperdak L, Lee B, Hrudey SE, Le XC, Li XF, Mandal R, Martin JW, Schopflocher D, Lyon ME, Cheung PY, Ackah F, Graydon JA, Reichert M, Lyon AW, Jarrell J, Benadé G, Charlton C, Huang D, Bennett MJ, Kinniburgh DW (2022) Maternal and child biomonitoring strategies and levels of exposure in western Canada during the past seventeen years: the Alberta Biomonitoring Program: 2005–2021. Int J Hyg Environ Health 244:113990

Maervoet J, Vermeir G, Covaci A, Van Larebeke N, Koppen G, Schoeters G, Nelen V, Baeyens W, Schepens P, Viaene MK (2007) Association of thyroid hormone concentrations with levels of organochlorine compounds in cord blood of neonates. Environ Health Perspect 115:1780–1786

Magoni M, Donato F, Speziani F, Leonardi L, Orizio G, Scarcella C, Gaia A, Apostoli P (2016) Substantial decline of polychlorinated biphenyls serum levels 10years after public health interventions in a population living near a contaminated site in Northern Italy. Environ Int 95:69–78

Malisch R, Schächtele A, Rolaf van Leeuwen FX, Moy G, Tritscher A (2023) WHO-and UNEP-coordinated exposure studies 2000–2019: findings of polychlorinated biphenyls, polychlorinated Dibenzo-p-Dioxins, and polychlorinated dibenzofurans. In: Malisch R et al (eds) Persistent organic pollutants in human milk. Springer Nature

Marek RF, Thorne PS, DeWall J, Hornbuckle KC (2014) Variability in PCB and OH-PCB serum levels in children and their mothers in urban and rural U.S. communities. Environ Sci Technol 48:13459–13467

Marek RF, Thorne PS, Wang K, Dewall J, Hornbuckle KC (2013) PCBs and OH-PCBs in serum from children and mothers in urban and rural U.S. communities. Environ Sci Technol 47:3353–3361

Matovu H, Li ZM, Henkelmann B, Bernhöft S, De Angelis M, Schramm KW, Sillanpää M, Kato CD, Ssebugere P (2021) Multiple persistent organic pollutants in mothers' breastmilk: implications for infant dietary exposure and maternal thyroid hormone homeostasis in Uganda, East Africa. Sci Total Environ 770:145262

Melymuk L, Blumenthal J, Sáňka O, Shu-Yin A, Singla V, Šebková K, Pullen Fedinick K, Diamond ML (2022) Persistent problem: global challenges to managing PCBs. Environ Sci Technol 56:9029–9040

Mendaš G, Jakovljević I, Romanić SH, Fingler S, Jovanović G, Sarić MM, Pehnec G, Popović A, Stanković D (2024) Presence of polycyclic aromatic hydrocarbons and persistent organochlorine pollutants in human milk: evaluating their levels, association with Total antioxidant capacity, and risk assessment. Sci Total Environ 931:172911

Miniero R, di Domenico A, Abate V, Abballe A, Dellatte E, De Filippis S, De Luca S, Ferri F, Fulgenzi A, Iacovella N, Iamiceli A, Ingelido A, Marra V, Valentini S, De Felip E (2022) Time trends of PCDDs, PCDFs, DL-PCBs, and NDL-PCBs in Italian women from biomonitoring studies. Chemosphere 308(Pt 3):136244

Miniero R, Ferri F, Carraro V, Franchini S, Fulgenzi AR, Iacovella N, Ingelido AM, De Felip E (2018) Chemical indicators of exposure to polychlorodibenzo-p-dioxins, polychlorodibenzofurans and polychlorobiphenyls in breast milk samples from mothers residing in Trento, Italy, and neighboring country municipalities. J Environ Sci Health, Part B 53:510–518

Montazeri P, Güil-Oumrait N, Marquez S, Cirugeda L, Beneito A, Guxens M, Lertxundi A, Lopez-Espinosa MJ, Santa-Marina L, Sunyer J, Casas M, Vrijheid M (2023) Prenatal exposure to multiple endocrine-disrupting chemicals and childhood BMI trajectories in the INMA Cohort Study. Environ Health Perspect 131:107006

Mørck TA, Erdmann SE, Long M, Mathiesen L, Nielsen F, Siersma VD, Bonefeld-Jørgensen EC, Knudsen LE (2014) PCB concentrations and dioxin-like activity in blood samples from Danish school children and their mothers living in urban and rural areas. Basic Clin Pharmacol Toxicol 115:134–144

Mori C, Nakamura N, Todaka E, Fujisaki T, Matsuno Y, Nakaoka H, Hanazato M (2014) Correlation between human maternal-fetal placental transfer and molecular weight of PCB and dioxin congeners/isomers. Chemosphere 114:262–267

Muckle G, Ayotte P, Dewailly EE, Jacobson SW, Jacobson JL (2001) Prenatal exposure of the northern Québec Inuit infants to environmental contaminants. Environ Health Perspect 109:1291–1299

Nadal M, Mari M, Schuhmacher M, Domingo JL (2019) Monitoring dioxins and furans in plasma of individuals living near a hazardous waste incinerator: temporal trend after 20 years. Environ Res 173:207–211

Nadal M, Perelló G, Schuhmacher M, Cid J, Domingo JL (2008) Concentrations of PCDD/PCDFs in plasma of subjects living in the vicinity of a hazardous waste incinerator: follow-up and modeling validation. Chemosphere 73:901–906

Nakamura T, Nakai K, Matsumura T, Suzuki S, Saito Y, Satoh H (2008a) Determination of dioxins and polychlorinated biphenyls in breast milk, maternal blood and cord blood from residents of Tohoku, Japan. Sci Total Environ 394(1):39–51

Nakamura T, Nakai K, Matsumura T, Suzuki S, Saito Y, Satoh H (2008b) Determination of dioxins and polychlorinated biphenyls in breast milk, maternal blood and cord blood from residents of Tohoku, Japan. Sci Total Environ 394:39–51

Naqvi A, Qadir A, Mahmood A, Baqar M, Aslam I, Sajid F, Mumtaz M, Li J, Zhang G (2018) Quantification of polychlorinated biphenyl contamination using human placenta as biomarker from Punjab Province, Pakistan. Environ Sci Pollut Res Int 25:14551–14562

Nhu DD, Kido T, Hung NN, Thom LTH, Naganuma R, Son LK, Honma S, Maruzeni S, Nishijo M, Nakagawa H (2011) Dioxin levels in the breast milk and estradiol and androgen levels in the saliva of Vietnamese primiparae. Toxicol Environ Chem 4:824–838

Ohta S, Okumura T, Nishimura H, Nakao T, Shimizu Y, Ochiai F, Aozasa O, Miyata H (2004) Levels of PBDEs, TBBPA, TBPs, PCDDs/DFs, PXDDs/DFs and PBDDs/DFs in human milk of nursing women and dairy milk products in Japan. Organohal Comp 66:2857–2862

Oltramare C, Riou J, Bochud M, Zennegg M, Vernez D, Berthet A (2025) Human exposure to persistent organic pollutants in Switzerland: the role of diet, age, smoking, and body composition. Food Chem Toxicol 205:115711

Orloff KG, Dearwent S, Metcalf S, Kathman S, Turner W (2003) Human exposure to polychlorinated biphenyls in a residential community. Arch Environ Contam Toxicol 44:125–131

Parizek O, Gramblicka T, Parizkova D, Polachova A, Bechynska K, Dvorakova D, Stupak M, Dusek J, Pavlikova J, Topinka J, Sram RJ, Pulkrabova J (2023) Assessment of organohalogenated pollutants in breast milk from the Czech Republic. Sci Total Environ 871:161938

Park H, Kang J-H, Baek S-Y, Chang Y-S (2010) Relative importance of polychlorinated naphthalenes compared to dioxins, and polychlorinated biphenyls in human serum from Korea: contribution to TEQs and potential sources. Environ Pollut 158:1420–1427

Park H, Kim J, Chang YS (2013) Prevalence of low chlorinated dibenzo-p-dioxin/dibenzofurans in human serum. Chemosphere 90:1658–1663

Park H, Lee SJ, Kang JH, Chang YS (2007) Congener-specific approach to human PCB concentrations by serum analysis. Chemosphere 68:1699–1706

Park H, Park E, Chang YS (2014) Ten-year time trend of dioxins in human serum obtained from metropolitan populations in Seoul, Korea. Sci Total Environ 470–471:1338–1345

Park HY, Park JS, Sovcikova E, Kocan A, Linderholm L, Bergman A, Trnovec T, Hertz-Picciotto I (2009) Exposure to hydroxylated polychlorinated biphenyls (OH-PCBs) in the prenatal period and subsequent neurodevelopment in eastern Slovakia. Environ Health Perspect 117:1600–1606

Park JS, Bergman A, Linderholm L, Athanasiadou M, Kocan A, Petrik J, Drobna B, Trnovec T, Charles MJ, Hertz-Picciotto I (2008) Placental transfer of polychlorinated biphenyls, their hydroxylated metabolites and pentachlorophenol in pregnant women from eastern Slovakia. Chemosphere 70:1676–1684

Park JS, She J, Holden A, Sharp M, Gephart R, Souders-Mason G, Zhang V, Chow J, Leslie B, Hooper K (2011) High postnatal exposures to polybrominated diphenyl ethers (PBDEs) and polychlorinated biphenyls (PCBs) via breast milk in California: does BDE-209 transfer to breast milk? Environ Sci Technol 45:4579–4585

Parsons RE, Douglas P, Ashworth D, Hansell AL, Sepai O, Chadeau-Hyam M, Toledano MB (2025) Polychlorinated dibenzo-dioxin/furan and polychlorinated biphenyl concentrations in the human milk of individuals living near municipal waste incinerators in the UK: findings from the Breast milk, Environment, Early-life, and development (BEED) human biomonitoring study. Environ Res 267:120588

Peecher JS, Schecter AJ, Lu H, Quynh HT, Stromberg A, Weng J, Crandall R, Birnbaum LS (2022) Biomonitoring of polybrominated dioxins & furans, polychlorinated dioxins & furans, and dioxin like polychlorinated biphenyls in Vietnamese female electronic waste recyclers. J Occup Environ Med 64:742–747

Pesatori AC, Consonni D, Bachetti S, Zocchetti C, Bonzini M, Baccarelli A, Bertazzi PA (2003) Short- and long-term morbidity and mortality in the population exposed to dioxin after the "Seveso accident." Ind Health 41:127–138

Pieters R, Focant JF (2014) Dioxin, furan and PCB serum levels in a South African Tswana population: comparing the polluting effects of using different cooking and heating fuels. Environ Int 66:71–78

Pineda S, Lignell S, Gyllenhammar I, Lampa E, Benskin JP, Lundh T, Lindh C, Kiviranta H, Glynn A (2023) Exposure of Swedish adolescents to elements, persistent organic pollutants (POPs), and rapidly excreted substances—the Riksmaten adolescents 2016–17 national survey. Int J Hyg Environ Health 251:114196

Piskorska-Pliszczyńska J, Maszewski S (2014) Brominated dioxins: little-known new health hazards—a review. Bull Vet Inst Pulawy 58:327–335

Plaku-Alakbarova B, Sergeyev O, Williams PL, Burns JS, Lee MM, Hauser R, Korrick SA; Russian Children's Study (2021) Peripubertal serum levels of dioxins, furans and PCBs in a cohort of Russian boys: can empirical grouping methods yield meaningful exposure variables? Chemosphere 275:130027

Polevoy C, Arbuckle TE, Oulhote Y, Lanphear BP, Cockell KA, Muckle G, Saint-Amour D (2020) Prenatal exposure to legacy contaminants and visual acuity in Canadian infants: a maternal-infant research on environmental chemicals study (MIREC-ID). Environ Health 19:14

Porta M, López T, Gasull M, Rodríguez-Sanz M, Garí M, Pumarega J, Borrell C, Grimalt JO (2012) Distribution of blood concentrations of persistent organic pollutants in a representative sample of the population of Barcelona in 2006, and comparison with levels in 2002. Sci Total Environ 423:151–161

Pratt IS, Anderson WA, Crowley D, Daly SF, Evans RI, Fernandes AR, Fitzgerald M, Geary MP, Keane DP, Malisch R, McBride J, Morrison JJ, Reilly A, Tlustos C (2012) Polychlorinated dibenzo-p-dioxins (PCDDs), polychlorinated dibenzofurans (PCDFs) and polychlorinated biphenyls (PCBs) in breast milk of first-time Irish mothers: impact of the 2008 dioxin incident in Ireland. Chemosphere 88:865–872

Quinete N, Esser A, Kraus T, Schettgen T (2016) Determination of hydroxylated polychlorinated biphenyls (OH-PCBs) in human urine in a highly occupationally exposed German cohort: new prospects for urinary biomarkers of PCB exposure. Environ Int 97:171–179

Raab U, Albrecht M, Preiss U, Völkel W, Schwegler U, Fromme H (2013) Organochlorine compounds, nitro musks and perfluorinated substances in breast milk—results from Bavarian Monitoring of Breast Milk 2007/8. Chemosphere 93:461–467

Raab U, Preiss U, Albrecht M, Shahin N, Parlar H, Fromme H (2008) Concentrations of polybrominated diphenyl ethers, organochlorine compounds and nitro musks in mother's milk from Germany (Bavaria). Chemosphere 72:87–94

Ramos JJ, Gómara B, Fernández MA, González MJ (2007) A simple and fast method for the simultaneous determination of polychlorinated biphenyls and polybrominated diphenyl ethers in small volumes of human serum. J Chromatogr A 1152:124–129

Rantakokko P, Tuomisto JT, Viluksela M, Tuomisto J, Airaksinen R, Raitakari O, Karvonen AM, Pekkanen J, Kiviranta H (2020) Estimated PCDD/F TEQ and total TEQ concentrations in the serum of 7–10 year old Finnish children. Chemosphere 257:127137

Rawn DF, Ryan JJ, Sadler AR, Sun WF, Haines D, Macey K, Van Oostdam J (2012) PCDD/F and PCB concentrations in sera from the Canadian Health Measures Survey (CHMS) from 2007 to 2009. Environ Int 47:48–55

Rawn DFK, Sadler AR, Casey VA, Breton F, Sun WF, Arbuckle TE, Fraser WD (2017) Dioxins/furans and PCBs in Canadian human milk: 2008–2011. Sci Total Environ 595:269–278

Ritter R, Scheringer M, MacLeod M, Moeckel C, Jones KC, Hungerbühler K (2011) Intrinsic human elimination half-lives of polychlorinated biphenyls derived from the temporal evolution of cross-sectional biomonitoring data from the United Kingdom. Environ Health Perspect 119:225–231

Rose M, Fernandes A, Mortimer D, Baskaran C (2015) Contamination of fish in UK fresh water systems: risk assessment for human consumption. Chemosphere 122:183–189

Rosofsky A, Janulewicz P, Thayer KA, McClean M, Wise LA, Calafat AM, Mikkelsen EM, Taylor KW, Hatch EE (2017) Exposure to multiple chemicals in a cohort of reproductive-aged Danish women. Environ Res 154:73–85

Rovira J, Martínez MÁ, Mari M, Cunha SC, Fernandes JO, Marmelo I, Marques A, Haug LS, Thomsen C, Nadal M, Domingo JL, Schuhmacher M (2022) Mixture of environmental pollutants in breast milk from a Spanish cohort of nursing mothers. Environ Int 166:107375

Runkel AA, Križanec B, Lipičar E, Baskar M, Hrženjak V, Kodba ZC, Kononenko L, Kanduč T, Mazej D, Tratnik JS, Horvat M (2021) Organohalogens: a persisting burden in Slovenia? Environ Res 198:111224

Ryan JJ, Masuda Y (1994) Polychlorinated naphthalenes (PCNs) in the rice oil poisoning. Organohalogen Comp 21:251–254

Ryan JJ, Rawn DF (2014) Polychlorinated dioxins, furans (PCDD/Fs), and polychlorinated biphenyls (PCBs) and their trends in Canadian human milk from 1992 to 2005. Chemosphere 102:76–86

Sandau CD, Ayotte P, Dewailly E, Duffe J, Norstrom RJ (2002) Pentachlorophenol and hydroxylated polychlorinated biphenyl metabolites in umbilical cord plasma of neonates from coastal populations in Québec. Environ Health Perspect 110:411–417

Santa-Marina L, Irizar A, Barroeta Z, Abad E, Lertxundi A, Ibarluzea J, Parera J, Urbieta N, Arruti E, Jimeno-Romero A, Zubero MB (2023) Serum levels of PCDDs, PCDFs and dl-PCBs in general population residing far and near from an urban waste treatment plant under construction in Gipuzkoa, Basque Country (Spain). Environ Res 236(Pt 1):116721

SC (Stockholm Convention) (2019) Stockholm convention on persistent organic pollutants. List of substances. https://www.pops.int/TheConvention/ThePOPs/AllPOPs/tabid/2509/Default.aspx

Schächtele A, Malisch R, Hardebusch B, Rolaf van Leeuwen F-X, Moy G, Tritscher A, van Duursen M, van den Berg M, Šebková K, Klánová J, Kalina J (2023) WHO-and UNEP-coordinated exposure studies 2000–2019: findings of polybrominated substances (PBDE, HBCDD, PBB 153, PBDD/PBDF). In (Malisch et al. (Eds.): Persistent Organic Pollutants in Human Milk. Springer Nature

Schecter A, Birnbaum L, Ryan JJ, Constable JD (2006) Dioxins: an overview. Environ Res 101:419–428

Schecter A, Startin J, Wright C, Kelly M, Päpke O, Lis A, Ball M, Olson JR (1994) Congener-specific levels of dioxins and dibenzofurans in U.S. food and estimated daily dioxin toxic equivalent intake. Environ Health Perspect 102:962–967

Schettgen T, Alt A, Esser A, Kraus T (2015) Current data on the background burden to the persistent organochlorine pollutants HCB, p, p'-DDE as well as PCB 138, PCB 153 and PCB 180 in plasma of the general population in Germany. Int J Hyg Environ Health 218:380–385

Schettgen T, Gube M, Alt A, Fromme H, Kraus T (2011) Pilot study on the exposure of the German general population to non-dioxin-like and dioxin-like PCBs. Int J Hyg Environ Health 214:319–325

Schuhmacher M, Kiviranta H, Ruokojärvi P, Nadal M, Domingo JL (2009) Concentrations of PCDD/Fs, PCBs and PBDEs in breast milk of women from Catalonia, Spain: a follow-up study. Environ Int 35:607–613

Schuhmacher M, Mari M, Nadal M, Domingo JL (2019) Concentrations of dioxins and furans in breast milk of women living near a hazardous waste incinerator in Catalonia, Spain. Environ Int 125:334–341

Schulte E, Malisch R (1984) Calculation of the real PCB content in environmental samples. II. Gas chromatographic determination of the PCB concentration in human milk and butter. Z Anal Chem 319:54–59

Seo SH, Kwon SY, Choi SD, Chang YS (2021) Twenty-year trends and exposure assessment of polychlorinated dibenzodioxins and dibenzofurans in human serum from the Seoul citizens. Chemosphere 273:128558

Sexton K, Ryan AD, Adgate JL, Barr DB, Needham LL (2011) Biomarker measurements of concurrent exposure to multiple environmental chemicals and chemical classes in children. J Toxicol Environ Health A 74:927–942

Shaw SD, Berger ML, Harris JH, Yun SH, Wu Q, Liao C, Blum A, Stefani A, Kannan K (2013) Persistent organic pollutants including polychlorinated and polybrominated dibenzo- p-dioxins and dibenzofurans in firefighters from Northern California. Chemosphere 91:1386–1394

She J, Holden A, Sharp M, Tanner M, Williams-Derry C, Hooper K (2007) Polybrominated diphenyl ethers (PBDEs) and polychlorinated biphenyls (PCBs) in breast milk from the Pacific Northwest. Chemosphere 67:S307-317

Sjödin A, Hagmar L, Klasson-Wehler E, Björk J, Bergman A (2000) Influence of the consumption of fatty Baltic Sea fish on plasma levels of halogenated environmental contaminants in Latvian and Swedish men. Environ Health Perspect 108:1035–1041

Soechitram SD, Athanasiadou M, Hovander L, Bergman A, Sauer PJ (2004) Fetal exposure to PCBs and their hydroxylated metabolites in a Dutch cohort. Environ Health Perspect 112:1208–1212

Sprowles JLN, Monaikul S, Aguiar A, Gardiner J, Monaikul N, Kostyniak P, Schantz SL (2022) Associations of concurrent PCB and PBDE serum concentrations with executive functioning in adolescents. Neurotoxicol Teratol 92:107092

Stubleski J, Lind L, Salihovic S, Lind PM, Kärrman A (2018) Longitudinal changes in persistent organic pollutants (POPs) from 2001 to 2009 in a sample of elderly Swedish men and women. Environ Res 165:193–200

Sun S, Zhang B, Luo Y, Ma X, Cao R, Zhang Y, Gao Y, Chen J, Zhang H (2022) Accumulation characteristics of polychlorinated dibenzo-p-dioxins and dibenzofurans and polychlorinated biphenyls in human breast milk from a seaside city of North China. Environ Pollut 297:118794

Suzuki G, Michinaka C, Matsukami H, Noma Y, Kajiwara N (2020) Validity of using a relative potency factor approach for the risk management of dioxin-like polychlorinated naphthalenes. Chemosphere 244:125448

Svarcova A, Lankova D, Gramblicka T, Stupak M, Hajslova J, Pulkrabova J (2019) Integration of five groups of POPs into one multi-analyte method for human blood serum analysis: an innovative approach within biomonitoring studies. Sci Total Environ 667:701–709

't Mannetje A, Coakley J, Bridgen P, Brooks C, Harrad S, Smith AH, Pearce N, Douwes J (2013) Current concentrations, temporal trends and determinants of persistent organic pollutants in breast milk of New Zealand women. Sci Total Environ 458–460:399–407

Takahashi T, Eguchi A, Watanabe M, Todaka E, Sakurai K, Mori C (2022) Association between telomere length in human umbilical cord tissues and polychlorinated biphenyls in maternal and cord serum. Chemosphere 300:134560

Takser L, Mergler D, Baldwin M, de Grosbois S, Smargiassi A, Lafond J (2005) Thyroid hormones in pregnancy in relation to environmental exposure to organochlorine compounds and mercury. Environ Health Perspect 113:1039–1045

Thomas GO, Wilkinson M, Hodson S, Jones KC (2006) Organohalogen chemicals in human blood from the United Kingdom. Environ Pollut 141:30–41

Thomsen C, Haug LS, Stigum H, Frøshaug M, Broadwell SL, Becher G (2010) Changes in concentrations of perfluorinated compounds, polybrominated diphenyl ethers, and polychlorinated biphenyls in Norwegian breast-milk during twelve months of lactation. Environ Sci Technol 44:9550–9556

Todaka T, Hirakawa H, Kajiwara J, Onozuka D, Sasaki S, Miyashita C, Yoshioka E, Yuasa M, Kishi R, Iida T, Uchi H, Furue M (2011) Concentrations of polychlorinated dibenzo-p-dioxins, polychlorinated dibenzofurans, and polychlorinated biphenyls in blood and breast milk collected from pregnant women in Sapporo City, Japan. Chemosphere 85:1694–1700

Torres-Moreno AC, Mejia-Grau K, Puente-DelaCruz L, Codling G, Villa AL, Ríos-Marquez O, Patequiva-Chauta L, Cobo M, Johnson-Restrepo B (2023) Polybrominated diphenyl ethers (PBDEs), polychlorinated biphenyls (PCBs), organochlorine pesticides (OCPs) in human breast milk from Colombia: a probabilistic risk assessment approach. Chemosphere 339:139597

Turci R, Balducci C, Brambilla G, Colosio C, Imbriani M, Mantovani A, Vellere F, Minoia C (2010) A simple and fast method for the determination of selected organohalogenated compounds in serum samples from the general population. Toxicol Lett 192:66–71

Turrio-Baldassarri L, Abate V, Battistelli CL, Carasi S, Casella M, Iacovella N, Indelicato A, La Rocca C, Scarcella C, Alivernini S (2008) PCDD/F and PCB in human serum of differently exposed population groups of an Italian city. Chemosphere 73:228–234

Uehara R, Nakamura Y, Matsuura N, Kondo N, Tada H (2007) Dioxins in human milk and smoking of mothers. Chemosphere 68:915–920

Ulaszewska MM, Zuccato E, Davoli E (2011) PCDD/Fs and dioxin-like PCBs in human milk and estimation of infants' daily intake: a review. Chemosphere 83:774–782

Ulutas OK, Çok I, Darendeliler F, Aydin B, Çoban A, Henkelmann B, Schramm KW (2015) Blood levels of polychlorinated biphenlys and organochlorinated pesticides in women from Istanbul. Turkey. Environ Monit Assess 187:132

UNEP (2010) Supporting document for technical review of the implications of recycling commercial pentabromodiphenyl ether and commercial octabromodiphenyl ether. Switzerland, Geneva

UNEP (2012) Report of the Persistent Organic Pollutants Review Committee on the work of its eighth meeting. In: Risk profile on chlorinated naphthalenes. Rome, Italy

UNEP (2017) Draft guidance on preparing inventories of polychlorinated naphthalenes (PCNs). file:///C:/Users/herma/Downloads/UNEP-POPS-COP.8-INF-19.English.pdf

Vafeiadi M, Vrijheid M, Fthenou E, Chalkiadaki G, Rantakokko P, Kiviranta H, Kyrtopoulos SA, Chatzi L, Kogevinas M (2014) Persistent organic pollutants exposure during pregnancy, maternal gestational weight gain, and birth outcomes in the mother-child cohort in Crete, Greece (RHEA study). Environ Int 64:116–123

Van den Berg M, Birnbaum L, Bosveld AT, Brunstrom B, Cook P, Feeley M, Giesy JP, Hanberg A, Hasegawa R, Kennedy SW, Kubiak T, Larsen JC, van Leeuwen FX, Liem AK, Nolt C, Peterson RE, Poellinger L, Safe S, Schrenk D, Tillitt D, Tysklind M, Younes M, Waern F, Zacharewski T (1998) Toxic equivalency factors TEFs for PCBs, PCDDs, PCDFs for humans and wildlife. Environ Health Perspect 106:775–792

Van den Berg M, Birnbaum LS, Denison M, De Vito M, Farland W, Feeley M, Fiedler H, Hakansson H, Hanberg A, Haws L, Rose M, Safe S, Schrenk D, Tohyama C, Tritscher A, Tuomisto J, Tysklind M, Walker N, Peterson RE (2006) The 2005 World Health Organization reevaluation of human and Mammalian toxic equivalency factors for dioxins and dioxin-like compounds. Toxicol Sci 93:223–241

Van den Berg M, Denison MS, Birnbaum LS, DeVito MJ, Fiedler H, Falandysz J, Rose M, Schrenk D, Safe S, Tohyama C, Tritscher A, Tysklind M, Peterson RE (2013) Polybrominated dibenzo-p-dioxins, dibenzofurans, and biphenyls: inclusion in the toxicity equivalency factor concept for dioxin-like compounds. Tox Sci 133:197–208

Van den Berg M, Kypke K, Kotz A, Tritscher A, Lee S, Magulova K, Fiedler H, Malisch R (2017) WHO/UNEP global surveys of PCDDs, PCDFs, PCBs and DDTs in human milk and benefit-risk evaluation of breastfeeding. Arch Toxicol 91:83–86

SC (Stockholm Convention) (2023) All POPs listed in the Stockholm Convention.https://www.pops.int/TheConvention/ThePOPs/AllPOPs/tabid/2509/Default.aspx

Vandentorren S, Frery N, Bidondo ML, Eppe G, Scholl G, De Pauw E, Saoudi A, Oleko A, Focant JF (2011) Dioxins in human milk from different regions of France: pilot of the French longitudinal study of children (ELFE). Organohalogen Comp 73:1626–1629

Varghese A, Chirackal RM, Mano VC, Kesavachandran C, Prathish KP (2025) Association of human milk dioxin like-POPs levels with maternal characteristics and newborn anthropometric measurements: a cross-sectional study from Southern India. Expo Health

VDI (Association of German Engineers) (2009) Ambient air measurement—Indoor air measurement—measurement of polychlorinated biphynyls (PCBs)—GC/MS method for PCB 28, 52, 101,138, 153, 180

Venisseau A, Antignac JP, Main K, Wohlfahrt Veje C, Veyrand B, Vincent S, Skakkebaek N, Topari J, Le Bizec B (2011) Concentration levels of environmental persistent organic pollutants in breast milk from Danish and Finish populations. Organohalogen Comp 73:1586–1589

Venkatesan AK, Halden RU (2014) Contribution of polybrominated dibenzo-p-dioxins and dibenzofurans (PBDD/Fs) to the toxic equivalency of dioxin-like compounds in archived biosolids from the U.S. EPA's 2001 National Sewage Sludge Survey. Environ Sci Technol 48:10843–10849

Vieth B, Albrecht M, Bruns-Weller E, Fromme H, Fürst P, Knoll A, Pydde E (2011) Time trends and current levels of dioxin and dioxin-like PCBs in Human milk from Germany. Organohal Compd 73:1559–1562

Vigh É, Colombo A, Benfenati E, Håkansson H, Berglund M, Bódis J, Garai J (2013) Individual breast milk consumption and exposure to PCBs and PCDD/Fs in Hungarian infants: a time-course analysis of the first three months of lactation. Sci Total Environ 449:336–344

Vizcaino E, Grimalt JO, Fernández-Somoano A, Tardon A (2014) Transport of persistent organic pollutants across the human placenta. Environ Int 65:107–115

Vuong TP (2022) Research on the relationship between exposure to dioxins and cancer incidence in Vietnam. Toxics 10:384

Wang Q, Yuan H, Jin J, Li P, Ma Y, Wang Y (2018) Polychlorinated biphenyl concentrations in pooled serum from people in different age groups from five Chinese cities. Chemosphere 198:320–326

Wang S, Jin J, Guo C, Li Z, Xu T, Wen X, Hu J, Wang Y, Wei Y, Jin J (2022) Polychlorinated naphthalene concentrations in human serum caused by unintentional production and emissions, and potential effects of polychlorinated naphthalenes on thyroid hormones. Sci Total Environ 806(Pt 1):150546

Weistrand C, Noren K, Nilsson A (1997) Occupational exposure. Organochlorine compounds in blood plasma from potentially exposed workers. PCB, PCN, PCDD/F, HCB and methylsulfonyl metabolites of PCB. Environ Sci Pollut Res Int 4:2–9

WHO (World Health Organization) (2001) International Programme on Chemical Safety. Chlorinated naphthalenes. Concise international chemical assessment document 34. Geneva, Switzerland

Wiberg K, Rappe C, Haglund P (1992) Analysis of bromo-, chloro- and mixed bromo/chloro-dibenzo-p-dioxins and dibenzofurans in salmon, osprey, and human milk. Chemosphere 24:1431–1439

Wilhelm M, Wittsiepe J, Lemm F, Ranft U, Krämer U, Fürst P, Röseler SC, Greshake M, Imöhl M, Eberwein G, Rauchfuss K, Kraft M, Winneke G (2008) The Duisburg birth cohort study: influence of the prenatal exposure to PCDD/Fs and dioxin-like PCBs on thyroid hormone status in newborns and neurodevelopment of infants until the age of 24 months. Mutat Res 659:83–92

Witczak A, Pohoryło A, Aftyka A, Pokorska-Niewiada K, Witczak G (2022) Changes in polychlorinated biphenyl residues in milk during lactation: levels of contamination, influencing factors, and infant risk assessment. Int J Mol Sci 23:12717

Wittsiepe J, Fürst P, Schrey P, Lemm F, Kraft M, Eberwein G, Winneke G, Wilhelm M (2007) PCDD/F and dioxin-like PCB in human blood and milk from German mothers. Chemosphere 67:286–294

Wong L-Y, Millette MD, Uddin MS, Needham LL, Patterson DG, Turner W, Henderson A (2008) Serum dioxin levels in residents of Calcasieu and Lafayette parishes, Louisiana with comparison to the US population. J Expo Sci Environ Epidemiol 18:252–261

Wuthe J, Piechotowski I, Papke O, Zier B, Gabrio T, Kramer D, Kouros B, Schwenk M, Pfaff G (1996) First data on background levels of non-ortho and mono-ortho PCBs in blood of residents from southern Germany. Chemosphere 32:567–574

Xu P, Zheng Y, Wang X, Shen H, Wu L, Chen Y, Xu D, Xiang J, Cheng P, Chen Z, Lou X (2022) Breastfed infants' exposure to polychlorinated biphenyls, polychlorinated dibenzo-p-dioxins and dibenzofurans, and per- and polyfluoroalkyl substances: a cross-sectional study of a municipal waste incinerator in China. Chemosphere 309(Pt 2):136639

Yang L, Liu G, Shen J, Wang M, Yang Q, Zheng M (2021) Environmental characteristics and formations of polybrominated dibenzo-p-dioxins and dibenzofurans. Environ Int 152:106450

Želježić D, Herceg Romanić S, Klinčić D, Matek Sarić M, Letinić JG (2018) Persistent organochlorine pollutants in placentas sampled from women in Croatia and an evaluation of their DNA damaging potential in vitro. Arch Environ Contam Toxicol 74:284–291

Zhang D, Saktrakulkla P, Marek RF, Lehmler HJ, Wang K, Thorne PS, Hornbuckle KC, Duffel MW (2022) PCB sulfates in serum from mothers and children in urban and rural U.S. communities. Environ Sci Technol 56:6537–6547

Zhang L, Li J, Zhao Y, Li X, Yang X, Wen S, Cai Z, Wu Y (2011) A national survey of polybrominated diphenyl ethers (PBDEs) and indicator polychlorinated biphenyls (PCBs) in Chinese mothers' milk. Chemosphere 84:625–633

Zhang L, Pei Z, Lyu B, Li J, Zhao Y, Wu Y (2023) Polychlorinated dibenzo-p-dioxins and dibenzofurans and dioxin-like polychlorinated biphenyls in human milk from national human breast milk monitoring in 2016–2019 in China. Sci Total Environ 872:162243

Zhang L, Yin S, Li J, Zhao Y, Wu Y (2016) Increase of polychlorinated dibenzo-p-dioxins and dibenzofurans and dioxin-like polychlorinated biphenyls in human milk from China in 2007–2011. Int J Hyg Environ Health 219:843–849

Zhao G, Xu Y, Li W, Han G, Ling B (2007) PCBs and OCPs in human milk and selected foods from Luqiao and Pingqiao in Zhejiang, China. Sci Total Environ 378:281–292

Zietz BP, Hoopmann M, Funcke M, Huppmann R, Suchenwirth R, Gierden E (2008) Long-term biomonitoring of polychlorinated biphenyls and organochlorine pesticides in human milk from mothers living in northern Germany. Int J Hyg Environ Health 211:624–638

Zober MA, Ott MG, Päpke O, Senft K, Germann C (1992) Morbidity study of extruder personnel with potential exposure to brominated dioxins and furans. I Results of blood monitoring and immunological tests. Brit J Indust Med 49:532–544

Zota AR, Linderholm L, Park J-S, Petreas M, Guo T, Privalsky ML, Zoeller T, Woodruff TJ (2013) A temporal comparison of PBDEs, OH-PBDEs, PCBs, and OH-PCBs in the serum of second trimester pregnant women recruited from San Francisco General Hospital, California. Environ Sci Technoil 47:11776–11784

Chapter 3
Polycyclic Aromatic Compounds and Metals/Elements

Abstract This chapter summarizes information on polycyclic aromatic compounds (PACs) and metals/elements. The key representatives of the PAC group are unsubstituted polycyclic aromatic hydrocarbons, along with nitrated or oxygenated analogs. They primarily originate from human activities, such as the incomplete combustion of organic materials (referred to as pyrolysis) in industrial processes, heating with fossil fuels, motor vehicles, and both intentional and accidental fires. These compounds are found throughout the environment, and their carcinogenic effects are significant for humans. Metals are naturally present in the Earth's crust, and in addition to this natural occurrence, exposure from human activities such as mining, processing, and extensive use in various industrial applications must also be taken into account. Such activities have resulted in high contamination of environmental media, posing a considerable risk to human health. They influence various biochemical and physiological functions of living organisms, with severe health effects seen for many metals at elevated concentrations. Therefore, this chapter includes a brief overview of the chemical and physical characteristics, usage patterns of these substances, their primary health risks, and toxicokinetic properties, as well as references to further scientific literature. The primary goal is to detail their occurrence in human matrices, including blood, urine, and breast milk. In this context, comprehensive information is provided on polycyclic aromatic hydrocarbons (PAHs) and their derivatives, such as nitro-PAH, oxy-PAH, and alkyl-PAH. Additionally, substantial data on metals and elements like arsenic, cadmium, lead, nickel, and mercury are also included.

Keywords Polycyclic aromatic compounds · PAH · Nitro-PAH · Oxy-PAH · Metals · Lead · Arsenic · Cadmium · Mercury

Polycyclic aromatic compounds (PACs) are hydrocarbons made up of condensed benzene rings (unsubstituted PAHs, henceforth referred to as PAHs), arranged in either a linear or angular structure. The hydrogen atoms can be replaced with various polar and nonpolar groups, forming alkylated, nitrated, or oxygenated analogs of PAHs. When nitrogen, oxygen, or sulfur are included in the ring structure instead of carbon, the compounds are referred to as heterocyclic PAHs.

© The Author(s), under exclusive license to Springer Nature Switzerland AG 2026

H. Fromme, *Human Biomonitoring II*, https://doi.org/10.1007/978-3-032-10036-8_3

3.1 Polycyclic Aromatic Hydrocarbons

Polycyclic aromatic hydrocarbons (PAHs) are numerous organic compounds made up of two or more mostly aromatic fused rings that contain only carbon and hydrogen. Table 3.1 summarizes the chemical structures and essential characteristics of 16 PAHs classified as priority substances by the US Environmental Protection Agency. In the environment, PAHs typically exist as complex mixtures of individual substances in varying ratios and often include compounds with 2 to 7 rings. Many particle-bound and gaseous PAHs released into the atmosphere are persistent chemicals that can be altered to varying degrees by light and oxidants such as hydroxyl radicals, ozone, and nitrogen oxides, as well as by biological systems (Abdel-Shafy and Mansour 2016). PAHs, which are of particular interest due to their carcinogenic effects, typically consist of 4–6 rings. They are poorly soluble in water but readily dissolve in lipids.

Low-molecular PAHs (two and three rings) predominantly exist in the air in the vapor phase, while PAHs with five rings or more are primarily bound to particles (WHO 2010). PAHs with four rings are split between the vapor and particle phases. It is essential to consider differences in the respective atmospheric conditions and the type and composition of the particles (Goss and Schwarzenbach 1998). Depending on their molecular weight, PAHs have varying boiling points, ranging from 96 °C (Nap) to 550 °C (BghiPer) (Lee and Vu 2010).

PAHs primarily originate from human activities. They are formed during the incomplete combustion of organic materials (so-called pyrolysis) such as wood, coal, or oil. They also appear in gaseous and liquid distillation products, which can result from the carbonization of coal or the refining of crude oil through cracking. PAHs can also be found in food products, particularly if their production or preparation involves a pyrolytic process or if they are contaminated with PAH-containing impurities. Additionally, PAHs are formed during carburization, as occurs during the formation of crude oil or coal. The more completely a carbon- and hydrogen-containing fuel is burned with oxygen to produce carbon dioxide and water, the lesser the amount of each PAH formed as a byproduct.

The following primary sources of outdoor air pollution must be taken into account (WHO 2010).

- Industrial processes: Coking plants, aluminum smelters (especially the Söderberg process), petroleum refining, production of tar oil, asphalt, and bitumen, tire production, power plants, and waste incineration.
- Domestic area: Heating with coal, wood, oil stoves, open fires, and tobacco smoke.
- Road traffic: Internal combustion engines, especially diesel motor vehicles.
- Agricultural: Activities: Deliberate forest, steppe, bush, and bog fires.
- Natural processes: unintentional forest, steppe, bush, and moorland fires, volcanic eruptions.

Table 3.1 Chemical structure and important characteristics of PAHs

Substance	Abbreviation	Formula	CAS-No.	Structure
Naphthalene	Nap	$C_{10}H_8$	91-20-3	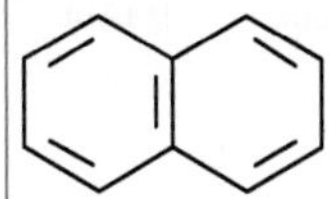
Acenaphthene	Ace	$C_{12}H_{10}$	208-96-8	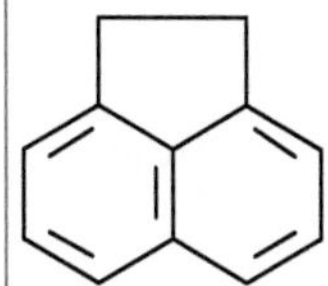
Acenaphthylene	Acy	$C_{12}H_8$	91-20-3	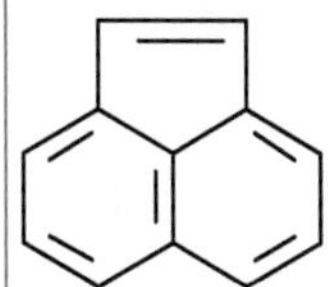
Fluorene	Fl	$C_{13}H_{10}$	86-73-7	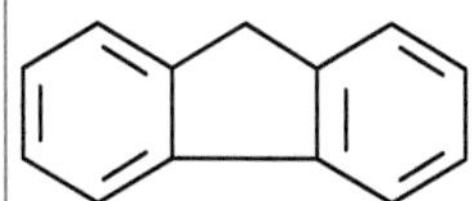
Anthracene	Ant	$C_{14}H_{10}$	120-12-7	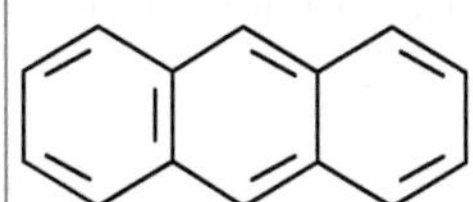
Phenanthrene	Phe	$C_{14}H_{10}$	85-01-8	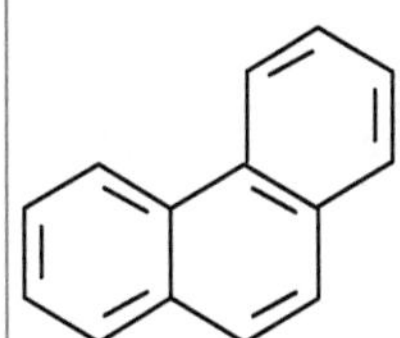
Fluoranthene	Flu	$C_{16}H_{10}$	206-44-0	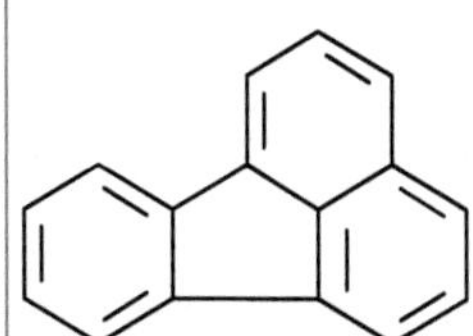
Pyrene	Pyr	$C_{16}H_{10}$	129-00-0	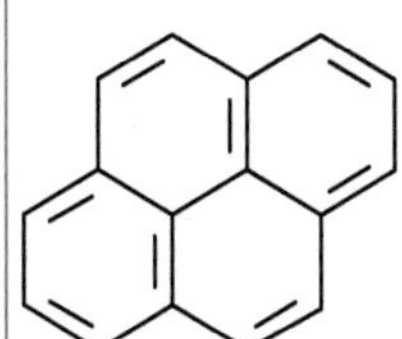

(continued)

Table 3.1 (continued)

Substance	Abbreviation	Formula	CAS-No.	Structure
Benzo[*a*]anthracene	BaAnt	$C_{20}H_{12}$	56-55-3	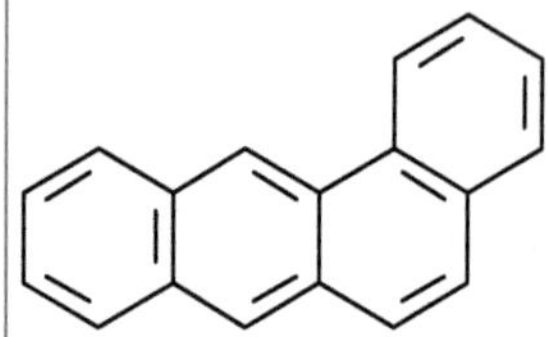
Chrysene	Chr	$C_{18}H_{12}$	218-01-9	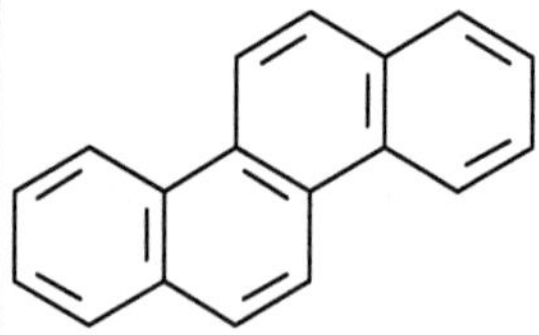
Benzo[*b*]fluoroanthene	BbFlu	$C_{20}H_{12}$	205-82-3	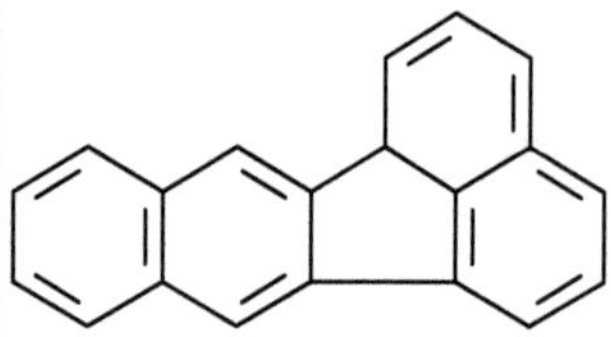
Benzo[*k*]fluoroanthene	BkFlu	$C_{20}H_{12}$	207-08-9	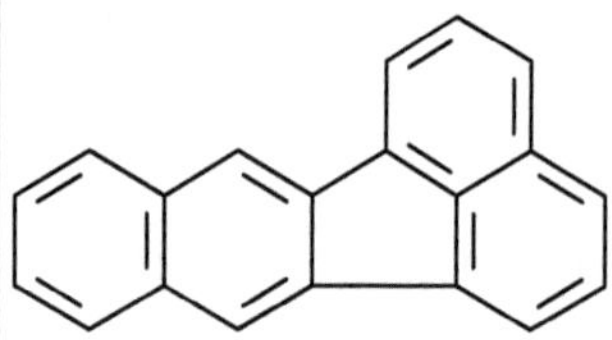
Benzo[*a*]pyrene	BaP	$C_{20}H_{12}$	50-32-8	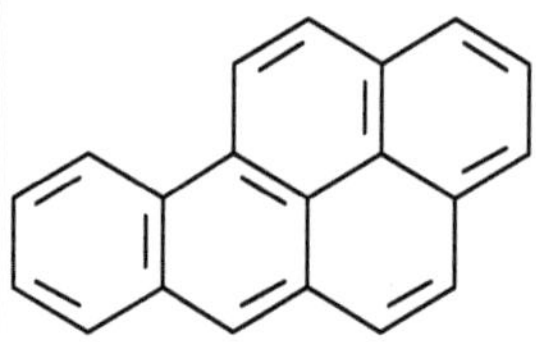
Dibenzo[*a,h*]anthracene	DBahAnt	$C_{22}H_{14}$	53-70-3	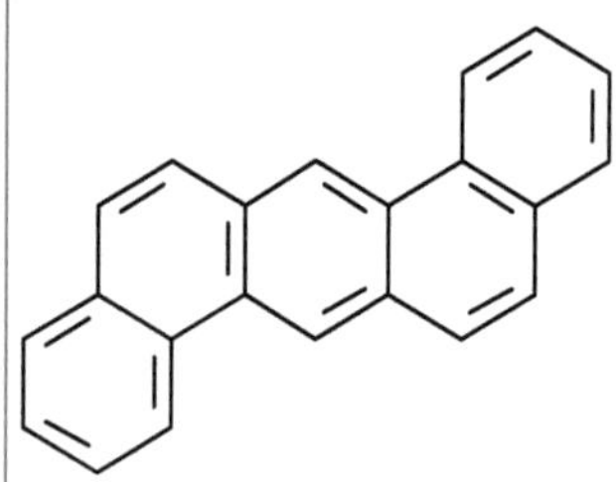

(continued)

Table 3.1 (continued)

Substance	Abbreviation	Formula	CAS-No.	Structure
Benzo[*g,h,i*]perylene	BghiPer	$C_{22}H_{12}$	191-24-2	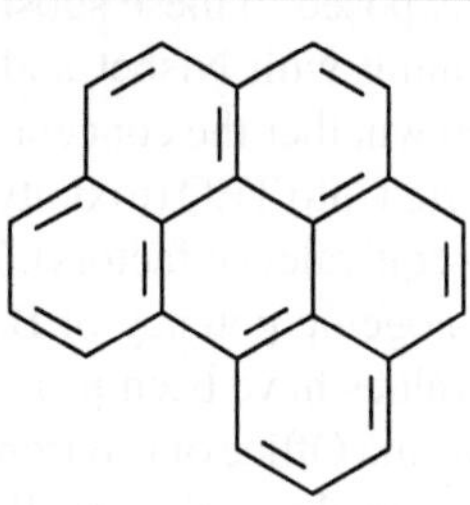
Ideno[*1,2,3-cd*]pyrene	IcdPyr	$C_{22}H_{12}$	193-39-5	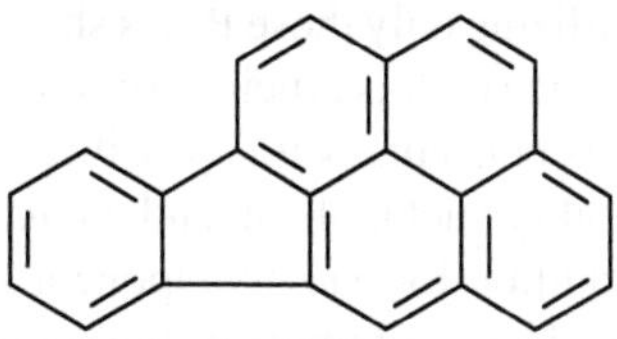

Estimating emission levels from the above sources is linked with significant uncertainties and varies notably between countries. In the European Union, the estimated total emissions in 2021 were 230 tons of BaP and 760 tons of PAHs (EEA 2023). For both BaP and PAHs, emissions from the commercial, institutional, and household sectors were the most critical emitter groups, accounting for 92 and 85% of total emissions, respectively. A global estimate for 2004 concludes that approximately 114,000 tons of ΣPAH16 were emitted in China and 32,000 tons in the USA (Zhang and Tao 2009). In 2011, annual emissions of approximately 324 tons of BaP and 30,000 tons of ΣPAH16 were still assumed in the USA (Zhang et al. 2017).

Multiple reviews concerning the occurrence, exposure, toxicology, and epidemiology of PAHs are currently accessible (WHO 2010; IARC 2010a, b, 2012a, b; HC 2016; US EPA 2017a, b; Santonicola et al. 2017; ECHA 2018; Oliveira et al. 2020; Lee and Vu 2010; Yost et al. 2021; Klotz 2021; Fromme et al. 2024). Unless otherwise specified, this literature serves as the foundation for the next two chapters.

3.1.1 *Health Aspects*

Animal studies demonstrate that certain PAHs impact the hematopoietic and immune systems, potentially affecting the reproduction and development of offspring. However, the carcinogenic effects are more prominent. Regarding the carcinogenic effects on humans following inhalation exposure, primarily epidemiological studies on workers, especially at coking plants, indicate an increased incidence of skin, lung, bladder, and gastrointestinal cancers. Since PAHs always occur in mixtures, the carcinogenic potential of individual PAHs cannot be accurately assessed from these studies for methodological reasons. Additionally, an increased incidence of skin, lung, bladder, liver, and stomach cancers is well-documented in laboratory animals exposed to these substances.

Beginning with Nisbet and LaGoy (1992), the scientific literature has broadly discussed whether the concentrations of the most significant carcinogenic PAHs can be converted into TEQ (toxicity equivalence concentration) using the concept of TEF (toxicity equivalence factors). The TEF value expresses the ratio of a single PAH to the carcinogenic potency of BaP, with BaP's value set to 1 by convention. To date, various values have been proposed (summary in Mueller et al. 2019). Additionally, the California Office of Environmental Health Hazard Assessment (OEHHA) derived potency equivalency factors (PEFs) for cancer induction relative to BaP for 21 PAHs, utilizing only those PAHs shown to be carcinogenic in bioassays (Collins et al. 1998). However, it is crucial to emphasize that considerable uncertainties exist surrounding this concept, as it is possible that different PAH mixtures vary in terms of bioavailability, metabolism, and antagonistic effects. This approach typically relies on dose addition for each component in a mixture to account for their differing potencies. Table 3.2 summarizes the different TEFs for PAHs.

The WHO International Agency for Research on Cancer has classified BaP as carcinogenic to humans (group 1) (IARC 2010a, b). Other PAHs, such as cyclopenta[cd]pyrene, dibenz[a,h]anthracene, and dibenzo[a,l]pyrene, are classified as probably carcinogenic to humans (group 2A). Additionally, anthracene, benz[j]aceanthrylene, benz[a]anthracene, benzo[b]fluoranthene, benzo[j]fluoranthene, benzo[k]fluoranthene, benzo[c]phenanthrene, chrysene, dibenzo[a,h]pyrene, dibenzo[a,i]pyrene, and indeno[1,2,3-cd]pyrene are classified as possibly carcinogenic to humans by IARC (group 2B) (IARC 2010a, b; Cattley et al. 2023).

3.1.2 Toxicokinetics

PAHs, with BaP serving as an example, can be absorbed by the organism during inhalation exposure, depending on whether they exist in gaseous or particulate form and on the type and size of the particles involved. In animal experiments, uptake rates of 10 and 22% have been reported in this context. After oral administration, absorption was observed in both experimental animals and human test subjects, influenced by conditions in the gastrointestinal tract and whether the substances were consumed alongside other foods (Vermillion Maier et al. 2022; Hummel et al. 2018). It has been demonstrated that Caco-2 cells in the intestinal wall impede the absorption of BaP by releasing resulting metabolites back into the intestinal lumen through cellular metabolism (Ramesh et al. 2004). Skin uptake can significantly contribute to occupational exposure, although the importance of dermal uptake in terms of environmental exposure remains unclear (Ruby et al. 2016). Regardless of the uptake site, absorbed PAHs are swiftly transported via blood and lymph to other

Table 3.2 PAHs and toxicity equivalence factors (TEFs)

Substance	CAS-No.	Abbreviation	US-EPA PAH[e]	TEFs
Acenaphthene	83-32-9	Ace	X	0.001[a]
Acenaphthylene	208-96-8	Acy	X	0.001[a]
Anthracene	120-12-7	Ant	X	0.01[a]
Benzo[*a*]anthracene	56-55-3	BaAnt	X	0.1[a]
Benzo[*a*]pyrene	50-32-8	BaP	X	1.0[a]
Benzo[*b*]fluoroanthene	205-99-2	BbFlu	X	0.1[a]
Benzo[*j*]fluoroanthene	205-82-3	BjFlu	X	0.1[c]
Benzo[*k*]fluoroanthene	207-08-9	BkFlu	X	0.1[a]
Benzo[*e*]pyrene	192-97-2	BePyr		0.01[c]
Benzo[*ghi*]perylene	191-24-2	BghiPer		0.01[a]
Benzo[*c*]fluorene	205-12-9	BcFl		20[d]
Chrysene	218-01-9	Chr	X	0.01[a]
Coronene	191-07-1	Cor		0.001[c]
Dibenzo[*a,c*]anthracene	215-58-7	DBacAnt		0.1[c]
Dibenzo[*a,h*]anthracene	53-70-3	DBahAnt	X	1.0[c]
Dibenzo[*a,e*]pyrene	192-65-4	DBaePyr		1.0[b]
Dibenzo[*a,i*]pyrene	189-55-9	DBahPyr		10[b]
Dibenzo[*a,h*]pyrenee	189-55-9	DBaiPyr		10[b]
Dibenzo[*a,l*]pyren	191-30-0	DBalPyr		10[b]
Fluoranthene	206-44-0	Flu	X	0.001[a]
Fluorene	86-73-7	Fl	X	0.001[a]
Indeno[1,2,3-*cd*]pyrene	193-39-5	IcdPyr	X	0,1[a]
Naphthalene	91-20-3	Nap	X	0,001[a]
Phenanthrene	85-01-8	Phe	X	0,001[a]
Pyrene	129-00-0	Pyr	X	0,001[a]

[a] Nisbet and LaGoy 1992; [b] Collins et al. 1998; [c] Malcom and Dobson 1994; [d] US EPA 2010; [e] Priority PAHs after US EPA (US Environmental Protection Agency)

tissues, particularly those with high lipid content. PAHs and PAH metabolites are capable of crossing the placental barrier and entering breast milk in humans.

The metabolism of PAHs is complex and will be illustrated using BaP as an example. In principle, BaP can be metabolized in all tissues; in addition to the high capacities in the liver, the metabolism rates in entry tissues, such as the lungs, skin, and digestive tract, as well as in reproductive tissues, are also considered relevant. In BaP metabolism, many metabolites are formed; for simplicity, only 3-OH-BaP, a metabolite often found in urine, is listed in Fig. 3.1. Three critical metabolic pathways can be roughly distinguished (Vondráček and Machala 2021). (1) The so-called epoxide-diol pathway. In this case, benzo[a]pyrene7,8-diol-9,10-epoxide is formed via BaP-dihydrodiol, and corresponding tetrols are subsequently excreted, or adducts are formed. (2) The reactive intermediate BaP-dihydrodiol (or similar PAH-trans-dihydrodiols) can be converted via BaP-7,8-dione to quinones, which can form DNA adducts directly or damage the DNA indirectly through the redox cycle mechanism by forming reactive oxygen species (ROS). (3) BaP can also be converted by peroxidase reactions to a reactive radical cation that induces ROS production.

As part of the epoxide-diol pathway, epoxides are formed in the first step of the reaction, which can be spontaneously rearranged or converted by enzymes such as hydrolases or oxygenases into phenols, quinones, dihydrodiols, tetrols, or other secondary products. Among the BaP-7,8-dihydrodiol-9,10-epoxides (BPDE), the (+)-BaP-7R,8S-diol-9S,10R-epoxide is the most potent carcinogen of these four isomers. It is produced in most tissues, and the corresponding covalent DNA adduct is predominantly found. The 3-OH-BaP was proposed as a marker of a detoxification reaction, while the benzo[a]pyrene tetrols were also formed as markers of an activating metabolic pathway. The reactive metabolites formed must be eliminated through various phase II reactions, such as the glutathione conjugation of diol epoxides or the sulfation and glucuronidation of phenols.

Animal experiments demonstrate that BaP and its metabolites are primarily excreted in bile or feces within a few days. The proportion of unmetabolized BaP in the feces varies significantly and is influenced by the ingested dose, among other factors. The excretion of PAHs and their metabolites through urine is comparatively lower. Table 3.3 summarizes the half-lives based on metabolite excretion after oral, inhalation, and dermal exposure to PAHs. After the oral administration of deuterium-labeled Nap, Fl, Phe, and Pyr to eight healthy adults, median elimination half-lives of 7.4 h for 1-OH-Nap, 9.8 h for 2-OH-Nap, 6.9 h for 1-OH-Pyr, 11.9 h for 2-OH-Fl, 9.2 h for 3-OH-Fl, and between 7.5 and 10.8 h for the Phe metabolites were observed (Choi et al. 2023).

Human biomonitoring of PAHs focuses on two main aspects. First, metabolites, especially hydroxy-PAHs, are identified as part of detoxification processes and can be effectively analyzed in urine and other biological materials. Monitoring procedures also rely on the understanding that toxic effects, particularly carcinogenic ones, result from interactions between reactive metabolites (in this case, diol epoxides) and essential cell structures. The known BaP intermediates induce this effect through covalent binding to cellular macromolecules, including proteins and DNA, forming protein or DNA adducts. These molecular lesions serve as the toxicological starting point for structural and functional changes, including tumor initiation.

3.1.3 Occurrence of Frequently Analyzed Metabolites in Urine

The most important metabolites of some PAHs are listed in Table 3.4, where hydroxyphenanthrenes, hydroxyfluorenes, hydroxynaphthalenes, and 1-hydropyrene are often measured in urine samples from the general population in scientific studies.

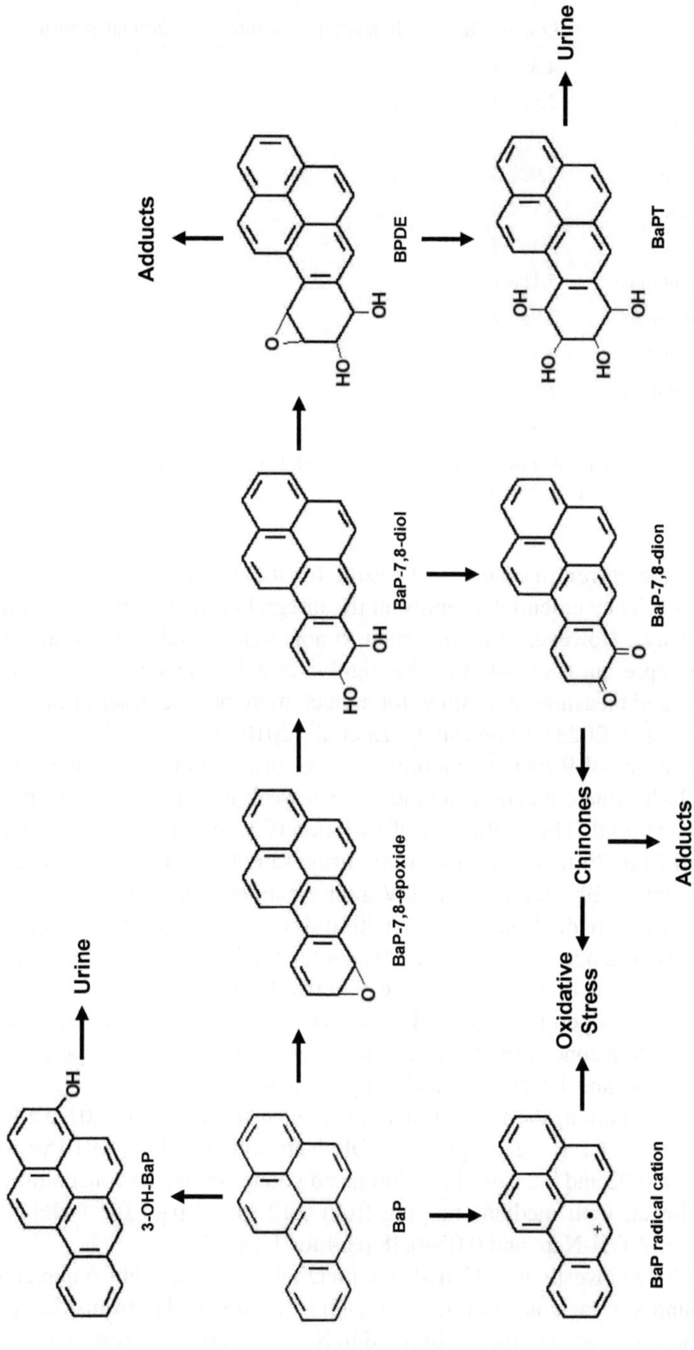

Fig. 3.1 Schematic representation of the most critical steps in the metabolism of benzo[*a*]pyrene (BPDE: benzo[a]pyrene-7,8-diol-9,10-epoxide; BaPT: benzo[a]pyrene-7,8,9,10-tetrol) (modified after Peterson et al. 2020; Vondráček and Machala 2021)

Table 3.3 Metabolite excretion after oral, inhalation, and dermal exposure to PAHs (half-lives in hours)

Metabolite	Oral intake	Inhalation exposure	Dermal exposure
1-Naphthol	4.3; 3.4	–	–
2-Naphthol	2.5; 2.4	9.4	–
1-Hydroxyfluorene	–	5.5	–
2-Hydroxyfluorene	2.9; 2.6	4.1	–
3-Hydroxyfluorene	6.1; 7.0	8.2	–
9-Hydroxyfluorene	3.1; 1.7	–	–
1-Hydroxyphenanthrene	5.1; 3.1	–	–
2-Hydroxyphenanthrene	3.9; 3.7	–	–
3-Hydroxyphenanthrene	4.1; 2.6	–	–
4-Hydroxyphenanthrene	3.5; 2.9	–	–
1-Hydroxypyrene	3.9; 4.4	6.0; 9.8	11.8

Modified after Brzeznicki et al. (1997), Viau (1999), St Helen et al. (2012), Li et al. (2008), Motorykin et al. (2015), Louro et al. (2022)

In this context, the current measurement results for these metabolites are summarized in Table 3.5. They essentially represent the integral of the internal load across all intake pathways. However, it is important to note that a single measurement in urine can only represent a snapshot of the highly variable exposure situation with PAHs, and repeated measurements allow for a much more precise description of the exposure (Zhou et al. 2023). In one study, Li et al. (2010) examined the variability of the concentrations of 9 PAH metabolites in spot urine samples, morning urine samples, and 24-h urine collection samples over a week in 8 adults who were not occupationally exposed. The coefficient of variation (CV) of all OH-PAH between subjects ranged from 26 to 153% in morning urine samples and 13–140% in 24-h urine samples, with a 29% reduction in CV after creatinine adjustment in morning urine samples. For individual subjects, the within-day variance was greater than the between-day variance, accounting for 72–89% of the total variance, possibly due to changes in daily routine, diet, and metabolic patterns. In another study from the US, urine samples from 19 people were analyzed over 44 consecutive days (Zhu et al. 2021). While OH-Nap concentrations fluctuated slightly during the study period, the excretion of OH-Phe and 1-OH-Pyr was highly variable.

In the adult population, the highest medians ranged from 0.26 to 8.01 μg/l for 1-OH-Nap and from 1.2 to 12.2 μg/l for 2-OH-Nap, followed by 1-OH-Pyr with medians between 0.02 and 8.2 μg/l. In children and young adults, the concentrations are generally lower, with medians ranging from 0.12 to 1.8 μg/l for 1-OH-Nap, 0.42–5.0 μg/l for 2-OH-Nap, and 0.05–0.48 μg/l for 1-OH-Pyr.

Guo et al. (2013) investigated 12 hydroxy metabolites across eight Asian countries. They found similar concentrations of 1-OH-Pyr in samples from Malaysia, Korea, and Japan compared to those observed in North America or Western Europe.

Table 3.4 PAHs and primary metabolites detectable by human biomonitoring

Substance	Metabolite	Abbreviation	CAS-No.
Naphthalene	1-Naphthol (1-Hydroxynaphthalene)	1-Nap	90-15-3
	2-Naphthol (2-Hydroxynaphthalene)	2-Nap	135-19-3
Benz[*a*]anthracene	1-Hydroxybenz[a]anthracene	1-OH-BaAnt	69847-26-3
	3 und 9-Hydroxybenz[a]anthracene	3/9-OH-BaAnt	4834-35-9
Benzo[*a*]pyrene	3-Hydroxybenzo[a]pyrene	3-OH-BaP	13345-21-6
Chrysene	2-Hydroxychrysene	2-OH-Chr	65945-06-4
	3-Hydroxychrysene	3-OH-Chr	63019-39-6
	4-Hydroxychrysene	4-OH-Chr	63019-40-9
	6-Hydroxychrysene	5-OH-Chr	37515-51-8
Fluoranthene	3-Hydroxyfluoroanthene	3-OH-Flu	7440-39-3
Fluorene	2-Hydroxyfluorene	2-OH-Flu	2443-58-5
	3-Hydroxyfluorene	3-OH-Flu	6344-67-8
	9-Hydroxyfluorene	9-OH-Flu	1689-64-1
Phenanthrene	1-Hydroxyphenanthrene	1-OH-Phe	2433-56-9
	2-Hydroxyphenanthrene	2-OH-Phe	605-55-0
	3-Hydroxyphenanthrene	3-OH-Phe	605-87-8
	4-Hydroxyphenanthrene	4-OH-Phe	7651-86-7
	9-Hydroxyphenanthrene	9-OH-Phe	484-17-3
Pyrene	1-Hydroxypyrene	1-OH-Pyr	5315-79-7

However, the concentrations in urine samples from China, India, and Vietnam were approximately 4–10 times higher. Among the analyzed compounds, the sum of OH-Nap constituted the highest share, comprising approximately 60–90% of the total OH-PAK, followed by OH-Phe at 3–16%, 2-OH-Flu at 3–20%, and 1-OH-Pyr at 2–8%. In a summary of 56 Chinese studies from 2010 to 2020 that included non-occupationally exposed subjects aged 45–55, the median concentrations in urine were 4.6 μg/l for 1-OH-Nap, 5.3 μg/l for 2-OH-Nap, 4.6 μg/l for four OH-Fl, 4.9 μg/l for up to seven OH-Phe, and 3.6 μg/l for 1-OH-Pyr (Zhang and Li 2023).

Various studies have demonstrated the influence of age on urine levels (Thai et al. 2020; Li et al. 2021; Hudson-Hanley et al. 2021; Huang et al. 2022; Murawski et al. 2020). In an Australian study, the concentrations of all OH-PAH in children aged 0–4 years and 5–14 years were consistently lower than in adolescents and adults (Thai et al. 2020). A Canadian study found that the median levels of 2-OH-Nap increased with age, showing levels of 3.0 μg/l (3–5 years), 3.6 μg/l (6–11 years), 4.8 μg/l (12–19 years), 5.4 μg/l (20–39 years), and 6.4 μg/l (40–59 years), only to decline to 3.1 μg/l in 60–79-year-olds (HC 2017).

Another factor influencing urine levels is exposure to tobacco smoke, which has been observed in various studies regarding 3-OH-BaP, 1-OH-Pyr, OH-Nap, and OH-Phe (summary in Burkhardt et al. 2023). For instance, in a Chinese study involving

Table 3.5 Median hydroxy-metabolite levels in the urine of the general population in μg/l

References	N	Age (years)	1-OH-Nap	2-OH-Nap	3-OH-Fl	2-OH-Fl	2-OH-Phe	3-OH-Phe	4-OH-Phe	1-OH-Phe	9-OH-Phe	1-OH-Pyr	Country, year
Adults													
Li et al. (2025)	2460	30[f]	1.17	2.90	0.09	0.17	0.23	–	–	0.11	0.03	0.04	China, 2022/23
Yang et al. (2026)	1311	27[a]	0.56	0.12	–	0.21	0.06	0.05	0.03	0.06	0.02	0.05	China, 2020–22
Yang et al. (2021a, b)	480	35[a]	1.53	3.16	–	0.33	0.06	0.17	0.10	0.03		0.14	China. 2019
Huang et al. (2022)	1154	1–99	1.77	4.64	0.99	0.97	0.42[b]		0.20	0.20[c]		0.27	China. 2019
Li et al. (2021)	480	5–77	4.94	4.17	< LOQ	< LOQ	0.026[b]		< LOQ	< LOQ[c]		0.297	China. 2018
Zhou et al. (2023)	378	32[a]	2.82	7.89	–	2.50	1.12	1.02	1.30	1.09	2.40	0.94	China
Urbancova et al. (2020)	330	18–44	0.36	4.66	–	0.23	1.15					0.12	Czech Republic. 2016/17
Thai et al. (2020)	24[d]	> 0	7.80	7.10	0.12	0.25	0.04	–	0.06	–	–	0.12	Australia, 2016/17
Kim et al. (2021)	1254	19–80	–	3.23	–	0.40	–	–		0.14	–	0.16	Korea, 2015–17
Fernández et al. (2021)	110	20–45	0.7	7.4	0.14	0.15	0.07	0.10	0.04	0.15	0.11	0.12	Spain, 2015
HC (2017)	312	40–59	0.89	4.60	0.09	0.20	0.07	0.09	0.02	0.14	0.04	0.09	Canada, 2014/15

(continued)

Table 3.5 (continued)

References	N	Age (years)	1-OH-Nap	2-OH-Nap	3-OH-Fl	2-OH-Fl	2-OH-Phe	3-OH-Phe	4-OH-Phe	1-OH-Phe	9-OH-Phe	1-OH-Pyr	Country, year
CDC (2019)	1802	> 20	1.29	4.18	0.07	0.16	0.11 [b]		–	0.09	–	0.11	USA. 2013/14
Tombolini et al. (2018)	747[e]	35–69	0.66	1.20	–	–	–	–	–	–	–	0.02	Italy, 2013/14
Joksić et al. (2022)	601	18–49		2.11	0.20		0.08	0.07	< LOQ	0.10	0.06		Slovenia, 2008–14
Guo et al. (2013)	84	29[a]	0.53	2.27	–	0.89	0.32	0.39	0.03	–	0.38	0.463	China, 2010–12
Guo et al. (2013)	23	51[a]	0.642	4.90	–	0.473	0.157	0.201	0.032	–	0.279	0.463	Vietnam, 2010–12
Guo et al. (2013)	34	32[a]	0.266	3.25	–	0.207	0.065	0.121	0.010	–	0.077	0.075	Japan, 2010–12
Guo et al. (2013)	38	37[a]	1.110	3.78	–	0.346	0.256	0.289	0.027	–	0.154	0.424	India, 2010–12
Guo et al. (2013)	29	30[a]	0.263	1.55	–	0.112	0.036	0.043	0.006	–	0.033	0.065	Malaysia, 2010–12
Guo et al. (2013)	38	37[a]	1.410	7.33	–	0.448	0.089	0.137	–	–	0.163	0.220	Kuwait, 2010–12
Berman et al. (2013)	249	20–74	–	–	–	–	0.095	0.141	0.032	0.219	–	0.204	Israel, 2011
Guo et al. (2013)	60	35[a]	2.060	5.87	–	0.165	0.049	0.078	0.007	–	0.066	0.103	Korea, 2006–07
Children/adolescents													

(continued)

Table 3.5 (continued)

References	N	Age (years)	1-OH-Nap	2-OH-Nap	3-OH-Fl	2-OH-Fl	2-OH-Phe	3-OH-Phe	4-OH-Phe	1-OH-Phe	9-OH-Phe	1-OH-Pyr	Country, year
Urbancova et al. 2020	330	Newborns	0.12	2.28	–	0.07	0.41					0.05	Czech Republic. 2016/17
Murawski et al. 2020	516	3–17	0.74	4.08	–	0.45	0.08	0.13	0.04	0.14	0.06	0.10	Germany, 2014–17
Kelishadi et al. (2017)	150	6–18	0.36	0.42	–	–	–	–	–	–	0.11	0.10	Iran, 2015/16
Fernández et al. (2021)	566	5–11	0.2	5.0	< LOQ	0.17	0.04	0.08	0.02	0.11	0.04	0.11	Spain, 2015
Yu et al. (2021)	167	8–11	0.65	1.45	–	0.49	0.22	–	0.14	0.25	0.19	0.40	China, 2015
HC 2017	510	6–11	0.57	3.60	0.08	0.08	0.04	0.08	0.02	0.12	0.03	0.10	Canada, 2014/15
CDC 2019	399	6–11	0.68	2.97	0.06	0.12	0.09[b]	–	0.07	–	0.13	USA. 2013/14	
Souza et al. (2022)	200	6–14	1.33	0.71	–	0.43	–	–	–	–	0.39	0.48	Brazil
Wallace et al. (2023)	1118	5,1[a]	0.53	3.23	–	–	0.06	0.06	–	0.17		0.08	USA. 2006–16
Dobraca et al. (2018)	431	6–8	1.30	1.52	0.06	0.15	0.03	0.07	0.008	0.08	–	0.08	USA, 2005–06
Sochacka-Tatara et al. (2018)	218	3	1.80	3.89	0.18	0.49	0.09	0.21	0.06	0.32	–	0.19	Poland, 2004–07

[a] Mean; [b] sum of 2-OH-Phe und 3-OH-Phe; [c] sum of 1-OH-Phe und 9-OH-Phe; [d] pooled from 1000 individual samples; [e] only nonsmoker; [f] median; *LOQ* limit of quantification, *2-OH-Nap* 2-Hydroxynaphthalene, *1-OH-Nap* 1-Hydroxynaphthalene, *2-OH-Nap* 2-Hydroxynaphthalene, *3-OH-Fl* 3-Hydroxyfluorene, *2-OH-Fl* 2-Hydroxyfluorene, *2/3-OH-Phe* 2/3-Hydroxyphenanthrene, *4-OH-Phe* 4-Hydroxyphenanthrene, *1/9-OH-Phe* 1/9-Hydroxyphenanthrene, *1-OH-Pyr* 1-Hydroxypyrene

1154 subjects, smokers exhibited higher median concentrations for the sum of 12 metabolites (22.5 μg/l) compared to nonsmokers (10.7 μg/l). However, the difference between the two groups was only significant for the sum of OH-Nap (Huang et al. 2022). In an Italian study with 1016 subjects, smokers also displayed significantly higher levels of 1-OH-Nap, 2-OH-Nap, and 1-OH-Pyr compared to nonsmokers, while 3-OH-BaP did not correlate with cotinine concentration (Tombolini et al. 2018). Additionally, a German study found a moderate but significant correlation between cotinine concentrations and the levels of 1-OH-Pyr, 1-Nap, and all five individual OH-Phe in 24-h urine samples (Burkhardt et al. 2023).

Burkhardt et al. (2023) examined 510 urine samples (24-h urine collection) from individuals aged 20–29 from the German Environmental Specimen Bank at nine time points to estimate the temporal trend of internal exposure. From 1995 to 2019, there was a significant decrease in urine concentrations of 3-OH-BaP by 19%, 1-OH-Pyr by 39%, 1-OH-Nap by 66%, and total OH-Phe by 16%, while 2-OH-Nap levels increased by approximately 70%. The authors attribute the decrease in concentrations of hydroxy metabolites and cotinine measured in parallel to smoking bans and regulations implemented in Germany to reduce tobacco smoke exposure. A comparison of results from the current German Environmental Health Study (GerES V; 2014–2017) with those of 3–17-year-olds in GerES II conducted in 1990/91 showed that urine concentrations of Phe and Pyr metabolites were still two to three times higher than those found in GerES IV (2003–2006) (Murawski et al. 2020). In GerES V, concentrations of 1-OH-Phe, 3-OH-Phe, and 1-OH-Pyr decreased further compared to GerES IV, while levels of 2/9-OH-Phe and 4-OH-Phe increased again. In an analysis of over 11,000 urine measurements from NHANES involving nonsmokers (> 6 years) between 2001 and 2014, concentrations of OH-Nap increased by 36% and OH-Pyr by 106%, while OH-Fl levels decreased by 55% and OH-Phe levels dropped by 37% during the same period (see Fig. 3.2) (Hudson-Hanley et al. 2021).

In a quasi-experimental study of 8 nonsmokers, Li et al. (2010) reported that urinary PAH metabolites increased after high inhalation and dietary exposure. On days with a low PAH diet, the total amount of inhaled naphthalene over each 24-h period correlated well with the amount of excreted OH-Nap. To a lesser extent, this was also true for fluorene and its metabolite in urine. Only naphthalene showed a significant correlation with its excreted metabolites on days with a high dietary intake of PAHs. In a Canadian pilot study involving nine women from a polluted region and ten from an "unpolluted" region during pregnancy, the following consistent predictors of urinary metabolite levels were also identified: 2-OH-Nap for past smoking, 3-OH-Fl and 3-OH-Phe for PAHs in outdoor air, and 1-OH-Pyr for the consumption of fried or grilled food and proximity to industrial areas (Nethery et al. 2012). In a Polish area with high environmental pollution, the concentrations of PAH metabolites in urine reflected external pollution only to a small extent (Sochacka-Tatara et al. 2018). In this study, higher urinary PAH metabolites (1-OH-Pyr, 2-OH-Fl, and 9-OH-Fl) were particularly observed in 218 three-year-old children exposed to tobacco smoke at home or in those whose homes used gas, coal, or oil for cooking and heating water. In New York City, 109 participants in a birth cohort wore a wristband as a passive collector for 48 h to determine their exposure and provided a urine sample at the end

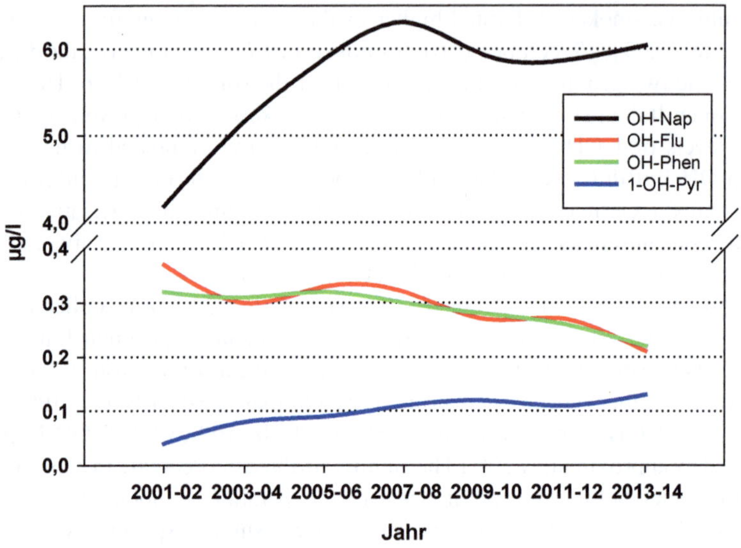

Fig. 3.2 Temporal development of the weighted geometric mean values of PAH metabolites in urine (after Hudson-Hanley et al. 2021)

of the collection period (Dixon et al. 2022). Significant correlations with personal external exposure were found for OH-Phe, OH-Fl, 1-OH-Nap, and 1-OH-Pyr in the urine.

3.1.4 Occurrence of Other Metabolites in Urine

Since the aforementioned monohydroxylated PAHs excreted in urine represent detoxified metabolites due to their nucleophilic character, carcinogenic PAH metabolites, such as benzo[a]pyrene, which indicate bioactivation (e.g., to the reactive dihydrodiole epoxides), would ideally be desirable (Klotz 2021). Additionally, while metabolites of low molecular weight PAHs, such as those with two or three rings, are preferentially excreted in urine, metabolites of relatively high molecular weight PAHs, like Chr and BaP, are preferentially excreted in feces (Ramesh et al. 2004). If the latter are to be documented as part of biomonitoring, only very sensitive and specific analytical methods can detect the small quantities of hydroxylated metabolites, such as 3-OH-BaP, in urine with adequate validity (MAK 2008). The following results on various other metabolites have been described in the scientific literature to date:

3-Hydroxy-benzo[a]pyrene

Lafontaine et al. (2006) reported mean excretions of 3-OH-BaP at 0.131 ng/24 h and 0.059 ng/ 24 h in 27 smokers and 27 nonsmokers, respectively. A 95th percentile of 0.032 ng/l (0.008 ng/g crea) was observed in 747 samples from Italian nonsmokers (Tombolini et al. 2018). Barbeau et al. (2011) analyzed the urine of 23 nonsmokers and 13 smokers (without workplace exposure). They found significant differences between the two groups, with medians of 0.009 nmol/mol crea and 0.023 nmol/mol crea, respectively. Fernández et al. (2021) found 3-OH-BaP above the limit of quantification of 50 ng/l in only 1% of the 110 samples from adults and 5% of the 566 samples from children. Using an ultra-sensitive analysis method (detection limit: 0.0006 ng/l), Richter-Brockmann et al. (2019) detected 3-OH-BaP in all samples at concentrations ranging from 0.037 to 0.270 ng/l in 7 nonsmokers and from 0.374 to 1.171 ng/l in 7 smokers. For instance, a geometric mean value of 0.057 ng/24 h (15% of values above the limit of quantification) or 0.030 ng/l (95th percentile: 0.08 ng/l) was reported for 3-OH-BaP in 24-h urine samples from 54 nonsmokers aged between 20 and 28 years in Germany in 2019 (Burkhardt et al. 2023).

In the Canadian CHMS from 2014 to 2015, 3-OH-BaP could only be quantified in individual samples exceeding the relatively high limit of quantification of 2.9 ng/l (HC 2017). In 480 samples from individuals in southern China, 3-OH-BaP was reported with mean values of 283 ng/l (urban areas) and 19 ng/l (rural areas), with only 3.8 and 6.7% of the results exceeding the limit of detection (Li et al. 2021). In a nationwide Brazilian study involving 200 children aged 6–14 years, 3-OH-BaP was detected in 43% of the samples at a limit of quantification of 150 ng/l, with a relatively high median of 111 ng/l calculated (Souza et al. 2022). When evaluating the results, it is important to note that the quantification limits of the analytical methods were quite high in some studies.

Luo et al. (2019) examined 15 urine samples for both conjugated and free forms of 3-OH-BaP. They found that the average concentration of free 3-OH-BaP was 0.19 ng/l, while the concentrations of 3-OH-BaP glucuronide ranged from < LOQ to 71.9 ng/l, and the concentrations of 3-OH-BaP sulfate ranged from < LOQ to 7.69 ng/l. Again, Park et al. (2025) were not able to find 3-OH-PaP above the limit of quantification of 18 ng/l in 124 urine samples from 28 child-mother-father triads in 2020/2021 in China.

Tetrole

The diol-epoxide pathway of BaP metabolism leads to the formation of benzo[a]pyrene-7,8,9,10-tetrol (BaPT), which is excreted in urine and serves as a biomarker for BaP exposure following metabolic activation (Peterson et al. 2020). Phenanthrene is also metabolized to tetrol (PheT), but in this instance, the concentrations in urine are significantly higher.

First, Zhong et al. (2011) examined the urine of 30 smokers and 30 nonsmokers, finding BaPT concentrations of 0.64 fmol/ml and 0.26 fmol/ml, respectively. In another study, significant differences were observed between two groups of 25 smokers and 25 nonsmokers, with mean BaPT concentrations (sum of two enantiomers) of 0.20 fmol/ml (0.03–0.61 fmol/ml) in smokers and 0.08 fmol/ml (0.009–0.25 fmol/ml) in nonsmokers (Hecht and Hochalter 2014). Hilton et al. (2017)

analyzed two tetrols, BaPT I-1 (benzo[a]pyrene-r-7,t-8,t-9,c-10-tetrahydrotetrol) and BaPT II-1 (benzo[a]pyrene-r-7,t-8,c-9,c-10-tetrahydrotetrol), in the urine of a volunteer who consumed a meal with high PAH content (grilled chicken) and in the urine of 30 smokers and 30 nonsmokers. The BaPT I-1 concentration increased from < 0.026 ng/l during the PAH-containing meal to 0.98 ng/l seven hours later, returning to baseline after about 24 h. BaPT II-1 showed a similar trend. Smokers had significantly higher concentrations of BaPT I-1, with a median of 0.069 ng/l (interquartile range: 0.031–0.10 ng/l) compared to nonsmokers, who had 0.043 ng/l (interquartile range: < 0.026–0.074 ng/l). BaPT II-1 was quantified in only 25% of the samples.

As part of a comprehensive epidemiological study in the US conducted from 2001 to 2006 involving 2613 smokers (who smoked fewer than 15 cigarettes per day), a mean value of 0.00143 fmol/ml of urine was found for PheT (Carmella et al. 2013). In a study of 428 nonsmoking women from three European countries (Poland, Italy, and Serbia), Nilsson et al. (2013) reported median (maximum) PheT concentrations ranging from 0.53 to 1.77 μg/g crea (0.07–24.8 μg/g crea). The medians from Poland and Italy, which recorded the highest and lowest concentrations, respectively, differed by a factor of approximately 3.3. In China, a 2009 study of 9 smokers and 40 nonsmokers found geometric mean values of 10.1 and 6.5 pmol/mg crea, respectively (Kensler et al. 2012).

3.1.5 Occurrence of Adducts in the BLOOD

Carcinogenic chemicals like BaP react with genetic material to form DNA adducts. Since the analytical determination of these adducts in target organs, such as the lungs, is generally impossible in epidemiological studies, more accessible tissues like blood must be used. The literature particularly describes the determination of DNA adducts in blood leukocytes (white blood cells) and protein adducts of human serum albumin or hemoglobin. The half-life in these test materials should be considered, which is 9–13 weeks in leukocytes, about 3–4 weeks for albumin adducts, and 17 weeks for hemoglobin adducts (Käfferlein et al. 2010). Käfferlein et al. (2010) summarize the various analytical techniques for determining protein adducts and the findings of studies on workplace exposure, as well as the general population not exposed at workplaces.

Recent measurements of BPDE-dG adducts formed by the reaction of BaP-trans-7,8-diol-anti-9,10-epoxide (BPDE) with cellular DNA are now available. As part of an epidemiological study conducted in the USA between 1998 and 2006, mean values of 0.21 and 0.23 adducts per 10^8 nucleotides were found in 581 maternal and 399 umbilical cord blood samples (Riley et al. 2024). For 150 mother–child pairs in China in 2002, the mean concentrations in umbilical cord blood were 0.33 adducts per 10^8 nucleotides, while maternal blood had a slightly lower mean of 0.29 adducts per 10^8 nucleotides (Tang et al. 2006). In a Polish birth cohort of 362 mother–child pairs, very low adduct levels in umbilical cord blood, but not in maternal blood, were significantly associated with BaP exposure levels during the second trimester of pregnancy (Jedrychowski et al. 2013). Here, at < 1.15 ng BaP/m^3, the average was 0.24 adducts per 10^8 nucleotides; at 1.16–5.7 ng/m^3, it was 0.25 adducts per 10^8 nucleotides; and at > 5.7 ng/m^3, it was 0.3 adducts per 10^8 nucleotides. In a more heavily contaminated region of Poland, with a mean BaP value of 1.3 ng/m^3, an average of 0.43 adducts per 10^8 nucleotides was found in 75 umbilical cord blood samples. In contrast, in a region contaminated with an average of only 0.9 ng BaP/m^3, the level was 0.21 adducts per 10^8 nucleotides (Sram et al. 2013). Kuang et al. (2021) examined 83 children with lung diseases, such as bronchial asthma and bronchitis, alongside 63 healthy subjects from 2016 to 2018. In the first group, with outdoor air exposure of 0.39 ng BaP/m^3 measured in 2015, the geometric mean of DNA adducts in the blood was significantly higher at 0.28 ng/μg DNA compared to the control group at 0.13 ng/μg DNA (0.21 ng/m^3 in outdoor air).

BPDE-dG adducts were also examined in the leukocytes of 42 umbilical cord blood samples from healthy individuals and 42 subjects with various congenital disabilities, and no significant difference in the exposure level (0.27–1.55 BPDE-dG/10^8 nucleotides and 0.27–11.8 BPDE-dG/10^8 nucleotides) was reported (Guo et al. 2019).

3.1.6 Occurrence in Breast Milk

The concentrations of PAHs in breast milk are shown in Table 3.6. For BaP, the concentrations vary globally from the detection limit of the respective method to 22.2 ng/g lipid. The highest levels were observed in a recent Chinese study that examined 98 samples from four industrialized districts in northwest China (Wang et al. 2018). In this study, the mean values of BaP concentrations ranged from 22.2 to 32.3 ng/g lw, and the total for all 15 PAHs measured fell between 260 and 374 ng/g lipid. PAHs with 2–4 rings (Acy, Ace, Fl, Ant, Flu, Phe, Pyr, BaAnt, Chr, BbFlu, and BkFlu) accounted for approximately 70% of the $\sum$PAH15. Two earlier Chinese studies, conducted in Hong Kong in 2005 and in Beijing in 2005/06, also reported very high mean concentrations in breast milk, particularly for the low molecular weight PAHs such as Phe, Flu, and Pyr (Tsang et al. 2011; Yu et al. 2011), whereas the levels of higher molecular weight PAHs like BaP were significantly lower compared to Wang et al. (2018) in northwest China.

In six European studies from the Czech Republic, Italy, Spain, and Portugal, the mean concentrations of higher molecular weight PAHs were significantly lower, similar to those in the USA, Colombia, and Ghana. For BaP, the mean concentration ranged from below the limit of quantification to 0.81 ng/g lw. Only the results from the study in Ghana, conducted in an industrialized region with many companies that recycle electronic waste, stood out (Asamoah et al. 2019). Very high concentrations of naphthalene (maximum: 4320 ng/g lw) were observed in this investigation.

Numerous factors influencing the topic are discussed in the literature. For instance, studies from Italy and Turkey reported higher concentrations in the breast milk of smoking mothers compared to that of nonsmoking mothers (Zanieri et al. 2007; Çok et al. 2012) and increased concentrations when exposure to secondhand smoke was

Table 3.6 Mean PAH levels in breast milk in ng/g l.w

References	N	Nap	Ace	Phe	Flu	Pyr	BaAnt	Chr	BbFlu	BkFlu	BaP	IcdPyr	BghiPer	DBahAnt	Country, year
He et al. (2024)	53	52.4	< LOQ	38.6	4.5	1.4	< LOQ	< LOQ	< LOQ	< LOQ	< LOQ	< LOQ	< LOQ	< LOQ	China, 2018/19
Mendas et al. (2024)	49	–	–	–	11.3	38.3	2.32	1.80	3.81	1.56	2.15	1.98	4.08	1.92	Croatia, 2018/19
Oliveira et al. (2020)	65[e]	157	3.64	7.21	< LOQ	< LOQ	< LOQ	< LOQ	< LOQ	< LOQ	< LOQ	< LOQ	6.95	10.3	Portugal, 2019/20
Acharya et al. (2019)	45	–	–	–	18.3	13.8	11.7	11.9	15.5	7.9	0.1	12.4	25.1	–	USA, 2015–18
Wang et al. (2018)	33[f]	–	13.0	29.3	14.0	13.4	7.26	7.79	19.1	17.7	22.2	26.0	14.6	15.9	China, 2016
Asamoah et al. (2019)	105[a]	1027	0.56	11.2	4.41	2.87	0.26	0.37	3.05	0.05	0.09	0.74	0.73	0.12	Ghana, 2014–16
Asamoah et al. (2019)	23[b]	78.7	7.27	1.62	2.31	5.63	< 0.001	< 0.001	2.42	< 0.001	< 0.001	0.48	< 0.001	0.27	Ghana, 2014–16
Torres-Moreno et al. (2022)	60	20.8	20.1	36.1	41.2	38.1	4.14	5.82	2.20	2.21	0.59	0.80	–	–	Colombia, 2014–16
Pulkrabova et al. (2016)	93	9.03	0.60	9.35	1.81	0.67	0.12	< 0.06	0.05	0.19	0.17	0.23	0.24	< 0.1	Czech Republic, summer 2013
Pulkrabova et al. (2016)	95	20.4	2.61	12.19	2.19	0.82	0.11	< 0.06	0.12	0.14	0.14	0.43	0.43	< 0.1	Czech Republic, winter 2013/14

(continued)

Table 3.6 (continued)

References	N	Nap	Ace	Phe	Flu	Pyr	BaAnt	Chr	BbFlu	BkFlu	BaP	IcdPyr	BghiPer	DBahAnt	Country, year
Luzardo et al. (2013)	23	1.04[d]	< 0.02[d]	5.33[d]	0.41[d]	1.06[d]	0.30[d]	0.17[d]	< 0.10[d]	< 0.10[d]	< 0.10[d]	< 0.2[d]	< 0.2[d]	< 0.2[d]	Spain, 2010
Santonicola et al. (2017)	30	–	4.09	10.02	7.32	24.5	0.34	0.84	0.78	6.33	0.81	7.77	5.23	4.36	Italy
Çok et al. (2012)	47	45.8	1.43	16.5	4.71	5.81	0.58	1.48	0.46	0.21	0.25	0.53	0.71	0.10	Turkey, 2009
Yu et al. (2011)	40	–	36.6	133	50.9	36.5	3.97	7.42	20.5	8.62	1.77	2.50	1.26	6.39	China, 2005/06
Tsang et al. (2011)	30	786	83	361	158	187	47	79	66		< LOQ	< LOQ	< LOQ	< LOQ	Hong Kong. 2005
Kim et al. (2008)	12	–	–	12.8	1.8	1.3	–	–	–	–	–	–	–	–	USA, 2005
Zanieri et al. (2007)	21[g]	5.56	6.48	0.80	0.53	0.78	0.33	0.28	0.26	<	<	<	<	<	Italy, 2004/05
Del Bubba et al. (2005)	10	4.70[c]	2.72[c]	0.55[c]	0.25[c]	0.62[c]	0.97[c]	< 0.03[c]	0.56[c]	0.11[c]	< 0.03[c]	< 0.02[c]	< 0.03[c]	< 0.03[c]	Italy
Kishikawa et al. (2003)	51	–	–	0.25	0.02	0.02	0.004	0.06	0.41	0.006	0.002	0.003	< 0.035	0.007	Japan

[a] Industrialized region with many companies that recycle electronic waste; [b] residential area; [c] in ng/g fresh weight; [d] medians in μg/l; [e] median; [f] only Chengguan province; [g] only nonsmoker; *LOQ* limit of quantification

reported (Wang et al. 2018). No statistically significant differences were found in the breast milk of primiparas and multiparas (Tsang et al. 2011; Asamoah et al. 2019). Even with environmental pollution from companies that recycle electronic waste in Ghana, or high outdoor air pollution containing PAHs in the Czech Republic, breast milk showed higher contamination compared to residential areas without known environmental contamination (Asamoah et al. 2019; Pulkrabova et al. 2016).

Only two studies have examined the correlations between breast milk and blood. In 2005, Tsang et al. (2011) analyzed 21 maternal and umbilical cord blood samples, along with 29 breast milk samples in Hong Kong. They found a significant correlation between the content of breast milk and umbilical cord blood, but not with the level of PAHs in maternal blood. Yu et al. (2011) also reported the same correlation between breast milk and umbilical cord blood in a study of 40 nonsmoking women from Beijing.

Only one study from Portugal has investigated hydroxy metabolites in breast milk (Oliveira et al. 2020). The medians (ranges) for the sum of 1-OH-Nap and 1-OH-Ace were 10.7 ng/g lw or 0.414 μg/l (0.413–21.5 μg/l); for 2-OH-Fl, it was 0.148 ng/g lw or 0.0053 μg/l (0.00529–0.166 μ/l), and for 1-OH-Phe, it was 11.5 ng/g lw or 0.418 μg/l (0.076–2.73 μg/l). In this study, the concentrations of $\sum$PAHs and $\sum$OH-PAHs showed a strong correlation, whereas the individual PAHs and OH-PAHs exhibited only a moderate correlation.

3.2 Polycyclic Aromatic Hydrocarbon Derivatives

Nitro-PAHs are derivatives of PAHs that contain at least one nitro-functional group (NO2) on the aromatic ring. At the same time, oxy-PAHs encompass compounds like polycyclic aromatic ketones and quinones. Both types are more polar than their unsubstituted analogs and are more hydrophilic, making them more mobile in environmental media (IPCS 2003). They are characterized by their persistence in the environment and greater reactivity due to electronegative atoms in their structure (Achten and Andersson 2015). In alkylated PAHs (alkyl-PAHs), a methyl group (CH_3) substitutes one or more hydrogen atoms. Compared to nitro-PAHs, oxy-PAHs such as 6H-benzo[c,d]pyrene-6-ones, 1,4-chrysenequinones, and 1-acenaphthenones, as well as alkyl-PAHs like methylfluorenes, methylphenantrenes, and methylnaphthalenes, are found in significantly higher concentrations in the environment (e.g., outdoor air) (Peng et al. 2023).

Nitro-PAHs are primarily produced by the incomplete combustion and pyrolysis of fossil fuels and biomass. They are emitted alongside parent PAHs from motor vehicles (diesel engine exhaust), industrial facilities, and waste incineration plants that use fossil fuels for heating, as well as from natural fires (IPCS 2003; Bandowe and Meusel 2017). Several nitro-PAHs are known to be significantly more abundant in diesel engine emissions compared to those from gasoline engines and other combustion sources (Abbas et al. 2018). Additionally, nitro-PAHs can form in the atmosphere through reactions of parent PAHs with oxidants such as hydroxyl radicals, nitrate radicals, or ozone (Idowu et al. 2019). The individual nitro-PAHs are distributed differently between the gas and particle phases (IPCS 2003). Oxy-PAHs are also emitted directly from combustion processes or formed secondarily through radical reactions of parent PAHs in the atmosphere (Lundstedt et al. 2007). Alkyl PAHs naturally occur in bitumen deposits, crude oil, coal, and volcanic eruptions. They can enter the environment through human activities related to fossil fuel combustion, emissions from transportation, and industrial processes (Wnorowski et al. 2022). Tobacco smoking also produces alkyl-PAHs, primarily methyl derivatives (Li et al. 2014).

A review article summarizes current knowledge on analytical methods, environmental occurrences, and the toxicity of PAH derivatives such as nitro-PAHs, oxy-PAHs, and alkyl-PAHs (Peng et al. 2023).

3.2.1 Health Aspects

Like unsubstituted PAHs, nitro-PAHs activate the Ah receptor, resulting in the formation of DNA adducts, changes in gene and protein expression, alterations in the cell cycle, inflammatory responses, and the production of reactive oxygen species (Idowu et al. 2019; Krzyszczak and Czech 2021). Their ability to form covalently bonded DNA adducts through various metabolic pathways makes nitro-PAHs highly potent carcinogens. In this context, the WHO International Agency for Research on Cancer has categorized 6-nitrochrysene and 1-nitropyrene in group 2A as probably carcinogenic to humans, while 3,7-dinitrofluoranthene, 3,9-dinitrofluoranthenes, 1,3-dinitropyrenes, 1,6-dinitropyrenes, 1,8-dinitropyrenes, 3-nitrobenzanthrones, 2-nitrofluorenes, and 4-nitropyrenes are assigned to group 2B as possibly carcinogenic to humans (IARC 2013).

The toxicological mechanisms of oxy-PAHs are complex and have not been fully elucidated (Lundstedt et al. 2007). Quinones, substances with specific toxic potential, can convert into electrophilic intermediates, forming adducts with proteins and DNA, which lead to genotoxic and cytotoxic effects. Additionally, reactive oxygen species are generated, resulting in oxidative stress within the cells.

In contrast to the parent compounds, the toxicological data on alkyl-PAHs are limited and primarily relate to methylnaphthalenes (Baird et al. 2007). Long-term studies demonstrated a dose-dependent increase in the incidence of pulmonary alveolar proteinosis in experimental animals. Various representatives of this substance class, such as 5-methylchrysene and 7,12-dimethylbenzo[a]anthracenes, displayed mutagenic, tumor-promoting, and carcinogenic properties (da Silva et al. 2021). The WHO International Agency for Research on Cancer has classified 5-methylchrysene as possibly carcinogenic to humans (group 2B), while 1,4-dimethylphenanthrenes, 1-methylchrysene, 2-methylchrysene, 3-methylchrysene, 4-methylchrysene, 6-methylchrysene, 2-methylfluoranthene, 3-methylfluoranthene, and 1-methylphenanthrene are classified as not classifiable regarding their carcinogenicity to humans (group 3) (IARC 2010a, b).

3.2.2 Toxicokinetics

The following brief overview primarily relies on the IPCS report (2003). The more thoroughly studied nitro-PAHs, such as 1-nitropyrene and 2-nitrofluorene, are rapidly absorbed following inhalation and oral exposure, with the resulting metabolites excreted in urine after conjugation with glutathione, sulfate, or glucuronic acid. The metabolism of nitro-PAHs in vivo is complex due to the numerous individual substances and occurs through various metabolic pathways. They can activate distinct pathways, leading to mutagenic compounds. Nitro-PAHs form DNA adducts, including C8-substituted deoxyguanosine adducts. In animal experiments, approximately 50–60% of 1-nitropyrene is eliminated in feces and 15–20% in urine.

Alkyl-PAHs are first biotransformed into alcohols through either ring epoxidation or oxidation of the alkyl group, followed by conjugation with water-soluble groups and the active excretion of the biotransformation products in urine (Peng et al. 2023).

In an experiment, 55 healthy adults were exposed to either clean control air or 300 $\mu g/m^3$ (as PM_{10}) of diesel engine exhaust for one hour. After exposure, significantly higher time-weighted average concentrations (median 138.7 ng/g crea vs. 21.7 ng/g crea) of 1-aminopyrene, the metabolite of 1-nitropyrene, were detected in urine (Laumbach et al. 2009).

3.2.3 Occurrence in Urine

The metabolites and their parent compounds examined in urine are shown in Table 3.7. Additionally, Table 3.8 summarizes the medians and ranges of certain nitro-PAH metabolites found in the urine of the general population. Studies were often conducted to assess exposure to diesel engine emissions (DME) from motorized road traffic as part of human biomonitoring. Most frequently, 1-AP was studied as a marker for DME in urine, with median/geometric mean values ranging from 29 to 160 ng/g crea. Only Ochirpurev et al. (2022) reported significantly lower geometric mean values in the urine of four office workers. The mean concentrations of the metabolites 1-AN and 2-AN were comparable across the studies conducted to date.

Various studies have shown a correlation between exhaust fumes from motorized road traffic and nitro-PAH concentrations in urine (Neophytou et al. 2014; Yang et al. 2021a, b). For example, Yang et al. (2021a, b) found significantly higher levels of all five measured metabolites in UK participants living within 100 m of a major road, although statistical significance was noted only for 2-AF. Additionally, in this study, levels of 2-AF were significantly linked to the traffic flow of nearby buses and heavy-duty vehicles, but not to motorcycles, taxis, or coaches. In the study involving 111 healthy volunteers from Beijing, China, conducted before, during, and after the Olympic Games, there was a 23% decrease in the sum of 1-AN and 2-AN due to measures implemented to enhance outdoor air quality before and during the Games; however, this change did not apply to 1-AP (Gong et al. 2015). Galaviz

Table 3.7 Parent compounds and common metabolites of PAH derivatives

Parent compound	CAS No.	Metabolites	Abbreviation
Nitro-PAHs			
1-nitropyrene	5522-43-0	6-hydroxy-1-nitropyrene	6-OHNP
		8-hydroxy-1-nitropyrene	8-OHNP
		6-hydroxy-*N*-acetyl-1-aminopyrenes	6-OHNAAP
		8-hydroxy-*N*-acetyl-1-aminopyrenes	8-OHNAAP
		1-aminopyrene	1-AP
2-nitrofluorene	607-57-8	2-aminofluorene	2-AF
9-nitrophenanthrene	954-46-1	9-aminophenanthrene	9-APhe
1-nitronaphthalene	86-57-7	1-aminonaphthalene	1-AN
2-nitronaphthalene	581-89-5	2-aminonaphthalene	2-AN
Alkyl-PAHs			
2-methylphenanthrene	832-69-9	2-phenanthrenecarboxylic acid	2-PHECA
2-methylnaphthalene	91-57-6	2-naphthoic acid	2-NAPCA

et al. (2017) reported two- to threefold higher mean urinary concentrations of 8-OHNP and 8-OHNAAP in Mexican residents who commuted on foot across the border to San Diego, USA, compared to San Diego residents, with both parameters linked to personal exposure to 1-NP in the previous 24 h.

Measurement results on the occurrence of metabolites of alkyl-PAH in urine are available only to a very limited extent. For instance, geometric mean values for the sum of the metabolites of 1-methyl naphthalene and 2-methyl naphthalene (11 metabolites in total) ranged from 34 to 453 ng/l for nonsmokers and from 1608 to 6990 ng/l for smokers, as reported in the analysis involving 30 smokers and 30 nonsmokers' urine (Li et al. 2014). Lin et al. (2021) detected metabolites of alkyl-naphthalene (2-NAPCA) and alkylphenanthrene (2-PHECA) in 314 morning urine samples collected repeatedly from 45 Los Angeles residents before, during, and after their ten-week stay in Beijing in the summer of 2014–2017 as part of a longitudinal study. They found that concentrations of 2-NAPCA and 2-PHECA increased from approximately 0.07 and 0.13 μg/g Crea to around 0.13 and 0.40 μg/g Crea after traveling from Los Angeles to Beijing, which is more polluted by particulate matter, and decreased significantly once the participants returned to Los Angeles. In another Chinese study, Phe and 2-methylphenanthrene (2-MePhe) were analyzed in personal PM2.5 samples, while 2-PHECA was examined in paired urine samples from 120 urban residents (Lin et al. 2022). The median 2-PHECA concentration was 0.22 μg/g crea, and after adjusting for seasonal effects, 2-PHECA levels were significantly associated with exposure to 2-MePhe but not Phe.

Table 3.8 Median (range) concentrations of nitro-PAH metabolites in urine samples of the general population

	N	6-OHNP	8-OHNP	6-OHNAAP	8-OHNAAP	1-AP	2-AF	9-APhe	1-AN	2-AN	Dimension	Country
He et al. (2021)	89	–	–	–	–	7.6 (2.5–59.8)	19.3 (0.21–139)	0.35 (0.089–127)	4.1 (0.025–865)	3.6 (0.039–56.6)	µg/l [c]	China
Yang et al. (2021a, b)	98	–	–	–	–	40[a]	350[a]	350[a]	260[a]	450[a]	ng/g crea	UK
Galaviz et al. (2017)	18	–	0.035[a] (0.016–0.13)	–	0.026[a] (0.087–0.10)	–	–	–	–	–	ng/g crea	USA
Gong et al. (2015)	111	–	–	–	–	190[a] (140, 270)[b]	–	–	470[a] (360, 630) [b]		ng/g crea	China
Miller-Schulze et al. (2016)	7	1.8 (<0.77–3.7)	0.58 (<0.24–1.0)	–	–	–	–	–	–	–	ng/g crea	Peru
Neophytou et al. (2014)	82	–	–	–	–	28.7	–	–	160		ng/g crea	USA
Ochirpurev et al. (2022)	4	–	–	–	–	0.18[a] (0.04, 0.79)[b]	–	–	–	–	pg/g crea	Korea
Toriba et al. (2007)	22	0.18 (0.03–0.42)	0.137 (0.03–0.28)	0.054 (<LOD–0.36)	0.070 (0.015–0.30)	–	–	–	-	-	nmol/mol crea	Japan

[a] Geometric mean; [b] geometric mean and 95% CI; [c] adjusted by specific gravity; *LOD* detection limit

3.3 Metals

Metals are commonly found as a natural component of the Earth's crust. In addition to this natural presence, exposure through human activities such as mining, processing, and extensive use in various industrial productions must also be considered. These activities have, in some cases, resulted in exceptionally high levels of environmental contamination (Tchounwou et al. 2012; Barthi and Sharma 2022). For instance, leaded gasoline (as tetraethyl lead) has caused widespread environmental contamination and human exposure globally since the 1920s. Following the ban between the 1980s and 2000s, decreasing concentrations were noted in ambient and indoor air as well as in human blood.

In the following, we will briefly discuss the occurrence and use of essential metals such as arsenic, cadmium, lead, nickel (see Table 3.9), and mercury (ATSDR 2007; EFSA 2010; ATSDR 2012; US DHHS 2012; EFSA 2012; IARC 2012a, b; NTP 2012; ECHA 2017; ECHA 2019; ECHA 2020; US EPA 2022; OSHA 2023):

Arsenic (As) rarely occurs as a pure metal but is often found in sulfur-containing minerals like arsenopyrite. There are three major groups of arsenic compounds: Inorganic arsenic in trivalent forms, such as arsenic trioxide, sodium arsenite, and arsenic trichloride, as well as pentavalent forms like arsenic pentoxide, arsenic acid, and arsenates. Additionally, organic arsenic compounds, including arsanilic acid, methylarsonic acid, and arsenobetaine, can be found in the environment. For centuries, arsenic and its compounds have been used commercially in various applications, including pharmaceuticals, wood preservatives, agricultural chemicals, animal feed, the mining industry, the production of nonferrous alloys, glass manufacturing, and the semiconductor industry. Today, their use is often restricted. Anthropogenic sources are estimated to contribute nearly 24,000 tons of arsenic emitted into the global atmosphere each year, while 7900 tons are attributed to natural sources.

Cadmium (Cd), along with its alloys and compounds, is utilized in various consumer and industrial materials. Cd-hydroxide is employed in electrodes for Ni–Cd batteries, while cadmium sulfide compounds serve as pigments in plastics, ceramics, rubber, artists' colors, and fireworks. Cadmium and its alloys are used in electroplated coatings on metals, and cadmium salts of organic acids were previously applied as heat and light stabilizers for plastics. Due to the semiconducting properties of cadmium sulfide, cadmium telluride is used in solar cells and various electronic devices. Global cadmium production slightly declined to an estimated 24,000 tons in 2018, with China and Korea being the leading primary producers, accounting for about 54%. Approximately 85% of cadmium was consumed in batteries, primarily in rechargeable nickel–cadmium batteries.

Lead (Pb) can be found in the environment in various oxidation states, primarily as + 2 inorganic compounds. Due to its physicochemical properties, such as high density, low melting point, flexibility, and relative inertness to oxidation, it was extensively used in the past in construction, plumbing, batteries, bullets and shots, weights, tin, fusible alloys, white paints, leaded gasoline, and radiation shielding. Pb

Table 3.9 Major sources and a brief overview of health effects (modified from Rehman et al. 2018; IARC 2012b)

Name	Major sources	Major uses	Health effects	IARC evaluation
Arsenic	Smelting, fossil fuel burning, agricultural pesticides, industrial waste	Pharmaceuticals, wood preservatives, agricultural chemicals, and applications in the mining, metallurgical, glass-making, and semiconductor industries	Arsenicosis, psychological effects, decreased mental performance, hypertension, cardiovascular disease risk, carotid atherosclerosis, diabetes mellitus, cancer of the lung, urinary bladder, and skin	Arsenic and inorganic arsenic compounds in Group 1 Dimethylarsinic acid and monomethylarsonic acid in Group 2B Arsenobetaine and other organic arsenic compounds in Group 3
Cadmium	Incineration fumes (from cadmium-containing products), phosphate fertilizers, color pigments, PVC products, cigarette smoking	Nickel–cadmium batteries, pigments, coatings and plating, stabilizers for plastics, nonferrous alloys, semiconductors and photovoltaic devices	Neurodegenerative disorders, end-stage renal disease, breast cancer, prostate cancer, demineralization of bones, diabetes	Cadmium and cadmium compounds in Group 1
Lead	Mining, fossil fuel burning, manufacturing of lead-acid batteries, oxide synthesis for paint and pigments		Neurotoxic effects on intelligence, decreased memory, hemolytic anemia, cardiovascular diseases, reproductive toxicity, lung cancer, bladder cancer	Lead in Group 2B Inorganic lead compounds in Group 2A Organic lead compounds in Group 3
Nickel	Chemical industries, food processing industries, forest fires, volcanic emissions, incineration of waste, combustion of coal	Electroplating, ceramics, pigments, production of ceramics, alloy steel and stainless steel, nickel–cadmium batteries in petrochemical metallurgical industry	Allergic contact dermatitis, oral hypersensitivity and risk of gingival hyperplasia, oral cancer, skin cancer, lung cancer, asthma, bronchitis, reproductive toxicity, carcinogenesis	Nickel compounds in Group 1

[a] Overall evaluation: Group 1: carcinogenic to humans; Group 2A: probably carcinogenic to humans; Group 2B: possibly carcinogenic to humans; Group 3: not classifiable as to their carcinogenicity to humans; Group 4: probably not carcinogenic to humans

is mainly utilized for batteries and lead alloys in ammunition, pipes, cable covering, building materials, solder, radiation shielding, and fishing weights. Additionally, lead is used in ceramic glazes and as a stabilizer in plastics. Organic lead compounds like tetramethyl lead and tetraethyl lead have historically been used as antiknock additives in gasoline but have been banned in most countries for decades. In 2018, lead production in the U.S. was estimated at 1.3 million tons, while in 2022, approximately 4.5 million tons of lead were extracted from global mining.

Nickel (Ni) salts are utilized in electroplating, ceramics, pigments, and as intermediates (e.g., catalysts, formation). Nickel oxide is employed in nickel catalysts within the ceramics industry, in manufacturing alloy steel and stainless steel, in producing nickel salts for specialty ceramics, and in creating nickel–cadmium and nickel–metal–hydride batteries. In contrast, nickel sulfide serves as a catalyst in the petrochemical industry or as an intermediate in the metallurgical industry. The total global usage of primary nickel in 2006 was 1.40 million tons. Estimates for natural nickel emissions into the atmosphere were 30,000 tons per year in the early 1990s.

Mercury (Hg) in the environment comes from both anthropogenic and natural sources in three distinct forms: Elemental (Hg^0), inorganic compounds (Hg^{2++} and Hg^{2+}), and organic compounds such as alkylmercury compounds. Exposure can occur during the mining, production, and transport of mercury, as well as during the mining and refining of gold and silver ores. Mercury is often found in thermometers, manometers, barometers, gauges, valves, switches, and batteries, and was previously used as an amalgam in dentistry, preservatives, heat transfer technologies, pigments, catalysts, and lubricating oils. Most exposure to methylmercury occurs through consumption of contaminated fish and shellfish. A global inventory of mercury emissions to air from anthropogenic sources in 2015 estimates total global emissions at about 2220 tons. The majority of these emissions originate in Asia (49%), followed by South America (18%), Sub-Saharan Africa (16%), Europe (9.1%), and North America (1.8%) (UNEP 2019). Emissions from artisanal and small-scale gold mining account for nearly 38% of the total global amount. Figure 3.3 illustrates the share of different sectors relative to the total emissions of that region.

3.3.1 Health Aspects

Heavy metals, particularly lead, cadmium, copper, chromium, zinc, and the semimetal arsenic, hold significant health implications. Heavy metals are typically defined as metals with a specific density greater than 5 g/cm^3 (Jaishankar et al. 2014). They impact the environment and various biochemical and physiological functions of living organisms in a dose-dependent manner, with severe health effects observed for many of these metals at higher concentrations. Numerous reviews on the subject are available from scientific organizations, as well as in more recent literature (e.g., Jaishankar et al. 2014; Nordberg et al. 2015; Pagano et al. 2019; Chen et al. 2019; CLS 2019; Kumar et al. 2020; Barthi and Sharma 2022).

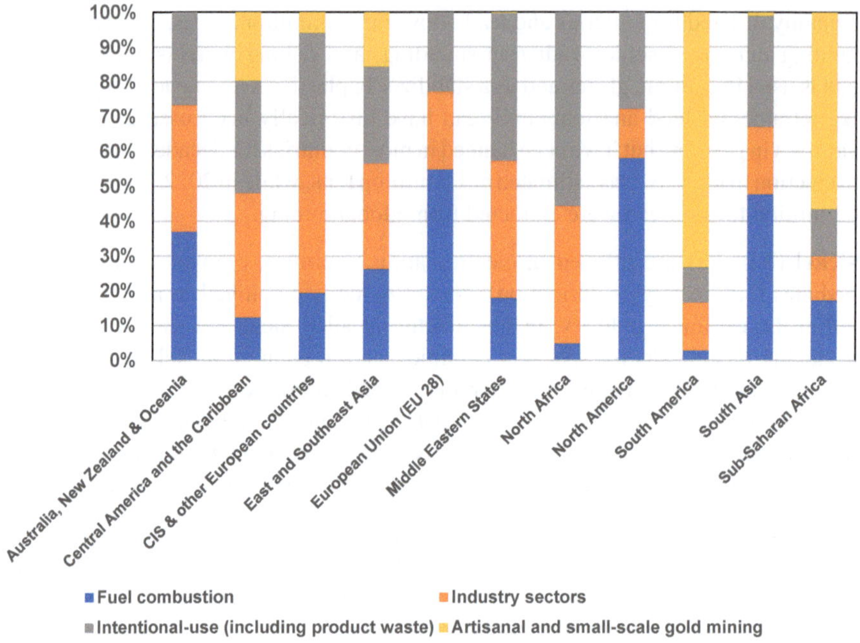

Fig. 3.3 Share of different sectors of the total mercury emission in WHO regions (modified from UNEP 2019)

In the following sections, we will briefly discuss the health effects of essential metals such as arsenic, cadmium, lead, nickel, and mercury (see Table 3.9), along with additional references mentioned in the second paragraph of this chapter.

Chronic high exposure to arsenic can lead to arsenical keratinosis, dark patchy skin hyperpigmentation, acne, cardiovascular disease, hypertension, and peripheral neuropathy. Moreover, sufficient evidence exists in humans regarding the carcinogenicity of mixed exposure to inorganic arsenic compounds. These compounds can cause cancers of the lung, urinary bladder, and skin. Positive associations have been observed between exposure to arsenic and inorganic arsenic compounds and cancers of the kidney, liver, and prostate.

High exposure to cadmium can lead to diseases affecting the kidneys, bones, and lungs. The primary critical effects include an increased excretion of low-molecular-weight proteins in the urine due to damage to the proximal tubular cells in the kidneys, as well as a heightened risk of osteoporosis. Additionally, a greater risk of lung cancer has been reported from inhalation exposure, and a positive correlation has been found between exposure to cadmium and cadmium compounds and cancer in the kidneys and prostate.

Lead accumulates in the bones and can be released during periods such as pregnancy and lactation. Chronic exposure is linked to impaired kidney function, cardiovascular issues (e.g., hypertension), effects on the nervous system and behavior, cognitive dysfunction later in life, reduced fetal growth, and decreased glomerular filtration rate. The most critical effect is exposure in utero and during early childhood, which impairs neurodevelopment in children. In cases of very high exposure, anemia, peripheral neuropathy, interstitial kidney fibrosis, and severe abdominal cramping have been observed. ECHA and IARC consider that the epidemiological data are inconsistent concerning the increased risk of cancer from inorganic lead compounds.

Nickel is the most common contact allergen. After a prior sensitization phase, an allergy can develop into allergic contact dermatitis. Chronic inhalation exposure to nickel in humans also leads to respiratory effects, including a typical form of asthma, decreased lung function, and bronchitis. At high doses, animal studies provide evidence of the reproductive toxicity of nickel salts taken orally or by inhalation. There is sufficient evidence in humans that nickel compounds and nickel metal can cause cancers of the lung, nasal cavity, and paranasal sinuses. Based on animal data, genotoxicity is likely due to indirect effects such as inhibition of DNA repair and the production of reactive oxygen species.

Mercury exposure at high levels can harm the brain, heart, kidneys, lungs, and immune systems of individuals of all ages. The primary target of mercury after long-term exposure is the nervous system, leading to emotional changes (e.g., nervousness, irritability, and mood swings), neuromuscular changes (e.g., weakness and muscle atrophy), and headaches. Mercury became infamous due to extensive contamination in Japan (Minamata) caused by a chemical factory that discharged methylmercury, a byproduct of acetaldehyde production, into the sea for decades. Consequently, adults developed various neurotoxic effects, including blurred vision, hearing impairment, olfactory and gustatory disorders, ataxic gait, hand clumsiness, speech disorders, as well as somatosensory and psychiatric disorders. Severe mental and motor development disorders have been observed in affected fetuses.

3.3.2 *Toxicokinetic*

The toxicokinetic characteristics are summarized below, primarily based on compilations from various scientific institutions (see also the references in the second paragraph of this chapter).

After inhaling inorganic As, roughly 30% is absorbed, with several studies in humans showing that arsenates and arsenites are effectively absorbed across the gastrointestinal tract. They are widely distributed in almost all tissues and accumulate in keratin-rich tissues such as skin, hair, and nails. Inorganic As^{V} is rapidly converted to AsIII in the body, followed by methylation steps to form various metabolites (see Fig. 3.4). After metabolization, excretion occurs mainly through urine as dimethylarsinic acid (DMA^{V}; 60–80% of total metabolites in humans) and monomethylarsonic acid (MMA^{V}), along with monomethylarsonic acid (MMA^{III}) and dimethylarsinous acid (DMA^{III}). Arsenic can cross the placenta and reach the fetus through breast milk.

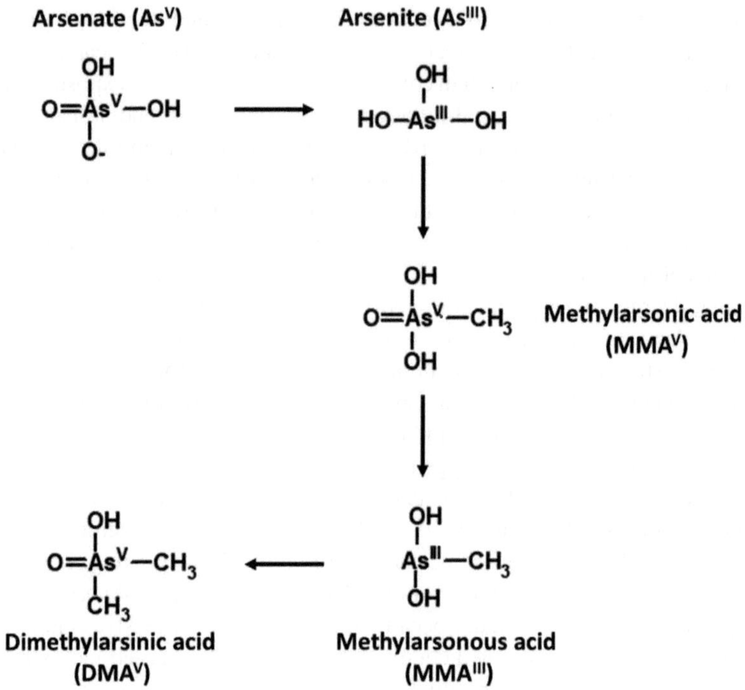

Fig. 3.4 Inorganic arsenic biotransformation pathway (according to ATSDR 2007)

In humans, approximately 2–50% of inhaled cadmium is absorbed, depending on factors such as solubility, particle size, respiratory rate, and deposition in the respiratory tract. In contrast, the uptake from the gastrointestinal tract is only 5%, but it is strongly dependent on dietary composition. Free cadmium or a cadmium-metallothionein complex is stored in the kidneys, and excretion is prolonged, having a biological half-life of 10–30 years in this organ. Cadmium can cross the placenta, but to a lesser extent.

The uptake of inorganic lead from the gastrointestinal tract in adults and children is assumed to be about 10 and 50%, respectively, for dietary intake and 6 and 30% for ingesting soil and dust particles. After inhalation, lead is absorbed very efficiently, while absorption through the skin is limited. Particular attention should be given to hand-mouth contact, which allows small children, for example, to absorb larger amounts of soil and dust particles. Following uptake into the body, lead is stored in soft tissues and bones and can be continually released from them. Studies indicate that reduced bone lead concentration correlates with a half-life of 6–37 years. Excretion primarily occurs through urine. There is fetal exposure via the placenta, and the blood–brain barrier is permeable to lead ions.

After inhalation exposure, approximately 20–35% of Ni is absorbed into the organism. In contrast, between 1 and 30% is absorbed in humans after oral intake, depending on the solubility of the administered nickel compound, the dosing vehicle, and the fasting state of the subject. The Ni ion binds to specific proteins in the blood and is rapidly distributed throughout the body. Excretion occurs predominantly through urine, with an elimination half-life of about 28 h.

The lungs absorb elemental Hg effectively after inhalation, but it is poorly absorbed from the gastrointestinal tract following oral intake. In contrast, inorganic Hg compounds are absorbed after ingestion at rates of approximately 2–38%. Methylmercury species are extensively and rapidly absorbed in humans after ingestion, with absorption rates of about 80%. Once absorbed, elemental Hg is converted to Hg(II) compounds in erythrocytes and various organs. Inorganic Hg binds to blood proteins and is distributed throughout most tissues, with the highest concentrations found in the kidneys. Blood concentrations initially decrease during a relatively rapid clearance phase, characterized by a half-life of around 1–3 days, followed by a slower phase with a half-life of 1–3 weeks. After forming water-soluble complexes with cysteine or glutathione, organic mercury can cross the blood–brain and placental barriers, accumulating in the blood, brain, and kidneys of the fetus. Methylmercury is excreted in breast milk.

Their toxicokinetic data suggest different biomarkers for determining exposure to the other elements. Arsenic accumulates in keratin-rich tissues such as hair, fingernails, or toenails and can be partly used as an exposure indicator. In contrast, arsenic is measured primarily in blood or urine as arsenic or metabolites. Concerning Cd, the concentration in blood tends to reflect more recent exposure, whereas urinary cadmium levels correlate well with long-term cadmium exposure. For lead, the concentration in whole blood lends itself as a marker. Serum and urine concentrations of Ni are valuable biomarkers for more recent exposure because they reflect the amount of nickel ingested in the last 24–48 h. Urinary Hg described the long-term exposure situation well, while Hg blood concentrations represent more recent exposure, mainly from the diet.

3.3.3 Occurrence in the Blood and Urine

Metals and elements are associated with different blood components, so the proper sampling method (whole blood, plasma, or serum) is crucial for biomonitoring studies. In a recent survey of 102 samples, 73 elements were analyzed in blood plasma, serum, and erythrocytes (Heitland and Köster 2021). The distribution between the two media indicated that for Pb, Cd, and Mn, 97–99% were found in the erythrocytes, while only 83% of Hg, 76% of As, 73% of Ni, and 54% of Co were in the cellular blood components.

Numerous studies in the scientific literature describe the internal load. Tables 3.10 and 3.11 summarize the medians of element concentrations in urine and blood from the general population in some recent and comprehensive studies. Significant concentration differences for Pb and Hg are evident between countries and regions, as well as within countries. Extensive and representative studies indicated 95th percentiles for Pb, Cd, and total Hg in blood at 24.1, 1.28, and 3.87 μg/l in the USA; 23.0, 1.7, and 3.8 µg/l in Canada; 38.8, 3.78, and 3.64 µg/l in China (90th percentile here); and 11.5, 1.55, and 9.26 µg/l in Japan (CDC 2023; HC 2021; Cao et al. 2021; Nakayama et al. 2019b).

Higher concentrations of various toxic elements have been consistently detected near specific industrial sources and mining activities (Barbieri et al. 2016; Hassan et al. 2023; Kwon et al. 2023). For example, a study involving 198 mother-infant pairs in an area with artisanal gold mining revealed elevated levels, with total Hg and Pb concentrations in cord blood and maternal blood measuring 6.0, 3.6, 11 and 14 µg/l, respectively (Berky et al. 2023).

Other important predictors include consumption of sea fish and rice for As, sea fish, shellfish, and canned food for Hg (Nakayama et al. 2019b; Ashrap et al. 2020; Vogel et al. 2021; Bartel-Steinbach et al. 2022; Hruba et al. 2023), and tobacco and marijuana smoking for Cd and Pb (Barbieri et al. 2016; Nakayama et al. 2019a; Ashrap et al. 2020; Vogel et al. 2021; Zhong et al. 2022; McGraw et al. 2023; Piasek et al. 2023; Hruba et al. 2023). Since inorganic Hg is transformed in the aquatic environment into highly toxic organic Hg compounds, such as methylmercury (MeHg), populations with high sea fish consumption are particularly exposed to MeHg. For example, median MeHg levels were found in maternal blood (5.15 µg/l) and cord blood (9.45 µg/l) in Japan (Iwai-Shimada et al. 2019). These levels are significantly higher than the median MeHg levels reported in the US, which are 0.50 µg/l for individuals older than 19 years and < 0.26 µg/l for 6–11 year olds (CDC 2023), or in Canada, where levels are 0.57 µg/l for those aged 20–79 years (HC 2021).

Indoors, lead-based paints can be a significant source of exposure because they mix with household dust through wear and aging and are easily absorbed, especially by young children, through hand-to-mouth contact. For instance, in the US in 2019, approximately 4.3 million children were estimated to live in homes with lead paint (Jacobs and Brown 2023). When using lead ammunition, recreational and sport shooters, as well as those occupationally exposed, such as police officers and military personnel, can be expected to experience significantly increased body burden (review in Laidlow et al. 2017). This risk is particularly high when activities take place in indoor firing ranges without adequate ventilation. For example, Demmler et al. (2009) studied 129 shooters and found median blood lead levels of 30 and 107 µg/l, with maximum values reaching 521 µg/l for individuals using airguns and large-caliber handguns, respectively. Additionally, lead from game meat shot with lead ammunition is a considerable source of lead exposure among groups that frequently consume game meat, including hunters and their families, as well as individuals with high consumption of this meat (Green and Pain 2012; Fachehoun et al. 2015; Wepler et al. 2022).

Additionally, elements transferred from maternal blood to the fetus warrant particular attention. For instance, a statistically significant correlation between umbilical cord and maternal blood was noted in a Belgian study involving 255 newborns and their mothers conducted from 2002 to 2006 (Baeyens et al. 2014). The authors also compiled the ratios between cord blood and maternal blood (C/M ratio) published in the scientific literature. They reported a C/M ratio ranging from 0.13 to 0.22 for Cd,

Table 3.10 Median element concentrations in urine in µg/l

References	N	Age (years)	Cd	Cr	Hg	Mo	Mn	Pb	Ni	Se	Sb	U	Country, year
Adults, general population													
HC (2021)	~ 2500	3–79	0.21	0.12	–	–	–	–	–	–	–	–	Canada, 2018/19
Zhong et al. (2022)	480		1.05	< 0.20	< 0.50	–	0.28	0.96	2.25	35.4	< 0.10	–	China, 2018
CDC 2023	~ 5000	≥ 20	0.18	< 0.19[b]	< 0.13	22.2	< 0.13	0.29	1.20[b]	191	0.042	0.004	USA, 2017/18
Cao et al (2021)	5859	3–79	0.34	0.38	0.22	–	–	0.89	–	–	–	–	China, 2017/18
Signes Pastor et al. (2023)	316	31[a]	0.06	–	–	42.4	–	0.20	0.86	29.2	0.02	–	USA
Liao et al. (2023)	1871	7–97	0.63	0.21	–	–	0.62	0.81	2.23	45.0	–	–	Taiwan, 2013–16
Goodrich et al. (2019)	56	26–40	< 0.06	0.71	0.06	48.0	1.54	0.21	3.98	45.3	–	< 0.01	USA, 2012–15
Ashrap et al. 2020	1040	18–40	0.12	–	0.59	62.9	1.2	0.27	5.50	–	0.08	–	Puerto Rico, 2011–17
Snoj Tratnik et al. (2019)	~ 1000	18–49	0.20	–	0.45	–	–	0.49	–	14.3	–	–	Slovenia, 2008–14
Dereumeaux et al. (2016)	990	18–47	0.12	0.33	–	–	–	–	1.50	–	0.05	< 0.01	France, 2011
Arrebola et al. (2019)	882	37[a]	0.27	–	–	–		–	–	–	–	–	Spain, 2009/10
Nisse et al. (2017)	1992	20–59	0.42	0.56	1.08	–	0.38	1.11	2.18	–	0.09		France, 2008–10

(continued)

Table 3.10 (continued)

References	N	Age (years)	Cd	Cr	Hg	Mo	Mn	Pb	Ni	Se	Sb	U	Country, year
Children													
CDC 2023	~ 800	6–11	< 0.04	0.29	< 0.13	70.4	< 0.13	0.26	–	175	0.065	0.005	USA, 2017/18
Schoeters et al. (2022)	428	14–15	0.29	–	–	–	–	–	–	–	–	–	Belgium, 2017/18
Vogel et al. (2021), Murawski et al. (2023)	~ 2200	3–17	0.08	0.43	0.07	–	–	–	–	–	0.06	–	Germany, 2015–17
Bennett et al. (2022)	237	2–5	0.02	–	–	91.9	–	–	–	–	–	0.009	USA, 2006–17

[a] Mean; [b] total population $\geq$ 6 years

Table 3.11 Median element concentrations in blood in µg/l

References	N	Age (years)	As	Cd	Cr	Cu	Hg	Mn	Pb	Se	Ni	Zn	Country, year
Adults, general population													
Hassan et al. (2023)	911	18–71	0.55	0.22	1.05	–	–	–	7.05	–	–		Israel, 2020/22
Heitland and Köster (2021)	102	19–66	0.47	0.36	0.04	934	0.41	8.5	11	106	0.38	5620	Germany, 2020
HC (2021)	4596	3–79	–	0.21	–	–	0.77	–	7.80	160	–	–	Canada, 2018/19
Cao et al. (2021)	5884	3–79	0.95	0.43	0.41	–	1.1	–	19.3	–	–	–	China, 2017/18
CDC 2023	~ 5700	≥ 20	–	0.27	< 0.41[b]	–	0.73	9.28	8.50	–	–	–	USA. 2017/18
Lv et al. (2021)	2501	65–99	–	–	0.70	–	1.43	10.7	35.1	102	< LOQ	–	China, 2017/18
Pineda et al. (2023)	1082	10–21	–	0.12	0.57	–	0.66	10.6	7.32	94.8	0.62	–	Sweden, 2016/17
Wu et al. (2023)	1939	49[c]	4.43	2.12	4.97	932	–	14.1	37.1	166	2.23	5535	China, 2015/16
Liu et al. (2022)	703	29[c]	–	0.06	1.16	1708	–	2.02	7.8	66.2	4.58	–	China, 2015/16
Ashrap et al. 2020	1183	18–40	0.21	0.12	–	1562	1.20	11.3	3.3	–	1.0	4752	Puerto Rico, 2011–17
Nakayama et al. (2019b)	17,997	31[c]	–	0.70	–	–	3.83	16.1	6.3	178	–	–	Japan, 2011–14
Snoj Tratnik et al. (2019)	~ 1000	18–49	0.77	0.29	–	940	1.20	13.3	17.5	104	–	6636	Slovenia, 2008–14

(continued)

Table 3.11 (continued)

References	N	Age (years)	As	Cd	Cr	Cu	Hg	Mn	Pb	Se	Ni	Zn	Country, year
Arrebola et al. (2019)	~ 900	37[a]	–	0.27	–	–	–	–	1.06	–	–	–	Spain, 2009/10
Nisse et al. (2017)	1992	20–59	2.03	0.37	0.57	0.29	1–65	7.66	18.4	–	1.28	5844	France, 2008–10
Iwai-Shimada et al. (2019)	649	31[c]	4.06	1.18	–	1289	5.42		10.8	176		4769	Japan, 2001–06
Children, adolescents													
CDC 2023	~ 1000	6–11	–	0.10	–	–	< 0.28	10.4	4.60	–	–	–	USA. 2017/18
Schoeters et al. (2022)	428	14–15	–	0.18	–	792	–	9.3	7.7	–	–	5350	Belgium, 2017/18
Vogel et al. (2021)	~ 2200	3–17	–	< 0.06	–	–	–	–	9.4	–	–	–	Germany, 2015–17
Cord blood													
Grzesik-Gąsior et al. (2023)	134	Cord blood	–	–	8.45	713	–	31.4	25.6	156	–	1822	Poland, 2020/21
Berky et al. (2023)	198	Cord blood	0.61	0.24	–	–	6.0	–	11.0	162	–	–	Peru, 2017/18
Zhang et al. (2023)	100	Cord blood	2.39	0.09	–	577	2.35	51.7	14.5	115	–	1.94	China, 2013–16
Baeyens et al. (2014)	241	Cord blood	0.54	0.07	–	600	–	31.2	8.60		–	–	Belgium, 2007–11
Iwai-Shimada et al. (2019)	594	Cord blood	3.68	0.53	–	511	9.96	–	9.89	191	–	2003	Japan, 2001–06

[a] Median; [b] total population ≥ 6 years; [c] mean; *LOQ* limit of quantification

0.54–0.98 for Pb, and 1.8–2.6 for Mn. Slightly higher ratios for Cd were observed in a Spanish study of 53 mother-infant pairs, while the Pb and Mn values remained comparable (Bocca et al. 2019). Median ratios were also reported in a large Japanese study, where the C/M for total Hg and MeHg was approximately 1.8 (Iwai-Shimada et al. 2019). Similarly, a Korean study of 347 cord blood samples yielded a ratio of 0.7 for Pb (Park et al. 2021). A recent Chinese study of 100 mother-infant pairs found a median C/M ratio of 0.2 for Cd, 1.3 for Pb, and 1.7 for total Hg (Zhang et al. 2023).

Several studies have investigated the time trend of internal exposure to toxic elements. Figure 3.5 presents the results of a German study showing blood Pb levels from 1981 to 2021 alongside a nationwide survey from the USA conducted between 1976 and 2018. These results highlight a significant decrease in blood Pb levels within the general population. Additionally, a multicenter birth cohort study in Korea measured various toxic elements in approximately 1500 cord blood samples and followed up with children at ages 2, 3, and 5 years (Jeong et al. 2017). The Pb concentration (geometric mean) was lowest at 9.2 μg/l in cord blood, rising to 15.3 μg/l after two years, and then decreasing steadily to 12.8 μg/l by five years of age. In contrast, the total Hg concentration was highest in umbilical cord blood at 5.10 μg/l and gradually decreased to 1.74 μg/l by the age of five. In the mother's blood, the geometric mean values for Pb and Hg at the end of pregnancy were 12.8 and 3.05 μg/l, respectively. Bartel-Steinbach et al. (2022) also analyzed total Hg in the urine of young adults aged 20–29 years in Germany from 1995 to 2018, finding an 86% reduction in daily total Hg excretion from 0.76 μg/l in 1995 to just 0.11 μg/l in 2018. The blood samples studied also demonstrated a decline from 1.76 μg/l in 2001 to 0.7 μg/l in 2020.

Table 3.12 compiles recent results on arsenic, arsenic species, and metabolites. Notably, the total arsenic and DMA values in Korea and Japan are significantly higher than in other regions. This is explained by the high consumption of sea and shellfish. Additionally, it must be considered that DMA is not only a metabolite of inorganic arsenic but can also form from arsenosugars and arsenolipids, which are also found in marine organisms (Luvonga et al. 2020).

3.3.4 Occurrence in Breast Milk

In the current scientific literature, several recent reviews on the presence of metals and elements in breast milk have been published (Rebelo and Caldas 2016; Cherkani-Hassani et al. 2017; Martín-Carrasco et al. 2023). Table 3.13 summarizes the recent findings on trace and toxic element concentrations in breast milk. Significant variation in concentrations can be expected among individuals (e.g., Björklund et al. 2012; Olujimi et al. 2023). The mean and median values differ widely across the respective studies, regions, and countries. For instance, the mean values for lead range from

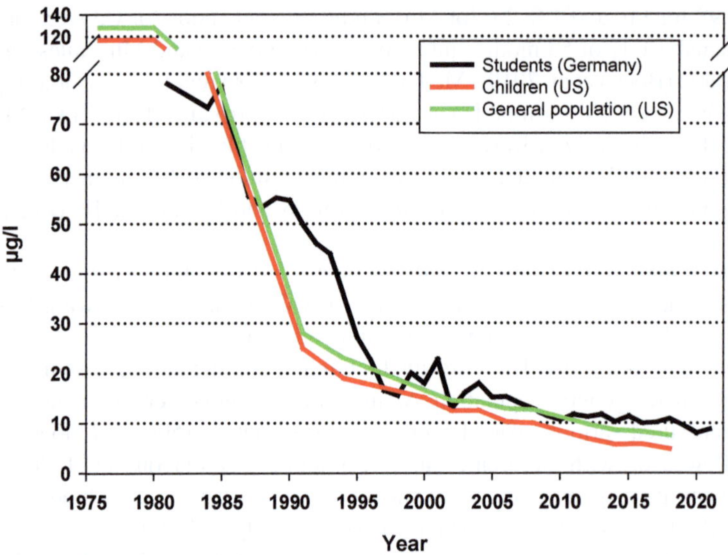

Fig. 3.5 Time trend of geometric mean levels of lead in blood (according to Lermen et al. 2021; Brody et al. 1994; Pirkle et al. 1994, 1998; CDC 2023) (In the U.S.: children 6–19 years in 1975–1983 and 6–11 years from 1988, general population > one year in 1976–1994 and > six years from 1999; Germany: students aged 20–29 years.)

0.2 to 15.6 µg/l in Europe, 1.5–41.9 µg/l in Asia, and 2.2–1185 µg/l in Africa. The average levels in Europe, Asia, and the Americas are comparable for arsenic and cadmium, while those in Africa are elevated.

The highest mean levels of arsenic and lead are reported to be 402 µg/l (170–1058 µg/l) and 1185 µg/l (366–5330 µg/l) in breast milk from a polluted area in Egypt where leaded gasoline is used (Mansour et al. 2015). The highest levels of cadmium were observed in Nigeria at 46 µg/l (3.3–198 µg/l) (Olujimi et al. 2023). In contrast, for other elements such as selenium or aluminum, for which fewer results have been published to date, the mean values in more recent studies range from 13 to 45 µg/l (Björklund et al. 2012; Grzunov Letinic et al. 2016; Alves Peixoto et al. 2019; Motas et al. 2021) and 7.3–57 µg/l (Chao et al. 2014; Mandiá et al. 2021; Motas et al. 2021; Freire et al. 2023), respectively.

Several authors have reported decreasing lead concentrations during breastfeeding (Ettinger et al. 2006; Chao et al. 2014; Samiee et al. 2019). For instance, in 100 samples collected in Iran, a decrease in mean Pb levels was observed, from 97 µg/l after two months of breastfeeding to 80 µg/l after four months, 43 µg/l after eight months, and then a slight increase to 50 µg/l (Samiee et al. 2019). Similarly, in 45 samples from Taiwan, lead, cadmium, and arsenic levels decreased from 13.2, 1.4, and 1.5 µg/l in colostrum to 2.9, 0.34, and 0.16 µg/l in mature milk around days 60–65 postpartum (Chao et al. 2014).

Table 3.12 Median arsenic concentrations in urine in µg/l

References	N	Age (years)	TotalAs	As III	As V	Arsenobetain	DMA	MMA	TRA	Country, year
HC (2021)	2531	3–79	–	0.42	< 0.14	1.2[a]	3.5	0.41	4.6	Canada, 2018/19
SLO CRP[b]	97	12–15	7.9	0.32	–	–	2.48	0.78	3.84	Slovenia, 2018
Schoeters et al. (2022)	428	14–15	–	0.32	< 0.1	2.17	8.07	0.60	4.23	Belgium, 2017/18
Choi et al. (2022)	2025	1–98	62.3	0.84	0.48	27.3	23.4	1.53	–	Korea, 2017/18
CDC 2023	333	6–11	5.09	< 0.12	< 0.79	< 1.16	2.91	< 0.2	–	USA, 2017/18
CDC 2023	1708	20 +	6.02	< 0.12	< 0.79	< 1.16	2.96	< 0.2	–	USA, 2017/18
Riksmaten[b]	300	10–21	27.8	0.31	0.24	9.59	3.79	1.48	6.20	Sweden, 2016/17
Takayama et al. (2021)	109	18–66	85.2	1.5	0.4	21.5	24.1	1.8	–	Japan, 2016
Murawski et al. (2023)	2250	3–17	8.0	–	–	0.7	–	–	4.1	Germany, 2015/17
Goodrich et al. (2019)	56	26–40	3.80	–	–	–	–	–	–	USA, 2012–15
Bennett et al. (2022)	237	2–5	10.8	–	–	–	–	–	–	USA, 2006–17
Yusa et al. (2018)	109	6–11	31.4	–	–	17.9	8.66	< 2.5	9.45	Spain, 2010
Signes Pastor et al. (2017)	400	4	–	0.35	9.71	3.97	0.44	–	Spain, 2007–10	
Kordas et al. (2016)	328	5–8	–	–	–	–	1.90	0.95	9.9	Uruguay
BEA[b]	300	14–16	32.1	0.45	0.35	10.7	5.70	1.31	8.47	Spain, 2009/10

[a] Sum of arsenobetaine and arsenocholine; [b] cited from Buekers et al. (2023) (values are geometric means); *TRA* sum of As III (arsenite), As V (arsenate), *MMA* monomethylarsonic acid, *DMA* dimethylarsinous acid

Table 3.13 Mean element concentrations in breast milk in μg/l

References	N	As	Cd	Cr	Cu	Hg	Mn	Ni	Pb	Zn	Country, year
Europe											
Nazlıcan et al. (2022)	34	1.6	0.37	8.3	–	2.6	–	–	12.1	–	Turkey, 2020
Çebi and Şengül 2022	50	< 0.7	< 0.09	3.4	59	–	9.3	2.0	2.3	380	Turkey, 2019
Mandiá et al. (2021)	70	1.37	0.15	3.5	250	0.31	1.7	2.4	0.3	1238	Spain, 2018/19
Szukalska et al. (2021)	52	1.4	0.06	0.62	457	–	7,2	–	0.86	3019	Poland
Ecsedi-Angyal et al. (2020)	27	0.41	0.19	–	–	< 0.4	–	–	1.7	–	Hungary, 2017
Kunter et al. (2017)	50	0.73	0.45	–	–	< LOQ	–	–	1.2	–	Cyprus, 2017
Motas et al. (2021)	50	0.9	0.40	16.1	367	5.6	10.7	25.3	5.2	1402	Spain
Freire et al. (2022)	83	1.7[b]	< 0.04	–	–	0.3[b]	–	–	0.2 [b]	–	Spain, 2015–18
Kılıç Altun et al. (2018)	42	< 1.0	–	–	540	–	–	–	< 1.0	2890	Turkey
Grzunov Letinic et al. (2016)	107	–	0.13	–	450	0.25	–	–	0.40	2600	Croatia, 2012/13
Klein et al. (2017)	23	3.9	–	–	187	–	1.6	–	1.0	750	Poland, 2012
Snoj Tratnik et al. (2019)	~ 400	0.16	< 0.2	–	360	0.16	–	–	0.21	2063	Slovenia, 2008–14
Winiarska-Mieczan 2014	320	–	2.1	–	137	–	–	–	6.3	1623	Poland, 2010
Dursun et al. (2016)	107	–	< 0.02	–	–	1.4[b]	–	–	14.6	–	Turkey, 2006/07
Vollset et al. (2019)	300	–	0.08	–	–	0.20	–	–	< 0.67	–	Norway, 2002–09
Asia											
Rahimi et al. (2025)	64	1.15	0.15	0.66	526	–	11.0	1.03	3.58	1436	Iran, 2024
Sharafi et al. (2023)	100	2.0	0.72	41.1	–	0.31	–	19.1	11.5	–	Iran, (2021)
Ahmad et al. (2023)	35	–	< LOQ	–	–	10.4	–	–	1.9	–	Pakistan, 2020/21

(continued)

Table 3.13 (continued)

References	N	As	Cd	Cr	Cu	Hg	Mn	Ni	Pb	Zn	Country, year
Zhang et al. (2024)	47	0.78	0.04	1.0	189	0.01	2.1	0.93	0.47	1432	China, 2019
Wang et al. (2025)	100[e]	0.35	0.09	<	–	< 0.1	–	–	< 0.7	–	China, 2016–19
Lin et al. (2023)	228[d]	–	–	–	–	0.76	–	–	6.5	–	Taiwan
Samiee et al. (2019)	100	0.50[b]	–	4.0[b]	–	–	–	–	41.9 [b]	–	Iran, 2016/17
Bassil et al. (2018)	74	2.4	0.9	–	–	–	–	–	18.2	–	Lebanon, 2015/16
Amiri et al. (2023)	100	–	–	7.6	183	2.3	38	45	20.0	–	Iran, PGA, 2015
Amiri et al. (2023)	100	–	–	1.4	170	0.8	4.4	7.4	4.6	–	Iran, urban, 2015
Al-Zurfi et al. (2016)	77	–	3.5	–	119	–	–	–	13.6	–	Iraq, 2014
Africa											
Philip-Slaboh et al. (2023)	72	0.65	3.24	–	–	3.08	–	–	12.2	–	Nigeria, 2020/21
Olujimi et al. (2023)	37	229	46	89	173	8.4	188	11	237	434	Nigeria
Ekeanyanwu et al. 2020	225	< LOQ	29	19	35	< LOQ	–	–	38	9	Nigeria, 2019
Klein et al. (2017)	6	6.7	–	–	131	–	11.6	–	2.2	1340	Namibia, 2013
Bansa et al. (2017)	114	27[a]	1.2[a]	–	–	7.6[a]	–	–	13.8 [a]	–	Ghana
Mansour et al. (2015)	32	176–402[c]	9–25 [c]	–	179–524[c]	–	64–381[c]	63–252[c]	322–1185[c]	622–2897[c]	Egypt
Americas											
Alves Peixoto et al. (2019)	156	–	–	–	618	–	4.5	–	–	2614	Brazil
Klein et al. (2017)	20	3.5	–	–	170	–	2.7	–	0.8	670	USA, 2013
Klein et al. (2017)	21	4.5	–	–	211	–	7.6	–	0.6	930	Argentina, 2012/13

[a] Geometric mean; [b] median; [c] range of means of three cites; [d] from 39 mothers; [e] pool from 3467 individual samples; *PGA* petrochemical and gas area, *LOQ* limit of quantification

Various studies have sought to establish a relationship between internal exposure and the influence of specific sources. For instance, environmental exposures, such as those from the petrochemical industry, mining, and regions with artisanal and small-scale gold mining, appear to be associated with higher concentrations of toxic elements, particularly in Africa and South America (Bansa et al. 2017; Molina et al. 2018; Bose-O'Reilly et al. 2020; Amiri et al. 2023). Furthermore, fish consumption seems linked to As levels and, to some extent, Hg levels (Björklund et al. 2012; Grzunov Letinic et al. 2016; Bassil et al. 2018; Vollset et al. 2019; Freire et al. 2022), while smoking behavior is associated with Cd levels (Chao et al. 2014; Grzunov Letinic et al. 2016; Bassil et al. 2018; Szukalska et al. 2021).

References

Abbas I, Badran G, Verdin A, Ledoux F, Roumié M, Courcot D, Garçon G (2018) Polycyclic aromatic hydrocarbon derivatives in airborne particulate matter: sources, analysis and toxicity. Environ Chem Lett 16:439–475

Abdel-Shafy HI, Mansour MSM (2016) A review on polycyclic aromatic hydrocarbons: source, environmental impact, effect on human health and remediation. Egypt J Petroleum 25:107–123

Acharya N, Gautam B, Subbiah S, Rogge MM, Anderson TA, Gao W (2019) Polycyclic aromatic hydrocarbons in breast milk of obese vs normal women: infant exposure and risk assessment. Sci Total Environ 668:658–667

Achten C, Andersson JT (2015) Overview of polycyclic aromatic compounds (PAC). Polycyclic Arom Compd 35:177–186

Ahmad N, Muhammad A, Zafar R, Afzal U, Aslam M (2023) Assessment of lead, cadmium, and mercury levels in the breast milk in Pakistani women. Environ Sci Pollut Res Int 30:85903–85909

Alves Peixoto RR, Bianchi Codo CR, Lacerda Sanches V, Guiraldelo TC, Ferreira da Silva F, Ribessi RL, Martins Marba ST, Cadore S (2019) Trace mineral composition of human breast milk from Brazilian mothers. J Trace Elem Med Biol 54:199–205

Al-Zurfi SKL, Ali BM, Abojassim AB, Albanon RA, Qazmooz HA (2016) Determination of some heavy metals in the blood and milk of nursing mothers: a comparative study between Wasit and Najyf provinces. Rasayan J Chem 9:404–412

Amiri L, Khalili Doroodzani A, Ostovar A, Dobaradaran S, Mohammadi A, Nabipour I, Raeisi A, Malekizadeh H, Farhadi A, Saeedi R, Afrashteh S, Nazmara S, Keshtkar M (2023) Lactational exposure of human infants to metal (loid)s: a comparison of industrial and urban inhabitants in North of the Persian Gulf. Biol Trace Elem Res

Arrebola JP, Ramos JJ, Bartolomé M, Esteban M, Huetos O, Cañas AI, López-Herranz A, Calvo E, Pérez-Gómez B, Castaño A (2019) Associations of multiple exposures to persistent toxic substances with the risk of hyperuricemia and subclinical uric acid levels in BIOAMBIENT.ES study. Environ Int 123:512–521

Asamoah A, Nikbakht Fini M, Essumang DK, Muff J, Søgaard EG (2019) PAHs contamination levels in the breast milk of Ghanaian women from an e-waste recycling site and a residential area. Sci Total Environ 666:347–354

Ashrap P, Watkins DJ, Mukherjee B, Boss J, Richards MJ, Rosario Z, Vélez-Vega CM, Alshawabkeh A, Cordero JF, Meeker JD (2020) Predictors of urinary and blood metal (loid) concentrations among pregnant women in Northern Puerto Rico. Environ Res 183:109178

ATSDR (2007) Toxicological profile for arsenic. https://www.atsdr.cdc.gov/ToxProfiles/tp2.pdf

ATSDR (2012) Toxicological profile for cadmium. https://www.atsdr.cdc.gov/toxprofiles/tp5-p.pdf

Baeyens W, Vrijens J, Gao Y, Croes K, Schoeters G, Den Hond E, Sioen I, Bruckers L, Nawrot T, Nelen V, Van Den Mieroop E, Morrens B, Loots I, Van Larebeke N, Leermakers M (2014)

Trace metals in blood and urine of newborn/mother pairs, adolescents and adults of the Flemish population (2007–2011). Int J Hyg Environ Health 217:878–890

Baird SJS, Bailey E, Vorhees DJ (2007) Evaluating human risk from exposure to alkylated PAHs in an aquatic system. Hum Ecol Risk Assess 13:322–338

Bandowe BAM, Meusel H (2023) Nitrated polycyclic aromatic hydrocarbons (nitro-PAHs) in the environment—a review. Environ Sci Technol 57:15314–15335

Bansa DK, Awua AK, Boatin R, Adom T, Brown-Appiah EC, Amewosina KK, Diaba A, Datoghe D, Okwabi W (2017) Cross-sectional assessment of infants' exposure to toxic metals through breast milk in a prospective cohort study of mining communities in Ghana. BMC Public Health 17:505

Barbeau D, Maître A, Marques M (2011) Highly sensitive routine method for urinary 3-hydroxybenzo[a]pyrene quantitation using liquid chromatography-fluorescence detection and automated off-line solid phase extraction. Analyst 136:1183–1191

Barbieri FL, Gardon J, Ruiz-Castell M, Paco VP, Muckelbauer R, Casiot C, Freydier R, Duprey JL, Chen CM, Müller-Nordhorn J, Keil T (2016) Toxic trace elements in maternal and cord blood and social determinants in a Bolivian mining city. Int J Environ Health Res 26:158–174

Bartel-Steinbach M, Lermen D, Gwinner F, Schäfer M, Göen T, Conrad A, Weber T, von Briesen H, Kolossa-Gehring M (2022) Long-term monitoring of mercury in young German adults: time trend analyses from the German Environmental Specimen Bank, 1995–2018. Environ Res 207:112592

Barthi R, Sharma R (2022) Effect of heavy metal: an overview. Mat Today: Proc 51:880–885

Bassil M, Daou F, Hassan H, Yamani O, Kharma JA, Attieh Z, Elaridi J (2018) Lead, cadmium and arsenic in human milk and their socio-demographic and lifestyle determinants in Lebanon. Chemosphere 191:911–921

Bennett DH, Busgang SA, Kannan K, Parsons PJ, Takazawa M, Palmer CD, Schmidt RJ, Doucette JT, Schweitzer JB, Gennings C, Hertz-Picciotto I (2022) Environmental exposures to pesticides, phthalates, phenols and trace elements are associated with neurodevelopment in the CHARGE study. Environ Int 161:107075

Berky AJ, Weinhouse C, Vissoci J, Rivera N, Ortiz EJ, Navio S, Miranda JJ, Mallipudi A, Fixen E, Hsu-Kim H, Pan WK (2023) In utero exposure to metals and birth outcomes in an artisanal and small-scale gold mining birth cohort in Madre de Dios. Peru. Environ Health Perspect 131:97008

Björklund KL, Vahter M, Palm B, Grandér M, Lignell S, Berglund M (2012) Metals and trace element concentrations in breast milk of first time healthy mothers: a biological monitoring study. Environ Health 14(11):92

Bocca B, Ruggieri F, Pino A, Rovira J, Calamandrei G, Martínez MÁ, Domingo JL, Alimonti A, Schuhmacher M (2019) Human biomonitoring to evaluate exposure to toxic and essential trace elements during pregnancy. Part A. concentrations in maternal blood, urine and cord blood. Environ Res 177:108599

Bose-O'Reilly S, Lettmeier B, Shoko D, Roider G, Drasch G, Siebert U (2020) Infants and mothers levels of mercury in breast milk, urine and hair, data from an artisanal and small-scale gold mining area in Kadoma/Zimbabwe. Environ Res 184:109266

Brody DJ, Pirkle JL, Kramer RA, Flegal KM, Matte TD, Gunter EW, Paschal DC (1994) Blood lead levels in the US population. Phase 1 of the Third National Health and Nutrition Examination Survey (NHANES III, 1988 to 1991). JAMA 272(4):277-283

Brzeznicki S, Jakubowski M, Czerski B (1997) Elimination of 1-hydroxypyrene after human volunteer exposure to polycyclic aromatic hydrocarbons. Environ Mol Mutagen 70:257–260

Buekers J, Baken K, Govarts E, Martin LR, Vogel N, Kolossa-Gehring M, Šlejkovec Z, Falnoga I, Horvat M, Lignell S, Lindroos AK, Rambaud L, Riou M, Pedraza-Diaz S, Esteban-Lopez M, Castaño A, Den Hond E, Baeyens W, Santonen T, Schoeters G (2023) Human urinary arsenic species, associated exposure determinants and potential health risks assessed in the HBM4EU Aligned Studies. Int J Hyg Environ Health 248:114115

Burkhardt T, Scherer M, Scherer G, Pluym N, Weber T, Kolossa-Gehring M (2023) Time trend of exposure to secondhand tobacco smoke and polycyclic aromatic hydrocarbons between 1995 and 2019 in Germany—showcases for successful European legislation. Environ Res 216(Pt 2):114638

Cao Z, Lin S, Zhao F, Lv Y, Qu Y, Hu X, Yu S, Song S, Lu Y, Yan H, Liu Y, Ding L, Zhu Y, Liu L, Zhang M, Wang T, Zhang W, Fu H, Jin Y, Cai J, Zhang X, Yan C, Ji S, Zhang Z, Dai J, Zhu H, Gao L, Yang Y, Li C, Zhou J, Ying B, Zheng L, Kang Q, Hu J, Zhao W, Zhang M, Yu X, Wu B, Zheng T, Liu Y, Barry Ryan P, Barr DB, Qu W, Zheng Y, Shi X (2021) Cohort profile: China National Human Biomonitoring (CNHBM)—a nationally representative, prospective cohort in Chinese population. Environ Int 146:106252

Carmella SG, Ming X, Olvera N, Brookmeyer C, Yoder A, Hecht SS (2013) High throughput liquid and gas chromatography-tandem mass spectrometry assays for tobacco-specific nitrosamine and polycyclic aromatic hydrocarbon metabolites associated with lung cancer in smokers. Chem Res Toxicol 26:1209–1217

Cattley RC, Kromhout H, Sun M, Tokar EJ, Abdallah MA, Bauer AK, Broadwater KR, Campo L, Corsini E, Houck KA, Ichihara G, Matsumoto M, Morais S, Mráz J, Nomiyama T, Ryan K, Shen H, Toyoda T, Vähäkangas K, Yakubovskaya MG, Yu IJ, DeBono NL, de Conti A, El Ghissassi F, Madia F, Mattock H, Pasqual E, Suonio E, Wedekind R, Benbrahim-Tallaa L, Schubauer-Berigan MK (2023) Carcinogenicity of anthracene, 2-bromopropane, butyl methacrylate, and dimethyl hydrogen phosphite. Lancet Oncol 24:431–432

CDC (2019) Fourth national report on human exposure to environmental chemicals. Updated tables, January 2019. Online: https://stacks.cdc.gov/view/cdc/75822

CDC (2023) National report on human exposure to environmental chemicals. In: Bomonitoring data tables for environmental chemicals. https://www.cdc.gov/exposurereport/data_tables.html

Çebi A, Şengül Ü (2022) Toxic metal and trace element status in the breast milk of Turkish new-born mothers. J Trace Elem Med Biol 74:127066

Chao HH, Guo CH, Huang CB, Chen PC, Li HC, Hsiung DY, Chou YK (2014) Arsenic, cadmium, lead, and aluminium concentrations in human milk at early stages of lactation. Pediatr Neonatol 55:127–134

Chen QY, DesMarais T, Costa M (2019) Metals and mechanisms of carcinogenesis. Ann Rev Pharmacol Toxicol 59:537–554

Cherkani-Hassani A, Ghanname I, Mouane N (2017) Assessment of cadmium levels in human breast milk and the affecting factors: a systematic review, 1971–2014. Crit Rev Food Sci Nutr 57:2377–2391

Choi JW, Kim M, Song G, Kho Y, Choi K, Shin MY, Kim S (2023) Toxicokinetic analyses of naphthalene, fluorene, phenanthrene, and pyrene in humans after single oral administration. Sci Total Environ 870:161899

Choi JW, Song YC, Cheong NY, Lee K, Kim S, Lee KM, Ji K, Shin MY, Kim S (2022) Concentrations of blood and urinary arsenic species and their characteristics in general Korean population. Environ Res 214(Pt 2):113846

CLS (Columbia Law School Health Justice Advocacy) (2019) Literature review: overview of childhood lead poisoning and its health effects. Online: https://web.law.columbia.edu/sites/default/files/microsites/clinics/health-advocacy/literature_review_health_effects_final_july_2019_1.pdf

Çok I, Mazmanci B, Mazmanci MA, Turgut C, Henkelmann B, Schramm KW (2012) Analysis of human milk to assess exposure to PAHs, PCBs and organochlorine pesticides in the vicinity Mediterranean city Mersin, Turkey. Environ Int 40:63–69

Collins JF, Brown JP, Alexeeff GV, Salmon AG (1998) Potency equivalency factors for some polycyclic aromatic hydrocarbons and polycyclic aromatic hydrocarbon derivatives. Regul Toxicol Pharmacol 28:45–54

da Silva Junior FC, Felipe MBMC, Castro DEF, Araújo SCDS, Sisenando HCN, Batistuzzo de Medeiros SR (2021) A look beyond the priority: a systematic review of the genotoxic, mutagenic, and carcinogenic endpoints of non-priority PAHs. Environ Pollut 278:116838

Del Bubba M, Zanieri L, Galvan P, Donzelli GP, Checchini L, Lepri L (2005) Determination of polycyclic aromatic hydrocarbons (PAHs) and total fats in human milk. Ann Chim 95:629–641

Demmeler M, Nowak D, Schierl R (2009) High blood lead levels in recreational indoor-shooters. Int Arch Occup Environ Health 82:539–542

Dereumeaux C, Saoudi A, Pecheux M, Berat B, de Crouy-Chanel P, Zaros C, Brunel S, Delamaire C, le Tertre A, Lefranc A, Vandentorren S, Guldner L (2016) Biomarkers of exposure to environmental contaminants in French pregnant women from the Elfe cohort in 2011. Environ Int 97:56–67

Dixon HM, Bramer LM, Scott RP, Calero L, Holmes D, Gibson EA, Cavalier HM, Rohlman D, Miller RL, Calafat AM, Kincl L, Waters KM, Herbstman JB, Anderson KA (2022) Evaluating predictive relationships between wristbands and urine for assessment of personal PAH exposure. Environ Int 163:107226

Dobraca D, Lum R, Sjödin A, Calafat AM, Laurent CA, Kushi LH, Windham GC (2018) Urinary biomarkers of polycyclic aromatic hydrocarbons in pre- and peri-pubertal girls in Northern California: predictors of exposure and temporal variability. Environ Res 165:46–54

Dursun A, Yurdakok K, Yalcin SS, Tekinalp G, Aykut O, Orhan G, Morgil GK (2016) Maternal risk factors associated with lead, mercury and cadmium levels in umbilical cord blood, breast milk and newborn hair. J Matern Fetal Neonatal Med 29:954–961

ECHA (2017) RAC; Opinion on Arsenic acid and its inorganic salts. https://echa.europa.eu/documents/10162/13641/opinion_arsenic_en.pdf/dd3eb795-108e-5d3a-6847-dddcc021a9dc

ECHA (2018) Note on reference dose-response relationship for the carcinogenicity of pitch, coal tar, high temperature and on PBT and vPvB properties. Risk Assessment Committee (RAC), European Chemical Agency. Agreed at RAC-43

ECHA (2019) Scientific report for evaluation of limit values for lead and its compounds at the workplace. https://echa.europa.eu/documents/10162/68cf7011-9c04-2634-efa6-b712f1b34a85

ECHA (2020) Scientific report for evaluation of limit values for cadmium and its inorganic compounds at the workplace. https://echa.europa.eu/documents/10162/2c23f940-fff8-59ab-43b1-05aadb30042e

Ecsedi-Angyal M, Tatár E, Óvári M, Kurin-Csörgei K, Záray G, Mihucz VG (2020) Determination of low-level arsenic, lead, cadmium and mercury concentration in breast milk of Hungarian women. Int J Environ Anal Chem 100:549–566

EEA (2023) Air quality in Europe 2022. Online: https://www.eea.europa.eu/publications/air-quality-in-europe-2022

EFSA (2010) Scientific opinion on lead in food. EFSA Panel on Contaminants in the Food Chain (CONTAM) EFSA J 8:1570

EFSA (2012) Scientific opinion on the risk for public health related to the presence of mercury and methylmercury in food. https://www.efsa.europa.eu/de/efsajournal/pub/2985

Ekeanyanwua CL, Alisia CS, Ekeanyanwu RC (2020) Levels of aflatoxin M1 and selected heavy metals (Pb, Cd, Cr, Cu, Zn, Fe, As, and Hg) in the breast milk of lactating mothers in South Eastern, Nigeria. Food Contr 107150

Ettinger AS, Téllez-Rojo MM, Amarasiriwardena C, Peterson KE, Schwartz J, Aro A, Hu H, Hernández-Avila M (2006) Influence of maternal bone lead burden and calcium intake on levels of lead in breast milk over the course of lactation. Am J Epidemiol 163:48–56

Fachehoun RC, Lévesque B, Dumas P, St-Louis A, Dubé M, Ayotte P (2015) Lead exposure through consumption of big game meat in Quebec, Canada: risk assessment and perception. Food Addit Contam Part A 32:1501–1511

Fernández SF, Pardo O, Hernández CS, Garlito B, Yusà V (2021) Children's exposure to polycyclic aromatic hydrocarbons in the Valencian Region (Spain): urinary levels, predictors of exposure and risk assessment. Environ Int 153:106535

Freire C, Iribarne-Durán LM, Gil F, Olmedo P, Serrano-Lopez L, Peña-Caballero M, Hurtado JA, Alvarado-González NE, Fernández MF, Peinado FM, Artacho-Cordón F, Olea N (2022) Concentrations and determinants of lead, mercury, cadmium, and arsenic in pooled donor breast milk in Spain. Int J Hyg Environ Health 240:113914

Freire C, Iribarne-Durán LM, Gil F, Olmedo P, Serrano-Lopez L, Peña-Caballero M, Hurtado-Suazo JA, Alvarado-González NE, Fernández MF, Peinado FM, Artacho-Cordón F, Olea N

(2023) Concentrations and predictors of aluminum, antimony, and lithium in breast milk: a repeated-measures study of donors. Environ Pollut 319:120901

Fromme H, Schröder K, Schober W (2024) VI-4 Organic compounds—polycyclic aromatic hydrocarbons (PAHs). In: Wichmann H-E, Fromme H, Zeeb H (eds) Handbook of environmental medicine. Ecomed-Verlag, Landsberg (in German)

Galaviz VE, Quintana PJ, Yost MG, Sheppard L, Paulsen MH, Camp JE, Simpson CD (2017) Urinary metabolites of 1-nitropyrene in US-Mexico border residents who frequently cross the San Ysidro Port of Entry. J Expo Sci Environ Epidemiol 27:84–89

Gong J, Zhu T, Kipen H, Rich DQ, Huang W, Lin WT, Hu M, Zhang JJ (2015) Urinary polycyclic aromatic hydrocarbon metabolites as biomarkers of exposure to traffic-emitted pollutants. Environ Int 85:104–110

Goodrich JM, Ingle ME, Domino SE, Treadwell MC, Dolinoy DC, Burant C, Meeker JD, Padmanabhan V (2019) First trimester maternal exposures to endocrine disrupting chemicals and metals and fetal size in the Michigan Mother-Infant Pairs study. J Dev Orig Health Dis 10:447–458

Grzesik-Gąsior J, Sawicki J, Pieczykolan A, Bień A (2023) Content of selected heavy metals in the umbilical cord blood and anthropometric data of mothers and newborns in Poland: preliminary data. Sci Rep 13(1):14077

Goss KU, Schwarzenbach RP (1998) Gas/solid and gas/liquid partitioning of organic compounds: critical evaluation of the interpretation of equilibrium constants. Environ Sci Technol 32:2025–2032

Green RE, Pain DJ (2012) Potential health risks to adults and children in the UK from exposure to dietary lead in gamebirds shot with lead ammunition. Food Chem Toxicol 50:4180–4190

Grzunov Letinić J, Matek Sarić M, Piasek M, Jurasović J, Varnai VM, Sulimanec Grgec A, Orct T (2016) Use of human milk in the assessment of toxic metal exposure and essential element status in breastfeeding women and their infants in coastal Croatia. J Trace Elem Med Biol 38:117–125

Guo L, Jiang X, Tian HY, Yao SJ, Li BY, Zhang RJ, Zhang SS, Sun X (2019) Detection of BPDE-DNA adducts in human umbilical cord blood by LC-MS/MS analysis. J Food Drug Anal 27:518–525

Guo Y, Senthilkumar K, Alomirah H, Moon HB, Minh TB, Mohd MA, Nakata H, Kannan K (2013) Concentrations and profiles of urinary polycyclic aromatic hydrocarbon metabolites (OH-PAHs) in several Asian countries. Environ Sci Technol 47:2932–2938

Hassan L, Shinar E, Groisman L, Rorman E, Kloog I, Jaffe E, Stoyanov E, Moser A, Novack V, Gat R, Grant-Sasson K, Novack L (2023) National blood bank services as a platform for national human biomonitoring—a proof-of-concept study. Chemosphere 328:138569

HC (2017) Fourth report on human biomonitoring of environmental chemicals in Canada. In: Results of the Canadian Health Measures Survey Cycle 4 (2014–2015), Ottawa, Canada. Online: https://www.canada.ca/en/health-canada/services/environmental-workplace-health/reports-publications/environmental-contaminants/fourth-report-human-biomonitoring-environmental-chemicals-canada.html

HC (2021) Sixth report on human biomonitoring of environmental chemicals in Canada. Minister of Health, Ottawa, ON. Available: www.canada.ca/en/health-canada/services/environmental-workplacehealth/reports-publications/environmental-contaminants/sixth-report-human-biomonitoring.html

HC (Health Canada) (2016) Guidelines for Canadian drinking water quality: guideline technical document—Benzo[a]pyrene. Online: https://www.canada.ca/en/health-canada/services/publications/healthy-living/guidelines-canadian-drinking-water-quality-guideline-technical-document-benzo-pyrene.html

He L, Lin Y, Day D, Teng Y, Wang X, Liu XL, Yan E, Gong J, Qin J, Wang X, Xiang J, Mo J, Zhang Y, Zhang JJ (2021) Nitrated polycyclic aromatic hydrocarbons and arachidonic acid metabolisms relevant to cardiovascular pathophysiology: findings from a panel study in healthy adults. Environ Sci Technol 55:3867–3875

He Q, Zhang X, Kuang L, Xu H, Hou Y, Huang F, Xu Z, Sun H, Deng W, Lin H, Hong H (2024) Residues of polycyclic aromatic hydrocarbons in breast milk collected from Jinhua (China) and the Influence of exercise habit: levels and the potential health effects. Expo Health 16:377–387

Hecht SS, Hochalter JB (2014) Quantitation of enantiomers of r-7,t-8,9,c-10-tetrahydroxy-7,8,9,10-tetrahydrobenzo[a]-pyrene in human urine: evidence supporting metabolic activation of benzo[a]pyrene via the bay region diol epoxide. Mutagenesis 29:351–356

Heitland P, Köster HD (2021) Human biomonitoring of 73 elements in blood, serum, erythrocytes and urine. J Trace Elem Med Biol 64:126706

Hilton DC, Trinidad DA, Hubbard K, Li Z, Sjödin A (2017) Measurement of urinary Benzo[a]pyrene tetrols and their relationship to other polycyclic aromatic hydrocarbon metabolites and cotinine in humans. Chemosphere 189:365–372

Hrubá F, Černá M, Chen C, Harari F, Horvat M, Koppová K, Krsková A, Laamech J, Li YF, Löfmark L, Lundh T, Lyoussi B, Mazej D, Osredkar J, Pawlas K, Pawlas N, Prokopowicz A, Rentschler G, Snoj Tratnik J, Sommar J, Spěváčková V, Špirić Z, Skerfving S, Bergdahl IA (2023) A regional comparison of children's blood cadmium, lead, and mercury in rural, urban and industrial areas of six European countries, and China, Ecuador, and Morocco. Int J Occup Med Environ Health 36:349–364

Huang S, Li Q, Liu H, Ma S, Long C, Li G, Yu Y (2022) Urinary monohydroxylated polycyclic aromatic hydrocarbons in the general population from 26 provincial capital cities in China: levels, influencing factors, and health risks. Environ Int 160:107074

Hudson-Hanley B, Smit E, Branscum A, Hystad P, Kile ML (2021) Trends in urinary metabolites of polycyclic aromatic hydrocarbons (PAHs) in the non-smoking U.S. population, NHANES 2001–2014. Chemosphere 276:130211

Hummel JM, Madeen EP, Siddens LK, Uesugi SL, McQuistan T, Anderson KA, Turteltaub KW, Ognibene TJ, Bench G, Krueger SK, Harris S, Smith J, Tilton SC, Baird WM, Williams DE (2018) Pharmacokinetics of [14C]-Benzo[a]pyrene (BaP) in humans: impact of Co-Administration of smoked salmon and BaP dietary restriction. Food Chem Toxicol 115:136–147

IARC (2010) Some non-heterocyclic polycyclic aromatic hydrocarbons and some related exposures. In: IARC monographs on the evaluation of carcinogenic risks to humans, vol 92. file:///C:/Users/herma/Downloads/mono92-2.pdf

IARC (2010) Some non-heterocyclic polycyclic aromatic hydrocarbons and some related exposures. In: IARC monographs on the evaluation of carcinogenic risks to humans. Lyon, France, vol 92

IARC (2012) Arsenic, metals, fibres, and dusts. In: IARC monographs on the evaluation of carcinogenic risks to humans, vol 100C, pp 219–309. Online: https://publications.iarc.fr/Book-And-Report-Series/Iarc-Monographs-On-The-Identification-Of-Carcinogenic-Hazards-To-Humans/Arsenic-Metals-Fibres-And-Dusts-2012

IARC (2012) Benzo[a]pyrene. In: IARC monographs on the evaluation of carcinogenic risks to humans. Lyon, France, vol 100F

IARC (2013) Diesel and gasoline engine exhausts and some nitroarenes. In: IARC monographs on the evaluation of carcinogenic risks to humans, vol 105. https://publications.iarc.fr/Book-And-Report-Series/Iarc-Monographs-On-The-Identification-Of-Carcinogenic-Hazards-To-Humans/Diesel-And-Gasoline-Engine-Exhausts-And-Some-Nitroarenes-2013

Idowu O, Semple KT, Ramadass K, O'Connor W, Hansbro P, Thavamani P (2019) Beyond the obvious: environmental health implications of polar polycyclic aromatic hydrocarbons. Environ Int 123:543–557

IPCS (International Programme on Chemical Safety) (2003) Selected nitro-and nitro-oxy-polycyclic aromatic hydrocarbons. Environmental Health Criteria 229. https://www.inchem.org/documents/ehc/ehc/ehc229.htm

Iwai-Shimada M, Kameo S, Nakai K, Yaginuma-Sakurai K, Tatsuta N, Kurokawa N, Nakayama SF, Satoh H (2019) Exposure profile of mercury, lead, cadmium, arsenic, antimony, copper, selenium and zinc in maternal blood, cord blood and placenta: the Tohoku Study of Child Development in Japan. Environ Health Prev Med 24:35

Jacobs DE, Brown MJ (2023) Childhood lead poisoning 1970–2022: charting progress and needed reforms. J Public Health Manag Pract 29:230–240

Jaishankar M, Tseten T, Anbalagan N, Mathew BB, Beeregowda KN (2014) Toxicity, mechanism and healtheffects of some heavy metals. Interdiscip Toxicol 7:60–72

Jedrychowski WA, Perera FP, Tang D, Rauh V, Majewska R, Mroz E, Flak E, Stigter L, Spengler J, Camann D, Jacek R (2013) The relationship between prenatal exposure to airborne polycyclic aromatic hydrocarbons (PAHs) and PAH-DNA adducts in cord blood. J Expo Sci Environ Epidemiol 23:371–377

Jeong KS, Ha E, Shin JY, Park H, Hong YC, Ha M, Kim S, Lee SJ, Lee KY, Kim JH, Kim Y (2017) Blood heavy metal concentrations in pregnant Korean women and their children up to age 5years: mothers' and Children's Environmental Health (MOCEH) birth cohort study. Sci Total Environ 605–606:784–791

Joksić AŠ, Tratnik JS, Mazej D, Kocman D, Stajnko A, Eržen I, Horvat M (2022) Polycyclic aromatic hydrocarbons (PAHs) in men and lactating women in Slovenia: results of the first national human biomonitoring. Int J Hyg Environ Health 241:113943

Käfferlein HU, Marczynski B, Mensing T, Brüning T (2010) Albumin and hemoglobin adducts of benzo[a]pyrene in humans–analytical methods, exposure assessment, and recommendations for future directions. Crit Rev Toxicol 40:126–150

Kelishadi R, Sobhani P, Poursafa P, Amin MM, Ebrahimpour K, Hovsepian S, Mansourian M, Najafi R, Hashemipour M (2018) Is there any association between urinary metabolites of polycyclic aromatic hydrocarbons and thyroid hormone levels in children and adolescents? Environ Sci Pollut Res Int 25:1962–1968

Kensler TW, Ng D, Carmella SG, Chen M, Jacobson LP, Muñoz A, Egner PA, Chen JG, Qian GS, Chen TY, Fahey JW, Talalay P, Groopman JD, Yuan JM, Hecht SS (2012) Modulation of the metabolism of airborne pollutants by glucoraphanin-rich and sulforaphane-rich broccoli sprout beverages in Qidong, China. Carcinogenesis 33:101–107

Kılıç Altun S, Dinç H, Temamoğulları FK, Paksoy N (2018) Analyses of essential elements and heavy metals by using ICP-MS in maternal breast milk from Şanlıurfa, Turkey. Int J Anal Chem 1784073

Kim MJ, Kim S, Choi S, Lee I, Moon MK, Choi K, Park YJ, Cho YH, Kwon YM, Yoo J, Cheon GJ, Park J (2021) Association of exposure to polycyclic aromatic hydrocarbons and heavy metals with thyroid hormones in general adult population and potential mechanisms. Sci Total Environ 762:144227

Kim SR, Halden RU, Buckley TJ (2008) Polycyclic aromatic hydrocarbons in human milk of nonsmoking U.S. Women. Environ Sci Technol 42:2663–2667

Kishikawa N, Wada M, Kuroda N, Akiyama S, Nakashima K (2003) Determination of polycyclic aromatic hydrocarbons in milk samples by high-performance liquid chromatography with fluorescence detection. J Chromatogr B 789:257–264

Klein LD, Breakey AA, Scelza B, Valeggia C, Jasienska G, Hinde K (2017) Concentrations of trace elements in human milk: comparisons among women in Argentina, Namibia, Poland, and the United States. PLoS ONE 12:e0183367

Klotz K (2021) Polycyclic aromatic hydrocarbons (PAHs)—evaluation of EKA and a BAR. In: Assessment values in biological material—translation of the German version from 2013. MAK Collect Occup Health Saf, Dec: Doc929

Kordas K, Queirolo EI, Mañay N, Peregalli F, Hsiao PY, Lu Y, Vahter M (2016) Low-level arsenic exposure: nutritional and dietary predictors in first-grade Uruguayan children. Environ Res 147:16–23

Krzyszczak A, Czech B (2021) Occurrence and toxicity of polycyclic aromatic hydrocarbons derivatives in environmental matrices. Sci Total Environ 788:147738

Kuang H, Dai Y, Ding X, Li Y, Cha C, Jiang W, Zhang H, Zhou W, Zeng Y, Pang Q, Fan R (2021) Association among blood BPDE-DNA adduct, serum interleukin-8 (IL-8) and DNA strand breaks for children with pulmonary diseases. Int J Environ Health Res 31:823–834

Kumar A, Kumar A, Chaturvedi AK, Shabnam AA, Subrahmanyam G, Mondal R, Gupta DK, Malyan SK, Kumar S, Khan S, Yadav KK (2020) Lead toxicity: health hazards, influence on food chain, and sustainable remediation approaches. Int J Environ Res Public Healt 17:E2179

Kunter İ, Hürer N, Gülcan HO, Öztürk B, Doğan İ, Şahin G (2017) Assessment of Aflatoxin M1 and heavy metal levels in mothers breast milk in famagusta, cyprus. Biol Trace Elem Res 175:42–49

Kwon JY, Lee S, Surenbaatar U, Lim HJ, Kim BG, Eom SY, Cho YM, Kim WJ, Yu BC, Lee K, Hong YS (2023) Association between levels of exposure to heavy metals and renal function indicators of residents in environmentally vulnerable areas. Sci Rep 13:2856

Lafontaine M, Champmartin C, Simon P, Delsaut P, Funck-Brentano C (2006) 3-Hydroxybenzo[a]pyrene in the urine of smokers and non-smokers. Toxicol Lett 162:181–185

Laidlaw MA, Filippelli G, Mielke H, Gulson B, Ball AS (2017) Lead exposure at firing ranges-a review. Environ Health 16:34

Laumbach R, Tong J, Zhang L, Ohman-Strickland P, Stern A, Fiedler N, Kipen H, Kelly-McNeil K, Lioy P, Zhang J (2009) Quantification of 1-aminopyrene in human urine after a controlled exposure to diesel exhaust. J Environ Monit 11:153–159

Lee BK, Vu VT (2010) Sources, distribution and toxicity of polycyclic aromatic hydrocarbons (PAHs) in particulate matter In: Villanyi V (ed) Air pollution. Sciyo

Lermen D, Weber T, Göen T, Bartel-Steinbach M, Gwinner F, Mueller SC, Conrad A, Rüther M, von Briesen H, Kolossa-Gehring M (2021) Long-term time trend of lead exposure in young German adults—evaluation of more than 35 Years of data of the German Environmental Specimen Bank. Int J Hyg Environ Health 231:113665

Li X, Yu X, Lian X, Kang L, Yang L, Ba F (2025) Maternal urinary levels of PAH metabolites, umbilical cord blood telomere length and anthropometric indices in newborns. Ecotoxicol Environ Saf 291:117767

Li X, Zhong Y, He W, Huang S, Li Q, Guo C, Ma S, Li G, Yu Y (2021) Co-exposure and health risks of parabens, bisphenols, triclosan, phthalate metabolites and hydroxyl polycyclic aromatic hydrocarbons based on simultaneous detection in urine samples from Guangzhou, south China. Environ Pollut 272:115990

Li Z, Romanoff LC, Lewin MD, Porter EN, Trinidad DA, Needham LL, Patterson DG Jr, Sjödin A (2010) Variability of urinary concentrations of polycyclic aromatic hydrocarbon metabolite in general population and comparison of spot, first-morning, and 24-h void sampling. J Expo Sci Environ Epidemiol 20:526–535

Li Z, Romanoff LC, Trinidad DA, Pittman EN, Hilton D, Hubbard K, Carmichael H, Parker J, Calafat AM, Sjödin A (2014) Quantification of twenty-one metabolites of methylnaphthalenes and polycyclic aromatic hydrocarbons in human urine. Anal Bioanal Chem 406:3119–3129

Li Z, Sandau CD, Romanoff LC, Caudill SP, Sjodin A, Needham LL, Patterson DG (2008) Concentration and profile of 22 urinary polycyclic aromatic hydrocarbon metabolites in the US population. Environ Res 107:320–331

Liao KW, Chen PC, Chou WC, Shiue I, Huang HI, Chang WT, Huang PC (2023) Human biomonitoring reference values, exposure distribution, and characteristics of metals in the general population of Taiwan: Taiwan environmental survey for Toxicants (TESTs), 2013–2016. Int J Hyg Environ Health 252:114195

Lin Y, Gao X, Qiu X, Liu J, Tseng C-H, Zhang JJ, Araujo JA, Zhu Y (2021) Urinary carboxylic acid metabolites as possible novel biomarkers of exposures to alkylated polycyclic aromatic hydrocarbons. Environ Int 147:106325

Lin Y, Zhang H, Han Y, Qiu X, Jiang X, Cheng Z, Wang Y, Chen X, Fan Y, Li W, Zhang J, Zhu T (2022) Field evaluation of a potential exposure biomarker of methylated polycyclic aromatic hydrocarbons: association between urinary phenanthrene-2-carboxylic acid and personal exposure to 2-methylphenanthrene. Environ Sci Technol Lett 9:166–172

Lin Y-C, Chang W-H, Li T-C, Iwata O, Chen H-L (2023) Health risk of infants exposed to lead and mercury through breastfeeding. Expo Health 15:255–267

Liu C, Huang L, Huang S, Wei L, Cao D, Zan G, Tan Y, Wang S, Yang M, Tian L, Tang W, He C, Shen C, Luo B, Zhu M, Liang T, Pang B, Li M, Mo Z, Yang X (2022) Association of both prenatal and early childhood multiple metals exposure with neurodevelopment in infant: a prospective cohort study. Environ Res 205:112450

Louro H, Gomes BC, Saber AT, Iamiceli AL, Göen T, Jones K, Katsonouri A, Neophytou CM, Vogel U, Ventura C, Oberemm A, Duca RC, Fernandez MF, Olea N, Santonen T, Viegas S, Silva MJ (2022) The use of human biomonitoring to assess occupational exposure to PAHs in Europe: a comprehensive review. Toxics 10:480

Lundstedt S, White PA, Lemieux CL, Lynes KD, Lambert IB, Oberg L, Haglund P, Tysklind M (2007) Sources, fate, and toxic hazards of oxygenated polycyclic aromatic hydrocarbons (PAHs) at PAH-contaminated sites. Ambio 36:475–485

Luo K, Gao Q, Hu J (2019) Determination of 3-hydroxybenzo[a]pyrene glucuronide/sulfate conjugates in human urine and their association with 8-hydroxydeoxyguanosine. Chem Res Toxicol 32:1367–1373

Luvonga C, Rimmer CA, Yu LL, Lee SB (2020) Organoarsenicals in seafood: occurrence, dietary exposure, toxicity, and risk assessment considerations—a review. J Agric Food Chem 68(4):943–960

Luzardo OP, Ruiz-Suárez N, Almeida-González M, Henríquez-Hernández LA, Zumbado M, Boada LD (2013) Multi-residue method for the determination of 57 persistent organic pollutants in human milk and colostrum using a QuEChERS-based extraction procedure. Anal Bioanal Chem 405:9523–9536

Lv Y, Wei Y, Zhou J, Xue K, Guo Y, Liu Y, Ju A, Wu B, Zhao F, Chen C, Xiong J, Li C, Gu H, Cao Z, Ji JS, Shi X (2021) Human biomonitoring of toxic and essential metals in younger elderly, octogenarians, nonagenarians and centenarians: analysis of the Healthy Ageing and Biomarkers Cohort Study (HABCS) in China. Environ Int 156:106717

MAK (2008) Polycyclic aromatic hydrocarbons (PAH) [in German]. https://onlinelibrary.wiley.com/https://doi.org/10.1002/3527600418.mb0223orgd0045

Malcolm HM, Dobsen S (1994) The calculation of an environmental assessment level (EAL) for atmospheric PAH using relative potencies. Department of the Environment, Report No DoE/HMIP/RR/94/041, ITE Project TO7061i1 Report to the Her Majesty's Inspectorate of Pollution, London

Mandiá N, Bermejo-Barrera P, Herbello P, López-Suárez O, Fraga JM, Fernández-Pérez C, Couce ML (2021) Human milk concentrations of minerals, essential and toxic trace elements and association with selective medical, social, demographic and environmental factors. Nutrients 13:1885

Mansour NA, Hassan NM, Abd et Aal SA, Ebrahem MH (2015) Quantitative analysis of lead, cadmium, heavy metals and other toxic elements in some guman breast samples. Asia J Chem 27:4443–4448

Martín-Carrasco I, Carbonero-Aguilar P, Dahiri B, Moreno IM, Hinojosa M (2023) Comparison between pollutants found in breast milk and infant formula in the last decade: a review. Sci Total Environ 875:162461

McGraw KE, Nigra AE, Klett J, Sobel M, Oelsner EC, Navas-Acien A, Hu X, Sanchez TR (2023) Blood and urinary metal levels among exclusive marijuana users in NHANES (2005–2018). Environ Health Perspect 131:87019

Mendaš G, Jakovljević I, Romanić SH, Fingler S, Jovanović G, Sarić MM, Pehnec G, Popović A, Stanković D (2024) Presence of polycyclic aromatic hydrocarbons and persistent organochlorine pollutants in human Milk: evaluating their levels, association with Total antioxidant capacity, and risk assessment. Sci Total Environ 931:172911

Miller-Schulze JP, Paulsen M, Kameda T, Toriba A, Hayakawa K, Cassidy B, Naeher L, Villalobos MA, Simpson CD (2016) Nitro-PAH exposures of occupationally-exposed traffic workers and associated urinary 1-nitropyrene metabolite concentrations. J Environ Sci 49:213–221

Molina CF, Arango CM, Sepúlveda H (2018) Mercury contamination in breast milk of nursing mothers in gold mining municipalities of Antioquia, Colombia. Biomedica 38:19–29

Motas M, Jiménez S, Oliva J, Cámara MÁ, Pérez-Cárceles MD (2021) Heavy metals and trace elements in human breast milk from industrial/mining and agricultural zones of Southeastern Spain. Int J Environ Res Public Health 18:9289

Motorykin O, Schrlau J, Jia Y, Harper B, Harris S, Harding A, Stone D, Kile M, Sudakin D, Simonich SLM (2015) Determination of parent and hydroxy PAHs in personal $PM_{2.5}$ and urine samples collected during Native American fish smoking activities. Sci Total Environ 505:694–703

Mueller A, Ulrich N, Hollmann J, Zapata Sanchez CE, Rolle-Kampczyk UE, von Bergen M (2019) Characterization of a multianalyte GC-MS/MS procedure for detecting and quantifying polycyclic aromatic hydrocarbons (PAHs) and PAH derivatives from air particulate matter for an improved risk assessment. Environ Pollut 255:112967

Murawski A, Lange R, Lemke N, Zimmermann P, Peisker J, Rucic E, Hahn D, Dębiak M, Kolossa-Gehring M (2023) German environmental survey for children and adolescents 2014–2017 (GerES V) (in German). https://www.umweltbundesamt.de/sites/default/files/medien/479/publikationen/uug_02-2023_deutsche_umweltstudie_zur_gesundheit_von_kindern_und_jugendlichen_2014-2017.pdf

Murawski A, Roth A, Schwedler G, Schmied-Tobies MIH, Rucic E, Pluym N, Scherer M, Scherer G, Conrad A, Kolossa-Gehring M (2020) Polycyclic aromatic hydrocarbons (PAH) in urine of children and adolescents in Germany—human biomonitoring results of the German Environmental Survey 2014–2017 (GerES V) Int J Hyg Environ Health 226:113491

Nakayama SF, Espina C, Kamijima M, Magnus P, Charles MA, Zhang J, Wolz B, Conrad A, Murawski A, Iwai-Shimada M, Zaros C, Caspersen IH, Kolossa-Gehring M, Meltzer HM, Olsen SF, Etzel RA, Schüz J (2019a) Benefits of cooperation among large-scale cohort studies and human biomonitoring projects in environmental health research: an exercise in blood lead analysis of the Environment and Child Health International Birth Cohort Group. Int J Hyg Environ Health 222:1059–1067

Nakayama SF, Iwai-Shimada M, Oguri T, Isobe T, Takeuchi A, Kobayashi Y, Michikawa T, Yamazaki S, Nitta H, Kawamoto T (2019b) Blood mercury, lead, cadmium, manganese and selenium levels in pregnant women and their determinants: the Japan Environment and Children's Study (JECS). J Expo Sci Environ Epidemiol 29:633–647

Nazlıcan E, Arıca E, Gören İE, Kılınçlı B, Mete B, Daglioglu N (2022) The risk estimation and assessment of heavy metal exposure by biomonitoring in the breast milk of mothers in the Cukurova Region, Turkey. Environ Sci Pollut Res Int 29:13963–13970

Neophytou A, Hart J, Chang Y, Zhang J, Smith T, Garshick E, Laden F (2014) Short-term traffic-related exposures and biomarkers of nitro-PAH exposure and oxidative DNA damage. Toxics 2:377

Nethery E, Wheeler AJ, Fisher M, Sjödin A, Li Z, Romanoff LC, Foster W, Arbuckle TE (2012) Urinary polycyclic aromatic hydrocarbons as a biomarker of exposure to PAHs in air: a pilot study among pregnant women. J Expo Sci Environ Epidemiol 22:70–81

Nilsson R, Antić R, Berni A, Dallner G, Dettbarn G, Gromadzinska J, Joksić G, Lundin C, Palitti F, Prochazka G, Rydzynski K, Segerbäck D, Souček P, Tekle M, Seidel A (2013) Exposure to polycyclic aromatic hydrocarbons in women from Poland, Serbia and Italy–relation between PAH metabolite excretion, DNA damage, diet and genotype (the EU DIEPHY project). Biomarkers 18:165–173

Nisbet ICT, Lagoy PK (1992) Toxic equivalency factors (TEFs) for polycyclic aromatic hydrocarbons (PAHs). Regul Toxicol Pharmacol 16:290–300

Nisse C, Tagne-Fotso R, Howsam M, Richeval C, Labat L, Leroyer A (2017) Blood and urinary levels of metals and metalloids in the general adult population of Northern France: the IMEPOGE study, 2008–2010. Int J Hyg Environ Health 220(2 Pt B):341–363

Nordberg GF, Fowler BA, Nordberg M (eds) (2015) Handbook on the toxicology of metals, 4th edn. Academic Press

NTP (National Toxicology Program) (2012) NTP monograph on health effects of low-level lead. Online: https://ntp.niehs.nih.gov/ntp/ohat/lead/final/monographhealtheffectslowlevellead_newissn_508.pdf

Ochirpurev B, Eom SY, Toriba A, Kim YD, Kim H (2022) Urinary 1-aminopyrene level in Koreans as a biomarker for the amount of exposure to atmospheric 1-nitropyrene. Toxicol Res 38:45–51

Oliveira M, Duarte S, Delerue-Matos C, Pena A, Morais S (2020) Exposure of nursing mothers to polycyclic aromatic hydrocarbons: levels of un-metabolized and metabolized compounds in breast milk, major sources of exposure and infants' health risks. Environ Pollut 266(Pt 3):115243

Olujimi O, Ajakore S, Abuganloye D, Arowolo T, Steiner O, Goessler W, Towolawi T (2023) Levels of toxic and trace metals in the breast milk of lactating mothers in Abeokuta, Ogun State, Nigeria. Toxicol Rep 11:168–173

OSHA (Occupational Safety & Health Administration) (2023) Online: https://www.osha.gov/

Pagano G, Thomas PJ, Di Nunzio A, Trifuoggi M (2019) Human exposures to rare earth elements: present knowledge and research prospects. Environ Res 171:493–500

Park J, Kim J, Kim E, Kim WJ, Won S (2021) Prenatal lead exposure and cord blood DNA methylation in the Korean Exposome Study. Environ Res 195:110767

Park S, Park NY, Lim JE, Shin MY, Kim S, Moon HB, Park J, Kho Y, Choi K (2025) Exposure assessment of PAHs on duplicate diet, indoor dust, and air measurements, and comparison with urinary biomonitoring: a family-based panel (CAFE-P) study. Environ Int 203:109726

Peng B, Dong Q, Li F, Wang T, Qiu X, Zhu T (2023) A systematic review of polycyclic aromatic hydrocarbon derivatives: occurrences, levels, biotransformation, exposure biomarkers, and toxicity. Environ Sci Technol 57:15314–15335

Peterson LA, Balbo S, Fujioka N, Hatsukami DK, Hecht SS, Murphy SE, Stepanov I, Tretyakova NY, Turesky RJ, Villalta PW (2020) Applying tobacco, environmental, and dietary-related biomarkers to understand cancer etiology and evaluate prevention strategies. Cancer Epidemiol Biomarkers Prev 29:1904–1919

Philip-Slaboh TP, Eleke C, Ezejiofor AN (2023) Comparison of toxic heavy metals in the breast milk of diabetic and non-diabetic postpartum mothers in Yenagoa, Nigeria. PLoS ONE 18(4):e0264658

Piasek M, Škrgatić L, Sulimanec A, Orct T, Sekovanić A, Kovačić J, Katić A, Branović Čakanić K, Pizent A, Brajenović N, Jurič A, Brčić Karačonji I, Kljaković-Gašpić Z, Tariba Lovaković B, Lazarus M, Stasenko S, Miškulin I, Jurasović J (2023) Effects of maternal cigarette smoking on trace element levels and steroidogenesis in the maternal-placental-fetal unit. Toxics 11:714

Pineda S, Lignell S, Gyllenhammar I, Lampa E, Benskin JP, Lundh T, Lindh C, Kiviranta H, Glynn A (2023) Exposure of Swedish adolescents to elements, persistent organic pollutants (POPs), and rapidly excreted substances—the Riksmaten adolescents 2016–17 national survey. Int J Hyg Environ Health 251:114196

Pirkle JL, Brody DJ, Gunter EW, Kramer RA, Paschal DC, Flegal KM, Matte TD (1994) The decline in blood lead levels in the United States. The National Health and Nutrition Examination Surveys (NHANES). JAMA 272(4):284–291

Pirkle JL, Kaufmann RB, Brody DJ, Hickman T, Gunter EW, Paschal DC (1998) Exposure of the U.S. population to lead, 1991–1994 Environ Health Perspect 106(11):745–750

Pulkrabova J, Stupak M, Svarcova A, Rossner P, Rossnerova A, Ambroz A, Sram R, Hajslova J (2016) Relationship between atmospheric pollution in the residential area and concentrations of polycyclic aromatic hydrocarbons (PAHs) in human breast milk. Sci Total Environ 562:640–647

Rahimi NR, Shahrokhi P, Soleimani F, Arfaeinia H, Dadipoor S, Nik ER, Mouseli A, Yazdi NB (2025) Breast milk contamination by toxic metals in active and passive smokers: A hidden threat to infant health. Food Chem Toxicol 204: 115677

Ramesh A, Walker SA, Hood DB, Guillén MD, Schneider K, Weyand EH (2004) Bioavailability and risk assessment of orally ingested polycyclic aromatic hydrocarbons. Int J Toxicol 23:301–333

Rebelo FM, Caldas ED (2016) Arsenic, lead, mercury and cadmium: toxicity, levels in breast milk and the risks for breastfed infants Environ Res 151:671–688

Rehman K, Fatima F, Waheed I, Akash MSH (2018) Prevalence of exposure of heavy metals and their impact on health consequences. J Cell Biochem 119(1):157–184

Richter-Brockmann S, Dettbarn G, Jessel S, John A, Seidel A, Achten C (2019) Ultra-high sensitive analysis of 3-hydroxybenzo[a]pyrene in human urine using GC-APLI-MS. J Chromatogr B 1118–1119:187–193

Riley KW, Guo J, Wang S, Factor-Litvak P, Miller RL, Andrews H, Hoepner LA, Margolis AE, Rauh V, Rundle A, Perera F, Herbstman JB (2024) Cohort profile: the Mothers and Newborns (MN) Cohort of the Columbia Center for Children's Environmental Health. Int J Epidemiol 53:dyae011

Ruby MV, Lowney YW, Bunge AL, Roberts SM, Gomez-Eyles JL, Ghosh U, Kissel JC, Tomlinson P, Menzie C (2016) Oral bioavailability, bioaccessibility, and dermal absorption of PAHs from soil—state of the science. Environ Sci Technol 50:2151–2164

Samiee F, Vahidinia A, Taravati Javad M, Leili M (2019) Exposure to heavy metals released to the environment through breastfeeding: a probabilistic risk estimation. Sci Total Environ 650(Pt 2):3075–3083

Santonicola S, Albrizio S, Murru N, Ferrante MC, Mercogliano R (2017) Study on the occurrence of polycyclic aromatic hydrocarbons in milk and meat/fish based baby food available in Italy. Chemosphere 184:467–472

Schoeters G, Verheyen VJ, Colles A, Remy S, Martin LR, Govarts E, Nelen V, Den Hond E, De Decker A, Franken C, Loots I, Coertjens D, Morrens B, Bastiaensen M, Gys C, Malarvannan G, Covaci A, Nawrot T, De Henauw S, Bellemans M, Leermakers M, Van Larebeke N, Baeyens W, Jacobs G, Voorspoels S, Nielsen F, Bruckers L (2022) Internal exposure of Flemish teenagers to environmental pollutants: results of the Flemish Environment and Health Study 2016–2020 (FLEHS IV). Int J Hyg Environ Health 242:113972

Sharafi K, Nakhaee S, Azadi NA, Mansouri B, Miri Kermanshahi S, Paknahad M, Habibi Y (2023) Human health risk assessment of potentially toxic elements in the breast milk consumed by infants in Western Iran. Sci Rep 13:6656

Signes-Pastor AJ, Carey M, Vioque J, Navarrete-Muñoz EM, Rodríguez-Dehli C, Tardón A, Begoña-Zubero M, Santa-Marina L, Vrijheid M, Casas M, Llop S, Gonzalez-Palacios S, Meharg AA (2017) Urinary arsenic speciation in children and pregnant women from Spain. Expo Health 9:105–111

Signes-Pastor AJ, Notario-Barandiaran L, Guill M, Madan J, Baker E, Jackson B, Karagas MR (2023) Prenatal exposure to metal mixtures and lung function in children from the New Hampshire birth cohort study. Environ Res 238(Pt 2):117234

Snoj Tratnik J, Falnoga I, Mazej D, Kocman D, Fajon V, Jagodic M, Stajnko A, Trdin A, Šlejkovec Z, Jeran Z, Osredkar J, Sešek-Briški A, Krsnik M, Kobal AB, Kononenko L, Horvat M (2019) Results of the first national human biomonitoring in Slovenia: trace elements in men and lactating women, predictors of exposure and reference values. Int J Hyg Environ Health 222(3):563–582

Sochacka-Tatara E, Majewska R, Perera FP, Camann D, Spengler J, Wheelock K, Sowa A, Jacek R, Mróz E, Pac A (2018) Urinary polycyclic aromatic hydrocarbon metabolites among 3-year-old children from Krakow, Poland. Environ Res 164:212–220

Souza MCO, Rocha BA, Ximenez JPB, Devóz PP, Santana A, Campíglia AD, Barbosa F Jr (2022) Urinary levels of monohydroxylated polycyclic aromatic hydrocarbons in Brazilian children and health risk assessment: a human biomonitoring-based study. Environ Sci Pollut Res Int 29:47298–47309

Sram RJ, Binkova B, Dostal M, Merkerova-Dostalova M, Libalova H, Milcova A, Rossner P Jr, Rossnerova A, Schmuczerova J, Svecova V, Topinka J, Votavova H (2013) Health impact of air pollution to children. Int J Hyg Environ Health 216:533–540

St Helen G, Goniewicz ML, Dempsey D, Wilson M, Jacob P 3rd, Benowitz NL (2012) Exposure and kinetics of polycyclic aromatic hydrocarbons (PAHs) in cigarette smokers. Chem Res Toxicol 25:952–964

Szukalska M, Merritt TA, Lorenc W, Sroczyńska K, Miechowicz I, Komorowicz I, Mazela J, Barałkiewicz D, Florek E (2021) Toxic metals in human milk in relation to tobacco smoke exposure. Environ Res 197:111090

Takayama Y, Masuzaki Y, Mizutani F, Iwata T, Maeda E, Tsukada M, Nomura K, Ito Y, Chisaki Y, Murata K (2021) Associations between blood arsenic and urinary arsenic species concentrations as an exposure characterization tool. Sci Total Environ 750:141517

Tang D, Li TY, Liu JJ, Chen YH, Qu L, Perera F (2006) PAH-DNA adducts in cord blood and fetal and child development in a Chinese cohort. Environ Health Perspect 114:1297–1300

Tchounwou PB, Yedjou CG, Patlolla AK, Sutton DJ (2012) Heavy metals toxicity and the environment. EXS 101:133–164

Thai PK, Banks APW, Toms LL, Choi PM, Wang X, Hobson P, Mueller JF (2020) Analysis of urinary metabolites of polycyclic aromatic hydrocarbons and cotinine in pooled urine samples to determine the exposure to PAHs in an Australian population. Environ Res 182:109048

Tombolini F, Pigini D, Tranfo G, Paci E, Carosi I, Marini F, Bauleo L, Ancona C, Forastiere F (2018) Levels of urinary metabolites of four PAHs and cotinine determined in 1016 volunteers living in Central Italy. Environ Sci Pollut Res Int 25:28772–28779

Toriba A, Kitaoka H, Dills RL, Mizukami S, Tanabe K, Takeuchi N, Ueno M, Kameda T, Tang N, Hayakawa K, Simpson CD (2007) Identification and quantification of 1-nitropyrene metabolites in human urine as a proposed biomarker for exposure to diesel exhaust. Chem Res Toxicol 20:999–1007

Torres-Moreno C, Puente-DelaCruz L, Codling G, Villa AL, Cobo M, Klanova J, Johnson-Restrepo B (2022) Polycyclic aromatic hydrocarbons (PAHs) in human breast milk from Colombia: spatial occurrence, sources and probabilistic risk assessment. Environ Res 204(Pt B):111981

Tsang HL, Wu S, Leung CK, Tao S, Wong MH (2011) Body burden of POPs of Hong Kong residents, based on human milk, maternal and cord serum. Environ Int 37:142–151

UNEP (2019) Global Mercury Assessment 2018, Geneva, Switzerland. Online: https://www.unep.org/resources/publication/global-mercury-assessment-2018

Urbancova K, Dvorakova D, Gramblicka T, Sram RJ, Hajslova J, Pulkrabova J (2020) Comparison of polycyclic aromatic hydrocarbon metabolite concentrations in urine of mothers and their newborns. Sci Total Environ 723:138116

US DHHS (2012) U.S. Department of Health and Human Services. In: National toxicology program monograph on health effects of low-level lead. Online: https://www.semanticscholar.org/paper/NTP-monograph-on-health-effects-of-low-level-lead./f765e1f6fc8e630289ce71515a5bf9b018e6495b

US EPA (2010) Development of a relative potency factor (RPF) approach for polycyclic aromatic hydrocarbon (PAH) mixtures. In: Support of summary information on the integrated risk information system (IRIS). EPA/635/R-08/012A

US EPA (2017a) Toxicological review of Benzo[a]pyrene. EPA/635/R-17/003Fa, Washington, DC, USA

US EPA (2017b) Toxicological review of Benzo[a]pyrene. In: Supplemental information. EPA/635/R-17/003Fb, Washington, DC, USA

US EPA (2022) Basic Information about Mercury. Online: https://www.epa.gov/mercury/basic-information-about-mercury

Vermillion Maier ML, Siddens LK, Pennington JM, Uesugi SL, Anderson KA, Tidwell LG, Tilton SC, Ognibene TJ, Turteltaub KW, Smith JN, Williams DE (2022) Benzo[a]pyrene (BaP) metabolites predominant in human plasma following escalating oral micro-dosing with [14C]-BaP. Environ Int 159:107045

Viau C (1999) The toxicokinetics of pyrene and its metabolites in rats. Toxicol Lett 108:201–207

Vogel N, Murawski A, Schmied-Tobies MIH, Rucic E, Doyle U, Kämpfe A, Höra C, Hildebrand J, Schäfer M, Drexler H, Göen T, Kolossa-Gehring M (2021) Lead, cadmium, mercury, and chromium in urine and blood of children and adolescents in Germany—human biomonitoring results of the German Environmental Survey 2014–2017 (GerES V). Int J Hyg Environ Health 237:113822

Vollset M, Iszatt N, Enger Ø, Gjengedal ELF, Eggesbø M (2019) Concentration of mercury, cadmium, and lead in breast milk from Norwegian mothers: association with dietary habits, amalgam and other factors. Sci Total Environ 677:466–473

Vondráček J, Machala M (2021) The role of metabolism in toxicity of polycyclic aromatic hydrocarbons and their non-genotoxic modes of action. Curr Drug Metab 22:584–595

Wallace ER, Buth E, Szpiro AA, Ni Y, Loftus CT, Masterson E, Day DB, Sun BZ, Sullivan A, Barrett E, Nguyen RH, Robinson M, Kannan K, Mason A, Sathyanarayana S, LeWinn KZ, Bush NR, Karr CJ (2023) Prenatal exposure to polycyclic aromatic hydrocarbons is not associated with behavior problems in preschool and early school-aged children: a prospective multi-cohort study. Environ Res 216(Pt 4):114759

Wang L, Liu A, Zhao Y, Mu X, Huang T, Gao H, Ma J (2018) The levels of polycyclic aromatic hydrocarbons (PAHs) in human milk and exposure risk to breastfed infants in petrochemical industrialized Lanzhou Valley, Northwest China. Environ Sci Pollut Res Int 25:16754–16766

Wang R, Yu J, Chen D, Li J, Qi K, Ma L, Lyu B, Shang X, Zhao Y (2025) Occurrences of heavy metal in breast milk in China: results of the third national breast milk survey. J Hazard Mater 486:137104

Wepler M, Schreckenberg J, Paul B, Fröba G, Muth CM (2022) Influence of ingestion of game meat on blood concentration of lead in Southern Germany: a pilot study. Bull Environ Contam Toxicol 110:21

WHO (2010) WHO guidelines for indoor air quality: selected pollutants. World Health Organization, Kopenhagen

Winiarska-Mieczan A (2014) Cadmium, lead, copper and zinc in breast milk in Poland. Biol Trace Elem Res 157:36–44

Wnorowski A, Harnish D, Jiang Y, Celo V, Dabek-Zlotorzynska E, Charland J-P (2022) Assessment and characterization of alkylated PAHs in selected sites across Canada. Atmosphere 13:1320

Wu S, Li L, Ji G, Xing X, Li J, Ma A, Wei Y, Zhao D, Huang H, Ma W, Wu B, Dong M, Liu T, Chen Q (2023) Association of multi-metals with the risk of hypertension and the interaction with obesity: a cross-sectional study in China. Front Public Health 11:1090935

Yang Z, Guo C, Li Q, Zhong Y, Ma S, Zhou J, Li X, Huang R, Yu Y (2021a) Human health risks estimations from polycyclic aromatic hydrocarbons in serum and their hydroxylated metabolites in paired urine samples. Environ Pollut 290:117975

Yang Z, Lin Y, Wang S, Liu X, Cullinan P, Chung KF, Zhang J (2021b) Urinary amino-polycyclic aromatic hydrocarbons in urban residents: finding a biomarker for residential exposure to diesel traffic. Environ Sci Technol 55:10569–10577

Yang J, Jiang J, Wu L, Xu P, Wu N, Yu R, Xiao Y, Zheng X, Zeng R, Zhou Y, Xie Y, Shen X (2026) Risk assessment and environmental influencing factors of urinary polycyclic aromatic hydrocarbons metabolites in pregnant women in Zunyi, China. J Environ Sci (China) 159:617–623

Yost EE, Galizia A, Kapraun DF, Persad AS, Vulimiri SV, Angrish M, Lee JS, Druwe IL (2021) Health effects of naphthalene exposure: a systematic evidence map and analysis of potential considerations for dose-response evaluation. Environ Health Perspect 129:76002

Yu Y, Peng M, Liu Y, Ma J, Wang N, Ma S, Feng N, Lu S (2021) Co-exposure to polycyclic aromatic hydrocarbons and phthalates and their associations with oxidative stress damage in school children from South China. J Hazard Mater 401:123390

Yu Y, Wang X, Wang B, Tao S, Liu W, Wang X, Cao J, Li B, Lu X, Wong MH (2011) Polycyclic aromatic hydrocarbon residues in human milk, placenta, and umbilical cord blood in Beijing, China. Environ Sci Technol 45:10235–10242

Yusà V, Pérez R, Sánchez A, Pardo O, Roca M (2018) Exposure and risk assessment to arsenic species in Spanish children using biomonitoring. Sci Total Environ 628–629:302–309

Zanieri L, Galvan P, Checchini L, Cincinelli A, Lepri L, Donzelli GP, Del Bubba M (2007) Polycyclic aromatic hydrocarbons (PAHs) in human milk from Italian women: influence of cigarette smoking and residential area. Chemosphere 67:1265–1274

Zhang J, Li J, Wang P, Chen G, Mendola P, Sherman S, Ying Q (2017) Estimating population exposure to ambient polycyclic aromatic hydrocarbon in the United States - Part I: model development and evaluation. Environ Int 99:263–274

Zhang T, Wang X, Luo ZC, Liu J, Chen Y, Fan P, Ma R, Ma J, Luo K, Yan CH, Zhang J, Ouyang F (2023) Maternal blood concentrations of toxic metal(loid)s and trace elements from preconception to pregnancy and transplacental passage to fetuses. Ecotoxicol Environ Saf 264:115394

Zhang X, Li Z (2023) Developing a profile of urinary PAH metabolites among Chinese populations in the 2010s. Sci Total Environ 857(Pt 2):159449

Zhang Y, Tao S (2009) Global atmospheric emission inventory of polycyclic aromatic hydrocarbons (PAHs) for 2004. Atmos Environ 43:812–819

Zhang C, Luo Y, Sun S, Tian T, Zhu M, Ahmad Z, Yang J, Jin J, Zhang H, Chen J, Geng N (2024) Accumulation characteristics of metals in human breast milk and association with dietary intake in northeastern China. Sci Total Environ 912:168515

Zhong Y, Carmella SG, Hochalter JB, Balbo S, Hecht SS (2011) Analysis of r-7,t-8,9,c-10-tetrahydroxy-7,8,9,10-tetrahydrobenzo[a]pyrene in human urine: a biomarker for directly assessing carcinogenic polycyclic aromatic hydrocarbon exposure plus metabolic activation. Chem Res Toxicol 24:73–80

Zhong Z, Li Q, Guo C, Zhong Y, Zhou J, Li X, Wang D, Yu Y (2022) Urinary heavy metals in residents from a typical city in South China: human exposure and health risks. Environ Sci Pollut Res Int 29:15827–15837

Zhou B, Deng Y, Zeng Q, Yang P, Hu X (2023) Associations between repeated measurements of urinary polycyclic aromatic hydrocarbon metabolites and thyroid hormones among reproductive-aged men. Environ Int 171:107698

Zhu H, Martinez-Moral MP, Kannan K (2021) Variability in urinary biomarkers of human exposure to polycyclic aromatic hydrocarbons and its association with oxidative stress. Environ Int 156:106720

Chapter 4
Perfluorinated and Polyfluorinated Compounds

Abstract Perfluorinated and polyfluorinated hydrocarbons (PFAS) are a significant group of over 4,700 distinct anthropogenic organic compounds in which fluorine atoms replace all or some hydrogen atoms in the carbon chain. Two notable examples of medium-chain and long-chain PFAS, mainly due to their previous application levels, are perfluorooctane sulfonate and perfluorooctanoic acid, which are highly persistent in the environment and are excreted from the human body with very long half-lives of several years. They have been classified as substances of very high concern due to their persistence, bioaccumulation, and toxic properties, and they are subject to various restrictions. Conversely, trifluoroacetic acid is a highly water-soluble, fully fluorinated alkanoic acid that is extremely mobile and less degradable in aquatic systems. It serves as a chemical process reagent for pharmaceuticals, biocides/crop protection agents, halogenated refrigerants, and blowing agents, and it is a potential breakdown product of fluorinated chemicals. In addition to the previously mentioned PFAS, several other substances or groups of substances, termed emerging PFAS, are used extensively and on a large scale, primarily as substitutes. They appear as residues in consumer goods such as carpets, upholstery, paper fibers, textiles, paper products, and cardboard, as well as in firefighting foams. These emerging PFAS are less persistent, can be degraded by atmospheric and biological processes, and are less toxic than the traditional long-chain PFAS. Thus, a brief description of the chemical-physical characteristics and usage patterns of various PFAS, their primary health risks and toxicokinetic properties, as well as references to further scientific literature, is provided. The main goal is to present detailed information about their occurrence in human matrices, such as blood, urine, and breast milk. In this context, detailed information is provided on perfluorinated and polyfluorinated compounds (PFAS) such as perfluorooctane sulfonate (PFOS) and perfluorooctanoic acid (PFOA), trifluoroacetic acid, emerging PFAS like fluorotelomer alcohols (FTOH), polyfluoroalkylphosphoric acid esters (PAP), and fluorotelomer sulfonate acids, as well as more recent substitutes such as GenX (HFPO-DA), ADONA, and 6:2-Cl-PFESA (F-53 B).

Keywords Perfluorinated compounds · PFAS · PFOS · PFOA · Trifluoroacetic acid · Emerging PFAS · FTOH · GenX

© The Author(s), under exclusive license to Springer Nature Switzerland AG 2026

H. Fromme, *Human Biomonitoring II*, https://doi.org/10.1007/978-3-032-10036-8_4

The term perfluorinated and polyfluorinated hydrocarbons (PFAS) encompasses anthropogenic organic compounds where fluorine atoms have replaced all or some hydrogen atoms in the carbon chain. Because the polar carbon–fluorine bond is one of the most stable bonds in organic chemistry, these substances exhibit greater thermal and chemical stability than similar hydrocarbon compounds. In total, more than 4700 distinct compounds are classified under PFAS. However, this category may also include compounds that are no longer produced or manufactured in small quantities, leading to estimates that approximately 6% of these substances are commercially relevant worldwide (Buck et al. 2021). Figure 4.1 provides an overview of the different classes. It is essential to note that these consist of solids, liquids, and gases with very different physical, chemical, and biological properties (Buck et al. 2011).

PFAS are primarily produced through two technical processes: electrochemical fluorination (ECF) and fluorotelomerization (ATSDR 2021). However, fluorination generates several by-products with shorter chain lengths (degradation of the alkyl chain) and branched alkyl chains. Consequently, the products formed in the ECF are complex mixtures of substances. This presents challenges for both analysis and risk evaluation. Fluorotelomerization yields linear alkyl chains that differ by two carbon atoms (Lehmler 2005). PFAS have been widely used since around the 1940s.

Due to the large number of existing PFAS and the introduction of new ones, the analysis will always be limited to a relatively small number of important and quantifiable individual substances. One way to gain insight into the total content of all PFAS is the approach first proposed by Miyake et al. (2007), which involves measuring inorganic and organic fluorine, followed by determining the extractable organic fluorine (EOF) content. In a 2004 Chinese study, a total of 30 whole blood samples from five cities were analyzed, revealing 15–67% of unexplained organofluorine (uEOF) using ten quantified PFAS (Yeung et al. 2008). In 130 whole blood samples from Swedish blood donors collected in 2018–2019, the mean uEOF was 60% in women and 41% in men (with 61 quantified PFAS), showing significant fluctuations (Aro et al. 2021). Kaiser et al. (2021) analyzed EOF in pooled serum samples, cord blood, and placental tissue, reporting values ranging from 2.85 to 7.17 ng fluorine/ml, from 1.2 to 2.10 ng fluorine/ml, and from 1.02 to 1.85 ng fluorine/g. They found that, on average, 24% of EOF in maternal serum, 9% in cord blood, and 51% in placental samples could not be identified. In the USA, Pennoyer et al. (2023) examined 20 serum samples and identified a proportion that could not be explained by the sum of the 44 PFAS measured, ranging from 15 to 86% or 0.9–7.5 ng fluorine/ml. Given that several hundred organofluorinated pharmaceuticals are prescribed, the authors sought to estimate their influence on fluorine content and found that an increase of 0.36 ng fluorine/ml could be expected in individuals reporting the use of organofluorinated drugs compared to those who did not.

Two studies have examined the time course of EOF/uEOF in blood samples. Yeung and Mabury (2016) assessed 52 PFAS in 122 plasma samples from students in two German cities from 1982 to 2009, reporting a uEOF of 0–48% and 43%, respectively. After 2000, a higher proportion of uEOF was found in one city. Additionally, Miaz

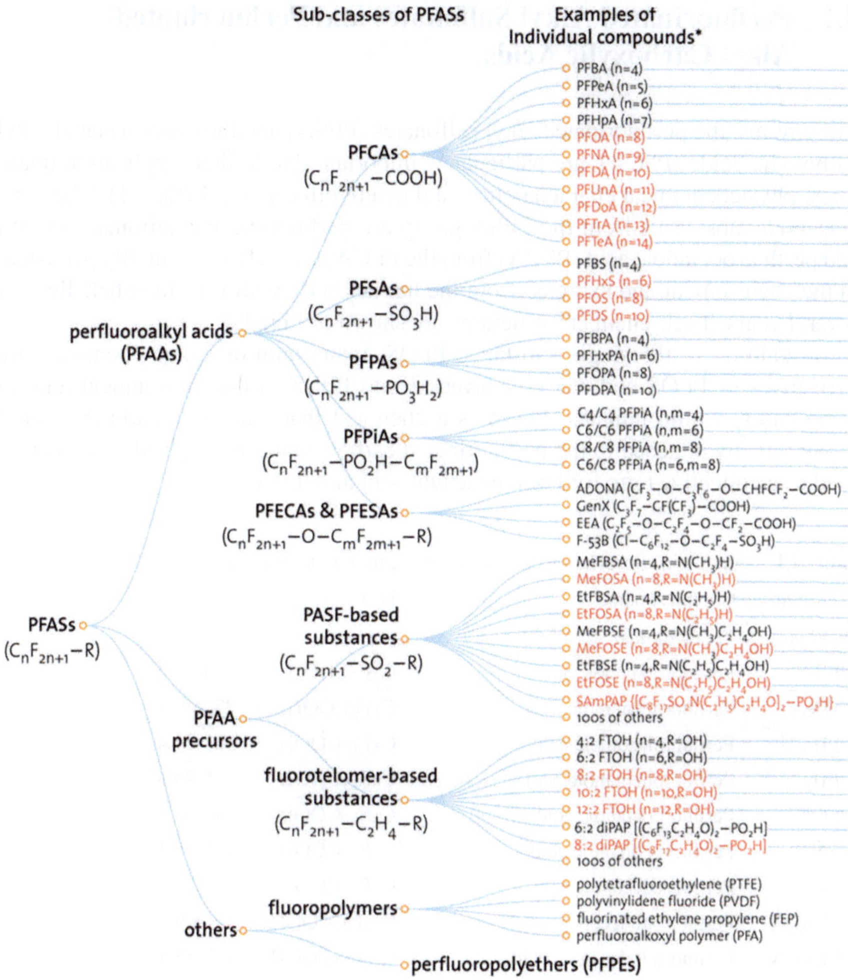

Fig. 4.1 Overview of the different classes of perfluorinated and polyfluorinated compounds (after Wang et al. 2017). PFAAS: Perfluoroalkyl acids; PFCA: Perfluoroalkyl carboxylic acids; PFSA: Perfluoroalkane sulfonic acids; PFPiA: Perfluoroalkyl phosphinic acids; PFECA: Perfluoroalkyl ether carboxylic acid; PFESA: Perfluoroalkyl ether sulfonic acid; PASF: Perfluoroalkane sulfonyl fluorides

et al. (2020) investigated the time trend of EOF between 1996 and 2016 in pooled Swedish serum samples from 622 mothers shortly after birth. A total of 61 PFAS were analyzed, with a uEOF of 67%, and the proportion of EOF attributable to the target PFAS decreased by 3.5% annually throughout the study period.

4.1 Perfluorinated Alkyl Sulfonates and Perfluorinated Alkyl Carboxylic Acids

Two groups, the perfluorinated alkyl sulfonates (PFSA) and the perfluorinated alkyl carboxylic acids (PFCA), are particularly important due to their application quantities, physicochemical characteristics, and health effects (see Table 4.1). The most discussed substances from the PFSA group are perfluorooctane sulfonate (PFOS) and perfluorooctanoic acid (PFOA) from the PFCA group; both are highly persistent in the environment and excreted from the human body with very long half-lives of several years. Their chemical structures are shown in Fig. 4.2.

In addition to the single substance PFOS, many anthropogenic chemicals are derivatives of PFOS and can be converted into PFOS in the environment and, in some cases, within humans. PFOA is a chemical that features a carboxylic acid group directly attached to the perfluorinated carbon chain. PFOA is also frequently used as a collective term for both the actual acid and its salts.

Table 4.1 Names and abbreviations of important medium to long-chain PFAS

Acronym	Chemical name	Formula	CAS No
Perfluoroalkyl carboxylic acid (PFCA)			
PFBA	Perfluorobutanoic acid	C_3F_7COOH	375-22-4
PFPeA	Perfluoropentanoic acid	C_4F_9COOH	2706-90-3
PFHxA	Perfluorohexanoic acid	$C_5F_{11}COOH$	307-24-4
PFHpA	Perfluoroheptanoic acid	$C_6F_{13}COOH$	375-85-9
PFOA	Perfluorooctanoic acid	$C_7F_{15}COOH$	335-67-1
PFNA	Perfluorononanoic acid	$C_8F_{17}COOH$	375-95-1
PFDA	Perfluorodecanoic acid	$C_9F_{19}COOH$	335-76-2
PFUnDA	Perfluoroundecanoic acid	$C_{10}F_{21}COOH$	2058-94-8
PFDoDA	Perfluorododecanoic acid	$C_{11}F_{23}COOH$	307-55-1
PFTrDA	Pentacosafluorotridecanoic acid	$C_{13}F_{25}COOH$	72629-94-8
PFTDA	Heptacosafluorotetradecanoic acid	$C_{14}F_{27}COOH$	376-06-7
Perfluoroalkane sulfonic acids (PFSA)			
PFBS	Perfluorobutane sulfonic acid	$C_4F_9SO_3H$	375-73-5 or 59933-66-3
PFPeS	Perfluoropentane sulfonic acid	$C_5F_{11}SO_3H$	2706-91-4
PFHxS	Perfluorohexane sulfonic acid	$C_6F_{13}SO_3H$	355-46-4
PFHpS	Perfluoroheptane sulfonic acid	$C_7F_{15}SO_3H$	375-92-8
PFOS	Perfluorooctane sulfonic acid	$C_8F_{17}SO_3H$	1763-23-1
PFNS	Perfluorononane sulfonic acid	$C_9F_{19}SO_3H$	474511-07-4
PFDS	Perfluorodecane sulfonic acid	$C_{10}F_{21}SO_3H$	335-77-3
PFUnDS	Perfluoroundecane sulfonic acid	$C_{11}F_{23}SO_3H$	749786-16-1
PFDoDS	Perfluorododecane sulfonic acid	$C_{12}F_{25}SO_3H$	79780-39-5

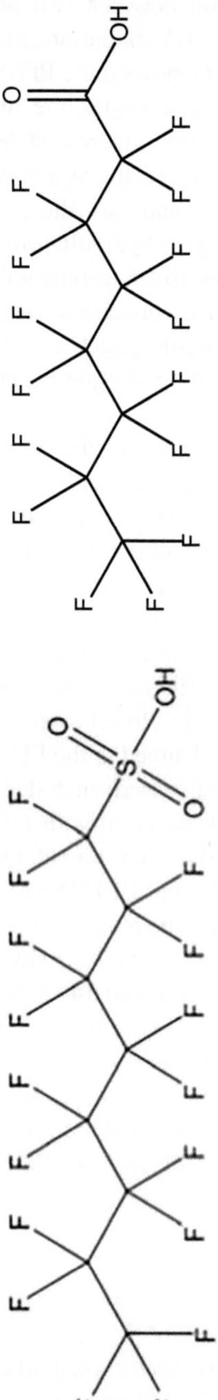

Fig. 4.2 Chemical structures of PFOS (left) and PFOA (right)

These substances have an amphiphilic character and can reduce the surface tension of a liquid or the interfacial tension between two phases. Due to their thermal and chemical stability, resistance to UV radiation and weathering, and their dirt, paint, grease, oil, and water-repellent properties, PFAS are used in various industrial and consumer products. Since the 1950s, the main applications for PFOS-type substances have included treating surfaces on textiles, leather, and carpets; coating paper and cardboard products; serving as a surfactant in cleaning and treatment agents; and being used in paints and varnishes, fire-fighting foam, the photographic industry, the aircraft industry in hydraulic oils, and the treatment of metal surfaces. This widespread use has led to environmental contamination. Conversely, PFOA compounds are mainly used as processing aids (emulsifiers) in manufacturing fluoroelastomers and fluoropolymers such as polytetrafluoroethylene (PTFE) and in producing nonstick kitchenware, photographic coatings, and food processing equipment.

According to the Stockholm Convention, PFOA (with exceptions) and PFHxS (without exceptions), along with their salts and related compounds, were classified in Annex A in 2019. This classification requires measures to prevent their production and use (SC 2019). PFOS and perfluorooctane sulfonyl fluoride (PFOS-F) are classified with permitted uses and specific exemptions in Annex B of the Convention, which mandates that parties take measures to restrict production and use, considering the listed permitted uses and specific exemptions. Since 2003, PFOS has only been allowed in the EU as a spray suppressant for nondecorative hard chrome plating (chromium VI) in closed-loop systems. PFOA, its salts, and PFOA-related compounds have been banned in the EU with exemptions since 2020. Long-chained C9-C14-PFCA and salts were included in the EU REACH Candidate List as substances of very high concern, with PFNA and PFDA classified as toxic for reproduction, persistent in the environment, and bioaccumulative. Furthermore, PFUnDA, PFDoDA, PFTrDA, and PFTDA are classified as very persistent and very bioaccumulative and may not be manufactured or placed on the market as independent substances in the EU starting in 2023 (ECHA 2017). Again, Commission Regulation (EU) 2024/2462 prohibits the use of PFHxA and related substances in textiles for consumers, food packaging, consumer mixtures like waterproofing sprays, cosmetics, and in some firefighting foams. In addition, Commission Regulation (EU) 2025/1988 stipulates that, with exceptions and transition periods, PFAS may no longer be placed on the market or used in firefighting foams in concentration of more than 1 mg/l for the sum of all PFAS from 2030.

4.1.1 Health Aspects

Results are now available for this substance class, particularly for PFOS and PFOA, from numerous individual epidemiological and animal studies compiled in various recent reviews (ECHA 2017; Rappozzo et al. 2017; EFSA 2018; Boesen et al. 2020; EFSA 2020; ATSDR 2021; Fenton et al. 2021; NASEM 2022; Costello et al. 2022; ECHA 2022; ECHA 2023; Crawford et al. 2023; Zhang et al. 2023;

OEHHA 2024; GC 2025; IARC 2025). Transferring data from animal experiments to humans presents challenges due to considerable differences in elimination half-lives, inadequate mechanistic data, species differences in toxicity mechanisms for some endpoints, and discrepancies in exposure levels between epidemiological and experimental studies (ATSDR 2021). Unless otherwise noted, these reviews will be referenced in this and the following subchapter.

EFSA (2020) summarized the health effects: PFOS and PFOA are linked to a reduced antibody response to childhood vaccinations against Haemophilus influenzae type b, diphtheria, and tetanus, while the evidence for other PFAS is less robust. Some studies indicate that PFOS and PFOA serum levels are associated with increased susceptibility to infections. Epidemiological studies provide clear evidence of a relationship between exposure to PFOS, PFOA, and PFNA and elevated serum cholesterol levels. There is insufficient evidence connecting these substances to liver disease, diabetes, obesity, metabolic syndrome, changes in kidney function, or an increased risk of cardiovascular disease. Epidemiological studies indicate that PFOS and PFOA may impact birth weight, but no connections were found between other PFAS and fertility or reproductive outcomes in both men and women. In a review paper, Steenland et al. (2020) conclude for PFOA that the epidemiological evidence remains limited and summarize the following associations: consistent evidence between PFOA and cholesterol, no evidence of association with heart disease; evidence of association with ulcerative colitis, but not for other autoimmune diseases; substantial evidence that PFOA affects immune response, but no clear evidence of association with infectious diseases; the proof of association with thyroid and kidney disease is suggestive but inconsistent; evidence of association with liver enzymes, but not with liver disease; minimal evidence of association with neurotoxicity.

Steenland and Winquist (2021) concluded in a review paper that the cancers most strongly associated with PFAS in humans are testicular and kidney cancer. They noted that contradictory evidence exists regarding the association with prostate cancer in several studies. Overall, they still regard the evidence linking cancer and PFAS as sparse. More recently, the Ronneby Register Cohort, which included over 60,000 people—among them, 5,702 with cancer and simultaneously high PFAS exposure (especially PFOS and PFHxS)—found no evidence of an overall increased cancer risk. However, there was a moderately increased risk of kidney cancer (Li et al. 2022). In a multiethnic cohort, an association was found between serum PFOA concentrations and the risk of renal cell carcinomas (RCC). A suggestive association was reported between higher PFOA concentrations and increased RCC risk among White participants and those with sera collected before 2002 (Rhee et al. 2023). In the second cohort, the same working group identified an exposure–response association between serum PFOA concentrations and the risk of RCC in 324 cases and 324 individually matched controls (Shearer et al. 2021).

Overall, the WHO's International Agency for Research on Cancer has classified PFOA as carcinogenic to humans (Group 1) based on limited evidence of cancer in humans (renal cell carcinoma and testicular cancer), sufficient evidence in experimental animals for its carcinogenicity, and strong evidence that PFOA exhibits multiple key characteristics of carcinogens in exposed humans, human primary cells, and experimental systems. PFOS was evaluated as possibly carcinogenic to humans (Group 2B) due to inadequate evidence in humans and limited evidence in experimental animals. However, strong evidence suggests that PFOS demonstrates multiple key characteristics of carcinogens in exposed humans, human primary cells, and experimental systems.

4.1.2 Toxicokinetics

Few data are available on the toxicokinetics of perfluorinated substances in humans. Animal studies indicate that PFAS are rapidly and almost completely absorbed following oral administration. Corresponding experiments with labeled substances demonstrate absorption rates of > 50% for PFHxS and > 95% for PFOA, PFBA, PFNA, PFDA, PFUnA, and PFDoDA (ATSDR 2021). Indirect evidence of oral intake comes from regions with contaminated drinking water, which show higher internal loads than control regions. Even after single and repeated inhalation exposure to PFOA in rats, there was significant uptake into the organism, supported by measurements from workers exposed to dust in fluoropolymer production facilities. Targeted studies on dermal uptake are still quite limited thus far. Available data on rodents suggest that the absorption of PFOA and PFOS through the skin is restricted and is only of minor importance as an exposure route.

Perfluorinated substances bind 94–98% to serum proteins (mainly albumin) following oral, intravenous, and intraperitoneal administration. Studies on protein binding revealed no significant species differences between rats, monkeys, and humans. A saturation effect was noted at very high concentrations in the blood solely for PFBS, but not for PFHxS, PFOS, or PFOA. A greater distribution in cellular blood components was observed in human volunteers only for PFHxA, not for PFOA, PFOS, or PFHxS (Jin et al. 2016). There is no accumulation in fatty tissue or fat fractions. In rodents, hamsters, and rabbits, PFAS primarily distribute in the liver, blood plasma, kidneys, and lungs (Hundley et al. 2006). Accumulation has also been reported in the testes, ovaries, and brain. However, measurements in the nervous tissue of deceased individuals indicate that levels here are very low (Maestri et al. 2006). The transfer of PFAS to the fetus during pregnancy is possible but decreases with increasing chain length.

The animal studies and in vitro tests available to date have not provided any evidence of metabolism for PFAS. In principle, PFAS can also be excreted in urine by humans, but due to the high reabsorption rate in the renal tubules for PFOA and PFOS, less than 0.1% of the administered dose is excreted through this route (ATSDR 2021). Menstruation can significantly contribute to the excretion of PFSA and PFCA due to blood loss (Lorber et al. 2015; Gomis et al. 2017). The elimination half-lives from blood described in the scientific literature so far are summarized for various PFSA and PFCA, as well as different species, in Table 4.2. They display apparent differences between species and genders for the individual substances. For human subjects or those exposed in the workplace, the half-lives of longer-chain substances span several years, whereas PFBS and PFBA typically last for days. One possible explanation for the gender-specific differences in excretion may involve specific kidney transporter systems. Notably, PFOA utilizes the so-called "organic anion transporter" (OAT), the expression of which is influenced by sex hormones (Worley and Fisher 2015). Experiments with rats have indicated that the renal clearance of PFOA can be stimulated by estradiol and decreased by ovariectomy (Kudo et al. 2002). In principle, perfluorinated alkyl sulfonates are excreted more slowly than perfluorinated alkyl carboxylic acids, and elimination rates decrease with increasing chain length while increasing with branched substances (ATSDR 2021). Although C7 and C8 compounds in this species are primarily eliminated renally, excretion via bile increases with chain length (e.g., for C10 and C11 compounds). However, there is significant reabsorption in the gastrointestinal tract.

Information on other PFAS is limited. For example, residents of the southern Swedish city of Ronneby were exposed to various PFAS through drinking water over several years (Li et al. 2022). In 114 participants, the calculated mean half-lives based on the decrease in serum levels between 2014 and 2018 were as follows: PFPeS 0.94 years, PFOA 2.47 years, L-PFOS 4.55 years, PFHxS 4.52 years, and PFHpS 4.55 years. Compared to L-PFOS, the half-life of the branched isomer 1 m-PFOS was longer at 5.0 years, while 3/4/5-PFOS and 2/6-PFOS had shorter half-lives of 3.43 and 2.69 years, respectively.

In a recent study, Abraham et al. (2024) administered a mixture of fifteen ^{13}C-labeled PFAS to a healthy volunteer and measured their concentrations in plasma, urine, and feces over 450 days. All substances were rapidly and almost completely excreted, with half-lives of 0.5 days for PFPeA, 1.5 days for PFHxA, 4.18 days for PFBA, 51 days for PFBS, and 152 days for PFHpA among the short-chain PFAS. In contrast, the longer-chain PFAS exhibited significantly longer mean half-lives of 5.5 years for PFOA, 3.6 years for PFNA, and 3.3 years for PFOS.

4.1.3 Occurrence in the Blood and Urine

Several comprehensive reviews have been conducted on the occurrence of PFAS in the blood of the general population (EFSA 2020; ATSDR 2021; ECHA 2023; Li et al. 2024; Adlard et al. 2024).

Typically, PFAS are detected in blood plasma or serum, and occasionally, in whole blood. Comparative studies have shown similar concentration levels of PFOS, PFOA, and PFHxS in serum and plasma samples (Ehresman et al. 2007). It was previously believed that whole blood levels account for approximately 50% of serum or plasma levels, although results vary between studies. For instance, Ehresman et al. (2007) evaluated samples across a wide concentration range and identified a median plasma/whole blood ratio (Pl/Wb ratio) of 2.3 for PFOS (range: 1.8–3.3 for EDTA whole blood and 1.8–2.9 for heparinized whole blood). The median Pl/Wb ratio for PFOA was 2.0, and for PFHxS, it was either 2.4 or 2.1, depending on the anticoagulant used. Kärrman et al. (2006) reached different conclusions when assessing whole blood and plasma for PFAS in five subjects, reporting a Pl/Wb ratio of 1.4 (PFOA), 1.2 (PFOS), 1.2 (PFHxS), 1.0 (PFNA), and only 0.2 for FOSA. Comparable findings

Table 4.2 Mean/median elimination half-life from plasma/serum (modified according to Olsen et al. 2007; Huang et al. 2019; ATSDR 2021)

Spezies	PFOS		PFOA		PFHxS		PFBS		PFBA		PFNA		PFHxA	
	♀	♂	♀	♂	♀	♂	♀	♂	♀	♂	♀	♂	♀	♂
Rat	24–83 Days	7.5–82 Days	3.2–16 Days	2–13 Days	2 Days	16–27 Days	1.3–7.4 h	3,3–4,7 h	1,6–1,8 h	7–9,2 h	1,4 Days	39–41 Days	2,1–2,6 h	2,4–2,5 h
Mouse	20–38 Days	20–42 Days	17 Days	19 Days	2.1–27 Days	16–31 Days	–	–	2,8–3,1 h	5,2–16,3 h	26–68 Days	34–56 Days	–	–
Rabbit	–	–	7 h	5.5 h	–	–	–	–	–	–	–	–	–	–
Dog	–	–	8–13 Days	20–30 Days	–	–	–	–	–	–	–	–	–	–
Monkey	179 Days	170 Days	–	20 Days	87 Days	141 Days	3.5 Days	4 Days	1,7 Days		–	–	–	–
Human	3.4–5.4 Years		2.3–3.9 Years		5.3–15.5 Years		28 Days		2.3–3.4 Days		2.5–4.3 Years		5.1 Days	

were also noted in a Chinese study involving 60 subjects, which showed the Pl/Wb ratio for PFOS at 1.2, PFHxS at 1.5, and PFOA at 1.2 (Jin et al. 2016). A Norwegian study reported median serum/plasma ratios of 0.9–1.3 in 61 blood samples for PFOS, PFOA, PFHxS, PFNA, and PFUnDA, while the ratios of serum and plasma to whole blood were 1.9–2.3 and 1.7–2.1, respectively (Poothong et al. 2017). This study determined that the relative median composition of the sums of PFOS and PFOA was 60% for whole blood, 64% for plasma, and 70% for serum. Figure 4.3 illustrates the percentages for the three blood fractions. Liu et al. (2023) analyzed 162 blood samples in China and found that the mean rate of C4-C7 PFCA in plasma increased from 76 to 82% with increasing chain length, then decreased with further increases in chain length up to C13.

4.1.3.1 PFOS and PFOA Isomers in Blood

In the production of PFAS, both linear and branched isomers, along with various fluorinated impurities, form during electrochemical fluorination (ECF), the primary process for producing PFOS. In contrast, the end product of telomerization predominantly consists of linear isomers (Langlois and Oehme 2006; Vyas et al. 2007). The technical mixture of branched isomers resulting from ECF is dependent on the batch and varies between 10 and 30% (Kärrman et al. 2007b). The increased polarity of the branched structure seems to contribute to their differing environmental behavior, as linear PFAS preferentially sorb in soil and sediments. Meanwhile, branched isomers tend to remain in water (Schulz et al. 2020).

As early as 2001, Hansen et al. (2001) assumed that PFOS and PFOA are also present in the blood of mammals in various isomers. Currently, no data on the toxicokinetic behavior of the different isomers or any specific effects are available. In serum and plasma samples collected from 1997 to 2003 in Sweden, Great Britain, and Australia, linear PFOS (L-PFOS) was the most common, accounting for 55% to 70% of the total content (Kärrman et al. 2007b; Berg et al. 2014; Lamichhane et al. 2021). Interestingly, in a standard product, according to ECF, the proportion of the linear isomer is 76% to 79% higher than the proportion observed in the blood, suggesting different behavior in the environment or the organism. Göckener et al. (2020) reported a decrease in the median proportion of branched PFOS (Br-PFOS) in the total content, which dropped from 37 to 27% in samples from the German Environmental Specimen Bank between 2009 and 2019. A Swedish study examining the temporal development of contamination in pooled samples from 1996 to 2017 also found that Br-PFOS represented 31% of the total PFOS concentrations in 2017 (Miaz et al. 2020). In the USA, results from 3759 participants in the NHANES 2013–2018 and 1303 women from the Study of Women's Health Across the Nation Multi-Pollutant Study, conducted from 1999 to 2000, indicated a weighted proportion of L-PFOS at 69% and 71% in both studies (Kang et al. 2023). In Chinese studies of 321 umbilical cord blood samples from 2013 and 145 serum samples from women, Br-PFOS comprised 25% and approximately 89% of the total content, respectively (Zhang et al. 2017; Han et al. 2023b).

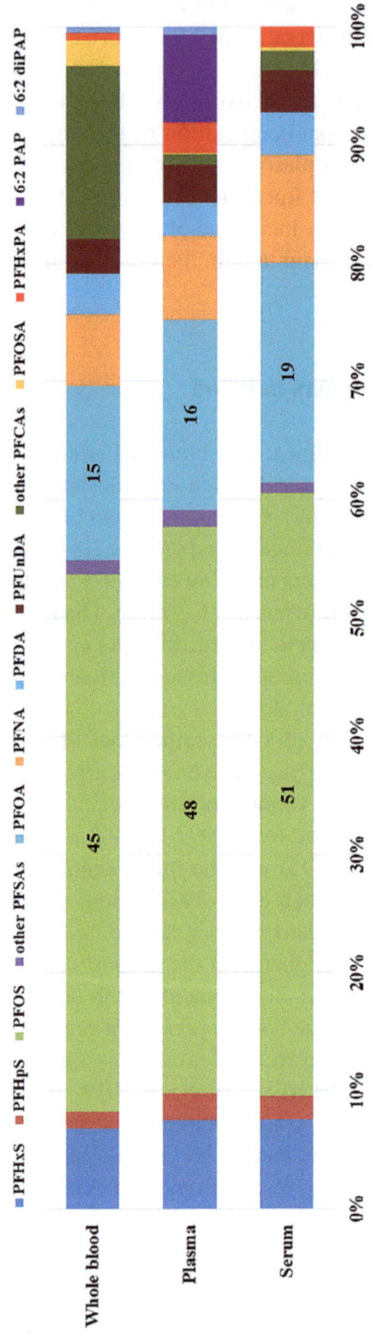

Fig. 4.3 Median relative profiles of PFASs in human serum, plasma, and whole blood for substances with detection frequency > 50%. The total PFAS serum, plasma, and whole blood concentrations were 10, 10, and 6.3 μg/l, respectively (Poothong et al. 2017)

The situation is different for PFCA. In the analysis of 16 pooled serum samples, approximately 98% of PFOA was found in its linear form (L-PFOA), with only about 2% existing as branched isomers (De Silva and Mabury 2006). This finding also applied to PFNA and PFUnDA. According to ECF, only 80% of L-PFOA was detected in a standard PFOA product. The authors attribute the high proportion of L-PFOA in the blood to the degradation of FTOH and possibly also of fluorotelomer olefins, which may significantly influence L-PFOA concentration. In the 2015/16 US NHANES among individuals older than 12, L-PFOA represented a median of only 4% during those years (CDC 2023). In China, L-PFOA accounted for 99% of the total content in the analysis of 321 umbilical cord blood samples conducted in 2013 (Zhang et al. 2017). In a recent review of eight studies, Schulz et al. (2020) concluded that the proportion of L-PFOA in serum ranged from 94 to 98%.

4.1.3.2 Concentrations in the Blood of Adults

A significant amount of data is now available on the general population's internal exposure to perfluorinated substances in serum or plasma, without any suspicion of contamination. Table 4.3 summarizes the findings of more recent studies. Many studies observed a clearer or less distinct difference in concentrations between the sexes, with higher levels in male test subjects.

In the 2000s, the median levels from European countries reached up to 18.5 μg/l for PFOS and 5.8 μg/l for PFOA. However, a more recent study from Germany in 2016/17 reported medians (95th percentile) of 2.1 μg/l (6.4 μg/l) for PFOS and 1.1 μg/l (2.4 μg/l) for PFOA among 158 blood donors (Fromme et al. 2017). Even lower values were observed in a study by Rovira et al. (2019) involving 39 pregnant women from Spain at the time of birth, where the mean value (95th percentile) for PFOA was 0.14 μg/l (0.93 μg/l), with only four values exceeding the quantification limit of 0.1 μg/l, and 2.20 μg/l (4.23 μg/l) for PFOS. The median and 95th percentiles in the last study were 0.4 μg/l and 0.8 μg/l for PFNA, while PFDA was detected in only three samples, with a maximum of 1.0 μg/l, and PFDoDA was not detected above the quantification limit in any sample.

The studies from East Asia are comparable to those in Europe and North America. For instance, Iwata et al. (2024) report medians (maxima) of 2.9 μg/l (39 μg/l) for PFOS, 1.6 μg/l (45 μg/l) for PFOA, 1.4 μg/l (36 μg/l) for PFNA, 0.49 μg/l (15 μg/l) for PFDA, and 1.1 μg/l (11 μg/l) for PFUnDA in the analysis of over 25,000 blood samples from the prospective, nationwide Japan Environment and Children's Study. Higher medians were observed in the analysis of 10,855 participants over 18 years of age from the nationwide China National Human Biomonitoring Study for PFOS and PFOA at 5.2 μg/l and 4.3 μg/l (Wu et al. 2023). In this study, the PFNA, PFDA, and PFUnDA medians were 0.99 μg/l, 0.58 μg/l, and 0.42 μg/l, respectively. Significantly lower concentrations were reported in the blood of 419 Lebanese women during pregnancy from 2018 to 2021 (Mahfouz et al. 2023). The medians (95th percentiles) were 0.09 μg/l (57.8 μg/l) for PFOS, 0.09 μg/l (51.3 μg/l) for PFOA, and 0.05 μg/l (2.3 μg/l) for PFHxS.

Table 4.3 Median (range) PFAS concentrations in the serum/plasma of the adult general population in µg/l

References	N	Age (years)	PFOS	PFOA	PFHxS	Sex	Country, year
Europe							
Cserbik et al. (2024)	98	34[a]	1.7 (0.3–5.9)	0.6 (0.2–1.8)	0.2 (0.06–1.3)	Female	Spain, 2021
Hron et al. (2024)	224 225	>18 >18	1.3 1.5	0.7 0.9	0.4 0.5	Female Male	Germany, 2020/21
Jensen et al. (2025)	187 110	32[a] 34[a]	4.1 (1.2–11.8) 6.9 (1.1–22.0)	0.8 (0.16–2.4) 1.3 (0.05–2.5)	0.3 (0.04–5.7) 0.8 (0.08–9.8)	Female Male	Danmark, 2018–21
Thépaut et al. (2021)	100	24–72	4.7 + (1.9–12.8)[b]	1.5 + (0.6–3.4)[b]	0.71 + (0.32–2.3)[b]	Female	Norway, 2016/17
	44	25–72	8.0 + (2.9–21.2)[b]	2.0 + (1.0–3.9)[b]	1.1 + (0.46–2.3)[b]	Male	
Rovira et al. (2019)	39	44	2.2[a] (0.10–4.23)[b]	0.14[a] (<0.10–0.93)[b]	–	Female	Spain, 2016/17
Fromme et al. (2017)	158	18–66	2.1 (0.6–17.9)	1.1 (0.3–3.7)	0.50 (<0.20–11.6)	Female/male	Germany, 2016
Ingelido et al. (2020)	250	20–51	5.6 (0.56–120)	1.8 (0.32–28)	2.5 (0.03–9.1)	Female/male	Italy, 2015/16
Sochorová et al. (2017)	300	18–65	2.4 (0.07–51)	0.76 (0.03–9.0)	0.18 (0.008–1.4)	Female/male	Czech Rep., 2015
Li et al. (2018a)	242		4.2 (<0.5–55)	1.6 (<0.4–5.0)	0.84 (<0.5–60)	female/male	Sweden, 2014–16
Lamichhane et al. (2021)	44	24–46	7.24 (3.02–19.5)	4.74 (1.22–11.3)	0.23 (0.15–0.38)	Female	Finland, 2013–15
Poothong et al. (2017)	61	20–66	5.2 (1.3–20.2)	1.9 (0.38–25)	0.78 (0.23–2.3)	Female/male	Norway, 2013/14
Colles et al. (2020)	205	50–65	7.6 (1.5–46)	2.9 (0.24–25)	1.61 (0.10–12.4)	Female/male	Belgium, 2014
De Filip et al. (2015)	549	20–40	2.4 (0.34.58.9)	1.6 (<0.1–9.99)	–	Female	Italy, 2011/12

(continued)

Table 4.3 (continued)

References	N	Age (years)	PFOS	PFOA	PFHxS	Sex	Country, year
Cariou et al. (2015)	100	20–46	3.1 (0.32–24.5)	1.0 (0.31–7.31)	0.62 (<0.3–31)	Female	France, 2010–13
Dereumeaux et al. (2016)	277	18- > 40	3.0 (2.1–7.9)[b]	1.5 (1.1–3.7)	0.73 (0.49–2.1)	Female	France, 2011
Beck et al. (2019)	981		7.7 (5.7–10.4) [d]	1.7 (1.1–2.4)[d]	0.36 (0.24–0.50)[d]	Female	Denmark, 2010–12
Bjermo et al. (2013)	270	22–75	11.2 (3.9–25.4)[b]	2.3 (0.76–5.0) [b]	1.95 (0.7–10.1) [b]	Female/male	Sweden, 2010/11
Bartolomé et al. (2017)	755	18–65	7.6 (3.8–19.3) [b]	2.0 (1.1–5.5) [b]	0.82 (<0.3–2.8) [b]	Female/male	Spain, 2009/10
Bach et al. (2016)	1507		8.3 (6.0–10.8)[c]	2.0 (1.5–2.6)[c]	0.5 (0.4–0.6)[c]	Female	Danmark, 2008–13
Hardell et al. (2014)	186	50–79	8.3 (1.7–49)	1.9 (0.35–8.4)	0.87 (0.15–3.0)	Male	Sweden, 2007–11
Fromme et al. (2010)	44	21–43	3.2 (0.49–9.13)	2.4 (0.74–8.73)	0.5 (<0.2–0.60)	Female	Germany, 2007–09
Góralczyk et al. (2015)	253	20–44	8.0 (1.6–21)	2.7 (0.67–9.8)	1.9 (0.34–102)	Female	Poland
Góralczyk et al. (2015)	176	20–44	18.4 (8.2–40)	4.8 (0.94–12.6)	1.2 (0.12–3.8)	Male	Poland
Hölzer et al. (2008)	153	18–69	5.2 + (1.0–70.7)	2.8 + (0.7–9.2)	0.6 + (<0.1–2.1)	Female	Germany, 2006
	103	18–69	9.7 + (1.7–92.5)	5.8 + (1.1–15.3)	2.2 + (0.7–5.4)	Male	
Fromme et al. (2007)	168	14–67	10.9 (2.5–30.7)	4.8 (1.5–16.2)	–	Female	Germany, 2005
	188	14–67	13.7 (2.1–55)	5.7 (0.5–19.1)	–	Male	
North America							

(continued)

Table 4.3 (continued)

References	N	Age (years)	PFOS	PFOA	PFHxS	Sex	Country, year
Baumert et al. (2024)	126	55[a]	2.0 + (1.7–2.3[g])	2.2 + (2.1–2.3[g])	0.63 + (0.55–0.72[g])	Female/male	USA, 2020/21
Schlezinger et al. (2025)	72	18–65	4.2+	1.8+	1.3+	Male	Canada, 2019/20
Botelho et al. (2025)	977	> 19	2.1 (9.3[f])	1.2 (3.5[f]	0.8 (3.3[f])	Female	USA, 2017–20
	952	> 19	3.4 (11.6[f])	1.5 (3.5[f])	1.4 (4.9[f])	Male	
HC (2021)	518	20–79	2.3 (12[f])	1.0 (2.5[f])	0.59 (2.1[f])	Female	Canada, 2018/19
	501	20–79	3.6 (13[f])	1.3 (3.3[f])	1.1 (4.4[f])	Male	
Yu et al. (2020)	517	20–74	2.4 (9.8 [f])	2.4 (5.5 [f])	0.9 (3.6 [f])	Female	USA, 2016/18
	513	20–74	4.0 (10.0 [f])	2.0 (5.3 [f])	1.5 (4.8 [f])	Male	
Eick et al. (2020)	506	32[a]	1.9 (1.2–3.2)	0.76 (0.5–1.1)	0.33 (0–2–0.6)	Female	USA, 2014–18
Starling et al. (2020)	583	28[a]	2.4 (<0.1–15.6)	1.1 (0.1–15.4)	0.7 (<0.1–10.9)	Female	USA, 2009–14
Asia							
Tian et al. (2024)	164	30[a]	3.4	2.3	1.0	Female	China, 2020/21
Liu et al. (2023)	162	27–69	– (0.49–14)	– (0.86–11)	– (0.36–3.2)	Female/male	China, 2020
Mahfouz et al. (2023)	419		0.09 (0.07–61.3)	0.09 (0.07–51.2)	0.05 (0.04–2.4)	Female	Lebanon, 2018–21
Iwata et al. (2024)	25,040	31[a]	2.9 (<LOQ-39)[e]	1.6 (<LOQ-45)[e]	0.3 (<LOQ-7.3)[e]	Female	Japan, 2017/18
Wu et al. (2023)	10,855	18–79	5.2	4.3	0.13	Female/male	China, 2017/18
Luo et al. (2022)	2257	29[a]	9.6 (1.1–159)	11.9 (3.0–165)	0.5 (<0.02–4.5)	Female	China, 2013–16
Kim et al. (2019)	94	2–82	4.2 (0.7–15)	3.2 (1.3–9.8)	0.9 (<LOQ-3.2)	Female/male	South Korea, 2014/15
Jiang et al. (2014)	141		4.3 (0.9–15.8)	4.0 (1.8–32.9)	0.4 (0.08–3.7)	Female	China, 2012

(continued)

Table 4.3 (continued)

References	N	Age (years)	PFOS	PFOA	PFHxS	Sex	Country, year
Hsu et al. (2013)	29	20–65	6.0 (3.5–25.7)	2.5 (1.6–5.98)	–	Female	Taiwan, 2011
	30	20–65	10.2 (4.67–22.6)	10.2 (4.7–22.6)	–	Male	
Nakayama et al. (2020)	339	32[a]	4.5	2.1	0.46	Female	Japan, 2009/10
Lee et al. (2019)	1069	18–70	8.99[+]	2.30[+]	0.19[+]	Female	South Korea, 2009/10
	805	18–70	11.6[+]	3.41[+]	1.00[+]	Male	
Itoh et al. (2019)	701	31[a]	6.2 (1.2–30.3)	2.0 (<0.2–12.4)	0.31 (<0.2–1.8)	Female	Japan, 2002–05

[+]Geometric mean value; [#]calculated from the whole blood values (multiplication by 2); [a]Mean value; [b]5th and 95th percentile; [c]Interquartile range; [d]25th and 75th percentile; [e]LOQ from 0.1 to 0.3 μg/l; [f]95th percentile; [g]95% confidence interval

Comprehensive studies from North America show medians of a comparable order of magnitude for PFAS. In the USA, as part of the representative NHANES 2017–2018, a total of 1616 serum samples from people older than 19 years were analyzed, and medians (95. percentile) of 4.70 μg/l (15.1 μg/l) for PFOS (total), 1.47 μg/l (3.87 μg/l) for PFOA, 1.20 μg/l (3.80 μg/l) for PFHxS, 0.40 μg/l (1.40 μg/l) for PFNA, and 0.20 μg/l (0.60 μg/l) for PFDA were reported (CDC 2023). A median of 3.20 μg/l and a 95th percentile of 11.0 μg/l were reported for L-PFOS.

So far, results from various current studies are available that attempt to map the temporal development of pollution levels in the population. Measurements of archived samples from the German Environmental Specimen Bank examined 258 plasma samples from eight time points between 1982 and 2010, focusing on students aged 20 to 29 years (Schröter-Kermani et al. 2013). In a follow-up study, Göckener et al. (2020) analyzed 100 plasma samples from students of the same age at five time points between 2009 and 2019. The results of both studies are graphically shown in Fig. 4.4. While the median PFOS content in the second half of the 1980s was still about 25 μg/l, the concentrations significantly dropped in 1992, especially after 1998, reaching 1.8 μg/l in 2019. For PFOA, a relatively consistent concentration level of around 4.0 μg/l was noted from the 1980s until about 2007, after which it slowly decreased to 9 μg/l in 2019. PFHxS concentrations rose to a median of approximately 2.4 μg/l until around 2000 and 2005, then declined to 0.5 μg/l in 2019. PFNA exhibited a wavelike pattern in a low concentration range, decreasing from 2006 to 0.4 μg/l in 2019. In southern Sweden, Noren et al. (2021) analyzed PFAS levels in Swedish adolescents aged 16 to 21 years in serum samples taken at five time points (1213 individuals, 83% male). PFOS, PFOA, and PFHxS declined by 12.5%, 6.5%, and 6.7% per year, respectively, while PFNA and PFDA increased until 2009 and only then began to decrease. In Denmark, a temporal development of PFOS and PFOA levels similar to that of Germany was also reported for 1533 serum samples from pregnant women collected from 2008 to 2013 as part of the Aarhus Birth Cohort (Bjerregaard-Olesen et al. 2016). Additionally, in a longitudinal study in Uppsala, Sweden, blood samples from 579 participants aged 70 were analyzed over three periods between 2001 and 2014 (Stubleski et al. 2016). While the internal levels of PFOS and PFOA decreased during this time, increases were observed for PFHxS and PFNA.

In the USA, measurement results are available from the NHANES study, which has examined a representative sample of approximately 1562 to 2233 participants nearly every two years since 1999 (CDC 2023). Figure 4.4 illustrates the development of the medians over time. The trend for PFOA concentration is comparable to that in Europe, while PFOS shows an overall higher level of contamination. Nonetheless, these levels have also significantly decreased in recent years. The PFHxS levels show a slight decline in the NHANES, dropping from 2.1 μg/l in 1999/2000 to 1.10 μg/l in 2017/18. A study conducted between 2000 and 2010 on 616 blood donors aged 20–69 from six regions in the USA also yielded similar results (Olsen et al. 2017).

The age profile was examined in several studies. For instance, Eriksson et al. (2017) studied the age progression in 54 pooled plasma samples from 4920 individuals in Australia in 2011. They noted a continuous increase in median PFOS

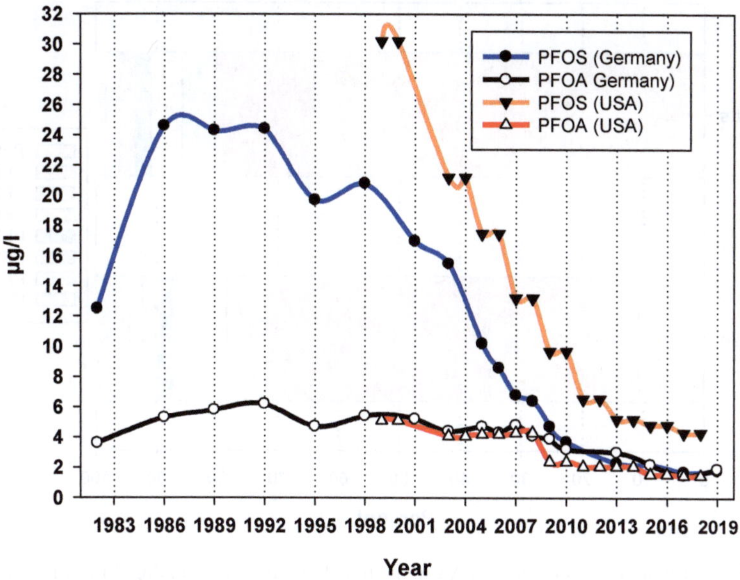

Fig. 4.4 Median PFOS and PFOA levels in the blood of the general population (USA: CDC 2023; Germany: years 1982–2008 and 2010 from Schröter-Kermani et al. 2013; years 2009 and 2013–2019 from Göckener et al. 2020)

concentration with age, rising from 2.9 µg/l for subjects aged up to 4 years to 5.6 µg/l for those over 60 years. The median PFOA concentration decreased from 3.2 µg/l at 0–4 years to 2.7 µg/l at 31–45 years, then increased steadily to 3.4 µg/l in the age group over 60 years. For PFNA, only slight differences in concentration were found among the age groups, while for PFHxS, an increase was observed up to the age group of 16–30 years, followed by a plateau. Utilizing European studies, EFSA (2020) calculated the relative contribution of individual PFAS to the total concentrations in blood samples (Fig. 4.5). In this study, PFOS, PFOA, PFHxS, PFNA, PFDA, PFUnDA, and PFHpS accounted for 97% and 93% of the total for adults and children, respectively. However, while in adults, PFOS and PFOA contributed 64% and 16%, in children, PFOA was dominant at 37%, with PFOS at only 35%. EFSA indicated that differences in transplacental transfer and exposure pathways might explain these variations.

4.1.3.3 Concentrations in the Blood of Children

Table 4.4 summarizes the results of various PFAS in serum samples from children and adolescents. In seven studies, L-PFOS and Br-PFOS were measured in blood samples. The proportion of L-PFOS to the total content averages 66% (49–73%).

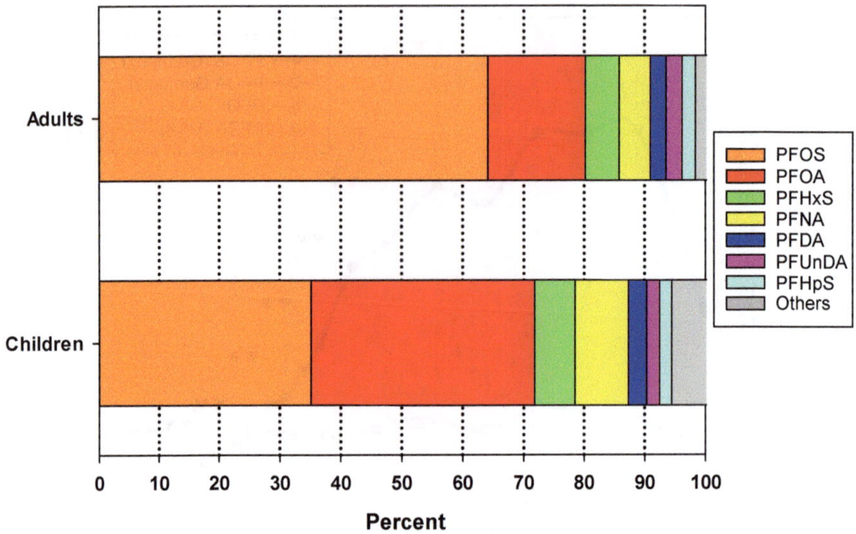

Fig. 4.5 Distribution of individual PFAS in the blood of adults and children in Europe (modified according to EFSA 2020)

Four nationally representative studies involving children and adolescents originate from Germany, Sweden, the USA, and Canada. As part of the 2014–2017 German Environmental Survey (GerES V), blood samples were collected from 1109 participants aged 3–17 (Duffek et al. 2020). In the USA, between 313 and 402 serum samples from individuals aged 12 to 19 were analyzed as part of the National Health and Nutrition Examination Survey (NHANES) 2017–2018 (CDC 2023). Additionally, the 2013–2014 NHANES studied the 3–11 age group. As part of the fifth Canadian Health Measures Survey (CHMS), 474–527 serum samples from individuals aged 3–19 were analyzed in 2016–2017 (HC 2019). The PFOS and PFOA medians and 95th percentiles for the three studies mentioned are presented in Fig. 4.6 for the age groups of 3–5 years and 6–11 years. In both age groups, PFOS levels are higher in the USA, whereas PFOA levels are elevated in Germany.

Moreover, the Riksmaten Adolescents 2016–2017 survey, a cross-sectional national population-based study of Swedish adolescents, analyzed 1082 blood samples from subjects 11–21 years old (Nyström et al. 2022).

Significantly fewer results have been reported for newborns and infants. In a German study, PFAS were initially collected from 44 pregnant women, as well as from the umbilical cord and their breastfed children aged six and 19 months (Fromme et al. 2010). The median plasma values for PFOS and PFOA increased from 1.0 μg/l and 1.4 μg/l in umbilical cord serum to 3.0 μg/l and 6.9 μg/l after six months, then decreased to 1.9 μg/l and 4.6 μg/l. The study's findings, along with Norwegian research involving 51 children aged 36 months (Granum et al. 2013), were modeled using a pharmacokinetic model, resulting in the concentration curves depicted in Fig. 4.7 (Verner et al. 2016). It was demonstrated that the median ratio of the internal

Table 4.4 Medians (95th percentiles) of perfluorinated substances in serum/plasma of children and adolescents in µg/l

References	N	Age (years)	PFHpA	PFOA	PFNA	PFDA	PFHxS	L-PFOS	Br-PFOS	Total-PFOS	Country, year
Lindfeldt et al. (2023)	19	4	< 0.03	1.59	–	–	2.45	2.21	1.22	3.39	Sweden, 2020/21
Lindfeldt et al. (2023)	16	8	< 0.03	0.95	–	–	1.78	1.96	1.06	2.98	Sweden, 2020/21
Lindfeldt et al. (2023)	44	12	< 0.03	1.15	–	–	2.03	2.60	1.20	3.82	Sweden, 2020/21
CDC (2023)	313–402	12–19	< 0.1[c] (0.2)	1.17 (2.37)	0.40 (1.20)	0.20 (0.40)	0.80 (3.40)	1.90 (5.70)	0.70 (2.00)	2.60 (7.30)	USA, 2017/18
Nyström et al. (2022)	1095	11–21	< 0.02	1.23	0.38	0.16	0.40	2.00	0.93	2.93[a]	Sweden, 2016/17
HC (2019)	474–527	12–19	–	1.0 (1.9)	0.37 (–)	0.13 (0.34)	0.58 (3.6)	–	–	1.8 (3.9)	Canada, 2016/17
Kornher et al. (2025)	381	11–17	–	0.32	0.07	–	–	–	–	0.54	Ecuador, 2016
Duffek et al. (2020)	1109	3–17	< 0.25 (< 0.25)	1.27 (3.24)	< 0.5 (0.66)	< 0.25 (0.35)	0.38 (1.26)	–	–	2.41 (6.00)	Germany, 2014–17
van Beijsterveldt et al. (2022)	307–310	2	–	1.80	0.20	0.06	0.40	0.64	0.66[a]	1.30	Netherlands, 2013–18
Kang et al. (2018)	150	3–18	< 0.01 (0.10)	1.88 (4.07)	0.94 (2.25)	0.06 (0.32)	0.79 (1.50)	–	–	5.68 (12.1)	Korea, 2012–14
Oh et al. (2024)	541	2–5	0.21 (0.87)	2.38 (6.81)	0.85 (2.60)	0.16 (0.48)	0.64 (3.39)	–	–	2.45 (11.0)	USA, 2009–17

(continued)

Table 4.4 (continued)

References	N	Age (years)	PFHpA	PFOA	PFNA	PFDA	PFHxS	L-PFOS	Br-PFOS	Total-PFOS	Country, year
Willhelm et al. (2015)	101	8–10	–	3.56 (6.13)	0.68 (1.09)	–	0.75 (1.43)	–	–	3.30 (5.85)	Germany, 2009/10
Bao et al. (2014)	225	12–15	–	0.52	0.83	0.94	1.3	–	–	29	Taiwan, 2009/10
Harris et al. (2017)	653	6–10	–	4.4 (7.9 ^{b})	1.5 (3.8^{b})	0.3 (0.6^{b})	1.9 (7.0^{b})	4.4 (9.9^{b})	1.7 (4.0^{b})	6.2 (13.7^{b})	USA, 2007–10
Schecter et al. (2012)	300	0–12	–	2.85	1.20	< 0.20	1.20	–	–	4.10	USA, 2009

[a]Quantified from reported data; [b]90th percentile; [c]Data from 2013/14 survey

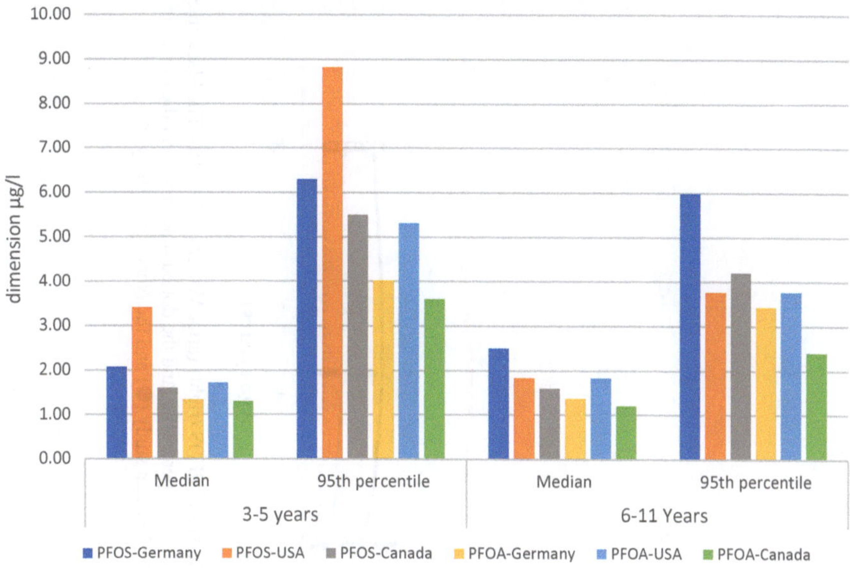

Fig. 4.6 Medians and 95th percentiles in two age groups in studies from Germany (Duffek et al. 2020), the USA (CDC 2023), and Canada (HC 2019)

load in maternal plasma to the child's plasma for PFOA peaks at 4.5 in the first year of life, then declines steadily as the child grows. For PFOS, the median ratio reaches its highest point of three between 1 and 1.5 years of age. Gyllenhammar et al. (2018) reported median levels of 13 µg/l PFOS, 7.2 µg/l PFOA, 2.0 µg/l PFHxS, and 0.34 µg/l PFNA in 107 Norwegian children aged 2 to 4 months who were examined between 1996 and 1999. The infant-to-maternal blood levels ratio was 0.7 for PFOS, 2.6 for PFOA, 1.1 for PFHxS, and 0.9 for PFNA. Van Beijsterveldt et al. (2022) also studied 276 children aged three months and 24 months, finding decreasing medians of 1.48 µg/l and 1.30 µg/l for PFOS, 2.40 µg/l and 1.80 µg/l for PFOA, and 0.43 µg/l and 0.40 µg/l for PFHxS, indicating a continuing trend. In 54 Finnish children, PFOS, PFOA, and PFNA concentrations decreased significantly between ages 1, 6, and 10.5 for PFHxS, but not remarkably between ages 1 and 6 (Koponen et al. 2018).

Breast milk can be a significant source during infancy. For example, a retrospective study involving 101 children aged one year, from whom blood samples were taken at the end of the 1990s, revealed differences in concentration between children who had been breastfed for at least four months and those fed with formula (Abraham et al. 2020). The mean values in the breastfed group (N = 80) and the formula group (N = 21) were 15.2 µg/l and 6.8 µg/l for PFOS, 16.8 µg/l and 3.8 µg/l for PFOA, 2.1 µg/l and 1.7 µg/l for PFHxS, and 0.6 µg/l and 0.2 µg/l for PFNA. These findings are supported by Mondal et al. (2014), who examined the blood of children from a region in the USA that was heavily contaminated with PFOA and found higher concentrations in 369 breastfed children compared to 254 non-breastfed children.

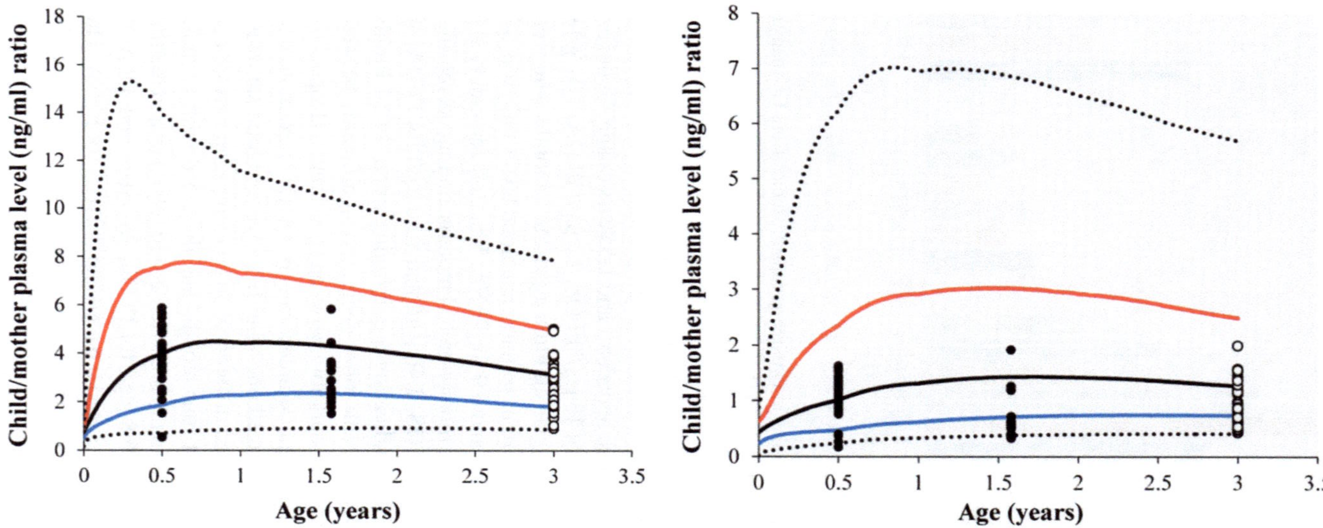

Fig. 4.7 Temporal trend of the ratios of the infant to maternal serum concentrations of PFOA (left) and PFOS (right) (after Verner et al. 2016). The black line represents the 50th percentile, the blue line represents the 5th percentile, the red line represents the 95th percentile, and the dotted lines represent the minimum and maximum values. Dots represent ratios of measured maternal and child PFAS levels from the German study (●) and the Norwegian study (○)

PFOS's geometric mean was 13.5 versus 12.7 μg/l, and for PFHxS, it was 4.0 versuss 3.5 μg/l. Very high concentrations of PFOA were reported in children due to existing environmental pollution. However, the concentrations in breastfed children (48.6 μg/l) were approximately twice as high as those in non-breastfed children (21.7 μg/l).

Eriksson et al. (2017) examined 54 pooled plasma samples from 100 children and adolescents aged 0 to 15 years from Australia. They found a decrease in the mean values between 2002 and 2013: for PFOS, from 10.9 μg/l to 1.8 μg/l; for PFOA, from 8.5 μg/l to 2.3 μg/l; and for PFHxS, from 9.2 μg/l to 1.2 μg/l. For PFNA, there was initially an increase from 2002 to 2006, rising from 0.35 μg/l to 0.70 μg/l, followed by a decrease to 0.48 μg/l in 2013. In Sweden, PFAS were analyzed in the serum of children aged 4, 8, and 12 years from 2008 to 2022. Declining temporal trends were observed for PFOS, PFOA, PFNA, and PFDA with average annual reductions of – 3.4%, – 5.8%, – 4.3%, and – 3.9%, respectively, except for PFHxS (Lindfeldt et al. 2023).

4.1.3.4 Concentrations in Umbilical Cord Blood

McAdam and Bell (2023) conducted a systematic review of 35 studies to evaluate the determinants of PFAS concentrations in the blood of pregnant mothers and newborns shortly after birth. The most robust and consistent evidence among pregnant mothers was linked to parity, breastfeeding, maternal race, country of origin, and household income. Conversely, the findings regarding smoking status, alcohol consumption, and prepregnancy body mass index do not appear to be significant determinants of PFAS levels in pregnant mothers or newborns. Additionally, Liu et al. (2020) provided an overview of studies on the presence of PFAS in amniotic fluid, placenta, and fetal organs.

The results of recent studies published in the scientific literature on umbilical cord plasma/serum measurements are summarized in Table 4.5. In North America and Europe, there is a decreasing trend in the concentrations of PFOS and PFOA in umbilical cord blood, while in China, the medians for both substances show no trend over time. Extremely high medians (95th percentile) of 55 μg/l (185 μg/l) for PFOS, 45 μg/l (77 μg/l) for PFOA, and 2.0 μg/l (81 μg/l) for PFHxS were reported in a recent study from Lebanon (Mahfouz et al. 2023). An investigation from eastern China also reported very high PFOA concentrations, with a median of 32 μg/l and a 95th percentile of 110 μg/l (Han et al. 2018). This last study was conducted in a highly industrialized area with over 400 chemical plants, some of which produce PFOA-containing chemicals such as firefighting foams. However, direct workplace exposure of the parents was excluded. Nonetheless, the concentrations in umbilical cord blood are also decreasing, at least in Europe and North America.

The median and mean concentration ratios of umbilical cord blood to maternal plasma or serum (see Table 4.6) range from 0.1 to 0.8 (mean: 0.4) for PFOS and from 0.3 to 0.9 (mean: 0.7) for PFOA. The mean ratios for PFNA and PFHxS are both 0.5. Numerous studies have found a significant correlation between maternal and umbilical cord blood concentrations. Compared to first-time mothers and women

Table 4.5 Median concentrations of perfluorinated substances in cord blood in µg/l

References	N	PFBA	PFPeA	PFHxA	PFHpA	PFOA	PFNA	PFDA	PFBS	PFHxS	PFOS	Country, year
Europe												
Kaiser et al. (2023)	136	< 0.49	-	< 0.12	< 0.15	0.62	< 0.16	< 0.08	< 0.29	< 0.19	< 0.38	Austria, 2017–19
Dufour et al. (2018)	214	–	–	–	< 0.05	0.68	0.15	< 0.15	–	0.16	0.73	Belgium, 2013–16
Colles et al. (2020)	269	–	–	–	–	1.27	0.21	–	< 0.20	0.37	1.11	Belgium, 2013/14
Quaak et al. (2016)	59	–	–	–	–	0.87	0.14	0.05	–	0.15	1.60	Netherlands, 2011–13
Cariou et al. (2015)	100	–	–	–	–	0.86	< 0.3	< 0.3	–	0.34	1.10	France, 2010–13
Richterová et al. (2018)	322	–	–	–	–	0.88	0.24	–	–	0.08	0.64	Slovakia, 2010–12
Fromme et al. (2010)	33	–	–	–	–	1.40	< 0.4	< 0.4	–	0.20	1.0	Germany, 2007/08
Gützkow et al. (2012)	123	–	–	–	–	0.88	0.12	0.04	–	0.20	1.52	Norway, 2007/08
Manzano–Salgado et al. (2015)	66	–	–	–	–	1.90	0.32	–	–	< 0.10	1.86	Spain, 2003–06
Needham et al. (2011)	12	–	–	–	–	3.1	0.37	0.10	–	9.1	6.6	Faroe Islands, 2000
North America												
MacDonald et al. (2022)	62	–	–	–	–	0.23	0.09	–	–	0.12	0.51	Canada, 2013–16

(continued)

Table 4.5 (continued)

References	N	PFBA	PFPeA	PFHxA	PFHpA	PFOA	PFNA	PFDA	PFBS	PFHxS	PFOS	Country, year
Morello–Frosch et al. (2017)	65	–	–	–	ng	ng	0.29	–	< 0.02	–	2.27	USA, 2010/11
Arbuckle et al. (2013)	100											Canada, 2005–08
Asia												
Zhao et al. (2025)	350	–	–	< 0.01	< 0.01	2.09	0.28	0.09	< 0.01	0.35	0.77	China, 2021–23
Song et al. (2025)	119	0.08	–	0.03	0.03	3.23	0.27	0.10	0.01	0.17	0.94	China, 2019–21
Yuan et al. (2025)	38	<	–	0.72	0.004	2.08	0.53	0.06	<	0.55	0.13	China
Mahfouz et al. (2023)	59	–	–	–	0.31	44.9	< 0.05	< 0.05	–	1.98	54.6	Lebanon, 2018–21
Gao et al. (2019)	132	0.12	0.04	0.18	0.04	1.88	0.38	0.21	0.12	0.11	2.35	China, 2015/16
Pan et al. (2017)	100	–	–	–	–	1.41	0.26	0.20	–	0.46	4.38	China, 2014
Cao et al. (2018)	337	–	–	–	–	1.25	0.11	0.10	–	< 0.02	1.01	China, 2013–15
Yang et al. (2016a)	157	–	–	–	–	1.15	0.20	0.10	–	0.18	1.18	China, 2013
Guo et al. (2021)	490	–	–	–	0.21	3.28	0.40	0.72	0.11	0.09	1.95	China, 2013
Han et al. (2018)	396	–	–	–	0.09[a]	31.8[a]	0.42[a]	0.20[a]	0.19[a]	0.31[a]	1.33[a]	China, 2010–13
Zhang et al. (2017)	321	0.05	–	–	0.05	1.23	0.15	0.12	–	3.87	2.99	China, 2010
Liu et al. (2011)	50	–	–	–	–	1.12	0.32	0.23	–	0.06	1.47	China, 2009

[a]Geometric mean; ng: not given

Table 4.6 Umbilical cord blood to maternal plasma/serum concentration ratios (medians or means)

References	N	PFOS	PFOA	PFNA	PFHxS	Country, year
Yuan et al. (2025)	38	0.3	0.8	1.3	0.5	China
Rovira et al. (2019)	40	0.4	0.3	–	–	Spain, 2016/17
Gao et al. (2019)	132	0.5	0.9	0.7	0.5	China, 2015/16
MacDonald et al. (2022)	62	0.5	0.8	0.6	0.5	Canada, 2013–16
Pan et al. (2017)	100	0.4	0.7	0.4	0.5	China, 2014
Yang et al. (2016a)	157	0.3	0.7	0.4	0.5	China, 2013
Han et al. (2018)	369	0.3	0.8	0.5	1.0	China, 2010–13
Cariou et al. (2015)	100	0.4	0.9	–	–	France, 2010–13
Morello-Frosch et al. (2016)	34–64	0.8	0.8	0.4	–	USA, 2010/11
Fisher et al. (2016)	315–865	0.1	0.3	–	0.2	Canada, 2008–11
Zhang et al. (2013)	32	0.6	0.5	0.3	0.3	China, 2010
Liu et al. (2011)	50	0.5	0.9	–		China, 2009
Fromme et al. (2010)	33	0.3	0.7	–	–	Germany, 2007/08
Gützkow et al. (2012)	123	0.3	0.8	0.4	0.7	Norway, 2007/08
Kato et al. (2014)	71	0.4	0.9	0.6	0.5	USA, 2003–06
Manzano-Salgado et al. (2015)	66	0.3	0.7	0.4	–	Spain, 2003–06
Needham et al. (2011)	12	0.3	0.7	0.5	0.7	Faroe Islands, 2000

who breastfeed for a short period or not at all, lower PFAS concentrations were observed in the umbilical cord blood of multiparous women and those who breastfed for a longer duration. A systematic review indicated a U-shaped trend in the placental transfer rate for PFCA with chain lengths ranging from C6 to C14, showing the lowest rates for PFOA and PFUnDA and the highest for PFHxA (Appel et al. 2022). When comparing PFCA and PFSA transfer rates for compounds with the same chain length in the C6-C8 range, PFSA compounds such as PFHxS, PFHpS, and PFOS are transferred less efficiently than PFCA.

4.1.3.5 Concentrations in the Blood in Special Exposure Situations

There is now substantial evidence in the scientific literature that proximity to industrial plants where fluorochemicals are produced or used can lead to increased exposure. However, extinguishing foams (known as AFFF, Aqueous-Film-Forming Foams), primarily used by the military and fire departments, also predominantly contained PFOS. In Germany, for example, telomer-based PFAS (e.g., 6:2 fluorotelomer sulfonate) was used until 2000 and continued to be used from 2011 onwards. Due to the long-standing use of these foaming agents during exercises at military and civilian airports, refineries, and chemical industry sites, PFOS-related contamination is commonly found there. Drinking water serves as a significant pathway for exposure. Between 2013 and 2015, Hu et al. (2016) conducted a spatial analysis of measured PFAS concentrations in drinking water in the USA and discovered that the drinking water supply for 6 million US citizens exceeded the limit of 70 ng/L for PFOS and PFOA recommended by the US EPA. Currently, various proven techniques for removing PFAS from drinking water exist, such as granular activated carbon filtration, ion exchange resins, reverse osmosis, and fluorographene-based sorbents attached to magnetic nanoparticles, which can work effectively, particularly for longer-chain PFAS like PFOA and PFOS (Wee and Aris 2023).

Against this backdrop, three notable exposure pathways can be identified beyond general population and occupational exposure. One concern is direct exposure in the vicinity of PFAS production facilities (e.g., through inhalation, via contamination of food and drinking water), after using AFFF products (primarily through drinking water), and from soil contamination with PFAS-containing materials.

4.1.3.6 Exposure in the Surroundings of PFAS Producers and Processors

For the first time, an unusual contamination situation was reported in the catchment area of the Little Hocking water supply in Ohio, USA. This region is located near an active PFOA emitter that has been operational since 1951. The results of drinking water measurements revealed high levels of PFOA in the wells and at the transfer point to the distribution network, with concentrations ranging from 1.9 to 18.6 μg/l over several years. Consequently, an initial study was conducted between 2004 and 2005 on 371 subjects, finding an exceedingly high median PFOA serum level of 329 μg/l among individuals without workplace-related exposure. In comparison, the median level was 775 μg/l among 18 participants who were simultaneously employed at the PFOA production company (Emmett et al. 2006). The 291 subjects who solely used water from the central drinking supply had a median level of 374 μg/l (ranging from 7 to 1950 μg/l), while 26 individuals who exclusively consumed packaged or spring water had a median level of 320 μg/l. The lowest blood concentrations were observed in the population group that only consumed packaged, cistern, or spring water, with a median of 71 μg/l among these ten individuals. A significant increase in blood PFOA levels was noted, corresponding to the number of meals that included locally harvested vegetables and fruits. The study revealed no gender differences in internal exposure; however, a specific age dependency was observed, with higher PFOA levels noted in age groups over 60 and under 6 years.

Moreover, at the "Fayetteville Works" in North Carolina, fluorochemicals were produced from 1980 to 2017, and the process wastewater was discharged into the Cape Fear River. Approximately 90 miles downstream, raw water was used to provide drinking water to around 200,000 people in the city of Wilmington and Hanover County. Both the raw and finished water were contaminated with various fluoroethers, including hexafluoropropylene oxide dimer acid (HFPO-DA or GenX), perfluoro-3,5,7,9-tetraoxadecanoic acid (PFO4DA), perfluoro-3,5,7,9,11-pentaoxadodecanoic acid (PFO5DoA), and perfluoro-2-{[perfluoro-3-(perfluoroethoxy)-2-propanyl]oxy}ethanesulfonic acid (Nafion by-product2). A biomonitoring study conducted with 344 participants from 2017 to 2018 found medians (95th percentiles) of 8.6 µg/l (26.8 µg/l) for PFOS, 4.3 µg/l (11.0 µg/l) for PFOA, and 3.2 µg/l (8.5 µg/l) for PFHxS (Kotlarz et al. 2020). Nafion by-product2, PFO4DA, and PFO5DA were measured with medians (95th percentiles) of 2.7 µg/l (8.4 µg/l), 2.5 µg/l (12.8 µg/l), and 0.3 µg/l (1.0 µg/l), respectively, while HFPO-DA was not quantifiable above the LOD of 2 µg/l. In a subset of 80 subjects from Wilmington, Kotlarz et al. (2024) also identified 2,2,3,3-tetrafluoro-3-((1,1,1,2,3,3-hexafluoro-3-(1,2,2,2-tetrafluoroethoxy)propan-2-yl)oxy)propanoic acid (also known as Hydro-EVE) with a median of < 0.1 µg/l and a 95th percentile of 0.47 µg/l.

From 2017 to 2019, a median concentration of 44.4 µg/l (95th percentile: 190 µg/l, maximum: 1400 µg/l) for PFOA and 3.9 µg/l each for PFOS and PFHxS was observed in blood samples from 18,345 participants aged 14 to 39 years in the vicinity of a PFAS production company active since the 1960s in northern Italy, which led to significant groundwater contamination (Pitter et al. 2020). The median contamination level for PFOA in drinking water was 0.32 µg/l (95th percentile: 1.1 µg/l) prior to the implementation of granular activated carbon filters. Key predictors of serum levels included gender, municipality, length of stay in the affected area, and number of births.

In the Netherlands, PFOA blood levels were measured in 2016 among residents in two zones surrounding the DuPont/Chemours chemical plant, where PFOA was used from 1970 to 2002 and phased out in 2012 (Van der Poll et al. 2017). For residents who had lived near the plant for an extended period, the median was 10.2 µg/l (1.3–147 µg/l), while those living farther away from the plant or for a shorter duration had medians of 3.4 µg/l (0.9–14.1 µg/l) and 2.8 µg/l (0.1–45.7 µg/l), respectively.

Furthermore, PFOA was produced in an industrial plant in southeastern Germany from 1968 to 2003 and was utilized for manufacturing fluoropolymers until 2008. This usage resulted in extensive environmental contamination and human exposure, with contaminated drinking water being the primary pathway for exposure among the general population. As part of a study conducted in 2018, a total of 906 participants aged 7 to 85 years from areas with varying degrees of contamination were examined, with no known occupational exposure. The medians (95th percentiles) were 20.1 µg/l (58.5 µg/l) for PFOA, 1.8 µg/l (6.2 µg/l) for PFOS, 0.5 µg/l (1.3 µg/l) for PFHxS, and 0.54 µg/l (1.5 µg/l) for PFNA. In a control area in 2016, the median PFOA level for 158 individuals was significantly lower at 1.1 µg/l (95th percentile: 2.4 µg/l), while the other PFAS levels remained in a similar concentration range (Fromme et al. 2017). In a follow-up study conducted in 2022, a 57% reduction in PFOA was observed compared to the first study, which used the same subjects from the general population aged over 12 years (medians: 23.2 vs. 10.0 µg/l) (Lahne et al. 2024).

In the separately studied group of women of reproductive age and children under 12 years, PFOA levels were reduced by 60% (medians: 10.8 vs. 4.3 μg/l) and 73% (medians: 20.7 vs. 5.5 μg/l), respectively.

In China, 1,202 people aged 22 to 96 years were examined as part of the C8 Health Project in Shenyang, an industrial city near the country's largest fluoropolymer production facilities. Medians of 24.1 μg/l for PFOS, 6.0 μg/l for PFOA, and 1.1 μg/l for PFNA were found (Zeeshan et al. 2020).

4.1.3.7 Exposure After the Application of AFFF

From 2013 to 2016, high levels of PFAS were detected in municipal water systems downstream from an Air Force base, impacting approximately 80,000 people in Colorado, USA (Barton et al. 2020). AFFF was suspected as the source. In 2018, median concentrations of 9.7 μg/l for PFOS, 3.0 μg/l for PFOA, and 16.6 μg/l for PFHxS were found in the serum of 213 adults, which were about 1.7, 2.1, and 14 times higher than the national NHANES data at that time. Factors influencing PFHxS concentrations included place of residence, frequency of bottled water consumption, age, race/ethnicity, and smoking history.

Up to 150 μg/l were detected in drinking water from public water systems and private wells in New Jersey communities. In 124 individuals, the median (95th percentile) was found to be 3.89 μg/l (12.4 μg/l) for PFNA, which was five times higher than the NHANES data (Grabet et al. 2019). After implementing measures to reduce drinking water treatment, 120 participants aged 20 to 74 were selected from 2017 to 2020 during the first year, with samples taken at one-year intervals (Yu et al. 2021). On average, PFNA serum levels decreased by 12% after two years and by 27% after three years compared to the first year.

In southern Sweden, drinking water from a waterworks in Ronneby, located near a military airport, was found to be contaminated with approximately 8.0 μg/l PFOS, 0.1 μg/l PFOA, and 1.7 μg/l PFHxS due to the long-term use of PFAS-containing extinguishing foams (Xu et al. 2021). Among 3299 potentially exposed individuals, geometric mean values of 135 μg/l for PFOS, 6.8 μg/l for PFOA, and 114 μg/l for PFHxS were recorded from 2014 to 2015, while in the blood of 219 controls, only medians of 3.9 μg/l for PFOS, 1.5 μg/l for PFOA, and 0.84 μg/l for PFHxS were noted.

In Australia, AFFF products were used on Department of Defense bases for firefighting and training purposes from about 1970 until the early 2000s, leading to the contamination of drinking water in three residential areas (Lazarevic et al. 2023). A study involving 881 exposed individuals and 801 controls found median values of 5.2–5.7 μg/l for PFOS, 1.5–1.8 μg/l for PFOA, and 2.8–3.7 μg/l for PFHxS among those exposed. These levels were higher than those of the controls by a factor of approximately 1.6–1.9 for PFOS, 1.3–1.4 for PFOA, and 2.2–4.6 for PFHxS.

4.1.3.8 Pollution from Contaminated Soils

Following a large-scale application of a PFOA-contaminated so-called soil conditioner, elevated concentrations of this substance were found in surface water and drinking water, reaching up to 640 μg/l in a region of western Germany. As part of a cross-sectional study, the internal contamination levels in this region and a control area were investigated, involving 170 children and 521 adults (Hölzer et al. 2008). The geometric mean PFOA values were 23.4 μg/l for children, 25.3 μg/l for men, and 22.1 μg/l for women, which were 4.6-fold (children), 4.4-fold (men), and 8.3-fold (women) higher than in the corresponding groups in the control region. This also applied to the PFHxS concentrations in the blood. The differences in PFHxS levels were 53% (children), 14% (men), and 80% (women) higher than in the control region, based on geometric mean values. PFOS was only statistically significantly higher in the exposed women group. It was also shown that the level of drinking water consumption, as estimated by the exposed subjects for themselves or their children, was significantly associated with the PFOA levels in plasma. In a follow-up study conducted in the exposure area in 2017, only geometric mean values of 5.5 μg/l were found in men, 2.8 μg/l in women, and 2.3 μg/l in 16–17-year-old adolescents (Bacher et al. 2019).

In another case in southwest Germany in 2008, compost contaminated with perfluorinated compounds was applied to agricultural land, leading to groundwater and drinking water contamination (LGA 2018). Following measures to reduce exposure, a total of 120 people were exposed via drinking water (Group A), 98 people were exposed through soil but not through drinking water (Group B), and 83 individuals without exposure to drinking water or soil (Group C) were examined in 2018. The highest median values for PFOA in the blood were found in Group A at 15.6 μg/l, while only 2.3 μg/l were recorded in Group B and 1.7 μg/l in Group C. For PFOS and PFHxS, there was no correlation with contamination from drinking water sources. In the 2020 follow-up study, only medians of 12.7 μg/l were found in Group A, indicating an 18% decrease (LGA 2021).

4.1.3.9 Concentrations in Urine

Table 4.7 presents the concentrations of various PFSA and PFCA in the urine of the general population. In the initial studies conducted in Korea in 2012 and in the USA in 2013/14, PFOA and PFOS were detected in only a few samples above the relatively high limits of quantification (Kim et al. 2014, 2019; Calafat et al. 2019). Recent studies conducted in China reported significantly different median concentration levels. The highest median values to date were observed in a Korean study in 2012 (Kim et al. 2014). Compared to studies in East Asia, the medians from an Austrian survey involving a total of 85 children examined in 2020 were generally much lower. Two recent studies investigated other PFAS such as PFPeS, PFHpS, PFNA, PFDA, PFUnDA, and PFDoDA, where the medians, except for the study from China showing 4.2 ng/l for PFDS and 2.5 ng/l for PFDA, consistently fell below the limits of quantification (Hartmann et al. 2023; He et al. 2023). In the study by Kim et al. (2019), no correlation was found between PFAS levels in urine and the serum samples taken in parallel from 94 individuals aged 2 to 89 years. In a Chinese study conducted in two cities in 2020/21, urine and umbilical cord blood samples were collected from 80 neonates (Yao et al. 2023). Short-chain PFAS like PFBA and PFPeA exhibited the highest medians at 92.9 μg/l and 14.8 μg/l, respectively, followed by PFOA at 10.9 μg/l.

Overall, the study results suggest that urine measurements offer a straightforward approach for estimating exposure to shorter-chain PFAS. In contrast, blood analyses are more appropriate for longer-chain PFAS.

4.1.3.10 Occurrence in Breast Milk

Since PFAS have a strong binding to plasma proteins, their transfer from maternal blood to breast milk is significantly lower than that of lipophilic substances. Table 4.8 summarizes the results of various studies. Additionally, several recent reviews on the occurrence of PFAS in breast milk have been published (Jian et al. 2017; Lehmann et al. 2018; Hu et al. 2021; LaKind et al. 2023).

In a global study of 44 sample pools conducted between 2016 and 2018 as part of the UNEP program, representing 42 countries, PFOA was quantified in all samples, ranging from 6.2 to 37.4 pg/g. PFOS appeared in 36 samples, with concentrations varying from less than 6.2 to 212 pg/g, while PFHxS was detected in only four samples, reaching a maximum of 111 pg/g (Fiedler and Sadia 2021). When examining the mean concentrations, PFOS levels were highest in the Asia–Pacific region, followed by Eastern and Western Europe, and were lowest in African, Latin American, and Caribbean countries. For PFOA, the mean values were highest in Eastern and Western European countries, while they were lower in all other regions.

In Europe, medians of approximately 30–120 ng/l for PFOS and 26 to 100 ng/l for PFOA were still observed in older samples until around 2016. The results from more recent studies show lower levels. For instance, Serrano et al. (2021) documented medians (95th percentile) of < 0.9 ng/l (26.0 ng/l) for PFOS, 7.2 ng/l (55.1 ng/l) for PFOA, and < 0.66 ng/l (16.0 ng/l) for PFHxS in 34 Spanish breast milk samples. Similarly, more recent studies from the USA have also found lower concentrations compared to older studies; for example, 24 ng/l for PFOS, 17 ng/l for PFOA, and < 10 ng/l for PFHxS in 426 breast milk samples from the New Hampshire Birth Cohort Study, with levels in the samples from primiparous mothers being higher than in multiparous mothers (Criswell et al. 2023). Significant differences in medians, ranging from 35 ng/l to 91 ng/l for PFOS and 35 ng/l to 336 ng/l for PFOA, have also been reported in Asian countries. The highest median values (95th percentile) for PFOS were measured in 324 samples from China at 201 ng/l (not provided) (Chen et al. 2024), while for PFOA, the highest median in 117 Chinese samples reached 970 ng/l (5137 ng/l) (Tao et al. 2025). In a study involving milk collection from 1151 mothers across 21 Chinese cities, considerable differences were noted among the 12 provinces (Yao et al. 2023). For example, the median PFOA level in Chongqing was

Table 4.7 Perfluorinated substances in the urine in ng/l (medians, in brackets: percent of values above the limit of quantification)

References	N	Age (years)	PFBA	PFPeA	PFHxA	PFHpA	PFOA	PFBS	PFHxS	PFOS	Country, year
Dai et al. (2024)	180		630 (97)	2000 (86)	30 (53)	< 40 (6)	< 50 (24)	110 (61)	90 (74)	< 40 (22)	China, 2020–22
Yao et al. (2023)	80	16–47	93 (100)	14.8 (96)	< 1.0 (30)	< 0.4 (46)	10.9 (98)	< 1.0 (36)	< 2.0 (25)	< 1.0 (0)	China, 2020/21
He et al. (2023)	153	49[a]	1890 (99)	281 (97)	161 (100)	< 10 (48)	12.9 (88)	4.2 (60)	34 (79)	49.6 (97)	China, 2021
Hartmann et al. (2023)	85	6–11	–	3.5 (88)	4.4 (100)	1.5 (87)	2.9 (100)	–	< 0.8 (34)	< 1.3 (47)	Austria, 2020
Li et al. (2021a)	189	8–12	–	–	21 (100)	48 (99)	79 (100)	22 (87)	19 (78)	21 (83)	China, 2019
Li et al. (2021b)	53	4–6	–	–	0.82 (58)	0.59 (75)	2.5 (100)	–	–	2.2 (32)	China, 2017–18
Wang et al. (2018)	41	3–68	16 (98)	14 (93)	11 (95)	4.2 (95)	17 (98)	14 (95)	13 (93)	3.5 (93)	China, 2015
Kim et al. (2019)	94	2–82	–	1.7 (84)	< 48 (23)	< 41 (4)	< 96 (5)	< 58 (2)	< 67 (3)	< 190 (0)	Korea, 2015
Calafat et al. (2019)	2628	> 6	< 100 (13)	–	< 100 (0)	< 100 (1)	< 100 (0)	< 100 (0)	< 100 (0)	< 100 (0.1)	USA, 2013–14
Kim et al. (2014)	120	5–13	< 175 (0)	2300[a] (70)	700[a] (11)	1350[a] (36)	< 163 (0)	492[a] (1)	< 88 (0)	< 138 (0)	Korea, 2012
Kim et al. (2014)	57	20–29	1720[a] (1)	2390[a] (25)	1389[a] (5)	495[a] (7)	< 163 (0.1)	< 113 (0)	< 88 (0)	< 138 (0)	Korea, 2012

[a]Mean; [b]LOQ: 41–190 ng/l

Table 4.8 Concentrations of perfluorinated substances in breast milk in ng/l

References	N	PFOS		PFOA		PFHxS		Country, year
		% > LOQ	Median (Min–Max)	% > LOQ	Median (Min–Max)	% > LOQ	Median (Min–Max)	
Europe								
Parizek et al. (2023)	70[h]	68	< 3 (2–40[g])	77	< 3 (4–11[g])	0	< 3	Czech Rep., 2019–21
Fromme et al. (2022)	180	31	< 25 (<25–248)	53	< 25 (<25–326)	0	< 75	Germany, 2016/17
Serrano et al. (2021)	82	34	< 0.9 (<0.9–65)	84	7.2 (<0.9–252)	24	< 0.7 (<0.7–45)	Spain, 2015–18
Cerná et al. (2020)	232	99	14[c] (<2–83)	100	24[c] (<3–160)	–	–	Czech Rep., 2017
Awad et al. (2020)	10	25	39[d] (23–58)	25	42[d] (<2–81)	20	7[d] (<4–13)	Sweden, 2016
Abdallah et al. (2020)	16[a]	62	20 (<20–120)	100	100 (16–350)	31	< 40 (<40–87)	Ireland
Van Beijsterveldt et al. (2022)	133	93	30 (<15–40)	97	36 (<15–50)	–	–	Netherlands, 2013–18
Hartmann et al. (2024)	40	98	12 (<13–55)	48	< 6.2 (<6.2–83)		< 6.2 (<6.2–9.7)	Austria, 2014–16
Beser et al. (2019)	20	55	69 (<66–78)	85	138 (<133–180)	–	–	Spain
Motas Guzmàn et al. (2016)	67	–	–	60	26 (<10–211)	–	–	Spain, 2014
Cariou et al. (2015)	61	31	– (<40–376)	18	– (<50–308)	15	– (<6–22)	France, 2010–13

(continued)

Table 4.8 (continued)

References	N	PFOS		PFOA		PFHxS		Country, year
		% > LOQ	Median (Min–Max)	% > LOQ	Median (Min–Max)	% > LOQ	Median (Min–Max)	
Barbarossa et al. (2013)	21	90	57[d] (<15–288)	81	76[d] (<24–241)	–	–	Italy, 2010
Lankova et al. (2013)	50	100	30 (7–114)	100	44 (12–128)	16	– (<6–22)	Czech Rep., 2010
Raab et al. (2013)	302	100	50 (20–260)	2	< 80 (<80–290)	2	< 10 (<10–30)	Germany, 2007–09
Antignac et al. (2013)	48	90	79 (<5–330)	98	75 (<5–224)	100	50 (40–66)	France, 2007
Völkel et al. (2008)	57	100	120 (28–309)	11	< 200 (<200–290)	–	–	Germany, 2006
Bernsmann and Fürst (2008)	203	55	56 (<40–284)	56	90 (<80–610)	1	[b]	Germany, 2006
North America								
Zheng et al. (2021)	50	100	30 (6.4–187)	86	14 (<16–51)	90	6.7 (<6.1–17)	USA, 2019
Criswell et al. (2023)	426	92	24 (16–36 [f])	84	17 (12–27[f])	8	< 10	USA, 2014–19
Rwan et al. (2022)	664	96	30 (–)	99.5	34.1 (–)	63	6.7 (–)	Canada, 2008–11
Tao et al. (2008)	45	96	106 (<32–617)	89	36 (<30–161)	51	12 (<12–64)	USA, 2007
Asia, Africa								
Jonathan et al. (2024)	29	86	93 (<24–190)	86	72 (<20–93)	–	–	Ghana, 2020/21

(continued)

Table 4.8 (continued)

References	N	PFOS		PFOA		PFHxS		Country, year
		% > LOQ	Median (Min–Max)	% > LOQ	Median (Min–Max)	% > LOQ	Median (Min–Max)	
Yao et al. (2023)	1151	100	50 (2.6–2503)	100	336 (12–2059)	95	7.3 (<1.0–29)	China, 2020/21
Chen et al. (2014)	324	93	201 (<LOD-3896)	96	64 (<LOD-333)	84	25 (<LOD-420)	China, 2018–21
Mahfouz et al. (2023)	41	100	91 (0.32–188)	100	42 (0.29–80)	100	48 (0.34–85)	Lebanon, 2018–21
Macheka et al. (2022)	50	66	50 (<5–510)	94	210 (5–1750)	46	< 5 (<5–1180)	South Africa, 2020
Yang et al. (2025)	447	96	21 (<2.8–38)	96	23 (<2.1–40)	60	3.9 (<3.0–8.0)	China, 2018/19
Jin et al. (2020)	174	50	1.0 (<1.0–202)	100	31 (10–1102)	< 1	< 7.7 (<7.7–80)	China, 2018/19
Kim et al. (2023)	218	100	50 (17–512)	100	100 (33–1370)	88	31 (<17–584)	Korea, 2018
Han et al. (2023a)	100[a]	100	35 (9.4–637)	100	61 (16–4294)	61	2.5 (<LOD-119)	China, 2017–20
Awad et al. (2020)	10[e]	100	65[d] (16–177)	96	139[d] (64–308)	65	8[d] (<4–18)	China, 2016
Kang et al. (2016)	264	99	50 (31–77)[f]	99	72 (52–110) [f]		–	Korea, 2013
Al-sheyab et al. (2015)	79	94	50 (<10–178)	100	83 (24–1220)	–	–	Jordan, 2013
Tao et al. (2025)	117	100	36 (3–224)	100	970 (47–15,879)	100	5 (1–33)	China, 2010–13
Lee et al. (2018)	293	100	48 (15–380)	88	40 (<10–657)	35	< 10 (<10–133)	Korea, 2011
Fujii et al. (2012)	30	–	–	93	89 (<40–194)	–	–	Japan, 2010
Liu et al. (2011)	50	100	42 (9–198)	100	121 (25–1440)	–	–	China, 2009
Liu et al. (2010)	1237	100	49 (6–137)	88	35 (<1–814)	–	–	China, 2007

[a]Pooled samples; [b]Only two samples > LOQ (180 and 160 ng/l); [c]Geometric mean; [d]Arithmetic mean; [e]Subgroup living in Shanghai; [f]Interquartile range; [g]5th and 95th percentile: [h]Subgroup living in Ceske Budejovice; *LOD* Limit of determination

63 ng/l, whereas in Huantai it was 1171 ng/l. The authors attribute the differences to the distribution patterns of numerous fluorochemical industrial plants in China.

In addition to various fluorinated impurities, linear (L-PFOS) and branched PFOS isomers (Br-PFOS) are produced during the electrochemical fluorination process. Cerná et al. (2020) examined pooled samples from four time points (2006, 2010/11, 2014, and 2017) and found that the median proportion of L-PFOS in the total content ranged from 58 to 66%. Another study determined a proportion of 64% in an analysis of 50 Czech breast milk samples (Lankova et al. 2013). The proportion was also approximately 66% in 372 Dutch samples (van Beijsterveldt et al. 2022). In the nationwide Canadian Maternal-Infant Research on Environmental Chemicals Study, the median proportion of L-PFOS was similar at 70% (Rawn et al. 2022). Awad et al. (2020) analyzed ten samples each from three Chinese cities and one Swedish city, reporting a proportion of L-PFOS between 88 and 92% in China and 85% in Sweden. In contrast to most European studies, the proportion of L-PFOS in 100 pooled breast milk samples from the 2017–2020 National Human Milk Survey in China was significantly higher at 96% (Han et al. 2023b, a). This study examined a total of six Br-PFOS isomers, including perfluoro-1-methylheptanesulfonate (1 m-PFOS), (3 + 4)m-PFOS, perfluoro-5-methylheptanesulfonate (5 m-PFOS), and perfluoro-6-methylheptanesulfonate (iso-PFOS), with detection frequencies ranging from 8 to 33%. The concentrations of L-PFOA observed in studies to date are low, if detectable at all (Awad et al. 2020; van Beijsterveldt et al. 2022; Han et al. 2023a).

Some studies describe trends in breast milk concentrations over time. The most extensive time series on PFAS presence in breast milk originates from Sweden, where data for Stockholm spanning 1972 to 2006 and for Gothenburg from 2007 to 2015 were evaluated, typically using pooled samples (Nyberg et al. 2018). Figure 4.8 illustrates the time trend in Stockholm for three PFAS. For PFOS and PFOA, the highest mean concentrations were observed between 1984 and 2004 and from 1988 to 2004, respectively, before sharply declining to a plateau between 2013 and 2016. In the Czech Republic, a thorough study also examined 59 breast milk samples from 2006, 183 from 2010–2011, 164 from 2014, and 232 from 2017 (Cerná et al. 2020). In this case, the mean values declined over time from 45 ng/l, 31 ng/l, 29 ng/l to 20 ng/l for PFOS and from 75 ng/l, 59 ng/l, 35 ng/l to 23 ng/l for PFOA. In the USA, the medians were still 106 ng/l for PFOS and 36 ng/l for PFOA in 45 samples from 2009 (Tao et al. 2008), while a study of 50 samples in 2018 showed they had dropped to only 30 ng/l and 14 ng/l (Zheng et al. 2021). In contrast, Kim et al. (2023) reported increasing PFOA values in Korea, rising from an average of 41 ng/l in 2007 to 114 ng/l in 2018, whereas PFOS remained relatively stable, with average values of 61 ng/l in 2007 and 58 ng/l in 2018. Han et al. (2023a) compared data from the third National Human Milk Survey in China, conducted between 2017 and 2020, with the first survey in 2007, which used the same study protocol across the same 12 provinces (Liu et al. 2010). The mean concentrations of the sum of six PFAS (PFOS, PFOA, PFHxS, PFNA, PFDA, and PFUnDA) decreased from 220 ng/l to 173 ng/l in pooled milk samples between both surveys, though the difference

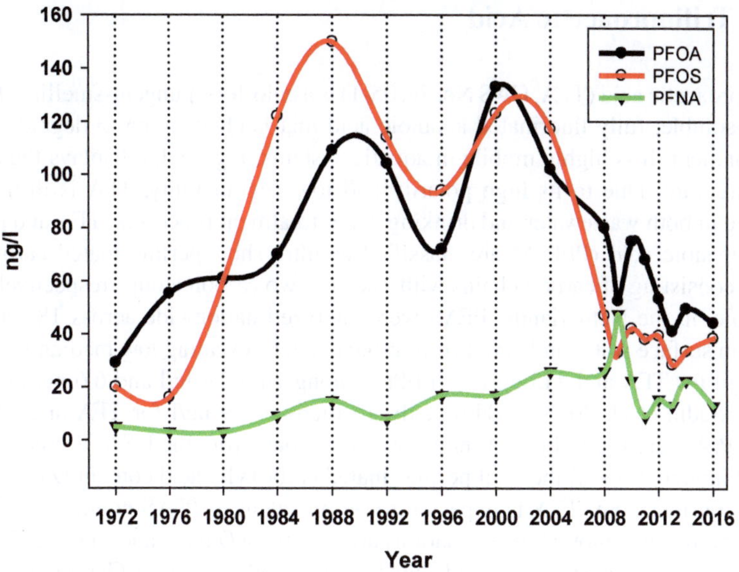

Fig. 4.8 Time trend of the concentrations of PFOS, PFOA, and PFNA in breast milk in Sweden in ng/l (after Nyberg et al. 2018)

was not statistically significant. Significant decreases in concentrations and some increases were observed within the individual provinces.

Several studies on PFOS have investigated the transfer efficiency (TE B/S) between breast milk and maternal serum. For instance, Kärrman et al. (2007a) found a TE B/S of 1% in 12 paired samples from Sweden, while Fromme et al. (2010) reported 1.3% in 38 pairs of samples from Germany. Additionally, a median TE B/S value of 0.9% was observed in 126 mothers from a highly contaminated region in southern Sweden (Blomberg et al. 2023). In China, Liu et al. (2011) reported a median TE B/S of 1.4% for PFOS in a study of 50 couples. Criswell et al. (2023) found a PFOS TE B/S of 0.9% in 294 paired samples in the USA. Furthermore, Han et al. (2023b) analyzed 145 paired maternal serum and human milk samples for six PFOS isomers in China and discovered that the TE B/S value was clearly below 1%.

Studies found a median TE B/S for PFOA of 9.3% in China (Liu et al. 2011), 1.8% in the USA (Criswell et al. 2023), and 2.8% in Sweden (Blomberg et al. 2023). The working group led by Liu et al. (2011) also determined the TE B/S for PFNA and PFDA to be 3.9% and 2.9%, respectively. These results confirm that longer-chain PFAS transfer into breast milk at low rates.

4.2 Trifluoroacetic Acid

Trifluoroacetic acid (TFA, CAS No. 76-05-1) is a colorless, pungent-smelling, highly water-soluble, fully fluorinated alkanoic acid that is challenging to degrade in the environment. It is highly mobile in aquatic systems. Figure 4.9 displays the chemical structure. Due to its high polarity and low degradability, TFA is difficult to remove in both wastewater and drinking water treatment processes. TFA and perfluoropropanoic acid (PFPrA) are classified as ultra-short perfluorinated carboxylic acids, consisting of carbon chains with one and two carbon atoms, respectively. For example, in the Netherlands, PFAS were analyzed nationwide across 18 samples, 11 from surface water and seven from groundwater, revealing low median values of 352 ng/l for TFA and 245 ng/l for PFPrA, along with 27 ng/l and 6.6 ng/l, respectively (Sadia et al. 2023). Additionally, medians of 79 ng/l for TFA and 6.9 ng/l for PFPrA were found in 80 drinking water samples from the USA, accounting for approximately 90% of the total perfluorinated carboxylic acid content (Zheng et al. 2023). Furthermore, TFA is expected to be prevalent in PFAS-contaminated sites where biotic and abiotic transformations are occurring (Zhang and Liang 2023). For instance, groundwater samples collected from five polluted sites in Germany revealed levels ranging from 800 to 2200 ng/l for TFA (Janda et al. 2019). Moreover, several recent reviews on the occurrence and fate of TFA in the environment have become available (Garavagno et al. 2024; Zhi et al. 2024).

TFA serves as a chemical process reagent and a synthetic building block for pharmaceuticals, veterinary pharmaceuticals, biocides/crop protection agents, halogenated refrigerants, and blowing agents (Solomon et al. 2016; Zhang and Liang 2023). Additionally, TFA may act as a potential breakdown product of fluorinated chemicals (Solomon et al. 2016). Moreover, halogenated hydrofluorocarbons (HCFCs) such as 1,1,1,2-tetrafluoroethane (R134a) and 2,3,3,3-tetrafluoropropene (R1234yf) can degrade in the atmosphere, producing TFA/trifluoroacetate. In 2020, total TFA emissions in Europe were projected to be just under 15,000 tons, with half attributed to HCFC conversion, which could potentially rise to around 50,000 tons annually by 2050 (Behringer et al. 2021). Furthermore, several PFAS have TFA as a transformation product, including pesticides, pharmaceuticals, and industrial chemicals (Arp et al. 2024). If TFA occurs naturally in the environment, it is critically discussed (Joudan et al. 2021).

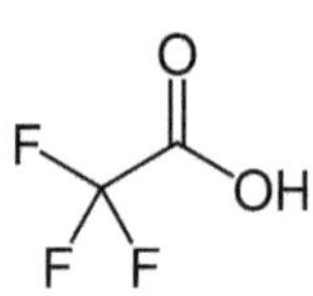

Fig. 4.9 Chemical structure of trifluoroacetic acid

4.2.1 Health Aspects

No studies on the health effects in humans are available. TFA has low acute toxicity, and in studies involving repeated oral administration to rats, the liver was the primary target organ, with increased liver weight due to hepatocellular hypertrophy caused by peroxisome proliferation (ECHA 2014; Dekant and Dekant 2023). An extended one-generation study noted no adverse effects on fertility, and no adverse effects were found in a developmental toxicity study.

4.2.2 Toxicokinetics

Data on toxicokinetics are limited. Summaries are available from Boutonnet et al. (1999) and Dekant and Dekant (2023). After uptake, TFA binds to albumin in humans at 44% to 55% and is present in erythrocytes at only 4% to 14%. EFSA (2014) reported a study in rabbits indicating half-lives of 16 to 34 h after intravenous administration and 17 h following oral administration. When injected intravenously into two human subjects, TFA was completely excreted in urine within 63 to 72 h, with a plasma half-life of 25 to 32 h. Indirect evidence arises from the use of halothane (2-bromo-2-chloro-1,1,1-trifluoroethane), a halogenated anesthetic. In six children aged 14 to 119 months who received halothane for 60 min, 11.4% was excreted in urine as TFA (Wark et al. 1990). The mean half-life of TFA in urine in this instance was reported to be 42 h. Furthermore, in an inhalation study, pregnant rats were exposed to R123. TFA, a major metabolite of the test substance, was detected in all urine samples from standard litters that nursed for 1 h from treated dams. In contrast, TFA was not detectable in urine samples from litters of untreated dams (Buschmann et al. 2001). The authors suggest that most, if not all, of the TFA in the urine originates from the mother's milk, where TFA could also be found.

4.2.3 Occurrence in the Blood and Urine

Few studies have examined TFA in the general population samples. Duan et al. (2020) initially analyzed 256 serum samples from individuals aged 19 to 87 in China in 2017. TFA was detected in 97% of the samples, with a median concentration of 8.5 μg/l (75th percentile: 12.6 μg/l), exhibiting increased levels with age. In another study conducted in 2020, four serum samples were analyzed from 81 participants (Zheng et al. 2023). The serum samples showed a median of 6.0 μg/l (<4.4–77 μg/l) in 74% of the samples. Additionally, PFPrA was found in 99% of the serum samples, with a median of 1.0 μg/l (<0.3–2.9 μg/l) in this study. Furthermore, a lower median level of 0.23 μg/l, with a range from < 0.005 to 2.48 μg/l, was reported in 55% of 66 cord blood samples collected in 2022 in China (Jia et al. 2023). This study indicated the presence of PFPrA in 73% of the samples, with concentrations ranging from <

0.004 to 4.53 μg/l (median: 0.27 μg/l). In the US, a median of 17 μg/l (<10–590 μg/l) was observed in 119 serum samples from Wilmington, NC, in the vicinity of a fluorochemical facility in 2010–2016 (Cheng et al. 2025).

In the US in 2020, Zheng et al. (2023) detected TFA in 31% of 81 urine samples above the LOD with a median of <3.5 μg/l and a maximum value of 290 μg/l. Moreover, Muir et al. (2025) analyted 70 pooled urine samples from 6040 Australians and reported a median of 24 μg/l (3.4–300 μg/l) for TFA. The concentrations showed no differences between the periods 2012/13 and 2020/21, so it must be assumed that exposure has remained constant. But higher levels were found in older than in younger groups.

4.3 Emerging PFAS

In addition to the medium-chain and long-chain PFCA and PFSA, various other substances or groups of substances are used extensively on a large scale, and results for these are available (Buck et al. 2011). Table 4.9 lists the most significant representatives, which are discussed below. Figure 4.10 shows the chemical structures of some emerging PFAS.

On one hand, these substances are primarily substitutes specifically developed for various applications, yet they have often been used for a long time and can be detected in the environment, such as sulfonamide ethanols and fluorotelomer alcohols (Jahnke et al. 2007). They appear as residues in consumer goods like carpets, upholstery, and paper fibers. In contrast to PFOS and PFOA, they are uncharged substances that are poorly soluble in water due to the perfluoroalkyl residues and are also relatively volatile. However, the nonpolar perfluorinated compounds are less persistent and can be degraded by atmospheric and biological processes (Ellis et al. 2004). In both the environment and organisms, the degradation of perfluorinated sulfonamide ethanols can lead to persistent PFOS via the sulfonamide. Simultaneously, fluorotelomer alcohols (FTOH) and polyfluoroalkylphosphoric acid esters (PAP) can be converted to PFOA (D'Eon and Maybury 2011a; b). In this context, degradation occurs following the oxidation of the telomer alcohol function through intermediate products, as well as PFOA or PFNA (Rand and Mabury 2017; Kabadi et al. 2018). Due to the aforementioned degradation mechanisms, these emerging PFAS are precursors that can be considered quasi-emission-remote sources of persistent PFOS and PFOA. In light of this, these transformation mechanisms must always be taken into account alongside the toxicological properties of the precursor substances.

Fluorotelomer alcohols contain carbon-bonded hydrogen atoms and a hydroxy group, in addition to the fluorinated carbon chain. They are named by indicating the number of fluorinated carbon atoms relative to those carrying a hydrogen atom, separated by a colon (e.g., 8:2 FTOH). Worldwide production is estimated to be between 6000 and 12,000 tons. Approximately 80% of the total produced quantity is utilized in fluoropolymer manufacturing. For instance, they are used in the impregnation of textiles and can be detected as residues on fabrics that have been treated appropriately. FTOH and fluorotelomer acrylates (FTAC) accounted for about 60% of the global

Table 4.9 Selection of important perfluorosurfactants and related compounds

Acronyms	Chemical names	Formula	CAS-No
Fluorotelomer alcohols (FTOH)			
4:2 FTOH	2-perfluorobutylethanol	$C_6H_5F_9O$	2043–47-2
6:2 FTOH	2-perfluorohexylethanol	$C_8H_5F_{13}O$	647–42-7
8:2 FTOH	2-perfluorooctylethanol	$C_{10}H_5F_{17}O$	678–39-7
10:2 FTOH	2-perfluorodecylethanol	$C_{12}H_5F_{21}O$	865–86-1
Perfluoroctansulfonamides			
FOSA	Perfluorooctane sulfonamide	$C_8F_{17}SO_2NH_2$	754-91-6
N-MeFOSA	N-methyl perfluorooctane sulfonamide	$C_9H_4F_{17}NO_2S$	31506-32-8
N-EtFOSA	N-ethyl perfluorooctane sulfonamide	$C_{10}H_6F_{17}NO_2S$	4151-50-2
N-MeFOSAA	N-methyl perfluorooctane sulfonamidoacetic acid	$C_{11}H_6F_{17}NO_4S$	2355-31-9
N-EtFOSAA	N-ethyl perfluorooctane sulfonamidoacetic acid	$C_{12}H_8F_{17}NO_4S$	2991-50-6
Perfluoroctansulfonamidethanols			
N-EtFOSE	N-ethyl perfluorooctane sulfonamidethanol	$C_{12}H_{10}F_{17}NO_3S$	1691-99-2
N-MeFOSE	N-methyl perfluorooctane sulfonamidethanol	$C_{11}H_8F_{17}NO_3S$	24448-09-7
Polyfluoroalkylphosphoric acid esters			
6:2-monoPAP	6:2 Fluoroalkyl phosphate monoester	$C_8H_6F_{13}O_4P$	57678-01-0
8:2-monoPAP	8:2 Fluoroalkyl phosphate monoester	$C_{10}H_6F_{17}O_4P$	57678-03-2
6:2/6:2-diPAP	6:2/6:2 Fluoroalkyl phosphate diester	$C_{16}H_9F_{26}O_4P$	57677-5-9
6:2/8:2-diPAP	6:2/8:2 Fluoroalkyl phosphate diester	$C_{18}H_9F_{30}O_4P$	943913-15-3
8:2/8:2-diPAP	8:2/8:2 Fluoroalkyl phosphate diester	$C_{20}H_9F_{34}O_4P$	678-41-1
Fluorotelomer sulfonic acids			
4:2 FTSA	4:2 Fluorotelomer sulfonic acid	$C_6H_5F_9O_3S$	757124-72-4

(continued)

Table 4.9 (continued)

Acronyms	Chemical names	Formula	CAS-No
6:2 FTSA	6:2 Fluorotelomer sulfonic acid	$C_8H_5F_{13}O_3S$	27619-97-2
8:2 FTSA	8:2 Fluorotelomer sulfonic acid	$C_{10}H_5F_{17}O_3S$	39108-34-4
Other substitutes			
ADONA	Ammonium 4,8-dioxa-3 H-perfluorononanoate	$C_7H_2F_{12}O_4$	958445-44-8
HFPO-DA (GenX)	Tetrafluoro-2-(heptafluoropropoxy)propanoic acid	$C_6HF_{11}O_3$	13252-13-6
EEA-NH4	Perfluoro(2-ethoxy-2-fluoroethoxy)acetic acid, ammonium salt	$C_6H_4F_{11}NO_4$	908020-52-0
FC-807	Ammonium salt bis[2-[N-ethyl(heptadecafluoroctane)sulfonylamino]ethyl]phosphate		30381-98-7
6:2-Cl-PFESA (F-53B) (major)	6:2 chlorinated polyfluorinated ether sulfonate	$C_8ClF_{16}O_4ClSO_4(K)$	73606-19-6
8:2-Cl-PFESA (F-53B) (minor)	8:2 chlorinated polyfluorinated ether sulfonate	$C_{10}ClF_{20}O_4ClSO_4(K)$	83,329–89-9
NVHOS	Tetrafluoro-2-(tetrafluoroethoxy)ethanesulfonic acid	$C_4H_2F_8O_4S$	801209-99-4

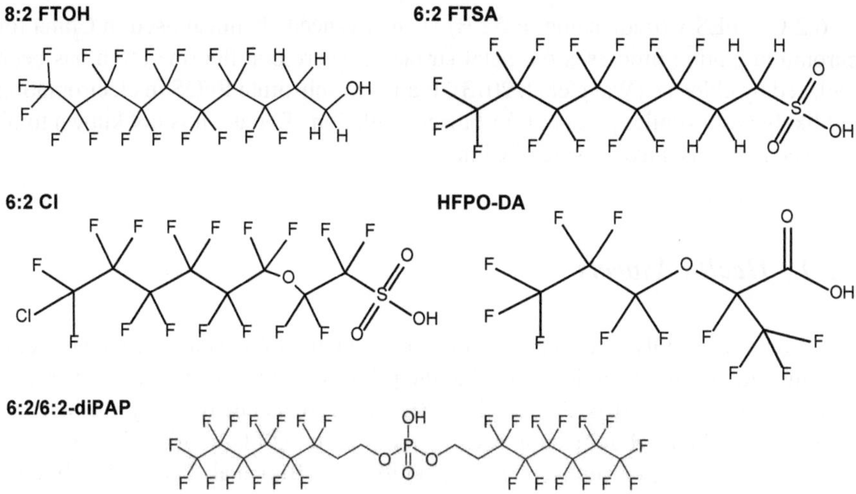

Fig. 4.10 Chemical structure of different emerging PFAS

market for fluorotelomer products in 2013. FTACs are utilized, for example, in the production of methacrylate polymers. Additionally, the use of FTOH in firefighting foam is well-known.

Polyfluoroalkyl phosphoric acid esters (PAP) are hydrophobic and possess mono-, di-, or trisubstituted fluorinated side chains, typically ranging from C8 to C12 in chain length, in addition to the phosphate group. In industrial products, di-PAPs, in particular, are prevalent, while mono- and tri-PAPs primarily exist as by-products. Due to their surface-active properties, they have been utilized in various applications since the 1960s, predominantly in paper products and cardboard, including packaging for consumer goods, cleaning agents, and paints.

Fluorotelomer sulfonate acids (FTSAs), such as 6:2 FTSA, are used as a substitute for PFOS or its salts in firefighting foams and in complex and decorative chromium plating processes involving electroplating. For a long time, particularly in the USA since the 1960s, 6:2-FTSA has been utilized in AFFF (Aqueous Film-Forming Foam) firefighting foams and as a wetting agent in various commercial products.

Tetrafluoro-2-(heptafluoropropoxy)propanoic acid (HFPO-DA) is synthesized through the condensation of two molecules of hexafluoropropylene oxide. GenX, the HFPO-DA ammonium salt, has been used as a replacement for PFOA in fluoropolymer production since around 2010. It is persistent in the environment and exhibits good water solubility. Consequently, HFPO − DA is commonly utilized as a lubricant in semiconductor fluid processing, telecommunication cables, aerospace, hydrogen energy, and lithium-ion batteries, as well as its traditional applications in nonstick coatings and firefighting foam (Guo et al. 2025). According to REACH, the annual production volume in the EU is between 10 and 100 tons.

Another substitute for PFOA in fluoropolymer production is ADONA. In Germany, for example, it has served as a processing aid in a plant in southeastern Bavaria since 2008, instead of PFOA. However, further information on the use of ADONA in other industrial plants and its production volumes is unavailable.

6:2-Cl-PFESA (trade name: F-53 B) is an advanced chemical used in China for chromium plating processes on metal surfaces, where one fluorine atom has been replaced by chlorine (Wang et al. 2013). It aims to substitute PFOS in electroplating but likely has a similar potential for bioaccumulation. To date, it is not known to be produced or consumed outside of China.

4.3.1 Health Aspects

Effect data for FTOH is available from experimental animal studies following oral administration. These studies show that the primary target organs are the liver and kidneys (Serex et al. 2014; Rice et al. 2020). This can also be partially attributed to the fact that 8:2 FTOH, for example, can be converted to PFOA within the organism. There is no evidence of mutagenicity or genotoxicity (Bull et al. 2014). Only limited results from animal studies exist for the other FTOH (Rice et al. 2020). Findings from epidemiological studies in humans have not yet been published.

The toxicological data for FTSA are also limited. In a study involving mice, an increase in liver weight, liver cell hypertrophy, and inflammatory markers in the liver was observed after 28 days of oral administration of 6:2 FTSA (Sheng et al. 2017).

To date, our understanding of the toxicology of PAPs is quite limited (Taxvig et al. 2014). In vitro studies have demonstrated that hormonal activity affects the production of steroid hormones.

For HFPO-DA, studies on animal oral feeding involving various exposure durations and species, as well as a 2-year study in rats, are available. These studies indicated minor effects on the liver and liver weight (hypertrophy), hematologic effects, and effects on the immune system and thyroid gland (Rae et al. 2015; Brase et al. 2021; US EPA 2021).

In a review article by 3 M, Gordon (2011) summarized results on several toxicological endpoints of ADONA. Effects on the liver, haematotoxic effects, and weight changes are reported. Overall, the data is still limited, and long-term studies are lacking.

Currently, the toxicological and epidemiological data on Cl-PFESA are extremely limited. Effects on the liver and thyroid hormones, lipid metabolism, and other endocrine effects are discussed (Wang et al. 2013; He et al. 2022).

4.3.2 Toxicokinetics

After a single administration of two doses of labeled 8:2 FTOH to rats, Fasano et al. (2006) observed a rapid increase in blood levels within just one hour and a plasma half-life of less than five hours. More than 70% of excretion occurred via feces. After oral administration of 8:2 FTOH to rats, Huang et al. (2019) also found that 22–41% of the substance is absorbed from the gastrointestinal tract and has a plasma half-life of 1.1–1.7 h, with PFOA and 7:3-fluorotelomer acid being the primary metabolites (Huang et al. 2019). Various studies are available on metabolism. In the first step, 6:2 FTOH is transformed by oxidation into 6:2-fluorotelomer aldehyde and then converted into saturated (6:2 FTCA) and unsaturated 6:2-fluorotelomer carboxylic acid (6:2 FTUCA) through further oxidation steps, which can then be excreted coupled with e.g., glutathione and sulfate or converted to PFBA or PFHpA via additional metabolic steps (Kabadi et al. 2018). Moreover, the transient intermediates were converted into other terminal metabolites like 5:3- and 4:3-fluorotelomer carboxylic acid, PFHxA, and PFBA (Kabadi et al. 2018; Rice et al. 2024). Based on their toxicokinetic considerations, the authors assume that 5:3 fluorotelomer carboxylic acid could be an important biomarker for assessing long-term exposure to 6:2 FTOH. A corresponding metabolic pathway is described for 8:2 FTOH (Li et al. 2016).

In a one-day inhalation study with 6:2 FTOH in rats, alongside PFBA, PFHxA, and PFHpA, 5:3 FTCA was particularly identified as the primary metabolite in plasma, with an elimination half-life of 5.4 h calculated for both sexes (Russell et al. 2015). In a five-day and a 23-day inhalation study, along with a 90-day oral study with 6:2 FTOH, the concentration profile of 5:3 FTCA in serum was significantly higher than in the one-day study, and the elimination half-life was estimated at 20–30 days. In a Swedish study, high concentrations of 5:3 FTCA were detected in the whole blood of 11 professional ski waxing technicians (median: 1.7 μg/l; maximum: 6.1 μg/l) who were occupationally exposed to concentrations of up to 92 $\mu g/m^3$ 8:2 FTOH (Nielsson et al. 2013). Based on the aforementioned study, Russell et al. (2015) estimate the elimination half-life of 5:3 FTCA to be 43 days.

D'eon and Mabury (2011a, b) administered a mixture of 4:2, 6:2, 8:2, and 10:2 diPAP or monoPAP intravenously and orally to two groups of four male rats. The bioavailability of diPAP increased with longer chain lengths, while monoPAP was not absorbed from the intestine. The plasma half-life of diPAP is reported to be 2.0–3.9 days. Twelve men who received a single dose of 700 μg 13C-labeled 8:2 diPAP orally with breakfast were examined for serum and urine for up to 27 days, showing gastrointestinal absorption of between 0.9 and 6% (Trier and Numata 20). Initially, the half-life of 8:2 diPAP was only 0.9–1.4 days at high 8:2 diPAP serum concentrations and increased to 5.7–15.4 days as serum levels decreased.

Animal studies in rats and mice following oral administration of HFPO-DA indicated rapid and nearly complete absorption (US EPA 2021; Rice et al. 2021). No metabolism of HFPO-DA was observed in both in vitro and in vivo studies (US EPA 2021). HFPO-DA, administered both orally and intravenously, exhibited rapid biphasic excretion of the unmetabolized substance via urine, with elimination half-lives ranging from 3 to 70 h in rats and mice after oral exposure (Gannon et al. 2016; Rice et al. 2021). Females demonstrated a slightly shorter half-life compared to males. Several studies involving rats and mice have shown that the ammonium salt of HFPO-dimeric acid can be transferred from the pregnant animal to the fetus (US EPA 2021). In a recent study, Abraham et al. (2024) administered ^{13}C-labeled

HFPO-DA once to a healthy volunteer, measuring plasma, urine, and feces concentrations over 450 days. After rapid absorption with a maximal plasma time of 1.8 h, 45% of the absorbed dose was excreted in urine within 6 days post-administration, resulting in an elimination half-life of 2.9 days.

After oral administration, ADONA is rapidly absorbed, not metabolized, and is excreted unchanged primarily through the urine (EFSA 2011; Gordon 2011). Elimination half-lives of 5.7 h (males) and 4.2 h (females) were observed in monkeys, while half-lives ranging from 21 to 36 days were determined in three exposed workers. In a recent study, Abraham et al. (2024) administered ^{13}C-labeled DONA once to a healthy volunteer and measured plasma, urine, and feces concentrations over 450 days. After rapid absorption with a maximum plasma time of 2.0 h, 16% of the absorbed dose was excreted in urine within 6 days post-administration. The elimination half-life was 35 days.

Toxicokinetic knowledge of Cl-PFESA remains limited. Yi et al. (2022) administered 6:2 Cl-PFESA to rats via gavage. They observed the highest levels in plasma and urine after 24 and 56 h, respectively, along with a parallel increase in the dechlorination product 6:2 H-PFESA. The urinary elimination half-life was 9.3 h. Shi et al. (2016) calculated median half-lives of 15.3 years based on internal exposure data from the Chinese population.

4.3.3 Occurrence of PAPs in the Blood

In various studies, different diPAPs were observed in serum samples, while other congeners, such as the 4:2/4:2-diPAP, 4:2/6:2-diPAP, and the monoPAP, were determined only very rarely. The following studies show a comparable and relatively low concentration level for the various PAPs.

In Germany, 150 serum samples from one city collected between 1995 and 2009 and 270 samples from another city from 1982 to 2009 were collected in the Environmental Specimen Bank and analyzed for seven PAP (Yeung et al. 2013). The levels (% above the limit of quantification) in all samples ranged from < 0.0007 to 0.095 μg/l (40%) for 4:2/4:2-diPAP, < 0.0007 to 2.38 μg/l (<20%) for 4:2/6:2-diPAP, < 0.0002 to 0.69 μg/l (40%) for 6:2/6:2-diPAP, < 0.0002 to 0.11 μg/l (<20%) for 6:2/8:2-diPAP, and < 0.0010 to 0.29 μg/l (approx. 30%) for 8:2/8:2-diPAP. In Sweden in 2012, levels of < 6.4 to 14.6 pg/g for 6:2/6:2-diPAP and < 0.5 to 2.3 pg/g for 8:2/8:2-diPAP were observed in three pooled serum samples of 30 individual samples each, whereby monoPAP could not be detected in any sample (Gebbing et al. 2015). Poothong et al. (2017) analyzed a total of 61 Norwegian serum samples collected in 2013 and 2014 and found 6:2-monoPAP, 6:2/6:2-diPAP, and 8:2/8:2-diPAP with concentrations of < 0.09 to 0.39 μg/l, < 0.018 to 0.14 μg/l, and 0.009 to 0.11 μg/l, respectively. In another study from Norway, a total of 144 serum samples were analyzed from 2014 to 2016, and 6:2/6:2-diPAP and 8:2/8:2-diPAP were quantified in only 28% and 38%, respectively (Thépaut et al. 2021). The medians (95th percentiles) for 6:2/6:2-diPAP were 0.03 μg/l (0.17 μg/l) for men and 0.02 μg/l

(0.26 μg/l) for women, while for 8:2/8:2-diPAP they were 0.02 μg/l (0.06 μg/l) for both men and women. The 95th percentiles for 6:2-monoPAP and 8:2-monoPAP were 0.24 and 0.17 μg/l in men and 0.22 and 0.20 μg/l in females. In this study, the main determining factors were dietary intake and the use of personal care products.

A nonsignificant increase in concentrations was only observed for 6:2/8:2-diPAP and 8:2/8:2-diPAP in a Swedish study conducted between 1997 and 2012 (Gebbing et al. 2015), while no changes in PAP concentrations over time or detection frequency were observed in a German study between 1982 and 2009 (Yeung et al. 2013). This finding was also confirmed in an Australian study, where no significant trend was noted for 6:2/6:2-diPAP and 6:2/8:2-diPAP in samples collected between 2002 and 2013 (Eriksson et al. 2017).

Lee & Mabury (2011) found 6:2/6:2-diPAP and 8:2/8:2-diPAP in 40 serum samples in the USA in 2009 with levels of < 0.02–0.39 μg/l and < 0.11–1.80 μg/l, respectively, with 50% and 78% of the measurement results being below the limit of quantification. In serum samples from 100 participants from the general population in the USA and 30 people working in office workplaces, 6:2/6:2-diPAP and 8:2/8:2-diPAP were detected in 47% and 58% of samples with mean values of 0.06 μg/l and 0.10 μg/l, respectively (Dagnino et al. 2016).

In Hong Kong in 2010, 20 whole blood samples showed concentrations ranging from < 1.52–13.2 pg/g (35% > LOQ) for 6:2/6:2-diPAP and < 1.52–5.52 pg/g (20% > LOQ) for 8:2/8:2-diPAP (Loi et al. 2013). Mean values of 0.18 μg/l for 6:2/6:2-diPAP and 0.08 μg/l for 8:2/8:2-diPAP were also reported for 162 plasma samples from pregnant Chinese women (Liu et al. 2023). In 339 Japanese serum samples collected from 2009 to 2010, 6:2/6:2-diPAP and 8:2/8:2-diPAP were observed in only 0.9% of the samples, with maxima of 0.34 μg/l and 0.21 μg/l, respectively (Nakayama et al. 2020).

Eriksson et al. (2017) examined 54 pooled serum samples from 4920 individuals aged between 0 and over 62 years in Australia in 2011. They found 6:2/6:2-diPAP in 56%, 6:2/8:2-diPAP in 59%, and 8:2/8:2-diPAP in only 28% of the samples, at concentrations of < LOQ–0.47 μg/l, < LOQ–0.96 μg/l, and < LOQ–0.12 μg/l.

4.3.4 Occurrence of FTOH, Perfluoro Octane Sulfonamides, Perfluorooctane Sulfonamide Ethanols, and Fluorotelomer Ethoxylates in the Blood

Very low concentrations of all substances have been reported in the blood of the general population. For example, 8:2 FTOH (as 8:2 FTOH sulfate) was found in serum samples from 100 individuals in the general population and 30 people working in office environments. This study observed concentrations in only 5% and 10% of the samples, within the range of < 0.05–0.08 μg/l, respectively (Dagnino et al. 2016).

In 2009, Lee and Maybury (2011) reported mean values of MeFOSAA at 0.36 μg/l (< 0.01–2.00 μg/l) (90% > LOQ) and for EtFOSAA at 0.05 μg/l (< 0.01–0.17 μg/l)

(82% > LOQ) in the serum of 40 blood donors from the USA. In plasma samples from two sites of the German Environmental Specimen Bank from 1982–2009, concentrations of MeFOSAA ranged from < 0.0031 to 10.8 μg/l, and for EtFOSAA, from 0.0063 to 14.3 μg/l (Yeung et al. 2013). Both substances could no longer be detected in this study after 2009. In Sweden, pooled serum samples collected between 1997 and 2012 showed a clear decreasing trend for MeFOSAA and EtFOSAA, while MeFOSA was not quantifiable in any samples, and EtFOSA was only quantifiable in samples from 2000 to 2004. Mean values of 0.32 μg/l for MeFOSA, 0.19 μg/l for EtFOSA, and 0.17 μg/l for MeFOSAA were also reported from 162 Chinese blood samples from women during pregnancy, while EtFOSAA could not be detected (Liu et al. 2023).

Fluorotelomer ethoxylates (polyethoxy-1H,1H,2H,2H-perfluoroalkanols; FTEOs), which function as anti-fogging agents, consist of repeating CF2 units linked to varying chain lengths of ethoxylates, resulting in a complex mixture of homologues (Frömel and Knepper 2010). They have been found in various environmental media. However, to date, only one human biomonitoring study has been published. Cheng et al. (2025) reported discovering seventeen 6:2 FTEO homologues in indoor dust samples from China, as well as in human serum collected from 90 women in 2019. Ten of the 17 targeted FTEOs were identified in the samples, with detection frequencies ranging from 7 to 78%, and cumulative concentrations varying from < LOD to 24.9 μg/l (median: 0.40 μg/l). The primary compounds included 6:2 FTEO8, 6:2 FTEO9, and 6:2 FTEO10, which had medians of 0.22 μg/l, 0.18 μg/l, and 0.20 μg/l, respectively.

4.3.5 Occurrence of FTSA in the Blood

In a German study, 100 plasma samples from students at the Environmental Specimen Bank, aged 20 to 29, were analyzed at five points between 2009 and 2019 (Göckener et al. 2020). 4:2 FTSA and 6:2 FTSA were not detected in any sample above the limit of quantification, and 8:2 FTSA was only found in one sample at 0.26 μg/l. In the analysis of pooled serum samples from Sweden between 1996 and 2017, 8:2 FTSA, in particular, was detected, showing a decreasing trend for 6:2 FTSA and 8:2 FTSA of 9.6% and 3.8%, respectively, during this period (Miaz et al. 2020).

In 2009, Lee & Mabury (2011) found 6:2 FTSA and 8:2 FTSA in 40 serum samples in the USA in 2009 at levels of < 0.011–0.030 μg/l and < 0.011–0.162 μg/l, respectively, with 69% and 13% of the results below the limit of quantification.

In China, a total of 50 maternal blood samples and 50 umbilical cord blood samples were examined in 2009. The levels of 6:2 FTSA ranged from < 0.003 to 0.048 μg/l (64% > LOQ), while 8:2 FTSA levels ranged from < 0.004 to 0.072 μg/l (34% > LOQ) (Yang et al. 2016b). In umbilical cord blood, the levels were < 0.003 to 0.090 μg/l for 6:2 FTSA and < 0.004 to 0.034 μg/l for 8:2 FTSA. In a comprehensive study conducted in 2015/16 with 1238 participants, 4:2 FTSA and 6:2 FTSA were detected in 0.5% and 7% of samples, respectively, with medians of < 0.024 and 0.035 μg/l.

Notably, 8:2 FTSA was not detected at all (Mi et al. 2020). In 339 Japanese serum samples, FTSA was not observed above the limits of quantification of 0.02–0.05 μg/l in any sample during 2009–2010 (Nakayama et al. 2020).

Eriksson et al. (2017) examined 54 pooled serum samples from 4920 people of the general population in Australia in 2011 and were able to measure 6:2 FTSA in 38% and 8:2 FTSA in 96% of the samples at concentrations of < 0.01–0.05 μg/l and < 0.01–0.04 μg/l.

4.3.6 Occurrence of HFPO-DA (GenX) and ADONA in the Blood

In Germany, HFPO-DA could not be detected above the LOQ of 0.25 μg/l in 100 samples from the Environmental Specimen Bank collected between 1982 and 2019 (Göckener et al. 2020). Even in 38 blood samples from adults in northern Italy, HFPO-DA could not be determined above the LOQ of 0.02 μg/l in any sample (Frigerio et al. 2022). In the USA, HFPO-DA was not found above the detection limit of 0.1 μg/l in any of the 30 serum samples taken from individuals living near a contaminated river used for drinking water production (Pritchett et al. 2019). HFPO-DA could also not be found above the limit of quantification of 0.05 μg/l in the blood of 39 children aged two years born in the USA from 2001 to 2010 and in 39 umbilical cord blood samples (Zhang et al. 2022). In the US-NHANES 2013–2014, HFPO-DA was measured in only 1.2% of the 2682 urine samples, with concentrations of < 0.1–0.4 μg/l (Calafat et al. 2019). In the US-NHANES 2017/18, HFPO-DA was not detected above the limit of 0.1 μg/l in any of the 1672 serum samples (CDC 2023). Meanwhile, in Canada, Schlezinger et al. (2025) reported a geometric mean of 0.45 μg/l in 71% of the samples above the LOD of 0.01 μg/l in 2019/20. Even in a comprehensive Chinese study conducted in 2015/16 with 1238 participants, HPFO-DA was detected in only 1.2% of the samples above the limit of quantification (LOQ) of 1.67 μg/l (Mi et al. 2020). Zhang et al. (2024) observed HFPO-DA in 43% of the samples above the LOD of 0.02 μg/l in 351 samples collected in 2017 in China. In 2019/20, Yao et al. (2020) reported results from 977 serum samples from Chinese individuals aged 12 days to 96 years in the vicinity of a fluorochemical plant, with HFPO-DA concentrations ranging from < 0.05 to 0.65 μg/l, where the substance could be determined in only 8.4% of the samples.

In Germany, ADONA was not found above the limit of quantification of 0.2 μg/l in 158 serum samples in 2016. In the vicinity of an industrial plant that used ADONA in fluoropolymer production, it was detected in three of 26 samples, with a maximum of 1.7 μg/l (Fromme et al. 2017). Additionally, ADONA was also not found above the limit of quantification of 0.25 μg/l in 100 samples from the Environmental Specimen Bank, collected between 1982 and 2019 (Göckener et al. 2020). Even in 38 blood samples from adults in northern Italy, ADONA could not be determined above the LOQ of 0.02 μg/l in any sample (Frigerio et al. 2022). Furthermore, in five Italian blood samples, ADONA was not detectable above the LOD of 0.006 μg/l (Belay et al. 2025). In the USA, ADONA was not found above the detection limit of 0.1 μg/l in any of the 30 serum samples from individuals living near a contaminated river used for drinking water production (Pritchett et al. 2019). In another study from the USA, ADONA could not be quantified above the LOQ of 0.1 μg/l in any of 50 serum and urine samples (Kato et al. 2018). ADONA could also not be found above the LOQ of 0.04 μg/l in the blood of 39 children aged two years born in the USA between 2001 and 2010, as well as in 39 umbilical cord blood samples (Zhang et al. 2022). In the US-NHANES 2017/18, ADONA was also not observed in any of the 1,672 serum samples above the detection limit of 0.1 μg/l, nor in 2328 urine samples collected in 2013/14 (CDC 2023). On the contrary, in a more recent Chinese study from 2020/21, ADONA was determined in 98% of samples from 169 women during pregnancy, with a median of 0.18 μg/l and a 75th percentile of 0.23 μg/l (Xu et al. 2022). In another Chinese investigation, ADONA was detectable in all 164 serum samples collected in 2020/21 with a median of 0.15 μg/l (Tian et al. 2024).

4.3.7 Occurrence of Cl-PFESA in the Blood and Urine

In Germany, 6:2 Cl-PFESA could not be detected above the LOQ of 0.25 μg/l in 100 samples from the Environmental Specimen Bank of students collected between 1982 and 2019 (Göckener et al. 2020). In 38 blood samples from adults in northern Italy, 6:2 Cl-PFESA was detected only with a median (95th percentile) of < 0.02 μg/l (0.02 μg/l) in 16% of the samples (Frigerio et al. 2022). Meanwhile, 8:2 Cl-PFESA could not be determined in any sample. In a study from the USA, Cl-PFESA was also undetectable above the LOQ of 0.10 μg/l in any of the 50 serum and urine samples (Kato et al. 2018). In the representative NHANES study 2017/18, among 1672 blood samples, the median of 6:2 Cl-PFESA was < 0.01 μg/l, and the 95th percentile was 0.10 μg/l, while it could not be detected in 2328 urine samples above the LOQ of 0.1 μg/l (CDC 2023).

The situation is different in China, where Cl-PFESA is produced and widely used under the commercial name F-53B, making it ubiquitous in the environment. Essentially, 6:2 Cl-PFESA is found in blood at lower concentrations than 8:2 Cl-PFESA. A preliminary Chinese study determined very high medians of 93.7 μg/l (2.0–357 μg/l) and significantly lower medians of 4.8 μg/l (1.9–5.9 μg/l) for 6:2 Cl-PFESA among 45 subjects with high fish consumption and eight participants from the general population (Shi et al. 2016). For 8:2 Cl-PFESA, medians were only 1.6 μg/l and 0.08 μg/l in both groups, respectively. This study reported a very low median concentration of 0.005 μg/l (0.0006–2.9 μg/l) for 6:2 Cl-PFESA in urine. Since this study was conducted in 2015, many additional results have been published in China, as summarized in a recent review by He et al. (2022) up to 2020. The studies by Mi et al. (2020) conducted in 2015/16 with 1238 participants and by Yao et al. (2020) in 2019/20 with 977 subjects living near a fluorochemical plant are particularly comprehensive. In both studies, medians of 1.75 μg/l and 2.31 μg/l (0.06–29.9 μg/

l) were reported for 6:2 Cl-PFESA, and 0.011 μg/l and 0.03 μg/l (<0.02–1.76 μg/l) for 8:2 Cl-PFESA. Zhang et al. (2024) reported a lower median of 0.52 μg/l for 6:2 Cl-PFESA and 0.07 μg/l for 8:2 Cl-PFESA in 351 samples collected in 2017 in China. He et al. (2022) summarize the available results as follows: exposure in blood can vary considerably spatially, dietary habits like high fish consumption could contribute to exposure, and no clear temporal trend was observed for 6:2 Cl-PFESA between 2015 and 2020; in general, men have a higher exposure than women, and older individuals have a higher exposure than younger ones. In a more recent Chinese study from 2020/21, medians (75th percentiles) of 2.38 μg/l (6.92 μg/l) were found for 6:2 Cl-PFESA and 0.16 μg/l (0.38 μg/l) for 8:2 Cl-PFESA in 169 women during pregnancy (Xu et al. 2022). Moreover, Tian et al. (2024) found medians of 2.37 μg/l for 6:2 Cl-PFESA and 0.16 μg/l for 8:2 Cl-PFESA in 164 serum samples collected 2020/21 in China.

In addition, a Chinese study in 2020–2021 also examined 80 cord blood samples from two cities and reported medians (95th percentile) of 0.59 μg/l (2.13 μg/l) for 6:2 Cl-PFESA and 0.02 μg/l (0.08 μg/l) for 8:2 Cl-PFESA (Yao et al. 2023). A slightly higher median of 1.35 μg/l and a range of 0.27–8.83 μg/l of 6:2 Cl-PFESA were reported in 119 cord blood samples collected 2019–2021 in China (Song et al. 2025).

4.3.8 Occurrence of Emerging PFAS in Breast Milk

To date, only minimal data are available on the occurrence of breast milk, partly from Europe and mainly from China.

In 2016, when 50 breast milk samples from Germany were examined, ADONA was not detected above the LOQ of 25 ng/l in any sample (Fromme et al. 2018). In Austria, Hartmann et al. (2024) reported a median (range) of < 5.8 ng/l (<5.8–9.6 ng/l) for ADONA in 9 of 40 samples, while HFPO-DA (GenX) was found in only 1 of the 40 samples with a median of 14 ng/l (LOD: 6.0). Furthermore, Yao et al. (2023) identified HFPO-DA in just 8.5% of the samples, with a 95th percentile of 5.5 ng/l. Awad et al. (2020) collected 30 breast milk samples in China in 2010 and 2016, along with 10 in Stockholm in 2016, finding ADONA in a range between < 3.0 and 8.0 ng/l in 20–30% of the samples from China, whereas in Sweden, no samples exceeded the detection limit of 3 ng/l. Tao et al. (2025) could not find ADONA above the LOD of 12 ng/l in 117 Chinese samples collected between 2010 and 13. In contrast, Yang et al. (2025) reported a median of 3.9 ng/l and a 95th percentile of 5.8 ng/l in 447 milk samples from Southern China collected in 2018/19 (LOD: 1.3 ng/l). Slightly higher ADONA-concentrations were found in 73 milk sample from China, with a mean value of 60 ng/l (12 % above the detection limit of 9.5 ng/l) (Lin et al. 2025).

Awad et al. (2020) collected 30 breast milk samples each from three Chinese cities between 2010 and 2016, and 10 samples from Stockholm in 2016. While 6: 2 Cl- PFESA and 8: 2 Cl- PFESA were not found above the detection limit of 2 ng/l, the mean values in the Chinese cities ranged from 126 to 173 ng/l and from 8 to 11 ng/l, respectively. In Stockholm, the total content of the 20 PFAS measured was determined by PFOS and PFOA, whereas in China, Cl- PFESA accounted for between 25 and 35%. As part of the third National Human Milk Survey from 2017 to 2020, 100 pooled samples out of 3531 breast milk samples from 25 Chinese provinces were analyzed, and 6: 2 Cl- PFESA and 8: 2 Cl- PFESA were quantified in 97% and 3% of the samples, respectively, with medians of 10. 2 ng/l (<5.0–531 ng/l) and < 5.0 ng/l (<5.0–27. 8 ng/l) (Han et al. 2023b). There were considerable differences in the mean values between the individual provinces, ranging from 2.16 ng/l to 136 ng/l for 6:2 Cl-PFESA. ADONA could not be detected in any breast milk sample above the detection limit of 12 ng/l. Additionally, the breast milk of 1151 mothers from 21 cities in China was examined in 2020–2021, with medians (95th percentiles) of 13.6 ng/l (104 ng/l) for 6:2 Cl-PFESA and < 1.0 ng/l (3.6 ng/l) for 8:2 Cl-PFESA reported (Yao et al. 2023). These concentrations were the third highest of all PFAS after PFOA and PFOS. Jin et al. (2020) detected medians (95th percentile) of 16 ng/l (58 ng/l) and < 1.0 ng/l (1.9 ng/l) for 6:2 Cl-PFESA and 8:2 Cl-PFESA in 174 breast milk samples collected in 2018–2019. Tao et al. (2025) reported medians (95th percentile) for 6:2 Cl-PFESA and 8:2 Cl-PFESA of 18 ng/l (84 ng/l) and < 1.5 ng/l (2.0 ng/l) in 117 Chinese samples collected from 2010 to 2013. Furthermore, Yang et al. (2025) reported a median (95th percentile) of 19 ng/l (29 ng/l) for 6:2 Cl-PFESA and 8:2 Cl-PFESA, respectively, in 447 milk samples collected in Southern China in 2018/19. In contrast, Hartmann et al. (2024) found 6:2 Cl-PFESA in only 1 of 40 samples (5.5 ng/l) from Austria above the LOD of 1.1 ng/l.

In two Swedish studies, linear and branched MeFOSAA and EtFOSAA were not found above the detection limits of 5 ng/l in breast milk (Nyberg et al. 2018; Awad et al. 2020). They also could not be determined in samples collected from three Chinese cities (Awad et al. 2020). In 40 samples collected in Austria from 2014 to 16, EtFOSAA and the two fluorotelomer sulfonate acids, 4:2-FTSA and 8:2-FTSA, were not detected in any sample above the LODs of 1.2–4.0 ng/l (Hartmann et al. 2024).

References

Abdallah MA, Wemken N, Drage DS, Tlustos C, Cellarius C, Cleere K, Morrison JJ, Daly S, Coggins MA, Harrad S (2020) Concentrations of perfluoroalkyl substances in human milk from Ireland: implications for adult and nursing infant exposure. Chemosphere 246:125724

Abraham K, Mielke H, Fromme H, Völkel W, Menzel J, Peiser M, Zepp F, Willich SN, Weikert C (2020) Internal exposure to perfluoroalkyl substances (PFASs) and biological markers in 101 healthy 1-year-old children: associations between levels of perfluorooctanoic acid (PFOA) and vaccine response. Arch Toxicol 94:2131–2147

Abraham K, Mertens H, Richter L, Mielke H, Schwerdtle T, Monien BH (2024) Kinetics of 15 per- and polyfluoroalkyl substances (PFAS) after single oral application as a mixture - A pilot investigation in a male volunteer. Environ Int 193:109047

Adlard B, Bonefeld-Jørgensen EC, Dudarev AA, Olafsdottir K, Abass K, Averina M, Ayotte P, Berner J, Byrne S, Caron-Beaudoin É, Drysdale M, Dumas P, Garcia-Barrios J, Gyllenhammar I, Laird B, Lemire M, Aker A, Lignell S, Long M, Norström K, Packull-McCormick S, Petersen MS, Ratelle M, Rautio A, Timmerman A, Toft G, Weihe P, Nøst TH, Wennberg M (2024)

Levels and trends of persistent organic pollutants in human populations living in the Arctic. Int J Circumpolar Health 83(1):2392405

Al-sheyab NA, Al-Qudah KM, Tahboub YR (2015) Levels of perfluorinated compounds in human breast milk in Jordan: the impact of sociodemographic characteristics. Environ Sci Pollut Res Int 22:12415–12423

Antignac JP, Veyrand B, Kadar H, Marchand P, Oleko A, Le Bizec B, Vandentorren S (2013) Occurrence of perfluorinated alkylated substances in breast milk of French women and relation with socio-demographical and clinical parameters: results of the ELFE pilot study. Chemosphere 91:802–808

Appel M, Forsthuber M, Ramos R, Widhalm R, Granitzer S, Uhl M, Hengstschläger M, Stamm T, Gundacker C (2022) The transplacental transfer efficiency of per- and polyfluoroalkyl substances (PFAS): a first meta-analysis. J Toxicol Environ Health B Crit Rev 25:23–42

Arbuckle TE, Kubwabo C, Walker M, Davis K, Lalonde K, Kosarac I, Wen SW, Arnold DL (2013) Umbilical cord blood levels of perfluoroalkyl acids and polybrominated flame retardants. Int J Hyg Environ Health 216:184–194

Aro R, Eriksson U, Kärrman A, Yeung LWY (2021) Organofluorine mass balance analysis of whole blood samples in relation to gender and age. Environ Sci Technol 55:13142–13151

Arp HPH, Gredelj A, Glüge J, Scheringer M, Cousins IT (2024) The Global Threat from the Irreversible Accumulation of Trifluoroacetic Acid (TFA). Environ Sci Technol 58(45):19925–19935

ATSDR (2021) Toxicological profile perfluoroalkyls. https://www.atsdr.cdc.gov/toxprofiles/tp200.pdf

Awad R, Zhou Y, Nyberg E, Namazkar S, Yongning W, Xiao Q, Sun Y, Zhu Z, Bergman Å, Benskin JP (2020) Emerging per- and polyfluoroalkyl substances (PFAS) in human milk from Sweden and China. Environ Sci Process Impacts 22:2023–2030

Bach CC, Bech BH, Nohr EA, Olsen J, Matthiesen NB, Bonefeld-Jørgensen EC, Bossi R, Henriksen TB (2016) Perfluoroalkyl acids in maternal serum and indices of fetal growth: the Aarhus Birth Cohort. Environ Health Perspect 124:848–854

Bacher S, Göen T, Stiegler H, Rauchfuss K, Hölzer J (2019) PFOA blood concentrations in the Arnsberg' cohort, Germany - 11 years follow-up. Abstracts of the 2019 Annual Conference of the International Society for Environmental Epidemiology, August 25-28 2019, Utrecht, the Netherlands. Environ Epidemiol 3, Suppl. 1.

Bao J, Lee YL, Chen PC, Jin YH, Dong GH (2014) Perfluoroalkyl acids in blood serum samples from children in Taiwan. Environ Sci Pollut Res Int 21:7650–7655

Barbarossa A, Masetti R, Gazzotti T, Zama D, Astolfi A, Veyrand B, Pession A, Pagliuca G (2013) Perfluoroalkyl substances in human milk: a first survey in Italy. Environ Int 51:27–30

Bartolomé M, Gallego-Picó A, Cutanda F, Huetos O, Esteban M, Pérez-Gómez B, Castaño A (2017) Perfluorinated alkyl substances in Spanish adults: geographical distribution and determinants of exposure. Sci Total Environ 603–604:352–360

Barton KE, Starling AP, Higgins CP, McDonough CA, Calafat AM, Adgate JL (2020) Sociodemographic and behavioral determinants of serum concentrations of per- and poly-fluoroalky substances in a community highly exposed to aqueous film-forming foam contaminants in drinking water. Int J Hyg Environ Health 223:256–266

Baumert BO, Wang H, Samy S, Park SK, Lam CN, Dunn K, Pinto-Pacheco B, Walker D, Landero J, Conti D, Chatzi L, Hu H, Goodrich JA (2024) Environmental pollutant risk factors for worse COVID-19 related clinical outcomes in predominately hispanic and latino populations. Environ Res 252(Pt 4):119072

Beck IH, Timmermann CAG, Nielsen F, Schoeters G, Jøhnk C, Kyhl HB, Høst A, Jensen TK (2019) Association between prenatal exposure to perfluoroalkyl substances and asthma in 5-year-old children in the Odense Child Cohort. Environ Health 18:97

Behringer D, Heydel F, Gschrey B, Osterheld S, Schwarz W, Warncke K, Freeling F, Nödler K, Henne S, Reimann S, Blepp M, Jörß W, Liu R, Ludig S, Rüdenauer I, Gartiser S (2021) Persistent degradation products of halogenated refrigerants and blowing agents in the environment:

nature, environmental concentrations and fate with special consideration of new halogenated substitutes with low global warming potential. Report prepared for the German Environment Agency. [in German] https://www.umweltbundesamt.de/publikationen/persistente-abbauprodukte-halogenierter-kaelte

Belay MH, Robotti E, Ghignone A, Fabbris A, Brandi J, Cecconi D, Masini MA, Dondero F, Marengo E (2025) Sensitive and accurate determination of 32 PFAS in human serum using online SPE-UHPLC-HRMS. J Hazard Mater 485:136780

Berg V, Nøst TH, Huber S, Rylander C, Hansen S, Veyhe AS, Fuskevåg OM, Odland JØ, Sandanger TM (2014) Maternal serum concentrations of per- and polyfluoroalkyl substances and their predictors in years with reduced production and use. Environ Int 69:58–66

Bernsmann T, Fürst P (2008) Determination of perfluorinated compounds in human milk. Prepared for DIOXIN 2008. Organoh Comp 70:718–721

Beser MI, Pardo O, Beltran J, Yusa V (2019) Determination of 21 perfluoroalkyl substances and organophosphorus compounds in breast milk by liquid chromatography coupled to orbitrap high-resolution mass spectrometry. Anal Chim Acta 1049:123–132

Bjermo H, Darnerud PO, Pearson M, Barbieri HE, Lindroos AK, Nälsén C, Lindh CH, Jönsson BA, Glynn A (2013) Serum concentrations of perfluorinated alkyl acids and their associations with diet and personal characteristics among Swedish adults. Mol Nutr Food Res 57:2206–2215

Bjerregaard-Olesen C, Bach CC, Long M, Ghisari M, Bossi R, Bech BH, Nohr EA, Henriksen TB, Olsen J, Bonefeld-Jørgensen EC (2016) Time trends of perfluorinated alkyl acids in serum from Danish pregnant women 2008–2013. Environ Int 91:14–21

Blomberg AJ, Haug LS, Lindh C, Sabaredzovic A, Pineda D, Jakobsson K, Nielsen C (2023) Changes in perfluoroalkyl substances (PFAS) concentrations in human milk over the course of lactation: A study in Ronneby mother-child cohort. Environ Res 15:115096

Boesen SAH, Long M, Wielsøe M, Mustieles V, Fernandez MF, Bonefeld-Jørgensen EC (2020) Exposure to perflouroalkyl acids and foetal and maternal thyroid status: a review. Environ Health 19:107

Botelho JC, Kato K, Wong LY, Calafat AM (2025) Per- and polyfluoroalkyl substances (PFAS) exposure in the U.S. population: NHANES 1999-March 2020. Environ Res 270:120916

Boutonnet JC, Bingham P, Calamari D, de Rooij C, Franklin J, Kawano T, Libre J-M, McCul-loch A, Malinverno G, Odom JM, Rusch GM, Smythe K, Sobolev I, Thompson R, Tiedje JM (1999) Environmental risk assessment of trifluoroacetic acid. Hum Ecol Risk Assess 1:59–124

Brase RA, Mullin EJ, Spink DC (2021) Legacy and emerging per- and polyfluoroalkyl substances: analytical techniques, environmental fate, and health effects. Int J Mol Sci 22:995

Buck RC, Franklin J, Berger U, Conder JM, Cousins IT, de Voogt P, Jensen AA, Kannan K, Mabury SA, van Leeuwen SP (2011) Perfluoroalkyl and polyfluoroalkyl substances in the environment: terminology, classification, and origins. Integr Environ Assess Manag 7:513–541

Buck RC, Korzeniowski SH, Laganis E, Adamsky F (2021) Identification and classification of commercially relevant per- and poly-fluoroalkyl substances (PFAS). Integr Environ Assess Manag 17:1045–1055

Bull S, Burnett K, Vassaux K, Ashdown L, Brown T, Rushton L (2014) Extensive literature search and provision of summaries of studies related to the oral toxicity of perfluoroalkylated substances (PFASs), their precursors and potential replacements in experimental animals and humans. External scientific report. EFSA supporting publication EN-572. https://efsa.onlinelibrary.wiley.com/doi/abs/https://doi.org/10.2903/sp.efsa.2014.EN-572

Buschmann J, Bartsch W, Dasenbrock C, Fuhst R, Pohlmann G, Preiss A, Berger-Preiss E (2001) Cross-fostering inhalation toxicity study with HCFC-123 in lactating Sprague-Dawley rats. Inhal Toxicol 13:671–687

Calafat AM, Kato K, Hubbard K, Jia T, Botelho JC, Wong LY (2019): Legacy and alternative per- and polyfluoroalkyl substances in the U.S. general population: Paired serum-urine data from the 2013–2014 National Health and Nutrition Examination Survey. Environ Int 131:105048

Cao W, Liu X, Liu X, Zhou Y, Zhang X, Tian H, Wang J, Feng S, Wu Y, Bhatti P, Wen S, Sun X (2018) Perfluoroalkyl substances in umbilical cord serum and gestational and postnatal growth in a Chinese birth cohort. Environ Int 116:197–205

Cariou R, Veyrand B, Yamada A, Berrebi A, Zalko D, Durand S, Pollono C, Marchand P, Leblanc JC, Antignac JP, Le Bizec B (2015) Perfluoroalkyl acid (PFAA) levels and profiles in breast milk, maternal and cord serum of French women and their newborns. Environ Int 84:71–81

CDC (2023) National Report on Human Exposure to Environmental Chemicals. Bomonitoring Data Tables for Environmental Chemicals https://www.cdc.gov/exposurereport/data_tables.html

Cerna M, Grafnetterova AP, Dvorakova D, Pulkrabova J, Malý M, Janos T, Vodrazkova N, Tupa Z, Puklova V (2020) Biomonitoring of PFOA, PFOS and PFNA in human milk from Czech Republic, time trends and estimation of infant's daily intake. Environ Res 188:109763

Chen HC, Cao WC, Liu XF, Liu X, Cheng QY, Zhou Y, Wen S (2014) Contamination levels of perfluorinated and polyfluoroalkyl compounds in breast milk and assessment of their exposure risk to infants. Se Pu 42(2):211–216 [in Chinese]

Cheng Y, Wang Y, Wang X, Lv Z, Zhou F, Huang B, Liu X, Chen D (2025a) 6:2 Fluorotelomer Ethoxylates in Human Serum and Residential Homes: A Growing Environmental Health Concern. Environ Sci Technol 59(10):5182–5190

Cheng L, Teagle S, Enders JR, Weed RA, Nichols HB, Knappe DRU, Hoppin JA (2025b) Historical blood serum samples from Wilmington, North Carolina: the importance of ultrashort-chain per- and polyfluoroalkyl substances. Environ Sci Technol 59(43):23125–23135

Colles A, Bruckers L, Den Hond E, Govarts E, Morrens B, Schettgen T, Buekers J, Coertjens D, Nawrot T, Loots I, Nelen V, De Henauw S, Schoeters G, Baeyens W, van Larebeke N (2020) Perfluorinated substances in the Flemish population (Belgium): levels and determinants of variability in exposure. Chemosphere 242:125250

Costello E, Rock S, Stratakis N, Eckel SP, Walker DI, Valvi D, Cserbik D, Jenkins T, Xanthakos SA, Kohli R, Sisley S, Vasiliou V, La Merrill MA, Rosen H, Conti DV, McConnell R, Chatzi L (2022) Exposure to per- and polyfluoroalkyl substances and markers of liver injury: a systematic review and meta-analysis. Environ Health Perspect 130:46001

Crawford L, Halperin SA, Dzierlenga MW, Skidmore B, Linakis MW, Nakagawa S, Longnecker MP (2023) Systematic review and meta-analysis of epidemiologic data on vaccine response in relation to exposure to five principal perfluoroalkyl substances. Environ Int 172:107734

Criswell RL, Wang Y, Christensen B, Botelho JC, Calafat AM, Peterson LA, Huset CA, Karagas MR, Romano ME (2023) Concentrations of Per- and Polyfluoroalkyl Substances in Paired Maternal Plasma and Human Milk in the New Hampshire Birth Cohort. Environ Sci Technol 57:463–472

Cserbik D, Casas M, Flores C, Paraian A, Haug LS, Rivas I, Bustamante M, Dadvand P, Sunyer J, Vrijheid M, Villanueva CM (2024) Concentrations of per- and polyfluoroalkyl substances (PFAS) in paired tap water and blood samples during pregnancy. J Expo Sci Environ Epidemiol 34(1):90–96

Dagnino S, Strynar MJ, McMahen RL, Lau CS, Ball C, Garantziotis S, Webster TF, McClean MD, Lindstrom AB (2016) Identification of biomarkers of exposure to FTOHs and PAPs in humans using a targeted and nontargeted analysis approach. Environ Sci Technol 50:10216–10225

Dai C, Peng L, Li Y, Li Z, Chen D, Wang F, Lin N (2024) Distribution of per- and polyfluoroalkyl substances in blood, serum, and urine of patients with liver cancer and associations with liver function biomarkers. J Environ Sci (China) 139:418–427

De Felip E, Abballe A, Albano FL, Battista T, Carraro V, Conversano M, Franchini S, Giambanco L, Iacovella N, Ingelido AM, Maiorana A, Maneschi F, Marra V, Mercurio A, Nale R, Nucci B, Panella V, Pirola F, Porpora MG, Procopio E, Suma N, Valentini S, Valsenti L, Vecchiè V (2015) Current exposure of Italian women of reproductive age to PFOS and PFOA: A human biomonitoring study. Chemosphere 137:1–8

D'eon JC, Mabury SA (2011a) Exploring indirect sources of human exposure to perfluoroalkyl carboxylates (PFCAs): evaluating uptake, elimination, and biotransformation of polyfluoroalkyl phosphate esters (PAPs) in the rat. Environ Health Perspect 119:344–350

D'eon JC, Mabury SA (2011b) Is indirect exposure a significant contributor to the burden of perfluorinated acids observed in humans? Environ Sci Technol 45:7974–7984

De Silva AO, Mabury SA (2006) Isomer distribution of perfluorocarboxylates in human blood: potential correlation to source. Environ Sci Technol 40:2903–2909

Dekant W, Dekant R (2023) Mammalian toxicity of trifluoroacetate and assessment of human health risks due to environmental exposures. Arch Toxicol 97:1069–1077

Dereumeaux C, Saoudi A, Pecheux M, Berat B, de Crouy-Chanel P, Zaros C, Brunel S, Delamaire C, le Tertre A, Lefranc A, Vandentorren S, Guldner L (2016) Biomarkers of exposure to environmental contaminants in French pregnant women from the Elfe cohort in 2011. Environ Int 97:56–67

Duan Y, Sun H, Yao Y, Meng Y, Li Y (2020) Distribution of novel and legacy per-/polyfluoroalkyl substances in serum and its associations with two glycemic biomarkers among Chinese adult men and women with normal blood glucose levels. Environ Int 134:105295

Duffek A, Conrad A, Kolossa-Gehring M, Lange R, Rucic E, Schulte C, Wellmitz J (2020) Per- and polyfluoroalkyl substances in blood plasma - Results of the German Environmental Survey for c hildren and adolescents 2014–2017 (GerES V). Int J Hyg Environ Health 228:113549

Dufour P, Pirard C, Seghaye M, Charlier C (2018) Association between organohalogenated pollutants in cord blood and thyroid function in newborns and mothers from Belgian population. Environ Pollut 238:389–396

ECHA (2017) Annex XV restriction report. Proposal for a restriction. Substance name: C 9 -C 14 PFCAs -including their salts and precursors. https://www.echa.europa.eu/documents/10162/2ec5dfdd-0e63-0b49-d756-4dc1bae7ec61

ECHA (2022) Annex XV restriction report. Proposal for a restriction. Substance name: Per- and polyfluoroalkyl substances (PFAS) in firefighting foams. https://echa.europa.eu/documents/10162/4524f49c-ae14-b01b-71d2-ac3fa916c4e9

ECHA (2023) Annex XV Restriction report. Pproposal for a restriction. Substance name: Per- and polyfluoroalkyl substances (PFASs). https://echa.europa.eu/documents/10162/f605d4b5-7c17-7414-8823-b49b9fd43aea

EFSA (2011) Scientific opinion of the panel on food contact materials, Enzymes, Flavourings and Processing Aids on the safety evaluation of the substance, 3H-perfluoro-3-[(3-methoxy-propoxy)propanoic acid], ammonium salt, CAS No. 958445-44-8, for use in food contact materials. EFSA J 9(6): 2182. Online: https://doi.org/10.2903/j.efsa.2011.2182/epdf

EFSA (2014) Reasoned opinion on the setting of MRLs for saflufenacil in various crops, considering the risk related to the metabolite trifluoroacetic acid (TFA). EFSA J 12:3585. https://www.efsa.europa.eu/en/efsajournal/pub/3585

EFSA (2018) Risk to human health related to the presence of perfluorooctane sulfonic acid and perfluorooctanoic acid in food. EFSA J 16:5194

EFSA (2020) Risk to human health related to the presence of perfluoroalkyl substances in food. EFSA J 18:6223

Ehresman DJ, Froehlich JW, Olsen GW, Chang S-C, Butenhoff JL (2007) Comparison of human whole blood, plasma, and serum matrices for the determination of perfluorooctanesulfonate (PFOA), and other fluorochemicals. Environ Res 103:176–184

Eick SM, Hom Thepaksorn EK, Izano MA, Cushing LJ, Wang Y, Smith SC, Gao S, Park JS, Padula AM, DeMicco E, Valeri L, Woodruff TJ, Morello-Frosch R (2020) Associations between prenatal maternal exposure to per- and polyfluoroalkyl substances (PFAS) and polybrominated diphenyl ethers (PBDEs) and birth outcomes among pregnant women in San Francisco. Environ Health 19:100

Ellis DA, Martin JW, De Silva AO, Mabury SA, Hurley MD, Sulbaek Andersen MP, Wallington TJ (2004) Degradation of fluorotelomer alcohols: a likely atmospheric source of perfluorinated carboxylic acids. Environ Sci Technol 15:3316–3321

Emmett EA, Shofer FS, Zhang H, Freeman D, Desai C, Shaw LM (2006) Community exposure to perfluorooctanoate: relationships between serum concentrations and exposure sources. J Occup Environ Med 48:759–770

Eriksson U, Mueller JF, Toms LL, Hobson P, Kärrman A (2017) Temporal trends of PFSAs, PFCAs and selected precursors in Australian serum from 2002 to 2013. Environ Pollut 220(Pt A):168–177

Fasano WJ, Carpenter SC, Gannon SA, Snow TA, Stadler JC, Kennedy GL, Buck RC, Korzeniowski SH, Hinderliter PM, Kemper RA (2006) Absorption, distribution, metabolism, and elimination of 8–2 fluorotelomer alcohol in the rat. Toxicol Sci 91:341–355

Fenton SE, Ducatman A, Boobis A, DeWitt JC, Lau C, Ng C, Smith JS, Roberts SM (2021) Per- and polyfluoroalkyl substance toxicity and human health review: current state of knowledge and strategies for informing future research. Environ Toxicol Chem 40:606–630

Fiedler H, Sadia M (2021) Regional occurrence of perfluoroalkane substances in human milk for the global monitoring plan under the Stocklm Convention on Persistent Organic Pollutants during 2016–2019. Chemosphere 277:130287

Fisher M, Arbuckle TE, Liang CL, LeBlanc A, Gaudreau E, Foster WG, Haines D, Davis K, Fraser WD (2016) Concentrations of persistent organic pollutants in maternal and cord blood from the maternal-infant research on environmental chemicals (MIREC) cohort study. Environ Health 15:59

Frigerio G, Cafagna S, Polledri E, Mercadante R, Fustinoni S (2022) Development and validation of an LC-MS/MS method for the quantitation of 30 legacy and emerging per- and polyfluoroalkyl substances (PFASs) in human plasma, including HFPO-DA, DONA, and cC6O4. Anal Bioanal Chem 414:1259–1278

Frömel T, Knepper TP (2010) Fluorotelomer ethoxylates: sources of highly fluorinated environmental contaminants part I: biotransformation. Chemosphere 80(11):1387–1392

Fromme H, Midasch O, Twardella D, Angerer J, Boehmer S, Liebl B (2007) Occurrence of perfluorinated substances in an adult German population in southern Bavaria. Int Arch Occup Environ Health 80:313–319

Fromme H, Mosch C, Morovitz M, Alba-Alejandre I, Boehmer S, Kiranoglu M, Faber F, Hannibal I, Genzel-Boroviczény O, Koletzko B, Völkel W (2010) Pre- and postnatal exposure to perfluorinated compounds (PFCs). Environ Sci Technol 44:7123–7129

Fromme H, Wöckner M, Roscher E, Völkel W (2017) ADONA and perfluoroalkylated substances in plasma samples of German blood donors living in South Germany. Int J Hyg Environ Health 220(2 Pt B):455–460

Fromme H (2018) Personal communication

Fromme H, Fuchs V, Albrecht M, Aschenbrenner B, Röhl C, Janitzki N, Herber-Jonat S, Wöckner M, Völkel W, Flemmer AW, Schober W (2022) Polychlorinated dioxins and dibenzofurans (PCDD/F), polybrominated dioxins and dibenzofurans (PBDD/F), polychlorinated biphenyls (PCB), polybrominated diphenyl ethers (PBDE), and per- and polyfluoroalkyl substances (PFAS) in German breast milk samples (LUPE 8). Sci Total Environ 825:154066

Fujii Y, Yan J, Harada KH, Hitomi T, Yang H, Wang P, Koizumi A (2012) Levels and profiles of long-chain perfluorinated carboxylic acids in human breast milk and infant formulas in East Asia. Chemosphere 86:315–321

Gannon SA, Fasano WJ, Mawn MP, Nabb DL, Buck RC, Buxton LW, Jepson GW, Frame SR (2016) Absorption, distribution, metabolism, excretion, and kinetics of 2,3,3,3-tetrafluoro-2-(heptafluoropropoxy)propanoic acid ammonium salt following a single dose in rat, mouse, and cynomolgus monkey. Toxicology 340:1–9

Gao K, Zhuang TF, Liu X, Fu JJ, Zhang JX, FuJ WLG, Zhang AQ, Liang Y, Song MY, Jiang GB (2019) Prenatal exposure to per- and polyfluoroalkyl substances (PFASs) and association between the placental transfer efficiencies and dissociation constant of serum proteins-PFAS Complexes. Environ Sci Technol 53:6529–6538

Garavagno MDLA, Holland RET, Khan MAH, Orr-Ewing AJ, Shallcross DE (2024) Trifluoroacetic acid: toxicity, sources. Sinks and Future Prospects. Sustainability 16(6):2382

GC (Government Canada) (2025) State of Per- and Polyfluoroalkyl Substances (PFAS). Environment and Climate Change Canada, Health Canada. https://publications.gc.ca/collections/collection_2025/eccc/En84-395-2025-eng.pdf

Gebbink WA, Glynn A, Berger U (2015) Temporal changes (1997–2012) of perfluoroalkyl acids and selected precursors (including isomers) in Swedish human serum. Environ Pollut 199:166–173

Göckener B, Weber T, Rüdel H, Bücking M, Kolossa-Gehring M (2020) Human biomonitoring of per- and polyfluoroalkyl substances in German blood plasma samples from 1982 to 2019 Environ Int 145:106123

Gomis MI, Vestergren R, MacLeod M, Mueller JF, Cousins IT (2017) Historical human exposure to perfluoroalkyl acids in the United States and Australia reconstructed from biomonitoring data using population-based pharmacokinetic modelling. Environ Int 108:92–102

Góralczyk K, Pachocki KA, Hernik A, Struciński P, Czaja K, Lindh CH, Jönsson BA, Lenters V, Korcz W, Minorczyk M, Matuszak M, Ludwicki JK (2015) Perfluorinated chemicals in blood serum of inhabitants in central Poland in relation to gender and age. Sci Total Environ 532:548–555

Gordon SC (2011) Toxicological evaluation of ammonium 4,8-dioxa-3H-perfluorononanoate, a new emulsifier to replace ammonium perfluorooctanoate in fluoropolymer manufacturing. Regul Toxicol Pharmacol 59:64–80

Graber JM, Alexander C, Laumbach RJ, Black K, Strickland PO, Georgopoulos PG, Marshall EG, Shendell DG, Alderson D, Mi Z, Mascari M, Weisel CP (2019) Per and polyfluoroalkyl substances (PFAS) blood levels after contamination of a community water supply and comparison with 2013–2014 NHANES. J Expo Sci Environ Epidemiol 29:172–182

Granum B, Haug LS, Namork E, Stølevik SB, Thomsen C, Aaberge IS, van Loveren H, Løvik M, Nygaard UC (2013) Pre-natal exposure to perfluoroalkyl substances may be associated with altered vaccine antibody levels and immune-related health outcomes in early childhood. J Immunotoxicol 10:373–379

Guo J, Zhang J, Wang Z, Zhang L, Qi X, Zhang Y, Chang X, Wu C, Zhou Z (2021) Umbilical cord serum perfluoroalkyl substance mixtures in relation to thyroid function of newborns: findings from Sheyang Mini Birth Cohort Study. Chemosphere 273:129664

Guo W, Hao W, Xiao W (2025) Emerging perfluorinated chemical GenX: environmental and biological fates and risks. Environ Health (Wash) 3(4):338–351

Gützkow KB, Haug LS, Thomsen C, Sabaredzovic A, Becher G, Brunborg G (2012) Placental transfer of perfluorinated compounds is selective–a Norwegian Mother and Child sub-cohort study. Int J Hyg Environ Health 215:216–219

Gyllenhammar I, Benskin JP, Sandblom O, Berger U, Ahrens L, Lignell S, Wiberg K, Glynn A (2018) Perfluoroalkyl acids (PFAAs) in serum from 2-4-month-old infants: Influence of maternal serum concentration, gestational age, breast-feeding, and contaminated drinking water. Environ Sci Techno 52:7101–7110

Han F, Liu J, Wang Y, Li J, Lyu B, Zhao Y, Wu Y (2023a) Penetration of perfluorooctanesulfonate isomers and their alternatives from maternal blood to milk and its associations with chemical properties and milk primary components. Environ Sci Technol 57:2457–2463

Han F, Wang Y, Li J, Lyu B, Liu J, Zhang J, Zhao Y, Wu Y (2023b) Occurrences of legacy and emerging per- and polyfluoroalkyl substances in human milk in China: results of the third national human milk survey (2017–2020). J Hazard Mater 443(Pt A):130163

Han W, Gao Y, Yao Q, Yuan T, Wang Y, Zhao S, Shi R, Bonefeld-Jorgensen EC, Shen X, Tian Y (2018) Perfluoroalkyl and polyfluoroalkyl substances in matched parental and cord serum in Shandong, China. Environ Int 116:206–213

Hansen KJ, Clemen LA, Ellefson ME, Johnson HO (2001) Compound-specific, quantitative characterization of organic fluorochemicals in biological matrices. Environ Sci Technol 35:766–770

Hardell E, Kärrman A, van Bavel B, Bao J, Carlberg M, Hardell L (2014) Case-control study on perfluorinated alkyl acids (PFAAs) and the risk of prostate cancer. Environ Int 63:35–39

Harris MH, Rifas-Shiman SL, Calafat AM, Ye X, Mora AM, Webster TF, Oken E, Sagiv SK (2017) Predictors of per- and polyfluoroalkyl substance (PFAS) plasma concentrations in 6–10 year old American children. Environ Sci Technol 51:5193–5204

Hartmann C, Jamnik T, Weiss S, Göß M, Fareed Y, Satrapa V, Braun D, Flasch M, Warth B, Uhl M (2023) Results of the Austrian Children's Biomonitoring Survey 2020 part A: Per- and polyfluorinated alkylated substances, bisphenols, parabens and other xenobiotics. Int J Hyg Environ Health 249:114123

Hartmann C, Kaiser AM, Moche W, Weiss S, Raffesberg W, Scharf S, Graf-Rohrmeister K, Thanhaeuser M, Haiden N, Uhl M (2024) Persistent organic pollutants in Austrian human breast milk collected between 2013 and 2016. J Xenobiot 14(1):247–266

HC (2019) Fifth report on human biomonitoring of environmental chemicals in Canada. Minister of Health. https://www.canada.ca/content/dam/hc-sc/documents/services/environmental-workplace-health/reports-publications/environmental-contaminants/fifth-report-human-biomonitoring/pub1-eng.pdf

HC (2021) Sixth report on human biomonitoring of environmental chemicals in Canada. Results of the Canadian Health Measures Survey Cycle 6 (2018–2019). https://www.canada.ca/en/health-canada/services/environmental-workplace-health/reports-publications/environmental-contaminants/sixth-report-human-biomonitoring.html

He Y, Lv D, Li C, Liu X, Liu W, Han W (2022) Human exposure to F-53B in China and the evaluation of its potential toxicity: An overview. Environ Int 161:107108

He A, Li J, Li Z, Lu Y, Liang Y, Zhou Z, Man Z, Lv J, Wang Y, Jiang G (2023) Novel insights into the adverse health effects of per- and polyfluoroalkyl substances on the kidney via human urine metabolomics. Environ Sci Techno 57:16244–16254

Hölzer J, Midasch O, Rauchfuss K, Kraft M, Reupert R, Angerer J, Kleeschulte P, Marschall N, Wilhelm M (2008) Biomonitoring of perfluorinated compounds in children and adults exposed to perfluorooctanoate-contaminated drinking water. Environ Health Perspect 116:651–657

Hron LMC, Wöckner M, Fuchs V, Fembacher L, Aschenbrenner B, Herr C, Schober W, Heinze S, Völkel W (2024) Monitoring of per- and polyfluoroalkyl substances (PFAS) in human blood samples collected in three regions with known PFAS releases in the environment and three control regions in South Germany. Arch Toxicol 98(11): 3727–3738

Hsu JY, Hsu JF, Ho HH, Chiang CF, Liao PC (2013) Background levels of persistent organic pollutants in humans from Taiwan: perfluorooctane sulfonate and perfluorooctanoic acid. Chemosphere 93:532–537

Hu L, Luo D, Wang L, Yu M, Zhao S, Wang Y, Mei S, Zhang G (2021) Levels and profiles of persistent organic pollutants in breast milk in China and their potential health risks to breastfed infants: A review. Sci Total Environ 753:142028

Hu XC, Andrews DQ, Lindstrom AB, Bruton TA, Schaider LA, Grandjean P, Lohmann R, Carignan CC, Blum A, Balan SA, Higgins CP, Sunderland EM (2016) Detection of poly- and perfluoroalkyl substances (PFASs) in U.S. drinking water linked to industrial sites, military fire training areas, and wastewater treatment plants. Environ Sci Technol Lett 3:344–350

Huang MC, Dzierlenga AL, Robinson VG, Waidyanatha S, DeVito MJ, Eifrid MA, Granville CA, Gibbs ST, Blystone CR (2019) Toxicokinetics of perfluorobutane sulfonate (PFBS), perfluorohexane-1-sulphonic acid (PFHxS), and perfluorooctane sulfonic acid (PFOS) in male and female Hsd:Sprague Dawley SD rats after intravenous and gavage administration. Toxicol Rep 6:645–655

Hundley SG, Sarrif AM, Kennedy GL (2006) Absorption, distribution, and excretion of ammonium perfluorooctanoate (APFO) after oral administration to various species. Drug Chem Toxicol 29:137–145

IARC (2025) Perfluorooctanoic Acid (PFOA) and Perfluorooctanesulfonic acid (PFOS). IARC monographs on the evaluation of carcinogenic risks to humans. Volume 135. file: ///C:/Users/herma/Downloads/Volume135.v2.pdf

Ingelido AM, Abballe A, Gemma S, Dellatte E, Iacovella N, De Angelis G, Marra V, Russo F, Vazzoler M, Testai E, De Felip E (2020) Serum concentrations of perfluorinated alkyl substances in farmers living in areas affected by water contamination in the Veneto Region (Northern Italy). Environ Int 136:105435

Itoh S, Araki A, Miyashita C, Yamazaki K, Goudarzi H, Minatoya M, Ait Bamai Y, Kobayashi S, Okada E, Kashino I, Yuasa M, Baba T, Kishi R (2019) Association between perfluoroalkyl substance exposure and thyroid hormone/thyroid antibody levels in maternal and cord blood: The Hokkaido Study. Environ Int 133(Pt A):105139

Iwata HS, Kobayash M, Itoh S, Itoh R, Mesfin Ketema N, Tamura C, Miyashita T, Yamaguchi K, Yamazaki H, Masuda Y, Ait Bamai Y, Saijo Y, Ito SF, Nakayama M, Kamijima R, Kishi (2024) The association between prenatal per-and polyfluoroalkyl substance levels and Kawasaki disease among children of up to 4 years of age: a prospective birth cohort of the Japan Environment and Children's Study. Environ Int 183:108321

Jahnke A, Ahrens L, Ebinghaus R, Temme C (2007) Urban versus remote air concentrations of fluorotelomer alcohols and other perfluorinated alkyl substances in Germany. Environ Sci Technol 41:745–752

Janda J, Nödler K, Brauch HJ, Zwiener C, Lange FT (2019) Robust trace analysis of polar (C2–C8) perfluorinated carboxylic acids by liquid chromatography-tandem mass spectrometry: method development and application to surface water, groundwater and drinking water. Environ Sci Pollut Res Int 26(8):7326–7336

Jensen IKB, Budtz-Jørgensen E, Lindh C, Roland CB, de Place KS, Bendix JM, Molsted S, Clausen TD, Stallknecht B, Mortensen OH, Løkkegaard E, Knudsen LE (2025) Serum concentrations of per- and poly-fluoroalkyl substances (PFAS) in Danish pregnant women-temporal trends during pregnancy, correlations with partners, associations with physical activity, and blood lipid concentrations. Environ Health 24(1):16

Jia J, Duan L, Dong B, Dong Q, Liu Y, Yu W, Yang L, Shi H (2023) Perfluoroalkyl and polyfluoroalkyl substances in cord serum of newborns and their potential factors. Chemosphere 313:137525

Jian JM, Guo Y, Zeng L, Liang-Ying L, Lu X, Wang F, Zeng EY (2017) Global distribution of perfluorochemicals (PFCs) in potential human exposure source-A review. Environ Int 108:51–62

Jiang W, Zhang Y, Zhu L, Deng J (2014) Serum levels of perfluoroalkyl acids (PFAAs) with isomer analysis and their associations with medical parameters in Chinese pregnant women. Environ Int 64:40–47

Jin H, Zhang Y, Jiang W, Zhu L, Martin JW (2016) Isomer-specific distribution of perfluoroalkyl substances in blood. Environ Sci Technol 50:7808–7815

Jin H, Mao L, Xie J, Zhao M, Bai X, Wen J, Shen T, Wu P (2020) Poly- and perfluoroalkyl substance concentrations in human breast milk and their associations with postnatal infant growth. Sci Total Environ 713:136417

Jonathan JWA, Essumang DK, Bentum JK, Kabotso DEK, Gborgblorvor IR, Eshun A, Hlorlewu ND, Davordzi E (2024) Exploring perfluoroalkyl substances contamination in human breast milk: first ghanaian study. Chemosphere 369:143769

Joudan S, De Silva AO, Young CJ (2021) Insufficient evidence for the existence of natural trifluoroacetic acid. Environ Sci Process Impacts 23:1641–1649

Kabadi SV, Fisher J, Aungst J, Rice P (2018) Internal exposure-based pharmacokinetic evaluation of potential for biopersistence of 6:2 fluorotelomer alcohol (FTOH) and its metabolites. Food Chem Toxicol 112:375–382

Kaiser AM, Forsthuber M, Aro R, Kärrman A, Gundacker C, Zeisler H, Foessleitner P, Salzer H, Hartmann C, Uhl M, Yeung LWY (2021) Extractable organofluorine analysis in pooled human serum and placental tissue samples from an AustriansSubpopulation-A mass balance analysis approach. Environ Sci Technol 55:9033–9042

Kaiser AM, Forsthuber M, Widhalm R, Granitzer S, Weiss S, Zeisler H, Foessleitner P, Salzer H, Grasl-Kraupp B, Moshammer H, Hartmann C, Uhl M, Gundacker C (2023) Prenatal exposure to per- and polyfluoroalkyl substances and pregnancy outcome in Austria. Ecotoxicol Environ Saf 259:115006

Kang H, Calafat AM, Karvonen-Gutierrez CA, Park SK (2023) Isomer-specific serum concentrations of perfluorooctane sulfonic acid among U.S. Adults: results from the National Health and

Nutrition Examination Survey (NHANES) and the Study of Women's Health Across the Nation Multi-Pollutant Study (SWAN-MPS) Environ Sci Technol 57:385–394

Kang H, Choi K, Lee HS, Kim DH, Park NY, Kim S, Kho Y (2016) Elevated levels of short carbon-chain PFCAs in breast milk among Korean women: current status and potential challenges. Environ Res 148:351–359

Kang H, Lee H-K, Moon H-B, Kima S, Lee J, Ha M, Hong S, Kim S, Choi K (2018) Perfluoroalkyl acids in serum of Korean children: Occurrences, related sources, and associated health outcomes. Sci Total Environ 645:958–965

Kärrman A, Ericson I, van Bavel B, Darnerud PO, Aune M, Glynn A, Lignell S, Lindström G (2007a) Exposure of perfluorinated chemicals through lactation: levels of matched human milk and serum and a temporal trend, 1996–2004, in Sweden. Environ Health Perspect 115:226–230

Kärrman A, Langlois I, van Bavel B, Lindström G, Oehme M (2007b) Identification and pattern of perfluorooctane sulfonate (PFOS) isomers in human serum and plasma. Environ Int 33:782–788

Kärrman A, van Bavel B, Järnberg U, Hardell L, Lindström G (2006) Perfluorinated chemicals in relation to other persistent organic pollutants in human blood. Chemosphere 64:1582–1591

Kato K, Wong LY, Chen A, Dunbar C, Webster GM, Lanphear BP, Calafat AM (2014) Changes in serum concentrations of maternal poly- and perfluoroalkyl substances over the course of pregnancy and predictors of exposure in a multiethnic cohort of Cincinnati, Ohio pregnant women during 2003–2006. Environ Sci Technol 48:9600–9608

Kato K, Kalathil AA, Patel AM, Ye X, Calafat AM (2018) Per- and polyfluoroalkyl substances and fluorinated alternatives in urine and serum by on-line solid phase extraction-liquid chromatography-tandem mass spectrometry. Chemosphere 209:338–345

Kim DH, Lee JH, Oh JE (2019) Perfluoroalkyl acids in paired serum, urine, and hair samples: Correlations with demographic factors and dietary habits. Environl Pollut 248:175–182

Kim DH, Lee MY, Oh JE (2014) Perfluorinated compounds in serum and urine samples from children aged 5–13 years in South Korea. Environ Pollut 192:171–178

Kim JH, Moon N, Lee JW, Mehdi Q, Yun MH, Moon HB (2023) Time-course trend and influencing factors for per- and polyfluoroalkyl substances in the breast milk of Korean mothers. Chemosphere 310:136688

Koponen J, Winkens K, Airaksinen R, Berger U, Vestergren R, Cousins IT, Karvonen AM, Pekkanen J, Kiviranta H (2018) Longitudinal trends of per- and polyfluoroalkyl substances in children's serum. Environ Int 121(Pt 1):591–599

Kornher K, Gould CF, Manzano JM, Baines K, Kayser G, Tu X, Suarez-Torres J, Martinez D, Peterson LA, Huset CA, Barr DB, Suarez-Lopez JR (2025) Associations of PFAS and pesticides with lung function changes from adolescence to young adulthood in the ESPINA study. Int J Hyg Environ Health 265:114526

Kotlarz N, McCord J, Collier D, Lea CS, Strynar M, Lindstrom AB, Wilkie AA, Islam JY, Matney K, Tarte P, Polera ME, Burdette K, DeWitt J, May K, Smart RC, Knappe DRU, Hoppin JA (2020) Measurement of novel, drinking water-associated PFAS in blood from adults and children in Wilmington. North Carolina. Environ Health Perspect 128(7):77005

Kotlarz N, McCord J, Wiecha N, Weed RA, Cuffney M, Enders JR, Strynar M, Knappe DRU, Reich BJ, Hoppin JA (2024) Measurement of Hydro-EVE and 6:2 FTS in Blood from Wilmington, North Carolina, Residents, 2017–2018. Environ Health Perspect 132(2):27702

Kudo N, Katakura M, Sato Y, Kawashima Y (2002) Sex hormone-regulated renal transport of perfluorooctanoic acid. Chem Biol Interact 139:301–316

Lahne H, Gerstner D, Völkel W, Schober W, Aschenbrenner B, Herr C, Heinze S, Quartucci C (2024) Human biomonitoring follow-up study on PFOA contamination and investigation of possible influencing factors on PFOA exposure in a German population originally exposed to emissions from a fluoropolymer production plant. Int J Hyg Environ Health 259:114387

LaKind JS, Naiman J, Verner MA, Lévêque L, Fenton S (2023) Per- and polyfluoroalkyl substances (PFAS) in breast milk and infant formula: a global issue. Environ Res 219:115042

Lamichhane S, Siljander H, Duberg D, Honkanen J, Virtanen SM, Orešič M, Knip M, Hyötyläinen T (2021) Exposure to per- and polyfluoroalkyl substances associates with an altered lipid composition of breast milk. Environ Int 157:106855

Langlois I, Oehme M (2006) Structural identification of isomers present in technical perfluorooctane sulfonate by tandem mass spectrometry. Rapid Commun Mass Spectrom 20:844–850

Lankova D, Lacina O, Pulkrabova J, Hajslova J (2013) The determination of perfluoroalkyl substances, brominated flame retardants and their metabolites in human breast milk and infant formula. Talanta 117:318–325

Lazarevic N, Smurthwaite KS, D'Este C, Lucas RM, Armstrong B, Clements AC, Trevenar SM, Gad I, Hosking R, Law HD, Mueller J, Bräunig J, Nilsson S, Lane J, Lal A, Lidbury BA, Korda RJ, Kirk MD (2023) Liver and cardiometabolic markers and conditions in a cross-sectional study of three Australian communities living with environmental per- and polyfluoroalkyl substances contamination. Environ Res 226:115621

Lee H, Mabury SA (2011) A pilot survey of legacy and current commercial fluorinated chemicals in human sera from United States donors in 2009. Environ Sci Technol 45:8067–8074

Lee S, Ahn RM, Kim JH, Han YD, Lee JH, Son BS, Lee K (2019) Study design, rationale and procedures for human biomonitoring of hazardous chemicals from foods and cooking in Korea. Int J Environ Res Public Health 16:2583

Lee S, Kim S, Park J, Kim HJ, Choi G, Choi S, Kim S, Kim SY, Kim S, Choi K, Moon HB (2018) Perfluoroalkyl substances (PFASs) in breast milk from Korea: time-course trends, influencing factors, and infant exposure. Sci Total Environ 612:286–292

Lehmann GM, LaKind JS, Davis MH, Hines EP, Marchitti SA, Alcala C, Lorber M (2018) Environmental chemicals in breast milk and formula: exposure and risk assessment implications. Environ Health Perspect 126:96001

LGA (Baden-Württemberg State Health Office) (2018) Results of the PFC blood control test in the district of Rastatt 2018. Report. [in German] https://sozialministerium.baden-wuerttemberg.de/fileadmin/redaktion/m-sm/intern/downloads/Downloads_Gesundheitsschutz/Bericht_PFC-BlutkontrolluntersuchungLKRastatt_2018.pdf

LGA (Baden-Württemberg State Health Office) (2021) Results of the PFC blood control test in the district of Rastatt 2020. Report. [in German] https://sozialministerium.baden-wuerttemberg.de/fileadmin/redaktion/m-sm/intern/downloads/Downloads_Gesundheitsschutz/Bericht_PFC-BlutkontrolluntersuchungLKRastatt_2020-Aug2021fin.pdf

Li Z-M, Guo L-H, Ren X-M (2016) Biotransformation of 8:2 fluorotelomer alcohol by recombinant human cytochrome P450s, human liver microsomes and human liver cytosol. Environ Sci: Processes Impacts 18:538

Li Y, Fletcher T, Mucs D, Scott K, Lindh CH, Tallving P, Jakobsson K (2018) Half-lives of PFOS, PFHxS and PFOA after end of exposure to contaminated drinking water. Occup Environ Med 75:46–51

Li H, Hammarstrand S, Midberg B, Xu Y, Li Y, Olsson DS, Fletcher T, Jakobsson K, Andersson EM (2022) Cancer incidence in a Swedish cohort with high exposure to perfluoroalkyl substances in drinking water. Environ Res 204(Pt C):112217

Li J, Li J, Ma Y, Chen B, Wang X, Jiao X, Jin Y, Shen Z, Yuan T, Yu X (2021a) Urine concentrations of perfluoroalkyl acids in children and contributions of dietary factors: a cross-sectional study from Shanghai, China. Environ Sci Pollut Res Int 28:20440–20450

Li N, Ying GG, Hong H, Deng WJ (2021b) Perfluoroalkyl substances in the urine and hair of preschool children, airborne particles in kindergartens, and drinking water in Hong Kong. Environ Pollut 270:116219

Li J, Duan W, An Z, Jiang Z, Li L, Guo M, Tan Z, Zeng X, Liu X, Liu Y, Li A, Guo H (2024) Legacy and alternative per- and polyfluoroalkyl substances spatiotemporal distribution in China: Human exposure, environmental media, and risk assessment. J Hazard Mater 480:135795

Lin BC, Zhu L, Xu H, Xiao YP, Xu JQ, Huang YF, Yang CZ, Wang MQ, Bai ZJ, Zhu W, Wei YH, Li HT, Chen YR (2025) Associations between per- and polyfluoroalkyl substances (PFAS)

exposure via infant feeding and the gut microbiota of preterm infants: a study of mother-preterm infant dyads. Ecotoxicol Environ Saf 303:118853

Lindfeldt E, Benskin JP, Plassmann M, Kruså M, Hedvall Kallerman P, Gyllenhammar I (2023) Levels of poly- and perfluoroalkyl substances (PFAS) in serum from children at 4, 8 and 12 years of age, in Uppsala, Sweden: results from year 2020–2021 and temporal trends for the time period 2008–2022. Report to the Swedish EPA

Liu D, Tang B, Nie S, Zhao N, He L, Cui J, Mao W, Jin H (2023) Distribution of per- and polyfluoroalkyl substances and their precursors in human blood. J Hazard Mater 441:129908

Liu J, Li J, Liu Y, Chan HM, Zhao Y, Cai Z, Wu Y (2011) Comparison on gestation and lactation exposure of perfluorinated compounds for newborns. Environ Int 37:1206–1212

Liu J, Li J, Zhao Y, Wang Y, Zhang Y, Lei Z, Wu Y (2010) The occurrence of perfluorinated alkyl compounds in human milk from different regions of China. Environ Int 36:433–438

Liu Y, Li A, Buchanan S, Liu W (2020) Exposure characteristics for congeners, isomers, and enantiomers of perfluoroalkyl substances in mothers and infants. Environ Int 144:106012

Loi EI, Yeung LW, Mabury SA, Lam PK (2013) Detections of commercial fluorosurfactants in Hong Kong marine environment and human blood: a pilot study. Environ Sci Technol 47:4677–4685

Lorber M, Eaglesham GE, Hobson P, Toms LM, Mueller JF, Thompson JS (2015) The effect of ongoing blood loss on human serum concentrations of perfluorinated acids. Chemosphere 118:170–177

Luo F, Chen Q, Yu G, Huo X, Wang H, Nian M, Tian Y, Xu J, Zhang J, Zhang J (2022) Exposure to perfluoroalkyl substances and neurodevelopment in 2-year-old children: a prospective cohort study. Environ Int 166:107384

MacDonald AM, Gabos S, Braakman S, Cheperdak L, Lee B, Hrudey SE, Le XC, Li XF, Mandal R, Martin JW, Schopflocher D, Lyon ME, Cheung PY, Ackah F, Graydon JA, Reichert M, Lyon AW, Jarrell J, Benadé G, Charlton C, Huang D, Bennett MJ, Kinniburgh DW (2022) Maternal and child biomonitoring strategies and levels of exposure in western Canada during the past seventeen years: the alberta biomonitoring program: 2005–2021. Int J Hyg Environ Health 244:113990

Macheka LR, Abafe OA, Mugivhisa LL, Olowoyo JO (2022) Occurrence and infant exposure assessment of per and polyfluoroalkyl substances in breast milk from South Africa. Chemosphere 288(Pt 2):132601

Maestri L, Negri S, Ferrari M, Ghittori S, Fabris F, Danesino P, Imbriani M (2006) Determination of perfluorooctanoic acid and perfluorooctanesulfonate in human tissues by liquid chromatography/single quadrupole mass spectrometry. Rapid Commun Mass Spectrom 20:2728–2734

Mahfouz M, Harmouche-Karaki M, Matta J, Mahfouz Y, Salameh P, Younes H, Helou K, Finan R, Abi-Tayeh G, Meslimani M, Moussa G, Chahrour N, Osseiran C, Skaiki F, Narbonne JF (2023) Maternal serum, cord and human milk levels of per- and polyfluoroalkyl substances (PFAS), association with predictors and effect on newborn anthropometry. Toxics 11:455

Manzano-Salgado CB, Casas M, Lopez-Espinosa MJ, Ballester F, Basterrechea M, Grimalt JO, Jiménez AM, Kraus T, Schettgen T, Sunyer J, Vrijheid M (2015) Transfer of perfluoroalkyl substances from mother to fetus in a Spanish birth cohort. Environ Res 142:471–478

McAdam J, Bell EM (2023) Determinants of maternal and neonatal PFAS concentrations: a review. Environ Health 22:41

Mi X, Yang YQ, Zeeshan M, Wang ZB, Zeng XY, Zhou Y, Yang BY, Hu LW, Yu HY, Zeng XW, Liu RQ, Dong GH (2020) Serum levels of per- and polyfluoroalkyl substances alternatives and blood pressure by sex status: Isomers of C8 health project in China. Chemosphere 261:127691

Miaz LT, Plassmann MM, Gyllenhammar I, Bignert A, Sandblom O, Lignell S, Glynn A, Benskin JP (2020) Temporal trends of suspect- and target-per/polyfluoroalkyl substances (PFAS), extractable organic fluorine (EOF) and total fluorine (TF) in pooled serum from first-time mothers in Uppsala, Sweden, 1996–2017. Environ Sci Process Impacts 22:1071–1083

Miyake Y, Yamashita N, Rostkowski P, So MK, Taniyasu S, Lam PKS, Kannan K (2007) Determination of trace levels of total fluorine in water using combustion ion chromatography for fluorine: a mass balance approach to determine individual perfluorinated chemicals in water. J Chromatogr A 1143:98–104

Mondal D, Weldon RH, Armstrong BG, Gibson LJ, Lopez-Espinosa MJ, Shin HM, Fletcher T (2014) Breastfeeding: a potential excretion route for mothers and implications for infant exposure to perfluoroalkyl acids. Environ Health Perspect 122:187–192

Morello-Frosch R, Cushing LJ, Jesdale BM, Schwartz JM, Guo W, Guo T, Wang M, Harwani S, Petropoulou SE, Duong W, Park JS, Petreas M, Gajek R, Alvaran J, She J, Dobraca D, Das R, Woodruff TJ (2016) Environmental chemicals in an urban population of pregnant women and their newborns from San Francisco. Environ Sci Technol 50:12464–12472

Motas Guzmàn M, Clementini C, Pérez-Cárceles MD, Jiménez Rejón S, Cascone A, Martellini T, Guerranti C, Cincinelli A (2016) Perfluorinated carboxylic acids in human breast milk from Spain and estimation of infant's daily intake. Sci Total Environ 544:595–600

Muir DCG, Freeling F, Nilsson S, Bugsel B, Bowles K, Hobson P, Toms L-M, Mueller JF (2025) Trifluoroacetic acid in Australian human urine samples. Environ Sci Technol Lett 12:1411–1417

Nakayama SF, Isobe T, Iwai-Shimada M, Kobayashi Y, Nishihama Y, Taniguchi Y, Sekiyama M, Michikawa T, Yamazaki S, Nitta H, Oda M, Mitsubuchi H, Sanefuji M, Ohga S, Mise N, Ikegami A, Suga R, Shimono M (2020) Poly- and perfluoroalkyl substances in maternal serum: Method development and application in Pilot Study of the Japan Environment and Children's Study. J Chromatogr A 1618:460933

NASEM (National Academies of Sciences, Engineering, and Medicine) (2022) Guidance on PFAS exposure, testing, and clinical follow-up. Washington, DC: The National Academies Press. https://doi.org/10.17226/26156.

Needham LL, Grandjean P, Heinzow B, Jørgensen PJ, Nielsen F, Patterson DG Jr, Sjödin A, Turner WE, Weihe P (2011) Partition of environmental chemicals between maternal and fetal blood and tissues. Environ Sci Technol 45:1121–1126

Nilsson H, Kärrman A, Rotander A, van Bavel B, Lindström G, Westberg H (2013) Biotransformation of fluorotelomer compound to perfluorocarboxylates in humans. Environ Int 51:8–12

Norén E, Lindh C, Glynn A, Rylander L, Pineda D, Nielsen C (2021) Temporal trends, 2000–2017, of perfluoroalkyl acid (PFAA) concentrations in serum of Swedish adolescents. Environ Int 155:106716

Nyberg E, Awad R, Bignert A, Ek C, Sallsten G, Benskin JP (2018) Inter-individual, inter-city, and temporal trends of per- and polyfluoroalkyl substances in human milk from Swedish mothers between 1972 and 2016. Environ Sci Process Impacts 20:1136–1147

Nyström J, Benskin JP, Plassmann M, Sandblom O, Glynn A, Lampa E, Gyllenhammar I, Moraeus L, Lignell S (2022) Demographic, life-style and physiological determinants of serum per- and polyfluoroalkyl substance (PFAS) concentrations in a national cross-sectional survey of Swedish adolescents. Environ Res 208:112674

OEHHA (Office of Environmental Health Hazard Assessment) (2024) Perfluorooctanoic Acid and Perfluorooctane Sulfonic Acid in Drinking Water. https://oehha.ca.gov/media/downloads/air/public-health-goal-document/pfoapfosphgfinaldraft040524.pdf

Oh J, Shin HM, Kannan K, Calafat AM, Schmidt RJ, Hertz-Picciotto I, Bennett DH (2024) Per- and Polyfluoroalkyl Substances (PFAS) in Serum of 2 to 5 year-Old Children: Temporal Trends, Determinants, and Correlations with Maternal PFAS Concentrations. Environ Sci Technol 58(7):3151–3162

Olsen GW, Mair DC, Lange CC, Harrington LM, Church TR, Goldberg CL, Herron RM, Hanna H, Nobiletti JB, Rios JA, Reagen WK, Ley CA (2017) Per- and polyfluoroalkyl substances (PFAS) in American Red Cross adult blood donors, 2000–2015. Environ Res 157:87–95

Olsen GW, Burris JM, Ehresman DJ, Froehlich JW, Seacat AM, Butenhoff JL, Zobel LR (2007) Half-life of serum elimination of perfluorooctanesulfonate, perfluorohexanesulfonate, and perfluorooctanoate in retired fluorochemical production workers. Environ Health Perspect 115:1298–1305

Pan YT, Zhu YS, Zheng TZ, Cui QQ, Buka SL, Zhang B, Guo Y, Xia W, Yeung LW, Li YY (2017) Novel chlorinated polyfluorinated ether sulfonates and legacy per-/polyfluoroalkyl substances:

placental transfer and relationship with serum albumin and glomerular filtration rate. Environ Sci Technol 51:634–644

Parizek O, Gramblicka T, Parizkova D, Polachova A, Bechynska K, Dvorakova D, Stupak M, Dusek J, Pavlikova J, Topinka J, Sram RJ, Pulkrabova J (2023) Assessment of organohalogenated pollutants in breast milk from the Czech Republic. Sci Total Environ 871:161938

Pennoyer EH, Heiger-Bernays W, Aro R, Yeung LWY, Schlezinger JJ, Webster TF (2023) Unknown organofluorine mixtures in U.S. adult serum:contribution from pharmaceuticals? Toxics 11:416

Pitter G, Da Re F, Canova C, Barbieri G, Zare Jeddi M, Daprà F, Manea F, Zolin R, Bettega AM, Stopazzolo G, Vittorii S, Zambelli L, Martuzzi M, Mantoan D, Russo F (2020) Serum levels of perfluoroalkyl substances (PFAS) in adolescents and young adults exposed to contaminated drinking water in the Veneto Region, Italy: a cross-sectional study based on a health surveillance program. Environ Health Perspect 128:27007

Poothong S, Thomsen C, Padilla-Sanchez JA, Papadopoulou E, Haug LS (2017) Distribution of novel and well-known poly- and perfluoroalkyl substances (PFASs) in human serum, plasma, and whole blood. Environ Sci Technol 51:13388–13396

Pritchett JR, Rinsky JL, Dittman B, Christensen A, Langley R, Moore Z, Fleischauer AT, Koehler K, Calafat AM, Rogers R, Esters L, Jenkins R, Collins F, Conner D, Breysse P (2019) Notes from the Field: targeted biomonitoring for GenX and other per- and polyfluoroalkyl substances following detection of drinking water contamination—North Carolina, 2018. MMWR 68:647–648

Quaak I, de Cock M, de Boer M, Lamoree M, Leonards P, van de Bor M (2016) Prenatal exposure to perfluoroalkyl substances and behavioral development in children. Int J Environ Res Public Health 13:511

Raab U, Albrecht M, Preiss U, Völkel W, Schwegler U, Fromme H (2013) Organochlorine compounds, nitro musks and perfluorinated substances in breast milk—results from Bavarian Monitoring of Breast Milk 2007/8. Chemosphere 93:461–467

Rae JMC, Craig L, Slone TW, Frame SR, Buxton LW, Kennedy GL (2015) Evaluation of chronic toxicity and carcinogenicity of ammonium2,3,3,3-tetrafluoro-2-(heptafluoropropoxy)-propanoate inSprague–Dawley rats. Tox Rep 2:939–949

Rand AA, Mabury SA (2017) Is there a human health risk associated with indirect exposure to perfluoroalkyl carboxylates (PFCAs)? Toxicology 375:28–36

Rappazzo KM, Coffman E, Hines EP (2017) Exposure to perfluorinated alkyl substances and health outcomes in children: a systematic review of the epidemiologic literatur. Int J Environ Res Public Health 14:691

Rawn DFK, Dufresne G, Clément G, Fraser WD, Arbuckle TE (2022) Perfluorinated alkyl substances in Canadian human milk as part of the maternal-infant research on environmental chemicals (MIREC) study. Sci Total Environ 831:154888

Rhee J, Chang VC, Cheng I, Calafat AM, Botelho JC, Shearer JJ, Sampson JN, Setiawan VW, Wilkens LR, Silverman DT, Purdue MP, Hofmann JN (2023) Serum concentrations of per- and polyfluoroalkyl substances and risk of renal cell carcinoma in the Multiethnic Cohort Study. Environ Int 180:108197

Rice PA, Aungst J, Cooper J, Bandele O, Kabadi SV (2020) Comparative analysis of the toxicological databases for 6:2 fluorotelomer alcohol (6:2 FTOH) and perfluorohexanoic acid (PFHxA). Food Chem Toxicol 138:111210

Rice PA, Cooper J, Koh-Fallet SE, Kabadi SV (2021) Comparative analysis of the physicochemical, toxicokinetic, and toxicological properties of ether-PFAS. Toxicol Appl Pharmacol 422:115531

Rice PA, Kabadi SV, Doerge DR, Vanlandingham MM, Churchwell MI, Tryndyak VP, Fisher JW, Aungst J, Beland FA (2024) Evaluating the toxicokinetics of some metabolites of a C6 polyfluorinated compound, 6:2 fluorotelomer alcohol in pregnant and nonpregnant rats after oral exposure to the parent compound. Food Chem Toxicol 183:114333

Richterová D, Fábelová L, Patayová H, Pulkrabová J, Lanková D, Rausová K, Šovčíková E, Štencl J, Hajšlová J, Trnovec T, Palkovičová Murínová Ľ (2018) Determinants of prenatal exposure to perfluoroalkyl substances in the Slovak birth cohort. Environ Int 121(Pt 2):1304–1310

Rovira J, Martínez MÁ, Sharma RP, Espuis T, Nadal M, Kumar V, Costopoulou D, Vassiliadou I, Leondiadis L, Domingo JL, Schuhmacher M (2019) Prenatal exposure to PFOS and PFOA in a pregnant women cohort of Catalonia, Spain. Environ Res 175:384–392

Russell MH, Himmelstein MW, Buck RC (2015) Inhalation and oral toxicokinetics of 6:2 FTOH and its metabolites in mammals. Chemosphere 120:328–335

Sadia M, Nollen I, Helmus R, Ter Laak TL, Béen F, Praetorius A, van Wezel AP (2023Feb 28) Occurrence, Fate, and Related Health Risks of PFAS in Raw and Produced Drinking Water. Environ Sci Technol 57(8):3062–3074

SC (Stockholm Convention) (2019) Stockholm Convention on Persistent Organic Pollutants. List of substances. https://www.pops.int/TheConvention/ThePOPs/AllPOPs/tabid/2509/Default.aspx

Schecter A, Malik-Bass N, Calafat AM, Kato K, Colacino JA, Gent TL, Hynan LS, Harris TR, Malla S, Birnbaum L (2012) Polyfluoroalkyl compounds in Texas children from birth through 12 years of age. Environ Health Perspect 120:590–594

Schlezinger JJ, Bello A, Mangano KM, Biswas K, Patel PP, Pennoyer EH, Wolever TMS, Heiger-Bernays WJ, Bello D (2025) Per- and poly-fluoroalkyl substances (PFAS) in circulation in a Canadian population: their association with serum-liver enzyme biomarkers and piloting a novel method to reduce serum-PFAS. Environ Health 24(1):10

Schröter-Kermani C, Müller J, Jürling H, Conrad A, Schulte C (2013) Retrospective monitoring of perfluorocarboxylates and perfluorosulfonates in human plasma archived by the German environmental specimen bank. Int J Hyg Environ Health 216:633–640

Schulz K, Silva MR, Klaper R (2020) Distribution and effects of branched versus linear isomers of PFOA. PFOS, and PFHxS: A Review of Recent Literature Sci Total Environ 733:139186

Serex T, Anand S, Munley S, Donner EM, Frame SR, Buck RC, Loveless SE (2014) Toxicological evaluation of 6:2 fluorotelomer alcohol. Toxicology 319:1–9

Serrano L, Iribarne-Durán LM, Suárez B, Artacho-Cordón F, Vela-Soria F, Peña-Caballero M, Hurtado JA, Olea N, Fernández MF, Freire C (2021) Concentrations of perfluoroalkyl substances in donor breast milk in Southern Spain and their potential determinants. Int J Hyg Environ Health 236:113796

Shearer JJ, Callahan CL, Calafat AM, Huang WY, Jones RR, Sabbisetti VS, Freedman ND, Sampson JN, Silverman DT, Purdue MP, Hofmann JN (2021) Serum concentrations of per- and polyfluoroalkyl substances and risk of renal cell carcinoma. J Natl Cancer Inst 113(5):580–587

Sheng N, Zhou X, Zheng F, Pan Y, Guo X, Guo Y, Sun Y, Dai J (2017) Comparative hepatotoxicity of 6:2 fluorotelomer carboxylic acid and 6:2 fluorotelomer sulfonic acid, two fluorinated alternatives to long-chain perfluoroalkyl acids, on adult male mice. Arch Toxicol 91:2909–2919

Shi Y, Vestergren R, Xu L, Zhou Z, Li C, Liang Y, Cai Y (2016) Human exposure and elimination kinetics of chlorinated polyfluoroalkyl ether sulfonic acids (Cl-PFESAs). Environ Sci Technol 50:2396–2404

Sochorová L, Hanzlíková L, Černá M, Drgáčová A, Fialová A, Švarcová A, Gramblička T, Pulkrabová J (2017) Perfluorinated alkylated substances and brominated flame retardants in serum of the Czech adult population. Int J Hyg Environ Health 220(2 Pt A):235–243

Solomon KR, Velders GJ, Wilson SR, Madronich S, Longstreth J, Aucamp PJ, Bornman JF (2016) Sources, fates, toxicity, and risks of trifluoroacetic acid and its salts: relevance to substances regulated under the Montreal and Kyoto Protocols. J Toxicol Environ Health B Crit Rev 19:289–304

Song Y, Du W, Hu J (2025) Association of per-and polyfluoroalkyl substances with thyroid hormones in the umbilical cord blood of neonates born by spontaneous delivery. Front Public Health 13:1528588

Starling AP, Liu C, Shen G, Yang IV, Kechris K, Borengasser SJ, Boyle KE, Zhang W, Smith HA, Calafat AM, Hamman RF, Adgate JL, Dabelea D (2020) Prenatal exposure to per- and polyfluoroalkyl substances, umbilical cord blood DNA methylation, and cardio-metabolic indicators in newborns: the Healthy Start Study. Environ Health Perspect 128:127014

Steenland K, Fletcher T, Stein CR, Bartell SM, Darrow L, Lopez-Espinosa MJ, Barry Ryan P, Savitz DA (2020) Review: Evolution of evidence on PFOA and health following the assessments of the C8 science panel. Environ Int 145:106125

Steenland K, Winquist A (2021) PFAS and cancer, a scoping review of the epidemiologic evidence. Environ Res 194:110690

Stubleski J, Salihovic S, Lind L, Lind PM, van Bavel B, Kärrman A (2016) Changes in serum levels of perfluoroalkyl substances during a 10-year follow-up period in a large population-based cohort. Environ Int 95:86–92

Tao L, Kannan K, Wong CM, Arcaro KF, Butenhoff JL (2008) Perfluorinated compounds in human milk from Massachusetts, U.S.A. Environ Sci Technol 42:3096–3101

Tao H, Zhang Y, Yao Q, Wang Y, Shi R, Cao Y, Lu P, Yuan T, Tian Y, Gao Y (2025) Per- and polyfluoroalkyl substances exposure through human milk, health risk assessment and physical growth in one-year-old children in Shandong, China. Environ Chem Ecotoxicol 7:539–546

Taxvig C, Rosenmai AK, Vinggaard AM (2014) Polyfluorinated alkyl phosphate ester surfactants—current knowledge and knowledge gaps. Basic Clin Pharmacol Toxicol 115:41–44

Thépaut E, Dirven HAAM, Haug LS, Lindeman B, Poothong S, Andreassen M, Hjertholm H, Husøy T (2021) Per- and polyfluoroalkyl substances in serum and associations with food consumption and use of personal care products in the Norwegian biomonitoring study from the EU project EuroMix. Environ Res 195:110795

Tian Y, Xu C, Zhang L, Shi D, Cappelli F, Yin S (2024) Maternal exposure to per- and polyfluoroalkyl substances: implications for intrahepatic cholestasis of pregnancy and adverse birth outcomes. Expo Health 16:1209–1223

Trier X, Numata J (2015) Uptake, elimination and toxicokinetic modeling of $^{13}C_4$–8:2 diPAPs given with food to twelve males. Poster session presented at Fluoros workshop 2015, Colorado, United States

US EPA (2021) Human health toxicity values for hexafluoropropylene oxide (HFPO) dimer acid and its ammonium salt (CASRN 13252-13-6 and CASRN 62037-80-3). https://www.epa.gov/system/files/documents/2021-10/genx-chemicals-toxicity-assessment_tech-edited_oct-21-508.pdf

van Beijsterveldt IALP, van Zelst BD, de Fluiter KS, van den Berg SAA, van der Steen M, Hokken-Koelega ACS 82022) Poly- and perfluoroalkyl substances (PFAS) exposure through infant feeding in early life. Environ Int 164:107274

Van Poll R, Jansen E, Janssen R (2017) PFOA measurements in blood Measurements in the serum of people living near DuPont/Chemours in Dordrech. RIVM Report 2017–0113. [in Dutch]. https://www.rivm.nl/bibliotheek/rapporten/2017-0113.pdf

Verner MA, Ngueta G, Jensen ET, Fromme H, Völkel W, Nygaard UC, Granum B, Longnecker MP (2016) A simple pharmacokinetic model of prenatal and postnatal exposure to perfluoroalkyl substances (PFASs). Environ Sci Technol 50:978–986

Völkel W, Genzel-Boroviczény O, Demmelmair H, Gebauer C, Koletzko B, Twardella D, Raab U, Fromme H (2008) Perfluorooctane sulphonate (PFOS) and perfluorooctanoic acid (PFOA) in human breast milk: results of a pilot study. Int J Hyg Environ Health 211:440–446

Vyas SM, Kania-Korwel I, Lehmler H-J (2007) Differences in the isomer composition of perfluorooctanesulfonyl (PFOS) derivatives. J Environ Sci Health Part A 42:249–255

Wang S, Huang J, Yang Y, Hui Y, Ge Y, Larssen T, Yu G, Deng S, Wang B, Harman C (2013) First report of a Chinese PFOS alternative overlooked for 30 years: its toxicity, persistence, and presence in the environment. Environ Sci Technol 47:10163101–10163170

Wang Y, Zhong Y, Li J, Zhang J, Lyu B, Zhao Y, Wu Y (2018) Occurrence of perfluoroalkyl substances in matched human serum, urine, hair and nail. J Environ Sci (China) 67:191–197

Wang Z, DeWitt JC, Higgins CP, Cousins IT (2017) A never-ending story of per- and polyfluoroalkyl substances (PFASs)? Environ Sci Technol 51:2508–2518

Wark H, Earl J, Chau DD, Overton J (1990) Halothane metabolism in children. Br J Anaesth 64:474–481

Wee SY, Aris AZ (2023) Revisiting the "forever chemicals", PFOA and PFOS exposure in drinking water. npj Clean Wat 6:5

Wilhelm M, Wittsiepe J, Völkel W, Fromme H, Kasper-Sonnenberg M (2015) Perfluoroalkyl acids in children and their mothers: association with drinking water and time trends of inner exposures—results of the duisburg birth cohort and bochum cohort studies. Int J Hyg Environ Health 218:645–655

Worley RR, Fisher J (2015) Application of physiologically-based pharmacokinetic modeling to explore the role of kidney transporters in renal reabsorption of perfluorooctanoic acid in the rat. Toxicol Appl Pharmacol 289:428–441

Wu B, Pan Y, Li Z, Wang J, Ji S, Zhao F, Chang X, Qu Y, Zhu Y, Xie L, Li Y, Zhang Z, Song H, Hu X, Qiu Y, Zheng X, Zhang W, Yang Y, Gu H, Li F, Cai J, Zhu Y, Cao Z, S Ji J, Lv Y, Dai J, Shi X (2023) Serum per- and polyfluoroalkyl substances and abnormal lipid metabolism: A nationally representative cross-sectional study. Environ Int 172: 107779.

Xu Y, Nielsen C, Li Y, Hammarstrand S, Andersson EM, Li H, Olsson DS, Engström K, Pineda D, Lindh CH, Fletcher T, Jakobsson K (2021) Serum perfluoroalkyl substances in residents following long-term drinking water contamination from firefighting foam in Ronneby. Sweden. Environ Int 147:106333

Xu C, Zhang L, Zhou Q, Ding J, Yin S, Shang X, Tian Y (2022) Exposure to per- and polyfluoroalkyl substances as a risk factor for gestational diabetes mellitus through interference with glucose homeostasis. Sci Total Environ 838(Pt 4):156561

Yang LL, Li J, Lai J, Luan H, Cai Z, Wang Y, Zhao Y, Wu Y (2016a) Placental transfer of perfluoroalkyl substances and associations with thyroid hormones: Beijing Prenatal Exposure Study. Sci. Rep 6.

Yang L, Wang Z, Shi Y, Li J, Wang Y, Zhao Y, Wu Y, Cai Z (2016b) Human placental transfer of perfluoroalkyl acid precursors: Levels and profiles in paired maternal and cord serum. Chemosphere 144:1631–1638

Yang MT, Yang F, Lan QY, Huang SY, Chen XY, Qiu YZ, Zhu HL (2025) Concentrations and exposure risks of per- and polyfluoroalkyl substances in human milk from Southern China during the first 400 days of lactation. Ecotoxicol Environ Saf 289:117442

Yao J, Pan Y, Sheng N, Su Z, Guo Y, Wang J, Dai J (2020) Novel perfluoroalkyl ether carboxylic acids (PFECAs) and sulfonic acids (PFESAs): Occurrence and association with serum biochemical parameters in residents living near a fluorochemical plant in China. Environ Sci Technol 54:13389–13398

Yao J, Dong Z, Jiang L, Pan Y, Zhao M, Bai X, Dai J (2023) Emerging and legacy perfluoroalkyl substances in breastfed Chinese infants: Renal clearance, body burden, and implications. Environ Health Perspect 131:37003

Yeung L, Miyake Y, Taniyasu S, Wang Y, Yu H, So MK, Jiang G, Wu Y, Li J, Giesy JP, Yamashita N, Lam PKS (2008) Perfluorinated compounds and total and extractable organic fluorine in human blood samples from China. Environ Sci Technol 42:8140–8145

Yeung LW, Robinson SJ, Koschorreck J, Mabury SA (2013) Part I. A temporal study of PFCAs and their precursors in human plasma from two German cities 1982–2009. Environ Sci Technol 47:3865–3874

Yeung LWY, Mabury SA (2016) Are humans exposed to increasing amounts of unidentied organofluorine? Environ Chem 13:102–110

Yi S, Yang D, Zhu L, Mabury SA (2022) Significant reductive transformation of 6:2 chlorinated polyfluorooctane ether sulfonate to form hydrogen-substituted polyfluorooctane ether sulfonate and their toxicokinetics in male sprague-dawley rats. Environ Sci Technol 56:6123–6132

Yu CH, Riker CD, Lu SE, Fan ZT (2020) Biomonitoring of emerging contaminants, perfluoroalkyl and polyfluoroalkyl substances (PFAS), in New Jersey adults in 2016–2018. Int J Hyg Environ Health 223:34–44

Yu CH, Weisel CP, Alimokhtari S, Georgopoulos PG, Fan ZT (2021) Biomonitoring: A tool to assess PFNA body burdens and evaluate the effectiveness of drinking water intervention for communities in New Jersey. Int J Hyg Environ Health 235:113757

Yuan KY, Gu YH, Pei YH, Yu SY, Li TZ, Feng T, Liu Y, Tian J, Miao X, Xiong J, Hu M, Yuan BF (2025) Comprehensive analysis of transplacental transfer of environmental pollutants detected in paired maternal and cord serums. J Hazard Mater 486:136970

Zeeshan M, Yang Y, Zhou Y, Huang W, Wang Z, Zeng XY, Liu RQ, Yang BY, Hu LW, Zeng XW, Sun X, Yu Y, Dong GH (2020) Incidence of ocular conditions associated with perfluoroalkyl substances exposure: Isomers of C8 Health Project in China. Environ Int 137:105555

Zhang L, Louie A, Rigutto G, Guo H, Zhao Y, Ahn S, Dahlberg S, Sholinbeck M, Smith MT (2023) A systematic evidence map of chronic inflammation and immunosuppression related to per- and polyfluoroalkyl substance (PFAS) exposure. Environ Res 220:115188

Zhang W, Liang Y (2023) The wide presence of fluorinated compounds in common chemical products and the environment: a review. Environ Sci Pollut Res Int 30(50):108393–108410

Zhang YZ, Zeng XW, Qian ZM, Vaughn MG, Geiger SD, Hu LW, Lu L, Fu C, Dong GH (2017) Perfluoroalkyl substances with isomer analysis in umbilical cord serum in China. Environ Sci Pollut Res Int 24:13626–13637

Zhang T, Sun H, Lin Y, Qin X, Zhang Y, Geng X, Kannan K (2013) Distribution of poly-and perfluoroalkyl substances in matched samples from pregnant women and carbon chain length related maternal transfer. Environ Sci Technol 47:7974–7981

Zhang M, Yu CH, Wang G, Buckley JP, Hong X, Pearson C, Adams WG, Fan ZT, Wang X (2022) Longitudinal trajectories and determinants of plasma per- and polyfluoroalkyl substance (PFAS) levels from birth to early childhood and metabolomic associations: A pilot study in the Boston Birth Cohort. Precis Nutr 1:e00004

Zhang L, Du J, Guo Q, Xu X, Li H, Zhong N, Zhang J, Li G, Shao B (2024) Serum levels of per- and poly-fluoroalkyl substances among middle-aged and elderly populations in Beijing and their association with dyslipidemia. Food Chem Toxicol 193:115066

Zhao N, Zhang X, Li Y, Zhang H, Yang E, Ding L, Liu Y (2025) Associations between in utero exposure of per- and polyfluoroalkyl substances (PFAS) mixture and anthropometry measures at birth. Environ Pollut 373:126093

Zheng G, Eick SM, Salamova A (2023) Elevated levels of ultrashort- and short-chain perfluoroalkyl acids in US homes and people. Environ Sci Techno 57:15782–15793

Zheng G, Schreder E, Dempsey JC, Uding N, Chu V, Andres G, Sathyanarayana S, Salamova A (2021) Per- and polyfluoroalkyl substances (PFAS) in breast milk: concerning trends for current-use PFAS. Environ Sci Technol 55:7510–7520

Zhi Y, Lu X, Munoz G, Yeung LWY, De Silva AO, Hao S, He H, Jia Y, Higgins CP, Zhang C (2024) Environmental occurrence and biotic concentrations of ultrashort-chain perfluoroalkyl acids: overlooked global organofluorine contaminants. Environ Sci Technol 58(49):21393–21410

Chapter 5
Flame Retardants

Abstract Under our current living conditions, the use of flammable materials, such as plastics, composites, and foams, in building envelopes and interior furnishings has increased significantly. Therefore, fire presents a serious health hazard, and various techniques have been developed to mitigate this risk. In addition to technical solutions, hundreds of chemicals are used today for fire protection. Notably, halogenated compounds, including brominated and chlorinated flame retardants and phosphorus compounds, were used in large quantities over decades, resulting in an exposure risk to humans. These are known as additive flame retardants, which are not chemically bound to polymers and can therefore be released into the environment through leaching, evaporation, or abrasion from products during production and processing. Some groups of flame retardants were chemically stable, persistent in the environment, and had a tendency to bioaccumulate, raising toxicological concerns. Thus, a brief description of the chemical-physical characteristics and usage patterns of various flame retardants, their primary health risks, and toxicokinetic properties, along with references to further scientific literature, is provided. The main objective is to detail their occurrence in human matrices, such as blood, urine, and breast milk. For this reason, detailed information is provided on polybrominated diphenyl ethers (PBDE), polybrominated biphenyls (PBB), organophosphate flame retardants, hexabromocyclododecane (HBCDD), and emerging brominated flame retardants like tetrabromobisphenol A (TBBPA), 1,2-bis-(2,4,6-tribromophenoxy)ethane (BTBPE), 2-ethyl-1-hexyl-2,3,4, 5-tetrabromobenzoate (EH-TBB), bis(2-ethyl-1-hexyl)tetrabromophthalate (BEH-TEBP), decabromodiphenylethane (DBDPE), and tetrabromoethylcyclohexane (DBE-DBCH).

Keywords Flame retardants · Polybrominated diphenyl ethers · PBDE · HBCDD · TBBPA · Emerging brominated flame retardants

Under our current living conditions, the use of flammable materials like plastics, composites, and foams in building envelopes and interior furnishings has significantly increased. Fire presents a major health hazard, especially in industrialized countries and densely populated areas. In response, various techniques have been developed

© The Author(s), under exclusive license to Springer Nature Switzerland AG 2026

H. Fromme, *Human Biomonitoring II*, https://doi.org/10.1007/978-3-032-10036-8_5

to minimize this risk. Beyond technical solutions, efforts are underway to enhance the ignition energy of polymers, decrease the fire rate, and limit the spread of flames and vapors.

Hundreds of substances are used today for chemical fire protection, with the following four classes of substances particularly utilized on an industrial scale for flame retardancy, either alone or in combination (ECHA 2023):

- Halogenated compounds (aromatic, aliphatic, and cycloaliphatic) act in the gas phase by radical quenching, i.e., interrupting the radical chain mechanism that carries the flame in the combustion process. These include, for example, brominated and chlorinated flame retardants.
- Phosphorus compounds (e.g., organophosphates, inorganic phosphates, polyphosphates, and their derivatives) are especially effective in oxygen-containing, charring polymers such as polyurethanes, epoxies, polyamides, and polyesters.
- Nitrogen-based substances (e.g., melamine, isocyanurates, and cyanuric acid derivatives) often work in conjunction with other flame retardants by providing cooling and dilution effects through the release of nitrogenous gases.
- Inorganic substances and metals such as antimony trioxide, aluminum trihydroxide, and boron salts serve as mineral fillers to enhance the product's density and thermal conductivity.

Annual global demand is projected to reach 2.39 million tonnes in 2019 (CPC 2020). The largest shares of global consumption are in China (27%), Western Europe (23%), and North America (21%). Figure 5.1 illustrates global consumption by type of flame retardant, with aluminum hydroxide and organophosphates comprising the most substantial shares by volume, at 38% and 18%, respectively. The primary sectors utilizing global flame retardants include electric and electronic equipment, cables and wires, at 38%; the building and construction sector, at 28%; automotive, trains, and airplanes, at 20%; and furnishing/upholstery and others, at 14% (ECHA 2023).

5.1 Polybrominated Diphenyl Ethers

Polybrominated diphenyl ethers (PBDEs) are produced through the bromination of diphenyl ether. In theory, 209 individual compounds (congeners) can be formed and categorized into ten homolog groups. The chemical structures of some significant congeners are displayed in Table 5.1. Due to their high molecular mass, these compounds have low vapor pressures and are lipophilic because of the absence of polar functional groups. Their solubility in water decreases sharply as the degree of bromination increases (Darnerud et al. 2001). Commercially, the three technical PBDE mixtures—penta-, octa-, and decabromodiphenyl ether—have primarily been produced, each consisting of various individual compounds with different levels of bromination. The homologous groups are summarized in Table 5.2.

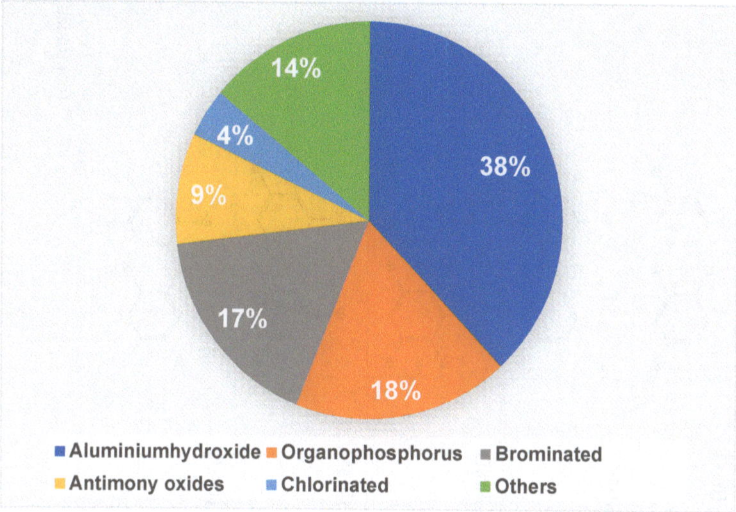

Fig. 5.1 Global consumption of various flame retardants in 2019 (from CPC 2020)

PBDEs have been extensively used for over 40 years as flame retardants in plastics, electronic devices such as televisions and computers, and building materials (NICNAS 2007). In 2019 alone, an estimated 53.6 million tons of e-waste (discarded products with a battery or plug, such as computers and mobile phones) were generated worldwide (ITU 2024). PBDEs are classified as additive flame retardants. They are added to polymers such as foams or textiles (e.g., carpets and upholstered furniture) in concentrations of 5 - 30%. Since they are not chemically bound, they can be released into the environment through leaching, evaporation, or abrasion from products during production and processing. Consequently, PBDEs and their degradation products can be detected in various environmental compartments, including sediments, sewage sludge, flowing water, and wastewater, as well as in marine mammals, fish, and bird eggs, for instance (Covaci et al. 2011; Besis and Samara 2012; McGrath et al. 2017). However, differences in the production and use of this group of substances have also led to significant regional variations in pollution levels (Frederiksen et al. 2009).

Tetra-, penta-, hexa-, and hepta-BDE mixtures are classified in Annex A of the Stockholm Convention (SC 2019) due to their environmental persistence, tendency to bioaccumulate, and high potential for long-range transport. Parties must take measures to prohibit the production and use of these chemicals, with the exception of recycling products containing these PBDEs. Commercial deca-BDE products are also included in Annex A, with an exemption for uses in automotive production. In the European Union, Regulation EU 2019/1021 prohibits the manufacture, placing on the market, and use of substances, mixtures, or articles, with a few exceptions. In the USA, some states have primarily banned penta- and octa-forms and partially deca-forms of PBDE, while producers decided to voluntarily phase out the production and use of deca-BDE by the end of 2012 (Sharkey et al. 2020). India has reportedly banned

Table 5.1 Important PBDE flame retardants

CAS No	Name	Abbreviation	Formula	Structure
41318–75-6	2,4,4′-Tribromodiphenyl ether	BDE 28	$C_{12}H_7Br_3O$	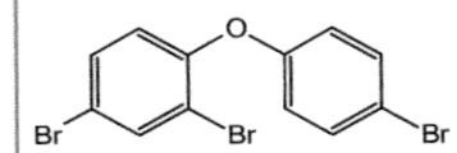
5436–43-1	2,2′,4,4′-Tetrabromodiphenyl ether	BDE 47	$C_{12}H_6Br_4O$	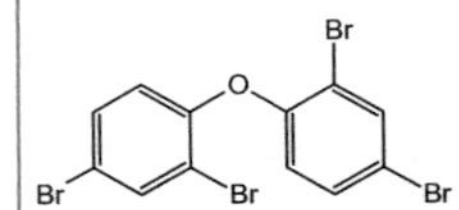
60328-60-9	2,2′,4,4′,5-Pentabromodiphenyl ether	BDE 99	$C_{12}H_5Br_5O$	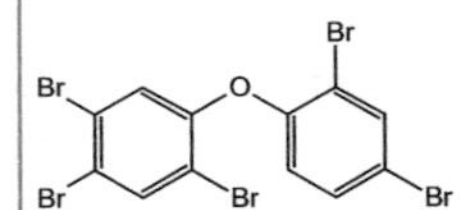
189084-64-8	2,2′,4,4′,6-Pentabromodiphenyl ether	BDE 100	$C_{12}H_5Br_5O$	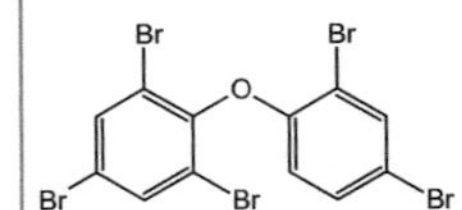
68,631–49-2	2,2′,4,4′,5,5′-Hexabromodiphenyl ether	BDE 153	$C_{12}H_4Br_6O$	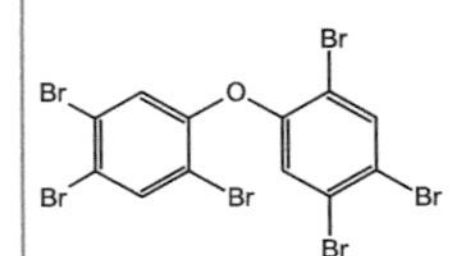

(continued)

Table 5.1 (continued)

CAS No	Name	Abbreviation	Formula	Structure
207122-15-4	2,2′,4,4′,5,6′-Hexabromodiphenyl ether	BDE 154	$C_{12}H_4Br_6O$	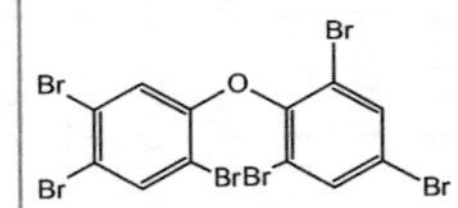
207122-16-5	2,2′,3,4,4′,5′,6-Heptabromodiphenyl ether	BDE 183	$C_{12}H_3Br_7O$	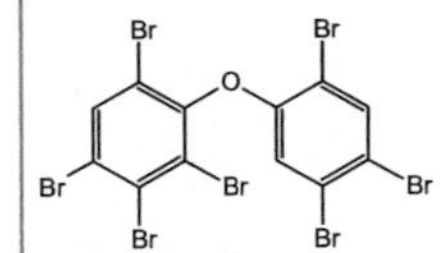
145538-74-5	Decabromodiphenyl ether	BDE 209	$C_{12}Br_{10}O$	

Table 5.2 List of homolog groups

Homologs	Acronym	Number of congeners	CAS No
Monobromodiphenylether	mono-BDE	1–3	101-55-3
Dibromodiphenylether	di-BDE	4–15	2050-47-7
Tribromodiphenylether	tri-BDE	16–39	49690-94-0
Tetrabromodiphenylether	tetra-BDE	40–81	40088-47-9
Pentabromodiphenylether	penta-BDE	82–127	32534-81-9
Hexabromodiphenylether	hexa-BDE	128–169	36483-60-0
Heptabromodiphenylether	hepta-BDE	170–193	68928-80-3
Octabromodiphenylether	octa-BDE	194–205	32536-52-0
Nonabromodiphenylether	nona-BDE	206–208	63936-56-1
Decabromodiphenylether	deca-BDE	209	1163-19-5

the trade, use, import, and export of commercial penta-BDE and commercial octa-BDE. Concurrently, China has also established concentration limits for congeners related to the penta- and octa-BDE commercial mixtures in newly manufactured products (Sharkey et al. 2020).

Numerous reviews discussing the environmental occurrence, exposure, toxicology, and epidemiology of PBDEs and other brominated flame retardants are now accessible (Lyche et al. 2015; Fromme et al. 2016a; ATSDR 2017; McGrath et al. 2017; Fromme and Roscher 2018; Klinčić et al. 2020; Wu et al. 2020; Xue et al. 2023; Renzelli et al. 2023; EFSA 2024a; Gomes et al. 2024). Unless stated otherwise, this literature served as a foundation in the subsequent two chapters.

5.1.1 Health Aspects

The focus is on the effects on the developing nervous and reproductive systems, as well as the endocrine system, thyroid gland, liver, and lipid metabolism. In animal experiments, neurological and behavioral changes, along with damage to the reproductive system, occurred after administering low-brominated PBDEs during early life stages and into adulthood. The toxic effects were generally more pronounced with low halogenated congeners like penta-BDEs compared to deca-BDEs, which can be attributed, among other factors, to the different kinetics. Effects on the male reproductive system, thyroid, and liver were also observed in adult animals subjected to long-term ingestion of low-brominated PBDEs. Meanwhile, administration of deca-BDEs resulted in impacts on the pancreas (diabetes), nervous system, immune system, and reproductive system. Several longitudinal and cross-sectional studies have reported significant associations between the levels of individual penta- and octa-BDEs in mothers' blood and the neuronal development of their children. In contrast, the relationship between concentrations in breast milk and a child's development yielded inconsistent results. While there is clear animal evidence for effects on the thyroid gland and reproduction, the epidemiological findings remain conflicting.

Individual studies on the carcinogenic effects in humans are available. However, most did not provide evidence of correlations with internal exposure to low-brominated PBDEs or yielded contradictory results. No human study results are available for BDE 209. There are well-founded suspicions of carcinogenicity from animal experiments for low-brominated BDEs and weak suspicions for decaBDE. For example, the US National Toxicology Program found clear evidence of carcinogenic activity in rats and mice for penta-BDE. However, there was only limited evidence of carcinogenicity in rats for deca-BDE (NTP 2016). The US EPA found suggestive evidence of carcinogenic potential for deca-BDE (US EPA 2008). The WHO's International Agency for Research on Cancer (IARC) has classified PBDE in Group 3 as not classifiable regarding its human carcinogenicity. It found insufficient evidence of carcinogenicity in humans and inadequate or limited evidence in experimental animals (IARC 1999). Nevertheless, the IARC is planning a re-evaluation.

5.1.2 Toxicokinetics

Human studies on the uptake of PBDEs are only available for BDE 209, for which an oral bioavailability of 0.29 was estimated based on a toxicokinetic model (Zhang et al. 2022). Animal studies on absorption after inhalation are limited and indicate good absorption, while absorption rates of approximately 70–85% were found for the most technically important PBDEs in animals after oral administration. For BDE 209, low absorption rates of 4% and 26% were observed in animal experiments, with some even below 1% without lipophilic vehicles to increase bioavailability (Huwe et al. 2008). In rodents, BDE 47, BDE 85, BDE 99, BDE 100, and BDE 153 were mainly distributed in adipose tissue, while BDE 209 was predominantly found in highly perfused tissues such as the liver.

Metabolization occurs primarily through oxidative hydroxylation at the phenyl rings, likely mediated by cytochrome P450, along with the cleavage of the ether bond and the formation of bromophenols. For example, the primary metabolic pathways for BDE 47 are illustrated in Fig. 5.2. BDE 47 is mainly converted to 5-OH-BDE 47, 6-OH-BDE 47, and 2,4-dibromophenol, and, to a lesser degree, to various other metabolites (Erratico et al. 2013). For BDE 209, multiple studies using labeled substances demonstrated reductive debromination to octa-BDE and nona-BDE following oral administration. It is presumed that this initial step is complemented by subsequent hydroxylation.

Table 5.3 presents estimates of mean half-lives for various PBDEs in humans, determined using different methods, primarily toxicokinetic models. Although the values may vary by a factor of up to 10 in some cases, it can be generally stated that congeners up to the hepta-BDEs have long half-lives, typically lasting for years, while octa- and nona-BDEs, especially deca-BDE, are excreted significantly faster, within a few days.

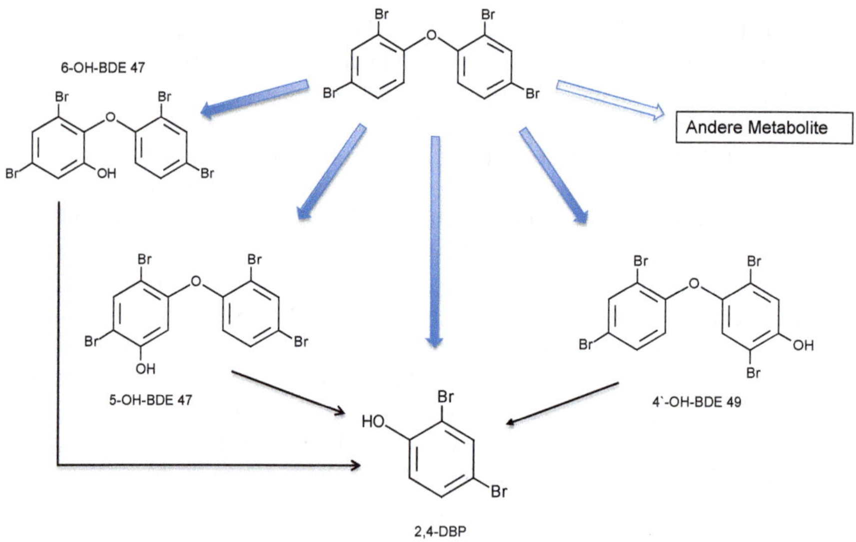

Fig. 5.2 Main metabolic pathways of BDE 47 in human liver microsomes (DBP: dibromo phenol) (after Erratico et al. 2013)

Table 5.3 Mean estimates of half-lives for various PBDEs in humans (in years, unless otherwise stated)

	Jakobsson et al. (2003)	Geyer et al. (2004)	Geyer et al. (2004)[a]	Trudel et al. (2011)	Wong et al. (2013)	Zhang et al. (2022)
BDE 28	–			3.0		–
BDE 47	–	1.8	3.0	1.4	0.4	–
BDE 99	–	2.8	5.4	0.8	8.2	–
BDE 100	–	1.6	2.9	1.8	2.0	–
BDE 153	1.9	6.5	11.7	7.4	3.5	–
BDE 154	0.7	3.3	5.8	1.3	–	–
BDE 183	0.3	–	–	2.7	–	–
OctaBDE	37–84 days	–	–	–	–	–
NonaBDE	17–35 days	–	–	–	–	–
BDE 209	14 days	–	–	4–7 days	-	15 and 18 days[b]

[a]For humans, back-calculated from data from the rat; [b]Predicted from two humans using PBK models without and with enterohepatic circulation

In principle, PBDEs and their hydroxyl metabolites can cross the placenta in animal experiments and enter breast milk.

5.1.3 Occurrence in the Blood of the Adult Population

In a meta-analysis, Meng et al. (2021) aggregated data from the literature between 2000 and 2019. They calculated medians of 58.5 ng/g lw in North America, 29.0 ng/g lw in Asia, and 7.3 ng/g lw in Europe for the sum of PBDEs (from BDE 28 to BDE 209). However, there are considerable differences among the individual congeners. For BDE 47, BDE 99, BDE 100, and BDE 153, the medians are higher in the USA than in the other regions, while BDE 209 has the highest values in Asia. Table 5.4 summarizes the concentrations of the individual PBDE congeners in the serum of the general, occupationally unexposed population from recent studies. These data support the general findings of the study by Meng et al. (2021). In Europe, even higher medians of BDE 47 and BDE 99 were found in the serum of 200 Danish women (Vorkamp et al. 2014), and for BDE 47 also in a recent study with pooled samples from Uppsala, Sweden (Gyllenhammar et al. 2023). In contrast, the medians of these congeners found in Germany, France, and Norway are significantly lower. Considerably higher levels can be found in studies from North America. For example, in Canada and the USA, elevated concentrations of BDE 47 and, more recently, BDE 153 were observed in serum. A specific exposure situation with BDE 47 is reported from California due to higher application quantities in flame retardants (Zota et al. 2018). In Asia, the concentration levels are comparable to those in Europe, although there are also regional hotspots in areas with larger e-waste recycling companies, for instance. Significantly fewer study results are available for BDE 209. While medians between 0.8 and 3.0 ng/g lw have been observed in recent European studies, they are substantially higher in China.

In many studies, no significant gender differences were observed in the serum of adults; however, lower levels were sometimes noted for individual congeners in the serum of women compared to men (Sjödin et al. 2008; Chen et al. 2023a). Most studies found no apparent influence of age on serum levels in adults, though only Anderson et al. (2008) and Qin et al. (2011) identified a positive association with age in groups with high fish consumption. Nevertheless, the results of various studies indicate, albeit not uniformly, that children and adolescents have a higher internal burden compared to adults (e.g., Thomsen et al. 2002a, 2007; Sjödin et al. 2008; Lunder et al. 2010; Wu et al. 2015; Knudsen et al. 2017a). In Australian samples for the sum of BDE 47, BDE 99, BDE 100, and BDE 153, an increase in the mean concentrations was observed, rising from 14 ng/g lw in the age group of 0–0.5 years to a peak of 51 ng/g lw at the age of 2.6–3 years, followed by a decrease to 9.9 ng/g lw in the group of 31-45-year-olds (Toms et al. 2008, 2009). After that, the concentrations remained stable with increasing age. Possible explanations could include PBDE

Table 5.4 Median PBDE blood levels in adults in ng/g lw

References	n	BDE 28	BDE 47	BDE 99	BDE 100	BDE 153	BDE 154	BDE 183	BDE 209	Country, year
Europe										
Gyllenhammar et al. (2023)	8[a]	0.21	1.53	0.39	0.26	1.31	0.23	0.17	2.98	Sweden, 2020
Charles et al. (2023)	58	–	1.02	0.52	0.49	1.36	0.18	0.32	–	Norway, 2015/16
Fillol et al. (2021)	742	< LOQ	0.22	0.08	0.07	0.75	< LOQ	< LOQ	1.14	France, 2014–16
Fromme et al. (2016b)	42	< LOQ	0.08	0.12	0.10	1.28	< LOQ	0.09	1.76	Germany, 2013/14
Fromme et al. (2015a)	70	0.02	0.38	0.16	0.13	0.97	0.02	0.12	–	Germany, 2013
Cequier et al. (2015)	48	0.33	0.49	0.13	0.14	0.82	0.07	0.09	–	Norway, 2012
Vorkamp et al. (2014)	200	–	3.40	2.50	< 0.70	1.70	< 0.37	< 0.36	7.50	Denmark, 2011
Darnerud et al. (2015)	30	0.06	0.36	0.05	0.18	0.72	0.12	0.04	0.90	Sweden, 2010
Sahlström et al. (2014)	24	0.04	0.56	0.08	0.10	0.95	–	–	0.68	Sweden, 2009/10
Frederiksen et al. (2010)	51	0.03	0.38	< 0.11	< 0.10	1.13	< 0.02	–	1.71	Denmark, 2007
Kalantzi et al. (2011)	61	0.01	0.16	0.09	0.11	0.51	0.02	0.03	1.18	Greece, 2007
Fromme et al. (2009)	47	0.71	1.80	0.75	0.58	2.37	< LOQ	0.72	–	Germany, 2005
Leijs et al. (2008)	18	0.20[b]	2.30[b]	2.40[b]	2.10[b]	2.70[b]	0.55[b]	1.40[b]	–	Netherlands, 2005
Karlsson et al. (2007)	5	< LOQ	3.90[b]	2.10[b]	0.92[b]	1.80[b]	< LOQ	< LOQ	10.4[b]	Sweden
Vizcaino et al. (2011)	174	–	2.30	0.35	–	2.10	1.50	–	< 0.7	Spain, 2004/05
García-Villarino et al. (2020)	155	0.16	0.26	1.63	–	2.52	2.60	–	1.02	Spain, 2004–07
Roosens et al. (2010)	[a]	–	1.34	0.29	0.20	1.24	1.26	0.32	–	Belgium, 2004/05
Gómara et al. (2007)	52	< 0.06	2.60	2.30	1.20	0.77	< 0.1	–	1.1	Spain, 2003
North America										
Eick et al. (2020)	506	< LOQ	10.9	4.1	< LOQ	9.59	< LOQ	< LOQ	< LOQ	USA, 2014–18

(continued)

Table 5.4 (continued)

References	n	BDE 28	BDE 47	BDE 99	BDE 100	BDE 153	BDE 154	BDE 183	BDE 209	Country, year
Zota et al. (2018)	103	–	28.8	7.44	5.07	8.62	–	–	–	USA, 2011–13
Makey et al. (2016)	51	0.50[c]	7.60[c]	1.70[c]	1.40[c]	5.30[c]	–	–	–	USA, 2011
Oulhote et al. 2016	745	–	9.0 [c]	1.1[c]	0.9[c]	1.4[c]	–	–	–	Canada, 2007–09
Woods et al. (2017)	272	0.90	18.0	4.20	3.50	4.20	0.10	0.07	–	USA, 2003–06
Foster et al. (2011)	97	2.18	23.3	8.85	3.66	3.80	1.24	1.26	–	Canada, 2004/05
Asia										
Wang et al. (2024b)	53	0.83	0.13	0.69	0.50	0.51	0.05	–	11.4	China, 2022
Ma et al. (2023)	104	0.56	4.39	0.10	0.11	0.17	0.18	0.35	–	China, 2020–22
Lin et al. (2023)	20	1.73	0.48	0.28	< LOQ	4.49	< LOQ	0.12	20.9	China, 2017/18
Yu et al. (2019)	76	0.53	0.95	0.55	0.28	3.32	0.42	0.90	6.12	China, 2016/17
Chen et al. (2023a)	33	6.56	1.24	4.46	3.33	1.74	1.42	2.52	–	China, 2016
Jin et al. (2020)	293	0.58	0.57	< LOQ	–	2.95	0.28	–	11.1	China
Wang et al. (2016b)	32	0.19	2.37	0.18	0.08	0.45	0.17	0.03	–	China, 2014
Choi et al. (2014)	198	< LOQ	1.72	0.49	0.12	0.72	0.08	< LOQ		Korè, 2011
Ho et al. (2015)	100	–	0.55[c]	0.33[c]	–	–	–	–	0.47[c]	China, 2010
Kim et alk. (2012)	720	< 0.10	1.69	0.71	< 0.41	2.07	< 0.22	< 1.13	< 3.69	korê, 2009/10

[a]Pooled samples; [b]Arithmetic mean; [c]Geometric mean; *LOQ* limit of detection

breast milk intake and the increased dust intake in small children (Toms et al. 2015). In the USA, the geometric mean values for BDE 47, for instance, also declined from 37.8 ng/g lw to 18.1 ng/g lw between the ages of two and nine years (Cowell et al. 2019); this trend was not observed for BDE 153.

Various studies have demonstrated a decrease in serum levels, except for BDE 209, in recent years (Schecter et al. 2005; Thomsen et al. 2007; Toms et al. 2009; Turyk et al. 2010; Darnerud et al. 2015; Fromme et al. 2016b). For instance, in Germany, the median Σ6-PBDE level dropped from 5.5 ng/g lw to 1.7 ng/g lw between 2005 and 2013/14 (Fromme et al. 2009, 2016b). In Norway, an increase from 0.5 ng/g lw in 1977 to 4.8 ng/g lw in 1998 was observed for the sum of the tri- to heptabrominated congeners, which then declined to 3.8 ng/g lw by 2003 (Thomsen et al. 2007). In the USA, an annual decrease of approximately 13% for BDE 47, 13% for BDE 99, 11% for BDE 100, and 11% for BDE 153 was also found in the examination of children's blood between 1998 and 2013, considering only samples collected during the postnatal period and direct environmental exposure (Cowell et al. 2019). A similar trend over time is also noted in a summary of various studies from northern China (Li et al. 2017).

5.1.4 *Occurrence in the Blood of Children*

Table 5.5 summarizes recent findings from studies on internal exposure in children and adolescents. Van der Schyff et al. (2023) provided an overview of PBDE results in the serum of 6- to 13-year-olds involved in the European Human Biomonitoring Initiative (HBM4EU) from France, Greece, Slovenia, and Norway. They reported medians (95th percentile) of < LOQ-0.36 ng/g lw (0.43–2.81 ng/g lw) for BDE 47, < LOQ-0.29 ng/g lw (0.1–1.5 ng/g lw) for BDE 153, and < LOQ-8.7 ng/g lw (< LOQ-159 ng/g lw) for BDE 209. Similar to the adults, it is primarily BDE 47, BDE 99, and BDE 100 that determine the overall PBDE levels. The highest medians are consistently found in North America. In China, the concentration level is comparable to that of Europe; in Australia, it is slightly higher. The higher application quantities explain the noticeable body burden in North America compared to Europe. In European studies, significantly fewer results are available for BDE 209, which measured similar concentrations ranging between 1.0 and 1.6 ng/g lw. Martinez et al. (2024) demonstrated that lower concentrations of low–chlorinated PBDEs in the blood of children in Canada were primarily related to indoor air quality. Specifically, lower blood levels of BDE 47 and BDE 100 were observed with the use of air purifiers.

5.1.5 *Occurrence in Umbilical Cord Serum*

Table 5.6 presents an overview of the median concentrations in umbilical cord serum. In most studies, BDE 47 and BDE 99 dominate the total concentrations. As expected,

Table 5.5 Median PBDE level in the blood of children and adolescents in ng/g lw

References	Age (years)	N	BDE 28	BDE 47	BDE 99	BDE 100	BDE 153	BDE 154	BDE 183	BDE 209	Country, year
Europe											
Fillol et al. (2021)	6–17	243	< LOQ	0.28	0.07	0.08	0.36	< LOQ	< LOQ	1.48	France, 2014–16
Knudsen et al. (2017a)	6–11	116	0.24	3.85	2.29	0.85	1.39	–	–	–	Denmark, 2011
Drobná et al. (2019)	6	91	0.02	0.18	0.08	0.05	0.18	0.01	0.10	< 0.49	Slovakia, 2010/11
Sahlström et al. (2014)	0.9–1,3	24	0.06	1.30	0.22	0.28	1.20	–	–	1.50	Sweden, 2009–10
Link et al. (2012)	9–11	29[b]	–	1.01	0.36	0.26	0.95	0.04	0.14	< LOQ	Germany, 2008–09
Caspersen et al. (2016)	3	99	0.20	1.80	1.10	0.08	0.90	0.10	–	–	Norway, 2003–08
Link et al. (2012)	9–11	8[b]	–	1.98	0.81	0.31	1.2	0.05	–	1.62	Germany, 2002–03
Carrizo and Grimalt (2007)	4	244	< LOQ	2.90	1.20	0.18	0.02	0.05	< LOQ	–	Spain
Fängström et al. (2005)	7	42	–	0.87	0.47	0.14	2.50	0.74	–	1.0	Faroe Islands, 2001–02
North America											
Martinez et al. (2024)	11.4[a]	170	–	4.42	1.68	1.08	3.87	–	< LOQ	< LOQ	Canada, 2019–22
Hoffman et al. (2024)	3–6	61	< LOQ	15.7	3.46	3.71	5.62	< LOQ	< LOQ	2.8	USA, 2014–16

(continued)

Table 5.5 (continued)

References	Age (years)	N	BDE 28	BDE 47	BDE 99	BDE 100	BDE 153	BDE 154	BDE 183	BDE 209	Country, year
Cowell et al. (2019)	9	160	–	18.1	4.4	4.0	5.5	–	–	–	USA, 2013
Jacobson et al. (2016)	1–5	80	–	14	4.0	2.0	3.0	< LOQ	–	< LOQ	USA, 2011–12
Darrow et al. (2017)	1.3–6	80	–	38.9	9.60	6.10	5.90	< LOQ	–	< LOQ	USA, 2011–12
Przybyla et al. (2016)	12–15	381	1.7	31.5	10.1	6.3	9.2	–	–	–	USA, 2003–04
Bradman et al. (2012)	7	272	2.1	46.9	10.7	10.9	11.5	1.1	< LOQ	–	USA, 2007–08
Lunder et al. (2010)	1.5–4	20	1.0	31.8	6.2	6.2	13.0	3.2	–	1.7	USA, 2006–07
Rose et al. (2010)	2–5	94	1.2	69.5	28.1	25.8	20.8	–	0.63	2.6	USA, 2003–05
Asia, Australia, Africa											
Chbihi et al. (2024)	5–11	93	< LOQ	0.65	1.68	< LOQ	< LOQ	< LOQ	–	–	Morocco, 2021
Shen et al. (2010)	5–11	22[b]	0.38 [a]	1.29[a]	0.98[a]	0.33[a]	3.05[a]	1.24[a]	1.26[a]	–	China, 2008
NICNAS 2007	5–15	[b]	–	9.35	4.15	2.5	4.3	0.45	–	–	Australia, 2002–05

[a]Mean; [b]Pooled samples; *LOQ* Limit of quantification

the medians in the USA are quite high. In Europe, however, the medians of BDE 209 range from < 2.1 to 1.87 ng/g lw; significantly higher concentrations can be observed in some regions of China, specifically in the prospective Wenzhou Birth Cohort.

Significant correlations were found between maternal and fetal serum for the tetra- to heptabrominated congeners (Mazdai et al. 2003; Bi et al. 2006; Kawashiro et al. 2008; Frederiksen et al. 2010; Terry et al. 2017), although other studies could not confirm this (Gómara et al. 2007; Vizcaino et al. 2011; Jakobsson et al. 2012). Besides BDE 209, PBDEs can cross the placental barrier. When only considering studies where samples were taken simultaneously, the median ratio between fetal and maternal serum is approximately 1.1 (0.2–2.2) for BDE 47 and roughly 1.0 (0.5–1.9) for BDE 153 (Mazdai et al. 2003; Bi et al. 2006; Frederiksen et al. 2010; Foster et al. 2011; Choi et al. 2014; Kim et al. 2021). Only Terry et al. (2017) and García-Villarino et al. (2020) reported significantly different levels in the ratio of fetal to maternal serum of 3.2 and 2.2 for BDE 47 and 2.51 and 0.1 for BDE 153.

5.1.6 Occurrence of OH-BDE in the Blood

Since Malmberg et al. (2005) detected hydroxylated metabolites in the blood plasma of rats after oral administration of PBDEs, OH-BDEs have been increasingly identified in human specimens. Table 5.7 provides an overview of commonly measured OH-BDEs. In principle, a distinction can be made between those produced by the degradation of PBDEs within the organism and those formed directly in the ecosystem (Malmvärn et al. 2005). Therefore, it is important to consider that various OH-BDEs are present in the aquatic system and can also lead to human exposure through food sources such as fish (Haraguchi et al. 2016).

Table 5.8 presents the medians of test results from the serum/plasma of the general population. While very low concentrations were noted in Asia, studies from the USA frequently report significantly higher levels, which can be attributed to differences in the use of PBDEs as flame retardants. Only one study, the Environment and Reproductive Health Study from the western USA, recruited subjects between 2006 and 2016 and reported lower medians (Ingle et al. 2021). Additionally, there were variations in the spectrum of OH-BDEs, even within countries, as many studies only identified a limited selection of substances. The highest levels were found for 6-OH-BDE 47, 5-OH-BDE 47, and 5′-OH-BDE 99. In some studies, PBDE concentrations were also measured simultaneously; however, except for Japanese studies (Fujii et al. 2014a; Haraguchi et al. 2016), where the medians of 6-OH-BDE 47 were significantly higher than the concentrations of the parent substances, an inverse relationship was observed in the USA and China.

Various studies have examined OH-BDE in umbilical cord blood, revealing medians of 1.0 ng/g lw and 5.6 ng/g ww in the USA during 2003–2004 and 2011, respectively. For 6-OH-BDE 47, the values were 5.7 ng/g lw and 19.1 ng/g ww for 5-OH-BDE 47, and 4.3 ng/g lw and 8.2 ng/g ww for 5′-OH-BDE-99 (Qiu et al.

Table 5.6 Median PBDE levels in umbilical cord blood in ng/g lw

References	n	BDE 28	BDE 47	BDE 99	BDE 100	BDE 153	BDE 154	BDE 183	BDE 209	Country, year
Europe										
Frederiksen et al. (2010)	40	0.05	< 0.07	< 0.29	< 0.27	0.51	< 0.05	-	< 2.4	Denmark, 2007
García-Villarino et al. (2020)	229	0.31	0.56	0.50	–	0.30	0.57	–	1.87	Spain, 2004–08
Vizcaino et al. (2011)	174	–	2.3	1.5	–	< 0.23	< 0.08	–	< 1.2	Spain, 2004–05
Gómara et al. (2007)	48	< 0.06	3.2	3.0	1.5	0.32	< 0.1	–	1.4	Spain, 2003–04
Lopez–Espinosa et al. (2015)	534	–	1.7	1.1	–	0.56	< 0.43	–	< 5.9	Spain, 2003–08
Hernik et al. (2013)	89	–	1.0[a]	0.6[a]	–	0.40[a]	–	–	–	Poland, 2002–05
Meijer et al. (2008)	12	–	0.5	0.1	0.1	0.9	0.3	–	–	Netherlands, 2001–02
Meironyté–Guvenius et al. (2003)	15	0.07	0.98	0.07	0.07	0.17	< 0.01	0.01	–	Sweden, 2000–01
Carrizo and Grimalt (2007)	92	0.25	2.8	1.3	1.0	< LOQ	< LOQ	0.05	–	Spain, 2000–01
North America										
Terry et al. (2017)	10	3.10	28.5	16.5	7.8	9.4	–	–	–	USA, 2013
Zota et al. (2018)	103	–	28.8	7.44	5.07	8.62	–	–	–	USA, 2011–13
Fisher et al. (2016)	1379	< LOQ	< LOQ	< LOQ	< LOQ	< LOQ	–	–	–	Canada, 2007–11
Foster et al. (2011)	97	5.93	50.5	17.9	6.79	5.55	3.11	3.88	–	Canada, 2004–05
Herbstman et al. (2007)	297	0.9	13.6	4.3	2.3	2.6	0.9	0.9	–	USA, 2004–05
Herbstman et al. (2010)	210	–	11.2	3.2	1.4	0.7	0.6	0.6	–	USA, 2001
Mazdai et al. (2003)	12	–	25	7.1	4.1	4.4	0.7	< LOQ	–	USA, 2001
Asia										
Zhao et al. (2019)	249	–	0.30[b]	–	–	1.50[b]	–	–	8.18[b]	China, 2016/17
Chen et al. (2018a)	222	2.92	3.93	7.03	3.03	3.14	1.46	2.55	–	China, 2010–12

(continued)

Table 5.6 (continued)

References	n	BDE 28	BDE 47	BDE 99	BDE 100	BDE 153	BDE 154	BDE 183	BDE 209	Country, year
Ding et al. (2017)	106	2.63[b]	3.96[b]	8.27[b]	3.31[b]	1.89[b]	0.87[b]	0.66[b]	–	China, 2010/11
Choi et al. (2014)	118	–	2.87	2.38	0.56	1.38	0.77	0.63	–	Korea, 2011
Xu et al. (2013)	45	0.37	1.95	1.48	0.44	1.86	0.17	0.71	< LOQ	China
Kim et al. (2015)	16	0.4	1.4	1.3	0.8	4.1	–	–	–	Korea, 2010
Li et al. (2024)	924	0.41	0.12	0.07	0.39	0.38	0.12	< LOQ	18.4	China, 2009/10
Kim et al. (2009)	108	0.35	6.1	2.1	0.80	3.49	0.54	2.04	–	Korea, 2007
Bi et al. (2006)	21	0.63	1.4	0.47	0.22	0.80	0.12	0.20	–	China

[a]Arithmetic mean; [b]Geometric mean; *LOQ* Limit of quantification

Table 5.7 Important hydroxylated PBDE metabolites

Name	Abbreviation
6-Hydroxy-2,2′,4,4′-tetrabromodiphenyl ether	6-OH-BDE 47
3-Hydroxy-2,2′,4, 4′-tetrabromodiphenyl ether	3-OH-BDE 47
5-Hydroxy-2,2′,4,4′-tetrabromodiphenyl ether	5-OH-BDE 47
4′-Hydroxy-2,2′,4,5′-tetrabromodiphenyl ether	4′-OH-BDE 49
4-Hydroxy-2,2′,3,4′-tetrabromodiphenyl ether	4-OH-BDE 42
5′-Hydroxy-2,2′,4,4′,5-pentabromodiphenyl ether	5′-OH-BDE 99
6′-Hydroxy-2,2′,4,4′,5-pentabromodiphenyl ether	6′-OH-BDE 99

2009; Chen et al. 2013). In China, the medians were found to be 7.7 ng/g ww for 6-OH-BDE 47, 4.4 ng/g ww for 3-OH-BDE 47, and 35.0 ng/g ww for 5′-OH-BDE-99 (Chen et al. 2016). Meanwhile, in Korea during 2008–2009, median concentrations of 26 ng/g ww were determined for 6-OH-BDE 47 (Wan et al. 2010). The three studies mentioned also analyzed maternal blood in parallel. The ratio of maternal to cord blood in two studies in the USA was 0.2 and 0.4 for 6-OH-BDE-47, 0.2 and 0.4 for 5-OH-BDE-47, and 0.4 and 1.0 for 5′-OH-BDE-99 (Qiu et al. 2009; Chen et al. 2013).

So far, only two studies have reported time trends in concentrations. For example, Zota et al. (2013) examined 25 and 36 pregnant women in California during 2008–2009 and 2011–2012, respectively. They found a decrease in the geometric mean values of $\sum$-PBDE by 60%, a 50% reduction for 5-OH-BDE 47 (from 0.028 to 0.004 ng/ml), and a 516% drop for 6-OH-BDE 47 (from 0.025 to 0.004 ng/ml). In Japan, however, an increase in mean 6-OH-BDE 47 concentrations from 0.200 ng/g ww to 0.610 ng/g ww was observed in 20 men between 1989 and 2011 (Fujii et al. 2014b).

There are few study results on breast milk to date. Chen et al. (2016) reported medians of 15 OH-BDEs in 30 breast milk samples from five large cities in southern China in 2012. The median results ranged from 1.8 to 13 pg/g ww and are shown in Fig. 5.3. The higher concentrations of low-brominated OH-BDEs are particularly striking. In Japan, Fujii et al. (2014a) examined nine breast milk samples from Okinawa collected between 2005 and 2006 and did not find 6-OH-BDE 47 above the limit of quantification of 50 pg/g ww in any sample.

5.1.7 *Occurrence in Breast Milk*

As part of a program coordinated by the WHO and UNEP, pooled breast milk samples collected from 80 countries worldwide between 2000 and 2019 were analyzed (Schächtele et al. 2023). For PBDEs, significant differences were observed in the concentrations, ranging from 0.3 to 352 ng/g lw for six congeners from BDE 47 to BDE 183 in 135 pooled samples (median: 1.6 ng/g lw). The results of more

Table 5.8 Median OH-PBDE levels in blood serum/plasma of the general population

References	N	3′-OH-BDE-28	6-OH-BDE-47	3-OH-BDE-47	5-OH-BDE-47	4′-OH-BDE-49	4-OH-BDE-42
ng/g lw							
Liu et al. (2017)	33	0.0006	0.007	0.010	0.005	0.007	0.008
Wang et al. (2016b)	32	< 0.0004	0.007	0.008	0.007	< 0.006	0.005
Schultz et al. (2020)	57	–	1.21	< LOQ	4.06	< LOQ	< LOQ
Wang et al. (2012)	54[b]	–	0.029	< 0.001	–	–	< 0.001
Wang et al. (2012)	62[c]	–	0.031	< 0.001	–	–	< 0.001
Xu et al. (2018)	25	–	0.073	0.022	–	–	–
Butt et al. (2016)	43	–	< 0.340	< 0.020	–	0.050	–
Ingle et al. (2021)	189	–	0.040	0.090	0.010	0.070	–
Qiu et al. (2009)	4	–	0.200	0.090	1.400	0.300	< LOQ
pg/g ww							
Dufour et al. (2017)		–	< 2.50	–	< 2.30	–	–
Zota et al. (2018)	103	–	5	–	4	–	–
Zota et al. (2013)	36	–	7	–	7	< 10	–
Chen et al. (2013)	20	–	2.50	–	8.29	2.50	–
Wan et al. (2010)	26	–	< 4.0	–	–	–	–
Haraguchi et al. (2016)	20[a]	–	0.183[d]	< 0.100	< 0.100	–	–

(continued)

Table 5.8 (continued)

References	N	3′-OH-BDE-28	6-OH-BDE-47	3-OH-BDE-47	5-OH-BDE-47	4′-OH-BDE-49	4-OH-BDE-42
Ingle et al. (2021)	189	–	0.060	0.140	0.020	0.100	–
Fujii et al. (2014a)	10	–	0.090	–	–	–	–
References	N	4-OH-BDE-90	5′-OH-BDE-99	6′-OH-BDE-99	6-OH-BDE-82	3-OH-BDE-154	Country, year
ng/g lw							
Liu et al. (2017)	33	< 0.006	0.005	–	< 0.004	0.005	China, 2014/15
Wang et al. (2016b)	32	< 0.006	< 0.003	–	0.004	0.008	China, 2014
Schultz et al. (2020)	57	–	1.380	< LOQ	–	–	USA, 2013–15
Wang et al. (2012)	54[b]	< 0.001	0.029	< 0.001	–	–	China, 2011
Wang et al. (2012)	62[c]	< 0.001	0.005	< 0.001	–	–	China, 2011
Xu et al. (2018)	25	–	–	< LOQ	–	–	China
Butt et al. (2016)	43	–	0.030	–	–	–	USA, 2008–10
Ingle et al. (2021)	189	–	–	–	–	–	USA, 2006–16
Qiu et al. (2009)	4	–	1.700	0.300	–	–	USA, 2003–04
pg/g ww							
Dufour et al. (2017)		–	< 2.00	–	–	–	Belgium, 2015
Zota et al. (2018)	103	–	–	–	–	–	USA, 2011–13
Zota et al. (2013)	36	–	–	–	–	–	USA, 2011/12
Chen et al. (2013)	20	–	8.20	–	–	–	USA, 2011
Wan et al. (2010)	26	–	–	–	–		Korea, 2008/09
Haraguchi et al. (2016)	20[a]	–	–	–	–	–	Japan, 2007–09
Ingle et al. (2021)	189	–	–	–	–	–	USA, 2006–16
Fujii et al. (2014a)	10	–	–	–	–	–	Japan, 2006

[a]Pooled samples; [b]Female; [c]Male; [d]Arithmetic mean; LOQ limit of quantification

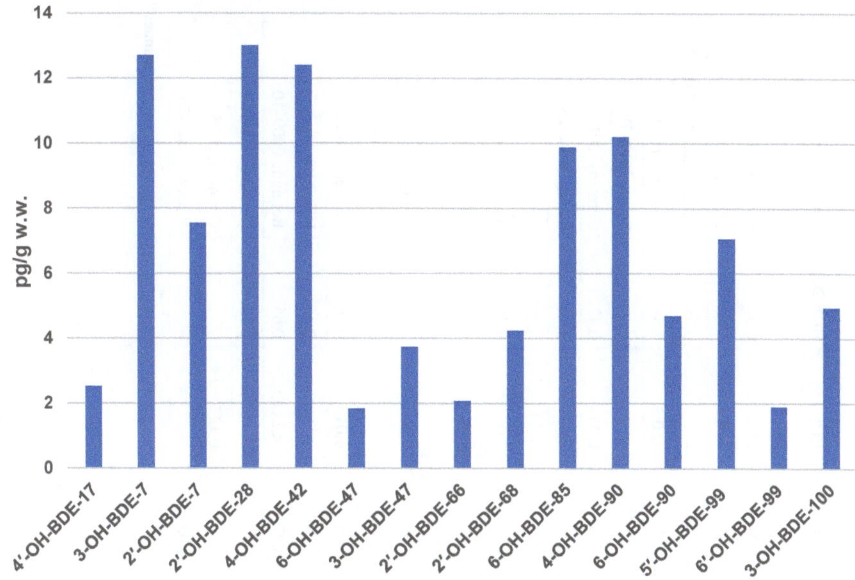

Fig. 5.3 Hydroxylated BDE in breast milk in pg/g ww (pg/g wet weight) (from Chen et al. 2016)

recent breast milk measurements published in the scientific literature are summarized in Table 5.9. When considering the sum of the median concentrations of the six congeners of BDE 28 to BDE 183, calculated from the above table, current levels in Europe and South America ranged from approximately 0.5 to 1.2 ng/g lw. Significantly higher levels were reported from North America, with approximately 28 ng/g lw in 2009–2012 and 11 ng/g in 2019. In studies from Asia, concentration levels comparable to those in Europe were found, while in Australia, the levels were slightly higher, at approximately 4.2 ng/g lw in 2014. In a study from Uganda, only BDE 47 was detected above the limits of quantification, with a median of just 0.39 ng/g lw. Significantly fewer results from studies are available for BDE 209. While medians between 0.06 ng/g lw and 1.4 ng/g lw have been reported from Europe and India, certain regions of China showed significantly higher medians of 2.2 and 2.5 ng/g lw. Overall, the differences in concentration between countries and within a country strongly reflect the historical use of individual congeners in flame retardancy.

Many studies identify BDE 47, BDE 99, and BDE 153 as the main congeners. While the share of BDE 47 in relation to the total of BDE 28 and BDE 180 (values from Table 5.9) ranges from 29 to 65% in North American studies, it varies in Europe, fluctuating between 4% in Ireland in 2017/18 and 60% in the UK in 2010/11. In both Europe and North America, the share of BDE 153 has increased, while that of BDE 47 has decreased. In 2019, the shares of BDE 47 and BDE 153 of the total content in the USA were 29% and 36%, respectively, whereas in the 2000s, they ranged between 55–65% and 6–20% in Canada and the USA. Studies from Asia show a significantly different pattern regarding the individual congeners, with the proportion of BDE 47

Table 5.9 Median PBDE-congeners level in breast milk samples in ng/g lw

References	n	BDE 28	BDE 47	BDE 99	BDE 100	BDE 153	BDE 154	BDE 183	BDE 209	Country, year
Europe										
Hedvall Kallerman et al. (2023)	30	0.02	0.17	0.11	0.04	0.54	0.04	0.02	0.07	Sweden, 2020/21
Jagić et al. (2024)	30	< 0.1	0.05	0.04	0.07	0.23	0.07	< 0.43	–	Croatia, 2019–22
Parizek et al. (2023)	231	–	0.17	0.15	–	0.15	–	–	0.75	Czech Republic, 2019–21
Wemken et al. (2020)	10[a]	< 0.06	0.05	< 0.2	< 0.2	0.71	< 0.12	< 0.3	1.4	Ireland, 2017/18
Rovira et al. (2022)	60	0.01	0.10	< 0.01	0.10	0.30	< 0.01	–	–	Spain, 2016–19
Fromme et al. (2022)	100	0.03	0.20	0.06	0.05	0.38	0.007	0.03	0.42	Germany, 2016/17
Milić et al. (2025)	33	ng	0.18	ng	ng	0.14	ng	ng	3.17	Serbia, 2014–16
Tao et al. (2017)	10	0.14	2.20	0.77	0.53	1.80	0.12		< 0.22	UK, 2014/15
Hartmann et al. (2024)	18	0.05	0.32	0.15	0.16	0.23	<	0.05	9.50	Austria, 2013
Herceg Romanic et al. (2023)	50[a]	0.02	0.22	0.07	0.08	0.24	0.01	0.03	–	Croatia, 2011–14
Iszatt et al. (2019)	257	0.16	1.10	0.26	0.26	0.50	0.03	–	–	Norway
Antignac et al. (2016)	22	0.33	3.12	0.54	0.42	0.77	0.05	0.03	–	Finland, 2011–14
Harrad and Abdallah (2015)	10	–	2.30	1.04	–	0.48	–	–	0.08	UK, 2010–11
Darnerud et al. (2015)	30	0.04	0.46	0.07	0.13	0.45	0.05	0.01	0.06	Sweden, 2010
Abdallah and Harrad (2014)	35	–	2.50	0.70	0.41	0.90	0.28	–	0.46	UK, 2010
Croes et al. (2012)	84	< LOQ	0.16	0.06	0.06	0.29	0.07	< LOQ	0.65	Belgium, 2009–10
Hoopmann et al. (2012)	2173		0.31	0.09	0.09	0.51	–	–	–	Germany, 2006–09
Raab et al. (2008)	42	–	0.45	0.18	0.14	0.56	–	0.12	–	Germany, 2005
Lenters et al. (2019)	1199	0.11	1.03	0.27	0.26	0.50	0.03	–	–	Norway, 2002–09

(continued)

Table 5.9 (continued)

References	n	BDE 28	BDE 47	BDE 99	BDE 100	BDE 153	BDE 154	BDE 183	BDE 209	Country, year
Americas										
Schreder et al. (2023)	50	< 0.3	3.17	1.44	1.73	3.90	0.40	< 0.30	–	USA, 2019
Torres–Moreno et al. (2023)	60	0.16	0.50	0.13	0.09	0.06	0.02	0.02	–	Colombia, 2014/15
Guo et al. (2016)	67	–	16.7	3.48	2.48	4.90	–	–	–	USA, 2009–12
Guo et al. (2016)	82	–	31.3	6.48	5.52	6.31	–	–	–	USA, 2003–05
Dunn et al. (2010)	40	1.5	13.4	2.0	2.5	4.9	0.1	–	–	USA, 2005/6
Ryan and Rawn (2014)	34	1.4	12.1	2.6	1.5	1.1			–	Canada, 2005
Siddique et al. (2012)	87	–	5.9	2.8	1.6	1.6	–	–	–	Canada, 2003–9
Daniels et al. (2010)	303	2.0	28	5.0	5.0	6.0	0.3	–	–	USA, 2004–06
Asia/Africa/Australia										
Matovu et al. (2021)	30	< LOQ	0.39	< LOQ	< LOQ	< LOQ	< LOQ	< LOQ	< LOQ	Uganda, 2018
Zhao et al. (2021)	105	0.06	0.08	0.03	0.03	0.39	0.01	0.06	0.28	China, 2018
Chen et al. (2019a)	111	0.09	0.07	0.02	0.02	0.25	0.02	0.03	2.20	China, 2014
Chen et al. (2018b)	25	0.03	0.24	0.06	0.08	0.63	0.03	0.05	0.69	Taiwan, 2013–16
Yang et al. (2016)	63*	0.03	0,02	–	–	0.14	–	0.02	2.50	China, 2014/15
Chen et al. (2015)	10	–	1.8	0.33	0.57	1.5	–	–	–	Australia, 2013
Zhang et al. (2014)	30	0.20	0.18	0.03	0.22	0.63	< LOQ	0.15	0.95	China, 2012
Lee et al. (2013b)	208	–	0.54	0.29	0.12	–	–	–	–	Korea, 2011
Zhang et al. (2017)	32[a]	0.22*	0.25*	0.08*	0.06*	0.50*	0.04*	0.32*	–	China, 2011
Shi et al. (2013)	103	0.12	0.12	0.03	0.02	0.29	0.01	0.06	1.95	China, 2011
Ben Hassine et al. (2012)	36	0.56	2.10	1.08	1.48	1.51	0.67	2.03	–	Tunisia, 2010
Tsai et al. (2021)	100	0.07	0.54	0.15	0.19	0.85	0.07	0.11	0.34	Taiwan, 2007–11
Coakley et al. (2013)	33	0.18	2.14	0.56	0.50	0.52	0.04	0.04	0.19	New Zealand
Toms et al. (2012)	4[a]	–	1.4	0.6	1.2	1.3	–	–	–	Australia, 2009
Devanathan et al. (2012)	10[a]	0.06	0.46	0.07	0.04	0.07	0.02	0.02	0.79	India, 2009

* Arithmetic means; [a]Pooled samples; *LOQ* Limit of quantification; *ng* Not given

being approximately between 10 and 27%, and that of BDE 153 between 34 and 67%. BDE 47 holds a considerable share of the total concentrations in Australia and is even more pronounced in India.

As seen in Fig. 5.4, Swedish studies indicate an increase in BDE 47 levels from the early 1970s, reaching a peak in the mid-1990s, followed by a significant decline. In contrast, a Swedish study reported a continuous rise in BDE 153 levels between 1997 and 2007 (Lignell et al. 2008). Gyllenhammar et al. (2021) also identified a decreasing temporal trend in Sweden from 1996 to 2017, with an average annual decrease of 9.0% for BDE 47, while no changes were noted for BDE 153 and BDE 209. In Germany, a notable increase in BDE 47 was observed between 1992 and 2002, rising from an average of 0.76 ng/g lw to 1.63 ng/g lw (Fürst 2006), before dropping to 0.45 ng/g lw in 2005 (Raab et al. 2008) and further to 0.20 ng/g lw in 2016/2017 (Fromme et al. 2022). A median reduction of 55% for BDE 47 and 29% for BDE 153 was reported in Irish breast milk samples between 2011 and 2016/18 (Wemken et al. 2020). In China, breast milk samples analyzed in 2005, 2011, and 2014 showed a significant decrease in BDE 47, but no changes were found for BDE 153 and BDE 209 during this period (Chen et al. 2019a). In the nationwide Chinese National Human Milk Survey from 2016 to 2019, the median levels of BDE 209 (0.27 ng/g lw) in 100 pooled milk samples sharply declined compared to the surveys conducted in 2007 and 2011 (around 0.55 ng/g lw) (Zhao et al. 2025). In the northwestern United States, a decrease in median PBDEs of approximately 70% was reported from 2003 to 2019 (Schreder et al. 2023). Conversely, no significant changes in low levels of contamination were observed in 16 breast milk samples collected in Croatia from 2010 to 2020 (Dvoršćak et al. 2024).

There are contradictory results in the literature regarding whether food, particularly animal products, can be a source of contamination (Shi et al. 2013; Zhao and Shi 2021). However, Vieth et al. (2005) identified a correlation between diet and PBDE levels, observing higher concentrations in the breast milk of mixed consumers compared to vegetarians. It is assumed that the concentrations of PBDEs remain statistically unchanged during the breastfeeding period, or that only slight fluctuations, for example, for BDE 153, can be observed (Vieth et al. 2005; Raab et al. 2008; Thomsen et al. 2010; Daniels et al. 2010; Harrad and Abdallah 2015; Yang et al. 2016; Chen et al. 2019a). Many studies have noted an increase in PBDE levels, or at least in some individual congeners, in breast milk as maternal age increases (Raab et al. 2008; Koh et al. 2010; Dunn et al. 2010; Glynn et al. 2011; Croes et al. 2012; Shi et al. 2013; Zhao and Shi 2021; Jagić et al. 2024).

The results of five studies on the relationship between maternal serum and breast milk levels (S/M ratio) are reported in a review ('t Mannetje et al. 2012). They demonstrate a clear trend of an increasing S/M ratio with a higher number of bromine atoms in the molecule. Mean S/M ratios of 0.69, 0.95, 0.73, 1.04, and 2.65 were found for BDE 47, BDE 99, BDE 100, BDE 153, and BDE 183. In 60 samples of breast milk and maternal serum from the eastern USA, S/M ratios between 1.0 and 1.1 were found for the congeners BDE 47, BDE 99, BDE 100, and BDE 153 (Guo et al. 2016). For BDE 209, an S/M of 10 was observed in a study with 89 paired samples from Japan (Inoue et al. 2006), while two smaller studies from the USA and Sweden

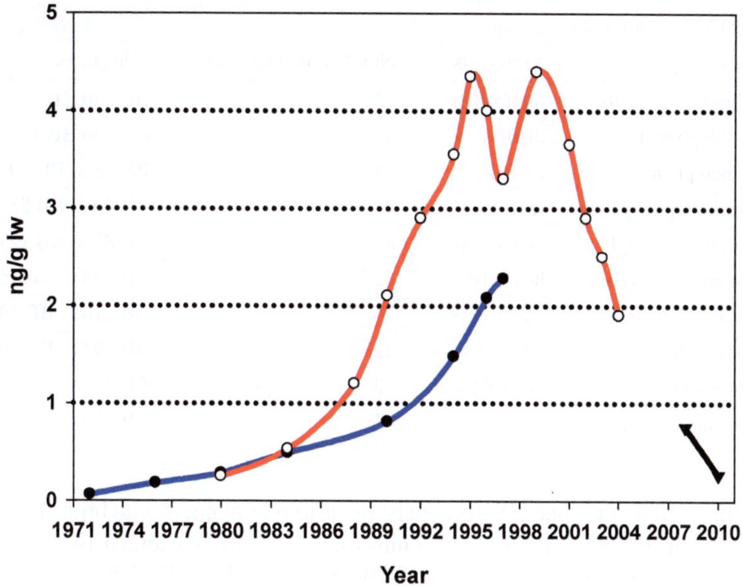

Fig. 5.4 Time course of the concentrations of BDE 47 in Swedish breast milk samples from two cities (blue: Meironyte et al. 1999, red: Fängström et al. 2008; black: Lignell et al. 2012)

determined an S/M of 40 and 17, respectively (Schecter et al. 2006; Darnerud et al. 2015). An animal study also investigated transplacental and lactational transport with repeated administration of isotope-labeled BDE 47 and BDE 209 (Shin et al. 2017). Both congeners and their debromination products increased rapidly in the offspring's blood, with exposure via breast milk being primarily responsible for this, rather than transplacental transport.

5.2 Polybrominated Biphenyls

Polybrominated biphenyls (PBBs) were additive flame retardants produced until the mid-1980s, and deca-BB continued to be produced until around 2000 (EFSA 2010). Because they are not chemically bound in the polymer, they can be released from products into the environment throughout their lifespan. Similar to polychlorinated biphenyls, they consist of two rings connected by a bridge, featuring different numbers and positions of bromine atoms, which theoretically results in 209 possible congeners. Table 5.10 provides a summary of the individual homolog groups. PBBs are lipophilic compounds that are chemically stable, persistent in the environment, and bioaccumulative. They have low vapor pressure and water solubility, both of which decrease with increasing bromination levels (EFSA 2010).

Table 5.10 List of homolog groups

Homologs	Acronyms	Number of congeners	Congeners
Monobromobiphenyls	mono-BBs	3	BB 1 to BB 3
Dibromobiphenyls	di-BBs	12	BB 4 to BB 15
Tribromobiphenyls	tri-BBs	24	BB 16 to BB 39
Tetrabromobiphenyls	tetra-BBs	42	BB 40 to BB 81
Pentabromobiphenyls	penta-BBs	46	BB 82 to BB 127
Hexabromobiphenyls	hexa-BBs	42	BB 128 to BB 169
Heptabromobiphenyls	hepta-BBs	24	BB 170 to BB 193
Octabromobiphenyls	octa-BBs	12	BB 194 to BB 205
Nonabromobiphenyls	nona-BBs	3	BB 206 to BB 208
Decabromobiphenyls	Deca-BB	1	BB 209

Commercially produced PBBs consist of various congeners with bromine content ranging from 76 to 85%. PBB 153 accounts for over 50% of the total in two commonly used technical products, followed by PBB 180 (ATSDR 2004). They were utilized in quantities of up to 10%, particularly in synthetic fibers and polymers in coatings, varnishes, and polyurethane foam.

The commercial product hexabromobiphenyl is listed under Annex A of the Stockholm Convention, and parties must take measures to eliminate its production and use (SC 2019).

5.2.1 *Health Aspects*

Various reviews provide summaries of the scientific literature on health effects, including toxicokinetics (Hakk and Letcher 2003; ATSDR 2004; EFSA 2010; IARC 2015).

In addition to workplace studies, epidemiological data are available, particularly from the so-called Michigan Cohort, which has been collecting health-related information since 1976, following the accidental contamination of cattle feed with PBBs and the subsequent exposure of residents. Overall, the partly methodologically limited and contradictory results of epidemiological studies indicate links between exposure to PBB and health effects, such as influences on neurological development, specific cancers, and effects on fertility and offspring.

In animal experiments, PBBs particularly cause effects on the liver, resulting in hypertrophy and fatty degeneration, impact the thyroid gland by decreasing thyroid hormones and increasing thyroid-stimulating hormone, and contribute to thyroid enlargement, including morphological changes. There are also indications of effects on neurobehavioral development and the immune system. In long-term studies with technical products, liver neoplasia and, in some cases, thyroid adenomas and leukemia were observed.

PBBs in which bromine atoms substitute the hydrogen atoms in the ortho position are referred to as ortho-PBBs, and those in which the hydrogen atoms in ortho positions are not substituted by bromine are referred to as nonortho-PBBs. This difference in molecular structure is relevant for interacting with different receptors that determine the toxicological properties of PBBs.

The WHO's International Agency for Research on Cancer classified PBBs as probably carcinogenic to humans (Group 2A) based on comparisons with PCBs (IARC 2015). IARC determined there was inadequate evidence for carcinogenicity in humans. At the same time, it found sufficient evidence for the carcinogenicity of Firemaster FF-1 in laboratory animals, limited evidence of PBB 153 in laboratory animals, and inadequate evidence for PBB 169 and Firemaster BP-6.

5.2.2 Toxicokinetics

Data on the toxicokinetics of this substance group are limited and primarily pertain to individual congeners. In animal studies, for instance, almost complete uptake after oral administration has been reported for BB 153, exceeding 90%. While hydroxylated metabolites were detected following metabolization in the urine of low-brominated PBBs, debromination and hydroxylation occasionally occurred with higher-brominated congeners. Excretion primarily occurs via feces. The elimination half-life of BB 153 from serum in rats is approximately 23 weeks, and based on toxicokinetic models and epidemiological data, it ranges from approximately 10 to 30 years for humans. For example, using 4768 serum samples from the Michigan cohort between 1974 and 2019, the adjusted median participant-specific half-life for PBB 153 was 12.2 years (Hood et al. 2023). Overall, the half-life of PBB 153 was extended by higher initial PBB level (~ + 1.5 years), younger age at exposure (~ + 5.4 years), higher BMI (~ + 1.0 years), and increased gravidity (~ + 7.3 years), while it was shortened by smoking status (~− 2.8 years) and breastfeeding (~− 3.5 years). The transfer from maternal blood to breast milk has been noted in animal studies.

5.2.3 Occurrence in Blood

Human samples were examined multiple times in relation to the previously mentioned Michigan accident. For example, Landrigan et al. (1979) reported median PBB levels of 2.5 μg/l (0–1150 μg/l) in 52 maternal serum samples from exposed residents and 1.0 μg/l (0–104 μg/l) in 58 umbilical cord blood samples. Even 40 years after the accident, median BB 153 levels of 38 ng/g lw (0.26 μg/l) and a 95th percentile (maximum) of 627 ng/g lw (38812 ng/g lw) can still be observed in the highly exposed group of workers and the population at that time (Chang et al. 2020). In contrast, BB 77, BB 101, and BB180 were found in only a few samples above the LOQ, with maxima of 0.34, 0.60, and 0.31 ng/g lw, respectively.

Many studies to date have focused on the analysis of BB 153. For instance, Abasilim et al. (2025) found 2008–2011 geometric mean values of 1.02 and 1.31 ng/g lw for 716 women and 1073 men in Florida, US. Again, Sjödin et al. (2019) investigated the occurrence of BB 153 in pooled serum samples from five representative US NHANES surveys conducted between 2005–2006 and 2013–2014. The adjusted geometric mean concentrations were 1.74 ng/g lw for women and 2.45 ng/g lw for men throughout the entire study period, with levels decreasing over time. Higher geometric mean values were observed for men compared to women, and the highest mean levels were noted in the 20–30 age group, while the lowest levels were found in the 12–19 age group. Sjödin et al. (2004) also examined a time trend by analyzing pooled samples from the USA over four measurement periods between 1985 and 2002. They reported a decrease in BB 153 from a median of 8.0 ng/g lw to 3.3 ng/g lw. On the contrary, Wattigney et al. (2022) reported a high median (95th percentile) of 15.4ng/g lw (62.7 ng/g lw) in the serum of 287 shorelane anglers along the Detroit river, a river with a long contamination history. Long et al. (2023) noted higher medians ranging from 2.0 to 28 ng/g lw for BB 153 in 59 plasma samples from subjects living in Greenland from 2013 to 2015, who had a high intake of marine mammals.

Significantly lower concentrations were found in Europe. For example, Frederiksen et al. (2010) reported medians of 0.18 ng/g lw (< 0.01–0.85 ng/g lw) and 0.07 ng/g lw (< 0.03–0.35 ng/g lw) in 51 maternal and 40 umbilical cord blood samples in Denmark in 2007. Again, Fillol et al. (2021) noted medians of 0.23 ng/g lw in 742 serum samples from adults in France during 2014–2016. Even in 394 serum samples collected from France between 1994 and 1999, the median was only 0.31 ng/g lw (Mancini et al. 2020). In the UK, an analysis of 20 serum samples from women and men conducted in 2011–2012 revealed a median of only 0.04 ng/g lw (< 0.01–0.91 ng/g lw). In the analysis of samples from Poland, Ukraine, and Greenland, medians below the LOQ and a 95th percentile of 0.32 ng/g lw and 9.22 ng/g lw were found in Poland and Ukraine (Lenters et al. 2013). Only in 99 serum samples from Greenland was a significantly higher median (95th percentile) of 1.1 ng/g lw (4.8 ng/g lw) reported in this study, likely due to specific dietary habits.

There is a significant risk of exposure to the general population in China due to the large number of e-waste dismantling facilities. Numerous studies have examined the exposure levels among the general population in this context. Ma et al. (2017) analyzed ten pooled serum samples from 452 individuals in China between 2014 and 2015, determining mean values for BB 153 of 0.05 ng/g lw (< LOQ-0.43 ng/g lw) and 0.16 ng/g lw (> LOQ-0.91 ng/g lw) in 10% and 20% of the samples, respectively. A maximum value of 0.82 ng/g lw was also noted for BB 153 in 13 of 155 serum samples collected between 2012 and 2016 (Bao et al. 2022a, b). In northern China, a median of 0.02 ng/g lw was reported in 21 serum samples from the general population gathered in 2009–2010 (Yang et al. 2013). Additionally, a total of 7 serum samples from a control group located 35 km away from an e-waste dismantling area were analyzed in 2017, revealing a median of only 0.03 ng/g lw (< LOQ-0.46 ng/g lw) for BB 153 (Wang et al. 2021a).

In addition to the frequently measured BB 153, other congeners are also increasingly being assessed in serum. For example, Bramwell et al. (2014) examined 20 samples from 2011 to 2012 and could not find the congeners BB 15, BB 49, BB 52, BB 80, BB 101, and BB-209 above the LOQs. Various studies are now available from China, where a larger number of congeners were assessed. In 155 samples from 31 blood donors between 2012 and 2016, the highest median of 254 ng/g lw was found for BB 1, followed by 141 ng/g lw for BB 10, 69.4 ng/g lw for BB 209, and 9.5 ng/g lw for BB 2 (Bao et al. 2022a, b). Wang et al. (2021a) also investigated a wide range of 35 congeners in 7 serum samples from a control group located 35 km away from an e-waste dismantling area in 2017. They found a similar distribution with medians of approximately 670 ng/g lw for BB 1, 55 ng/g lw for BB 10, 8.5 ng/g lw for BB 209, 5.2 ng/g lw for BB 7, and 4.9 ng/g lw for BB 18. In 2017, Wang et al. (2024a) collected serum samples from both an e-waste dismantling area and a control area. In the 44 samples from the control area, the total $\sum$35PBB concentrations ranged from 105 to 1980 ng/g lw (median 468 ng/g lipid). The highest medians were observed at 303 ng/g lw for BB 1 and 81.8 ng/g lw for BB 10. These congeners account for 78% and 18% of the total PBB concentration, respectively. In addition, the median results of the other congeners are shown in Fig. 5.5.

5.2.4 Occurrence in Breast Milk

To date, only limited results are available on PBBs in breast milk, which generally indicate low exposure levels. Pratt et al. (2013) analyzed 11 pooled breast milk samples from 109 Irish women, finding BB 153 with a median of 0.13 ng/g lw. Conversely, BB 15, BB 49, BB 52, BB 80, and BB 101 were not detected above the LOQs of 0.002 to 0.02 ng/g lw, and BB 209 was not found above the LOQs of 0.3–0.4 ng/g lw in any of the samples. A slightly lower median of 0.08 ng/g lw (0.06–0.79 ng/g lw) for BB 153 was observed in all six UK breast milk samples from 2011–2012, while BB 101 was detected in only one sample at 0.007 ng/g lw. Additionally, BB 15, BB 49, BB 52, and BB 80 were all below the LOQs of 0.001–0.11 ng/g lw, with BB 209 also below the LOQ of 0.05 (Bramwell et al. 2014). Aerts et al. (2019) analyzed 206 samples from Belgian primiparous mothers in 2014 and could find BB 153 in only one sample at 0.58 ng/g lw, above the LOQ of 0.1 ng/g lw. Fromme et al. (2022) identified BB 153 in 64 of the total 100 German milk samples, reporting a median of 0.043 ng/g lw and a range from 0.001 to 0.180 ng/g lw.

In Beijing, China, in 2014, 111 samples from 31 women were analyzed during the first three months postpartum, revealing a low median of 0.02 ng/g lw (< LOQ-0.13 ng/g lw) (Chen et al. 2019a).

In the analysis of 30 breast milk samples from Kampala, Uganda, in 2018, BB 153 was present above the LOQ in 17% of the samples, with values ranging from < 0.0001 to 0.06 ng/g lw (Matovu et al. 2021).

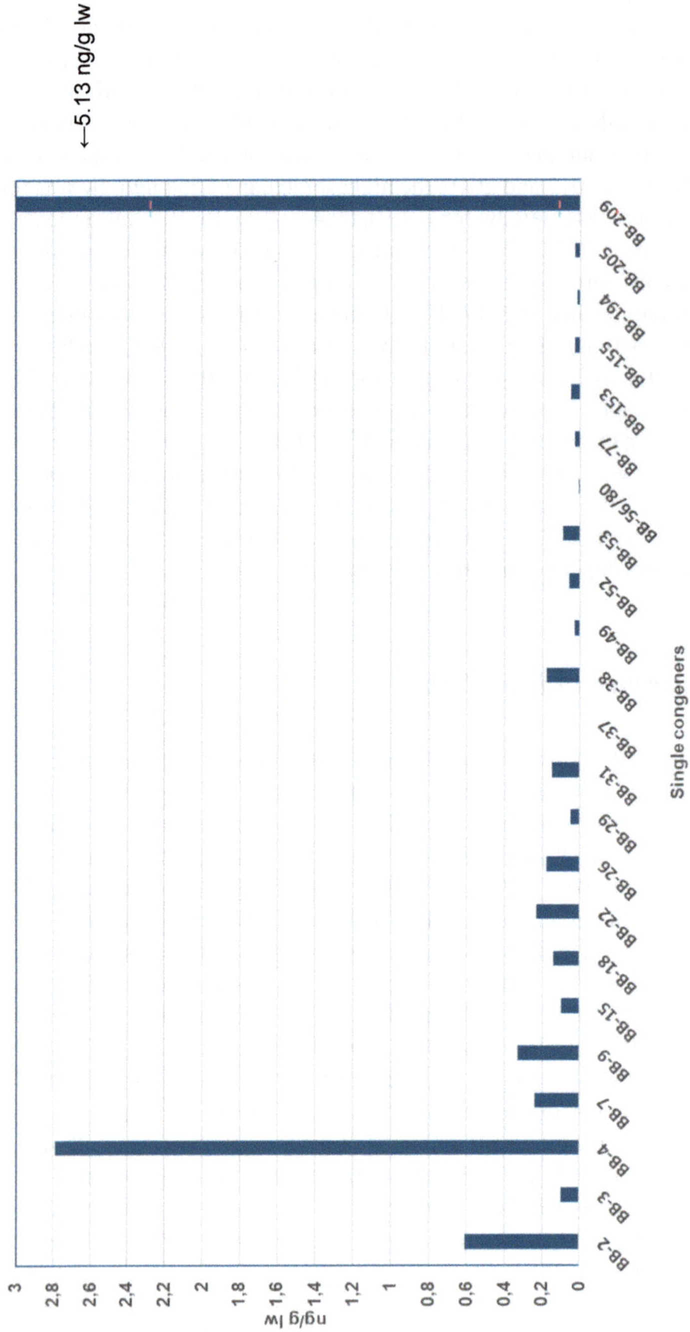

Fig. 5.5 Median concentrations of PBB congeners in ng/g lw in serum from the general population (after Wang et al. 2024a, b). Medians of BB 30, BB 75, BB 154, BB156, BB 169, BB 180, and BB 206 were under the limit of quantification

5.3 Hexabromocyclododecane

Hexabromocyclododecane (HBCDD, CAS No. 25637-99-4) was primarily used in polystyrene insulation materials for buildings and, to a lesser extent, in upholstery textiles and packaging plastics. The chemical structure is shown in Fig. 5.6. As of 2016, approximately 12,000 tons of HBCDD were utilized annually in the EU. In the EU, HBCDD has been included in the candidate list in Annex XIV of REACH and has been a substance targeted for elimination under Annex A of the Stockholm Convention since 2013. Under the POP Regulation, no products containing more than 100 mg HBCDD/kg may be manufactured or placed on the market in the EU since 2016, with a temporary exception for insulation materials made of expanded polystyrene.

5.3.1 Health Aspects

Many scientific publications and reviews on toxicological (including toxicokinetic) and epidemiological knowledge are available. The main ones include: HBCDD (HC 2011; NICNAS 2012; US EPA 2020; EFSA 2021) and various flame retardants (EFSA 2012a,b; Xiong et al. 2019; Fromme and Roscher 2020).

This is only briefly summarized here. In animal experiments, the main effects of HBCDD are observed in the liver, with increased liver weight and hepatocellular hypertrophy; the thyroid gland, characterized by increased thyroid weight, follicular hypertrophy, and hyperplasia; and significant changes in thyroid hormone levels along with neurodevelopmental effects on behavior. An increasing number of epidemiological studies have demonstrated a link between exposure to HBCDDs and

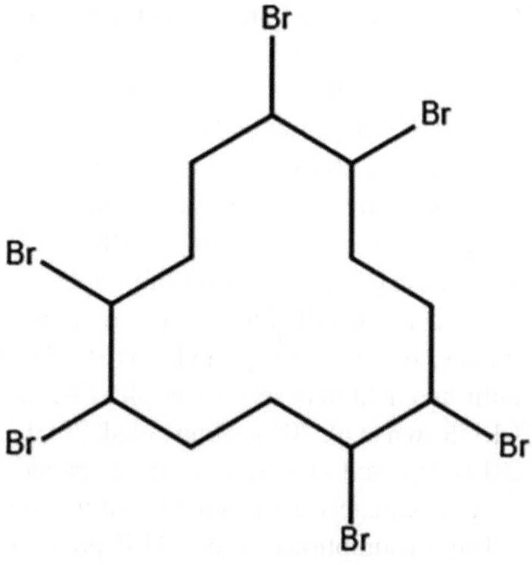

Fig. 5.6 Chemical structure of hexabromocyclododecane (α, β, γ)

factors such as birth weight and length, neuronal development, and thyroid dysfunction in children, as well as metabolic and hormonal consequences. The acute toxicity of HHCB is estimated to be low.

5.3.2 Toxicokinetics

The limited data on individual HBCDD congeners revealed nearly complete absorption in animal experiments after single or repeated oral administration, and only lower percentages after dermal application. Following distribution, especially in the liver, fatty tissue, and muscles, accumulation occurs in fatty tissue, with α-HBCDD accumulating approximately 100-fold and β-HBCDD around tenfold more than γ-HBCDD. In experimental animals, α-HBCDD is primarily metabolized through hydroxylation, ring opening, and glutathione conjugation, while γ-HBCDD is metabolized by debromination and decarboxylation, followed by ring opening and subsequent cleavage of the carbon chain. In animal experiments, about 50% of γ-HBCDD is excreted via feces and up to 30% via urine, with half-lives lasting several days. The total body half-life for Σ-HBCDD in humans was found to be 64 days (ranging from 23 to 219 days) (Geyer et al. 2004).

5.3.3 Occurrence in the Blood and Urine of Adults

The study results on HBCDD levels in blood samples are summarized in Table 5.11. Overall, since 2010, medians in Europe have remained in a low range of approximately < 0.1–12.6 ng/g lw for Σ-HBCDD. Drage et al. (2019) found slightly higher median levels in the serum of 59 adults from the UK in 2014 and in Norway during 2010–11, where the 95th percentile was only 1.8 ng/g lw, although some individuals had concentrations of up to 77 ng/g lw (Bjermo et al. 2017). In Canada (2007–2009) and Australia (2002–2014), mean levels of Σ-HBCDD were observed at 0.85 ng/g lw (0.33–8.9 ng/g lw) and 0.88 ng/g lw (< 0.5–36 ng/g lw), respectively, which are comparable to those in Europe (Rawn et al. 2014; Drage et al. 2017). This finding also applies to a Chinese study examining 42 pooled samples from Beijing in 2011 (Shi et al. 2013). Only Kim and Oh (2014) reported a high median of 8.55 ng/g lw for ∑-HBCDD in the serum of mothers and their children in Korea.

Typically, α-HBCDD is the predominant isomer found in all biota, including humans, followed by γ- and β-HBCDD. The dominance of α-HBCDD was also confirmed in human serum samples (Weiss et al. 2006; Roosens et al. 2009; Shi et al. 2013; Rawn et al. 2014; Drage et al. 2019; Tay et al. 2019). In contrast, Kim and Oh (2014) reported in a study of mothers and their children that γ-HBCDD accounted for approximately 75% and 86% of the total levels of the three congeners.

The consumption of fish and fish products appears to influence exposure. Thomsen et al. (2008) found a median of 4.1 ng/g lw and 2.6 ng/g lw in hobby fishermen and

Table 5.11 Levels of $\sum$-HBCDD in blood serum of the general adult population in ng/g lw

References	N	Median	Range	Group, Age (mean or range in years)	Country, year
Hedvall Kallerman et al. (2023)	3[a]	< 0.50	–	Pregnant women (27–40)	Sweden, 2021
Drage et al. (2017)	20[a]	–	< 0.1–7.9	(5- > 60)	Australia, 2014/15
Drage et al. (2019)	59	2.2	< 0.3–12.6	Females (23–42)	UK, 2014
Shi et al. (2013a)	42[a]	0.84	< LOQ-7.2		China, 2011
Fromme et al. (2016b)	42	< 19	< 19	Adults (20–68)	Germany, 2013–14
Tay et al. (2019)	61	< 0.11	< 0.11–3.4	Adults (20–66)	Norway, 2013
Dereumeaux et al. (2016)	277	< 0.3	< 0.3–1.5[b]	Women (18–47)	France, 2011
Bjermo et al. (2017)	170	0.1	< 0.5–77	Adults (50)	Sweden, 2010–11
Darnerud et al. (2015)	3[a]	< 0.49		Pregnant women (30)	Sweden, 2010
Rawn et al. (2014)	59[a]	1.0[d]	0,33–8,9	General population (6–79)	Canada, 2007–09
Roosens et al. (2009)	16	1.7	< 0.5–11.3	Adults (20–25)	Belgium, 2007
Kim and Oh (2014)	76	8.6[d]	< LOQ-166		Korea
Kalantzi et al. (2011)	61	1.3	< 1.0–38.8	Computer professionals and controls (40)	Greece, 2007
Thomsen et al. (2008)	41	4.1	< 1.0–52	Hobby fishermen (30–87)	Norway, 2004–05
Thomsen et al. (2008)	25	2.6	< 1.0–18	Women (9–78)	Norway, 2004–05
Meijer et al. (2008)	69	0.7	< LOQ–7.4	Pregnant women (24–42)	Netherlands, 2000–01
Weiss et al. (2006)	50	0.46	< 0.24–3.4	Fishermen's wives (62[c])	Sweden, 2000
Peters (2004)	91	< 0.08[e]	< 0.08–0.36[e]	Adults	The Netherlands

[a]Pooled samples; [B]95th percentile; [c]median; [d]mean; [e]values in pg/g serum; *LOQ* limit of quantification

their wives who lived near a contaminated lake. The serum levels were correlated with fish consumption.

An increase in HBCDD concentrations was observed in Sweden from 1996 to 2010, followed by a decrease after a plateau phase (Darnerud et al. 2015). Hedvall Kallerman et al. (2023) also reported an annual mean decrease in pooled serum samples from Sweden from 1996 to 2021 of 5.9%. In Australian serum samples, an increase was noted between 2002 and 2008–2010, followed by a slight decrease in serum concentrations until 2014–2015 (Drage et al. 2017).

5.3.4 Occurrence in the Blood and Urine of Children and Cord Blood

Studies on children's blood have been limited to date. Rawn et al. (2014) examined 57 pools of 6–11-year-olds and 12–19-year-olds from Canada during 2007/8 and found geometric mean values of 0.89 ng/g lw (0.51–8.9 ng/g lw) and 0.99 ng/g lw (0.5–3.3 ng/g lw) for Σ-HBCDD. Additionally, a median of 0.48 ng Σ-HBCDD/g lw (< 0.1–1.9 ng/g lw) was observed in the pooled serum samples of 800 Australian children aged 0–4 years in 2014/15 (Drage et al. 2019). This study found no significant trends with age or differences between genders. The results of this study are comparable to a mean level of 0.35 ng/g lw observed in four pooled Australian samples from 2013 collected from children aged 0–4 years (Drage et al. 2017). In Korea, Kim and Oh (2014) report very high mean values of 16.4 ng/g lw in the blood of 12 children, which were also higher than those in the maternal serum samples examined in parallel. Serum samples from 93 Moroccan children aged 5–11 collected in 2021 showed a median of 2.20 ng/g lw (0.51–4.56 ng/g lw) for Σ-HBCDD (Chbihi et al. 2024). In this study, α-HBCDD was the dominant congener with a median of 1.43 ng/g lw.

Meijer et al. (2008) examined 12 umbilical cord serum samples along with maternal serum. They found lower concentrations in four out of the 12 samples, with a median of 0.2 ng/g lw (0.2–4.3 ng/g lw) for Σ-HBCDD in cord blood. The authors reported the median ratio of cord to maternal blood (C/M-ratio) as 0.7. Furthermore, Yuan et al. (2025) analyzed 38 paired maternal and cord blood samples from China, revealing a median of 0.30 μg/l for Σ-HBCDD and a C/M-ratio of 1.4.

5.3.5 Occurrence in Breast Milk

As part of a program coordinated by the WHO and UNEP, 102 pooled breast milk samples from 72 countries collected worldwide between 2006 and 2019 were analyzed (Schächtele et al. 2023). In this study, α-HBCDD levels ranged from < 0.1 to 15 ng/g lw (median: 0.5 ng/g lw), while β-HBCDD and γ-HBCDD were found in nearly all samples at concentrations below the LOQ. Table 5.12 summarizes the recent results of HBCDD measurements in breast milk. In Europe, the medians for Σ-HBCDD range from 0.22 to 6.0 ng/g lw. Only a Spanish study involving 33 samples from La Coruña observed significantly higher levels at 27 ng/g lw, whereas samples from Barcelona showed similar concentrations in other European countries (Eljarrat et al. 2009). In North America, median concentrations are 0.4 and 1.0 ng/g lw, which are comparable to those measured in Europe, while the medians in Asia and Australia were slightly higher. In breast milk, α-HBCDD constitutes the largest portion of the total concentrations of the three congeners (Abdallah and Harrad 2011; Croes et al. 2012; Carignan et al. 2012; Pratt et al. 2013; Shi et al. 2013; Ryan and Rawn 2014; Harrad and Abdallah 2015; Fujii et al. 2018; Wemken et al. 2020; Parizek et al. 2023; Rawn et al. 2024; Zhao et al. 2025).

The time course of HBCDD concentrations between 1980 and 2004 in pooled samples from Sweden was examined (Fängström et al. 2008). This study reported an increase until approximately 1995, followed by fluctuating concentrations settling on a plateau range. In English, breast milk samples collected for the first time in 2010 and again in 2014/2015, a slight decrease was found in the median levels of Σ-HBCDD (3.8 vs. 2.9 ng/g lw), albeit not statistically significant (Tao et al. 2017). In Ireland, a decrease was observed between 2007 and 2016–2018 (Wemken et al. 2020). Kakimoto et al. (2008) analyzed pooled Japanese breast milk samples from 1973 to 2006 at 14 time points. While concentrations between 1973 and 1983 were below the LOQ, the Σ-HBCDD rose to a plateau between 1.4 and 2.5 ng/g lw, then again to 4.0 ng/g lw in 2006. In 2007 and 2008, the concentrations fell again to 1.4 and 1.7 ng/g lw (Kakimoto et al. 2012). In China, significantly higher HBCDD levels were found in 100 pooled samples during the third Chinese National Human Milk Survey from 2016 to 2019 (5.01 ng/g lw) compared to 2007 (around 1 ng/g lw) (Zhao et al. 2025).

5.4 Emerging Brominated Flame Retardants

In addition to polybrominated diphenyl ethers, the following emerging brominated flame retardants (EBFRs) are particularly technically significant: tetrabromobisphenol A (TBBPA), 1,2-bis-(2,4,6-tribromophenoxy)ethane (BTBPE), 2-ethyl-1-hexyl-2,3,4, 5-tetrabromobenzoate (EH-TBB), bis(2-ethyl-1-hexyl)tetrabromophthalate (BEH-TEBP), decabromodiphenylethane (DBDPE), and tetrabromoethylcyclohexane (DBE-DBCH). The chemical structures of the most essential EBFRs are shown in Table 5.13. The nomenclature was chosen based on Bergman et al. (2012).

TBBPA (CAS No. 79-94-7), TCBPA (tetrachlorobisphenol A, CAS No. 79–95-8), and TBBPS (tetrabromobisphenol S, CAS No. 39635-79-5), which are halogenated derivatives of bisphenol A, are reactive and additive flame retardants used in the production of epoxy and polycarbonate resins, acrylonitrile–butadiene–styrene resins, polystyrene, and printed circuit boards. Globally, over 1220,000 tons of TBBPA are produced annually, and in the EU, around 125,000 tons are consumed yearly. These substances are low in volatility and have high lipophilicity.

BTBPE (CAS No. 37853-59-1) is an additive flame retardant that has been produced on a large scale since the 1970s, particularly as a substitute for octa-BDE.

Table 5.12 Median HBCDD levels in breast milk in ng/g lw

References	N	α-HBCDD	β-HBCDD	γ-HBCDD	Σ-HBCDD (range)	Country, Year
Parizek et al. (2023)	231	0.50	< LOQ	< LOQ	–	Czech Republic, 2019–21
Zhao et al. (2025)	100[a]	4.91	< 0.03	0.05	5.01 (1.60–57.1)	China, 2016–19
Wemken et al. (2020)	16[a]	1.7	0.34	0.21	1.8 (0.83–1.6)	Ireland, 2016–18
Tao et al. (2017)	10	2.1	–	–	6.0 (1.0–22.4)	England, 2014–15
Huang et al. (2020)	38	4.23	< LOQ	0.65	5.67 (< LOQ-36.3)	China, 2014
Shi et al. (2013b)	103	2.27	< LOQ	< LOQ	2.4 (< LOQ-78.3)	China, 2011
Harrad and Abdallah (2015)	10[a]	4.2	0.4	0.8	–	England, 2010–11
Lankova et al. (2013)	50	< 2.0	–	–	–	Czech Republic, 2010
Lignell et al. (2012)	30	–	–	–	0.22 (0.07–1.0)	Sweden, 2010
Croes et al. (2012)	84	3.2	0.05	0.55	3.8	Belgium, 2009–10
Rawn et al. (2024)	298	0.27	< 0.001	< 0.001	0.30 (< 0.01–7.66)	Canada, 2008–11
Fujii et al. (2018)	64	1.9	0.15	0.08	2.2 (< LOQ-14)	Japan, 2008–10
Abdallah and Harrad (2011)	35	3.2	0.30	0.56	3.8 (1.0–22.3)	England
Devanathan et al. (2012)	7[d]	0.06	< 0.05	< 0.05	0.06 (< 0.05–0.95)	India, 2009
Toms et al. (2012)	[a]	2.1	< LOQ	0.4	2.5 (< LOQ-19)	Australia, 2009
Thomsen et al. (2010)	310	–	–	–	0.86 (< 0.2–31)	Norway, 2003–09
Shi et al. (2009)	24[a]	0.86/1.21[e]	< LOQ	< LOQ	–	China, 2007
Hammel et al. (2024)	40	0.24	< LOQ	< LOQ	–	Denmark, 2007
Pratt et al. (2013)	11[a]	2.6[b]	0.5[b]	0.5[b]	3.5[b] (1.7–5.9)	Ireland, 2007

(continued)

Table 5.12 (continued)

References	N	α-HBCDD	β-HBCDD	γ-HBCDD	Σ-HBCDD (range)	Country, Year
Eljarrat et al. (2009)	33	–	< LOQ	–	27 (3–188)	Spain, 2006–07
Roosens et al. (2010)	22[a]	–	–	–	0.60 (0.6–5.7)	Belgium, 2006
Ryan and Rawn (2014)	34	0.5	< LOQ	< LOQ	0.7 (< 0.1–28)	Canada, 2005
Antignac et al. (2008)	23	–	–	–	– (< LOQ-5.0)	France, 2005
Carignan et al. (2012)	43	0.71[c]	0.08[c]	0.20[c]	1.02[c] (3.36–8.1)	USA, 2004/05
Darnerud et al. (2011)	14	–	–	–	0.34 (< 0.23–1.4)	South Africa, 2004
Ryan and Rawn (2014)	7	0.4	–	0.5	0.4 (0.2–1.2)	USA, 2004
Glynn et al. (2011)	92	–	–	–	0.30 (< 0.2–4.4)	Sweden, 2000–04
Polder et al. (2008)	23	–	–	–	0.62 (0.24–1.7)	Russia, 2002

[a]Pooled samples; [b]Arithmetic mean; [c]Geometric mean; [d]Urban site of Kolkata; [e]Mean of urban and rural sites; *LOQ* Limit of quantification

It is used in polymers and foams, and probably also in toys. In 1998, 450-4500 tons were produced in the USA.

EH-TBB (CAS No. 183658-27-7) is an additive flame retardant used in the production of polyurethane foams. It was originally used as a replacement for penta-BDE. It is frequently combined with BEH-TBB in Firemaster 550 and Firemaster BZ-54 products.

BEH-TEBP (CAS No. 26040-51-7), a brominated phthalate, is an additive flame retardant specifically designed to replace penta-BDE in various applications, including polyurethane foam, PVC, neoprene rubber, and cable sheathing. It is incorporated into the products Firemaster 550 and Firemaster BZ-54 at rates of 15% and 30%, respectively, and, along with EH-TBB, contributes to the effects. Annual production in the USA ranged from 450 to 4500 tons between 1990 and 2006.

DBDPE (CAS No. 84852-53-9) is not produced in the EU, but approximately 2000 tons per year are consumed in the EU, showing an upward trend. It is a flame-retardant additive used in various polymers and textiles, with a range of applications comparable to deca-BDE.

DBE-DBCH (TBECH) (CAS No. 3322-93-8) is an additive flame-retardant additive primarily used in OSB boards, electrical cables, and both rigid and flexible insulation foams, as well as in furnishings and extruded polystyrene and polyurethane.

Table 5.13 Abbreviations and chemical structures of novel brominated flame retardants

CAS-Nr	Substanzname	Abkürzung	Formel	Struktur
79-94-7	Tetrabromobisphenol A	TBBPA	$C_{15}H_{12}Br_4O_2$	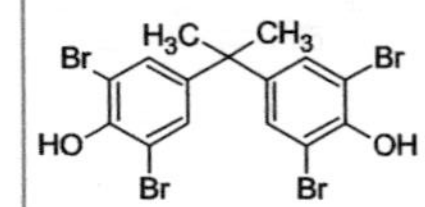
37853-59-1	1,2-Bis(2,4,6-tribromophenoxy)ethane	BTBPE	$C_{14}H_8Br_6O_2$	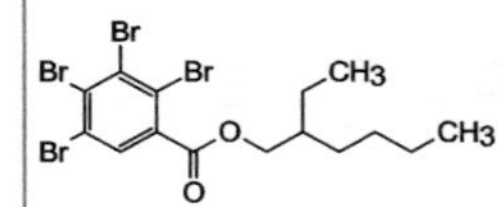
183658–27-7	2-Ethyl-1-hexyl-2,3,4,5-tetrabromobenzoate	EH-TBB	$C_{15}H_{18}Br_4O_2$	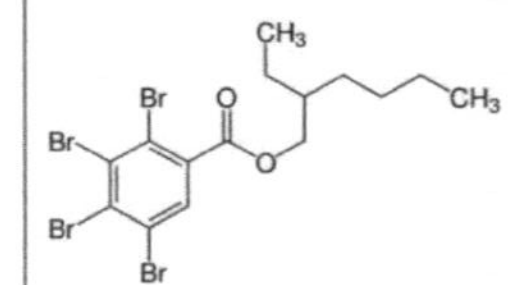
26040-51-7	Bis(2-ethyl-1-hexyl)-tetrabromophthalate	BEH-TEBP (TBPH)	$C_{24}H_{34}Br_4O_4$	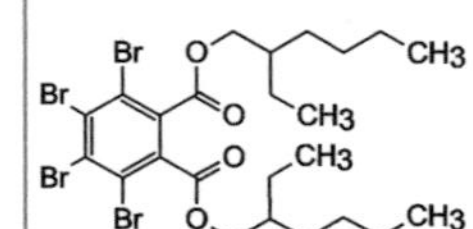

(continued)

Table 5.13 (continued)

CAS-Nr	Substanzname	Abkürzung	Formel	Struktur
84852-53-9	Decabromodiphenylethane [1,2-Bis(pentabromo-phenyl)-ethan]	DBDPE	$C_{14}H_4Br_{10}$	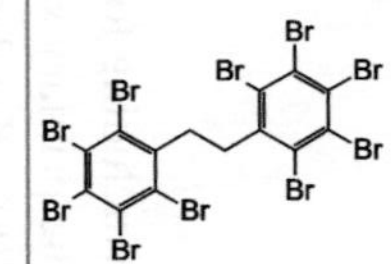
3322-93-8	1,2-Dibromo-4-(1,2-dibromoethyl)cyclohexane	DBE-DBCH (TBECH)	$C_8H_{12}Br_4$	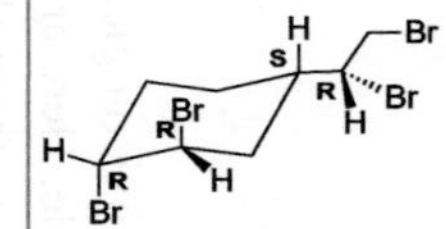

It consists mainly of α- and β-diastereoisomers, although other isomers may also be present.

5.4.1 Health Aspects

Numerous scientific publications and reviews on toxicological (including toxicokinetic) and epidemiological knowledge are available. Here are the primary sources: TBBPA (Geyer et al. 2004; Schauer et al. 2006; Kuester et al. 2007; Knudsen et al. 2014; Colnot et al. 2014; Sunday et al. 2022; EFSA 2024b), BTBPE (Nomeir et al. 1993; Hakk et al. 2004), EH-TBB (US EPA 2015; Knudsen et al. 2016; Hays and Kirman 2017), BEH-TEBP (US EPA 2015; Frederiksen et al. 2016; Knudsen et al. 2016, 2017b; Silva et al. 2016), DBDPE (Wang et al. 2010; ECHA 2011–2017; US EPA 2014; Knudsen et al. 2017a, b), and DBE-DBCH (Nguyen et al. 2017), along with various flame retardants (EFSA 2012a,b; Xiong et al. 2019; Fromme and Roscher 2020).

This is only briefly summarized here. For TBBPA, the mechanisms behind the various toxic effects, which often vary between studies, have yet to be adequately clarified. Nevertheless, effects on metabolic and hormonal processes, such as a decrease in thyroid hormones, seem plausible. Data on TCBPA are limited, but this substance acts as an estrogen receptor alpha agonist and an antagonist for androgen and progesterone receptors. Following repeated oral and inhalation administration or dermal application, BTBPE showed no evidence of significant toxic effects, except for potential methodologically induced inflammatory reactions in the lungs of the animals. Very little data is available on the toxicity of BEH-TEBP. In a long-term study on rats, oral administration of the mixture Firemaster® BZ 54 resulted in effects on the kidneys, particularly in females. DBDPE showed only slight systemic toxicity after repeated oral administration, along with increased absolute and relative liver weights at the highest dose level. Otherwise, no clinical signs or changes in liver and kidney function and no effects on reproduction and development were observed. For DBE-DBCH, results on chronic toxicity in animal experiments are still lacking; however, in vitro studies indicate an androgenic effect. Overall, the toxicological data for many substances are minimal, except for TBBPA. Acute toxicity is estimated to be low for all substances in this class.

5.4.2 Toxicokinetics

TBBPA: TBBPA is rapidly and almost completely absorbed by humans and animals after oral administration, and to a lesser extent after dermal application. Rapid metabolization occurs mainly in the tissues of the digestive tract and the liver, resulting in only small amounts of the parent substance reaching other organs and tissues. Regardless of the route of absorption, excretion primarily occurs via bile and feces, with only a small fraction excreted through urine. Terminal half-lives of less than 10 h have been observed in rats, and approximately 2–4 days in humans.

Information on TCBPA metabolism remains limited. Plattard et al. (2022) compared the metabolic kinetics in the liver of rats and humans using hepatocyte suspensions. They reported that TCBPA was quickly metabolized in human hepatocytes, resulting in the formation of glucuronidated and sulfated metabolites. Additionally, Hanioka et al. (2024) noted that the glucuronidation abilities toward TBBPA and TCBPA in the liver vary significantly across species. For example, Ho et al. (2017) analyzed 140 urine samples and observed geometric mean levels of 1.69 μg/l for TBBPA, 1.74 μg/l for TBBPA di-glucuronide, and 0.63 μg/l and 0.89 μg/l for TBBPA mono- and di-sulfate.

BTBPE: After oral administration, BTBPE is absorbed only minimally and is excreted almost entirely unchanged in the feces. Following oxidative debromination and ether cleavage, the remainder is excreted at low concentrations within three to four days, primarily in the feces and partially in the urine. The average half-life in the blood is reported to be 36 h.

EH-TBB: Limited information is available on absorption and distribution. Absorption through the skin is estimated to be low. It is a lipophilic substance that is rapidly metabolized by esterases in the intestinal wall and liver, resulting in the cleavage of the side chain to tetrabromobenzoic acid (TBBA), which is then excreted in the urine. In animal studies, the parent substance was not detected in the urine, but two sulfate and glycine conjugates of TBBA were found. Since TBBA does not occur in the environment and is not formed through the degradation of other substances, it can be considered a specific biomarker of EH-TBB exposure.

BEH-TEBP: In animal experiments, BEH-TEBP is not absorbed dermally and is absorbed by approximately 2–7% after oral administration. Accumulation in the liver and adrenal glands, along with transfer into breast milk, but not into the fetus, was observed. In vitro metabolism to monoethylhexyl tetrabromophthalate (TBMEHP) was noted. In animal studies, up to 98% of the administered dose was excreted unchanged in the feces within 72 h following a single dose.

DBDPE: Absorption following oral administration is low for both single and repeated doses, with only a small percentage of the doses absorbed. It is not detected in the body after topical application. Following oral administration, only a slight proportion of the administered dose was excreted in the urine, while the majority was eliminated unchanged in the feces.

DBE-DBCH: The technical DBE-DBCH consists of a mixture of α- and β-diastereoisomers. In vitro hydroxylation and debromination catalyzed by cytochrome P450 have been reported, leading to various metabolites. Studies indicate the potential for binding and activation of the androgen receptor, particularly for the γ- and δ-diastereoisomers.

5.4.3 *Occurrence in the Blood and Urine of Adults*

For the first time, TBBPA concentrations ranging from < 0.1 to 2.0 ng/g lw were measured in annual serum samples collected from 1982 to 2003 (Thomsen et al. 2007). However, TBBPA could no longer be detected in the last three years of the study. In France, from 2004 to 2006, a median concentration of 16.1 ng/g lw (0.23–93 ng/g lw) was found in 91 serum samples (Cariou et al. 2008). Tay et al. (2019) reported a median of < 0.28 ng/g lw (< 0.28–74 ng/g lw) in 61 Norwegian serum samples in 2013, where TBBPA was detected in only 36% of the samples. A comparison of 21 individual samples from Belgium in 2007 and 14 pooled samples from 1999 showed nearly identical medians of 0.08 and 0.09 μg/l for TBBPA (Dirtu et al. 2008). In contrast, Shi et al. (2013) reported a maximum concentration of only 6.58 ng/g lw in 42 pooled serum samples from China, with TBBPA detected in 36% of the samples above the LOQ. Moreover, Ho et al. (2020) observed geometric mean TBBPA levels of 3.65 μg/l (maximum 7.49 μg/l) in 140 plasma samples collected in China in 2012. Additionally, Li et al. (2020) analyzed 181 serum samples in China in 2015 and reported maximum values (detection rates) of 5.83 μg/l (30%) for TBBPA, 0.14 μg/l (29%) for TCBPA, and 10.8 μg/l (90%). Wang et al. (2021b) studied TCBPA in 150 serum samples in China and detected a mean level of 0.018 μg/l and a maximum level of 0.145 μg/l in 18.6% of the samples. Xu et al. (2022) collected 300 urine samples from three cities in southern China and reported detection rates ranging from 0.9% to 44%, with geometric means of TBBPA at 0.13 μg/l, 0.14 μg/l, and 0.71 μg/l. The maximum values varied from 0.14 μg/l to 33.7 μg/l, indicating significant differences in local usage of this substance. In Japan, a median TBBPA concentration of 0.001 ng/g ww (< 0.2–238 ng/g ww) was found in 3 out of 10 serum samples in 2006 (Fujii et al. 2014b). Conversely, high concentrations were observed in Korea, with a mean of 10.7 ng/g lw (< 0.07–74 ng/g lw) for TBBPA in serum from 12 women (Kim and Oh 2014).

Xu et al. (2022) collected 300 urine samples from three cities in southern China and analyzed TBBPS and TCBPA with the geometric means (and maximum values) from 0.07 to 0.25 μg/l (7.11–39.5 μg/l) and from 0.01 to 0.06 μg/l (0.01–0.76 μg/l), respectively. Jin et al. (2024) analyzed urine samples from 18 workers without occupational exposure and could not detect TCBPA above the LOQ of 0.3 μg/kg. In the US, Estill et al. (2024) analyzed urine samples from workers between 2015 and 2017, finding TBBPA in 10% of 103 preshift samples, with a geometric mean of 0.05 μg/l.

In a previous Swedish study, BTBPE could not be quantified above the LOQ (Karlsson et al. 2007). In 2012, Cequier et al. (2015) found BTBPE in 9% of 46 Norwegian samples, with a maximum value of 0.65 ng/g lw. In Canada, from 2008 to 2009, BTBPE was detected in only 4% of 102 serum samples, with a maximum value of 16 ng/g lw (Zhou et al. 2014a). In 9 of 12 Chinese serum samples collected in 2016, concentrations ranged between < 0.004 and 2.56 ng/g ww (Gao et al. 2016). Even in 30 serum samples from Peking, medians of only 1.1 ng/l (0.11–18.5 ng/l) were determined (Wang et al. 2018). Similarly, in 239 serum samples from China collected in 2020–2021, BTBPE was identified in 41% of the samples, with a geometric mean of 0.5 ng/g lw (Liu et al. 2022). In another Chinese study involving 22 serum samples collected in 2020–2021, BTBPE was detected in 6% of the samples, with a maximum value of 0.8 ng/g lw (Guan et al. 2023). In contrast, in a region with high production of brominated flame retardants in China, BTBPE was reported to have a median of 2.43 ng/g lw (< LOQ-126 ng/g lw) (Zhao et al. 2021).

In 2008–2009, EH-TBB was found in 57% of 102 Canadian serum samples, with a maximum value of 68 ng/g lw (median: 1.6 ng/g lw) (Zhou et al. 2014a). In a Chinese study involving 22 serum samples collected in 2020–2021, EH-TBB was detected in 27% of the samples, with a maximum of only 1,8 ng/g lw (Guan et al. 2023). Again, in a separate analysis of 239 serum samples from China collected in 2020–2021, EH-RBB was found in 21% of the samples, with a geometric mean of 0.1 ng/g lw (Liu et al. 2022). In the USA in 2011, the specific metabolite TBBA (tetrabromobenzoic acid) was found in 78% of urine samples from 52 adults, with a median of 5.3 ng/l (maximum: 341 ng/l) (Hoffman et al. 2014). The same research group examined the urine of 22 women in 2013–2014 and measured TBBA in a range of < 3.0 - 62.2 ng/l in 27% of the samples (Butt et al. 2014).

In Canada, BEH-TEBP (TBPH) was detected above the LOD of 7.3 ng/g lw in only 17% of 102 blood samples collected between 2008 and 2009, with a 95th percentile of 33 ng/g lw and a maximum of 164 ng/g lw (Zhou et al. 2014a). In contrast, a Chinese study involving 22 serum samples collected in 2020–2021 found BEH-TEBP present in 9% of the samples, with a lower maximum of just 1,7 ng/g lw (Guan et al. 2023). Additionally, Ma et al. (2017) reported no detection of BEH-TEBP in any of 20 pooled serum samples from 452 Chinese subjects in 2014/15.

Cequier et al. (2015) did not find DBDPE above the LOQ in any of the 46 Norwegian samples from 2012. Tay et al. (2019) detected DBDPE in 61 Norwegian blood samples from 2013, with a median of < 1.4 ng/g lw (<1.4–27 ng/g lw), and the substance was detected in only 2% of the samples.

In contrast, DBDPE was detected in all 43 Chinese samples in 2014, with a high median of 39.2 ng/g lw (Qiao et al. 2018). In another study from China, a median of 10.9 ng/g lw (< LOQ-33.2 ng/g lw) was reported in 10 serum samples from individuals in one city (Liang et al. 2016). Additionally, another Chinese study conducted with 22 serum samples collected in 2020–2021 found DBDPE in 68% of the samples, with a median of 53 ng/g lw (< LOQ-215 ng/g lw) (Guan et al. 2023). In a region with a high production of brominated flame retardants in China, DBDPE was observed, showing a median of 32.5 ng/g lw and a range from < LOQ to 1600 ng/g lw (Zhao et al. 2021). Furthermore, in 239 serum samples from China collected in 2020–2021, DBDPE was found in 71% of the samples, with a median of 45 ng/g lw (Liu et al. 2022).

In Canada, however, DBDPE was detected above the LOQ of 3.5 ng/g lw in only 6% of 102 blood samples from 2008 to 2009 (Zhou et al. 2014a).

5.4.4 Occurrence in the Blood and Urine of Children and Cord Blood

Studies on children's blood have been limited to date. In France, a median of 55 ng/g lw (2.1–649 ng/g lw) was found for TBBPA in 90 umbilical cord serum samples collected between 2004 and 2006 (Cariou et al. 2008). In the maternal serum samples measured in parallel, the median was only 30% of the median found in the umbilical cord serum. In contrast, unusually high concentrations were reported in Korea, with a mean of 72.5 ng/g lw (< 0.07–457 ng/g lw) in serum from 12 children (Kim and Oh 2014). TBBPA concentrations were 2–5 times higher in the infants than in their mothers' serum.

BTBPE was analyzed in 174 serum samples from an average of 10-year-old subjects living near a petrochemical complex in South China in 2015 (Guo et al. 2018). BTBPE was detected in 70% of the samples, with a median (range) of 0.83 ng/g lw (< LOQ-14 ng/g lw).

EH-TBB was analyzed in the urine of 23 children in the USA from 2013 to 2014, with the metabolite TBBA detected in 70% of the samples, ranging from < 3.0 to 84.9 ng/l (geometric mean: 7.4 ng/l), which was higher than in the urine of their mothers (Butt et al. 2014). Furthermore, Hoffman et al. (2024) identified TBBA in 43% of 125 urine samples from children aged 3 to 6 years living in the USA, reporting a 95th percentile of 118 ng/l. Although the detection frequency was low (7.1 ng/l), the maximum value recorded was approximately 23 times higher, indicating that exposure to EH-TBB varies significantly among the population.

In Europe, BEH-TEBP (TBPH) was not found above the detection limit of 0.005 ng/g ww in unspecified pooled umbilical cord blood samples (Sales et al. 2017). In contrast, a 2015 Chinese study conducted near a petrochemical complex analyzed BEH-TEBP in 174 samples from subjects with a median age of 10 years (Guo et al. 2018). In 83% of the samples, BEH-TEBP was found, with a median level of 6.6 ng/g lw, ranging from < LOQ to 78 ng/g lw.

DBDPE was analyzed in 174 serum samples from subjects averaging 10 years old living near a petrochemical complex in South China in 2015 (Guo et al. 2018). This study found high concentrations of DBDPE, with a median of 180 ng/g lw (< LOQ-690 ng/g lw) in 99% of the samples.

In contrast, a 2015 Chinese study near a petrochemical complex analyzed DBE-DBCH in 174 samples from subjects with a median age of 10 years (Guo et al. 2018). DBE-DBCH was detected in 100% of the samples, with a median (range) of 30 ng/g lw (1.0–110 ng/g lw).

5.4.5 Occurrence in Breast Milk

The most recent results of TBBPA levels in breast milk samples are shown in Table 5.14 and indicate generally low contamination levels. In European studies, median concentrations ranged from < 0.15 ng/g lw in the Czech Republic from 2019 to 2021 to 0.48 ng/g lw in France from 2004 to 2006. Slightly higher levels were observed in some studies from China and Japan, with medians of approximately 1.6 to 3.2 ng/g lw. One explanation could be that TBBPA is a substance with a high production volume, particularly in China. Information regarding TBBPA occurrence remains limited. So far, Rodríguez-Gómez et al. (2014) analyzed 10 breast milk samples and did not detect TBBPA in any of the samples above the LOQ of 0.13 μg/l.

Pratt et al. (2013) and Zhou et al. (2014a, b) could not determine BTBPE above the LOD of approximately 0.30 ng/g lw and 0.86 ng/g lw in 11 pooled samples from Ireland and 105 samples collected in 2008–2009 in Canada. Nevertheless, Rawn

Table 5.14 Median (range) TBBPA levels in breast milk in ng/g lw

References	N	Detection frequency (%)	Concentration	Country, Year
Parizek et al. (2023)	231	0	< 0.15[d]	Czech Republic, 2019–21
Rovira et al. (2022)	40	50	– (< 0.7–13)	Spain, 2016–19
Zhao et al. (2025)	100	97	0.05 (< 0.01–3.15)	China, 2016–19
Zhao et al. (2021)	105	99	0.27 (< 0.02–34.8)	China, 2018
Huang et al. (2020)	111	64	1.57 (<LOD-42)	China, 2014
Nakao et al. (2015)	19	95	1.9[b] (< 0.02–8.7)	Japan, 2012/13
Shi et al. (2013)	103	53	0.1 (<LOD-12.5)	China, 2011
Harrad and Abdallah (2015)	120	61	0.04 (0.03–0.17[b])	UK, 2010/11
Inthavong et al. (2017)	106	42	< 0.05 (< 0.05–15.1)	France, 2010
Abdallah and Harrad (2011)	34	36	< 0.04 (< 0.04–0.65)	UK
Lankova et al. (2013)	50	30	– (<2.0–688)	Czech Republic, 2010
Pratt et al. (2013)	11[a]	18	0.33[b] (< 0.29–0.17)	Ireland
Fujii et al. (2018)	64	97	3.2 (< 0.1–2200)	Japan, 2008–10
Carignan et al. (2012)	43	35	< LOD (< 0.03–0.55)	USA, 2004/05
Cariou et al. (2008)	77	44	0.48 (0.06–93.2)	France, 2004–06
Fujii et al. (2014b)	9	100	0.72 (0.39–2.22)	Japan, 2005/06

[a]Pooled samples; [b]Mean; [c]5th and 95th percentile; [d]Values in μg/l; *LOD* limit of detection

et al. (2024) detected BTBPE in 56% of 541 Canadian samples from 2008–2011 and reported a median of 0.006 ng/g lw (< LOQ-0.58 ng/g lw). In China, BTBPE was detected in 28% of 111 human milk samples in 2014, with a mean value of 0.009 ng/g lw (maximum: 0.09 ng/g lw) reported by Chen et al. (2019a).

In Canada, 105 samples were analyzed from 2008 to 2009, and EH-TBB was detected in 78% of the samples, with a median concentration of 0.41 ng/g lw (95th percentile: 5.3 ng/g lw) (Zhou et al. 2014a). Schreder et al. (2023) reported a median of < 0.30 ng/g lw and a maximum of 278 ng/g lw in an analysis of 50 samples collected in the USA in 2019, with only 6% of the samples exceeding the LOQ. Tao et al. (2017) found that during 2014/15 in the UK, EH-TBB concentrations ranged from < 0.01 to 0.48 ng/g lw, with 90% of the samples above the LOQ. In a study conducted in three European countries, 458 breast milk samples were collected from 2003 to 2014, and EH-TBB was found in 46% to 69% of the samples, with mean values between 0.04 and 0.24 ng/g lw (Čechová et al. 2017a).

Pratt et al. (2013) found that DBDPE was not above the LOD of 2.2 ng/g lw in 11 pooled samples from Ireland. Wemken et al. (2020) also analyzed Irish samples collected from 2016 to 2018 and found that DBDPE was present in 19% of the samples, exceeding the LOQ of 2.5 ng/g lw, with a maximum concentration of 4.6 ng/g lw. From 2008 to 2009, DBDPE was detected in only 9% of 105 samples collected in Canada, with a 95th percentile of 3.3 ng/g lw and a maximum of 25 ng/g lw (Zhou et al. 2014a). Furthermore, Rawn et al. (2024) detected DBDPE in just 9% of 541 Canadian samples from 2008 to 2011, reporting a mean concentration of 0.26 ng/g lw (maximum: 6.8 ng/g lw). In the US, Schreder et al. (2023) found a higher mean of 2.68 ng/g lw and a maximum of 134 ng/g lw in an analysis of 50 samples in 2019, with only 2% of the samples exceeding the LOQ. In China, DBDPE was detected in 88% of the 111 samples in 2014, with a higher median of 5.96 ng/g lw (maximum: 35.3 ng/g lw) (Chen et al. 2019a).

In Canada, 105 samples were analyzed from 2008 to 2009, and 32% of BEH-TEBP was found above the LOQ, with a 95th percentile of 4.0 ng/g lw and a maximum of 6.6 ng/g lw (Zhou et al. 2014a). In the UK, Tao et al. (2017) reported that from 2014 to 2015, BEH-TEBP was within a concentration range of < 0.1–0.73 ng/g lw, with 50% of samples exceeding the LOQ. In a study conducted across three European countries, 458 breast milk samples were collected between 2003 and 2014, revealing that BEH-TEBP was detected in 15–29% of the samples, with mean values ranging from 0.02 to 1.34 ng/g lw (Čechová et al. 2017a).

In 120 Dutch breast milk samples, DBE-DBCH was detected in 26% of the samples, with a median concentration of 0.04 ng/g lw and a 95th percentile of 0.08 ng/g lw (Čechová et al. 2017b). Tao et al. (2017) reported concentrations of α-DBE-DBCH and β-DBE-DBCH in 10 UK samples from 2014 to 2015, ranging from 0.3 to 1.1 ng/g lw and 0.43 to 10 ng/g lw, respectively.

5.5 Organophosphate Flame Retardants

Organophosphate flame retardants (OPFRs) are a large group of aliphatic, chloroaliphatic, and aromatic phosphoric acid triesters.

Since the 1940s, they have been widely used as flame retardants in plastic products, textiles such as upholstered furniture, building materials, hydraulic fluids, lubricants, plasticizers, stabilizers, wetting agents, and defoamers. In terms of flame retardancy, OPFRs account for about 20% of the substances in use. Additionally, they are increasingly being used as a substitute for brominated flame retardants.

The key representatives include tri-n-butyl phosphate (TnBP) and tri-(2-butoxyethyl) phosphate (TBOEP), which are the aliphatic, halogen-free structures; tri-(2-chloroethyl) phosphate (TCEP) and tri-(2-chloroisopropyl) phosphate (TCIPP), which represent the chlorine-containing structures; and triphenyl phosphate (TPhP) and tricresyl phosphate (TCrP), which represent the aromatic OPFRs. Figure 5.7 illustrates the chemical structure of some substances, and Table 5.15 offers an overview of names, abbreviations, and CAS numbers. Moreover, organophosphate diesters are either biotic or abiotic degradation products of triesters or are utilized in industrial processes and can be detected in various environmental matrices (Wang et al. 2020a; Liu et al. 2021).

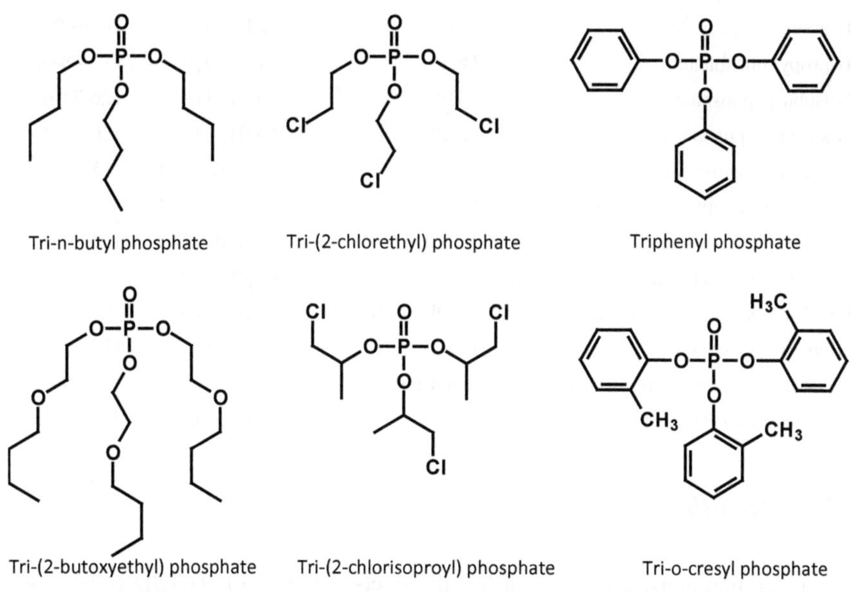

Fig. 5.7 Chemical structure of the most common organophosphate flame retardants

Fig. 5.8 Chemical structure of the stereoisomers *syn*-DDC-CO (top) and *anti*-DDC-CO (bottom) of Dechloran Plus (after ECHA 2020)

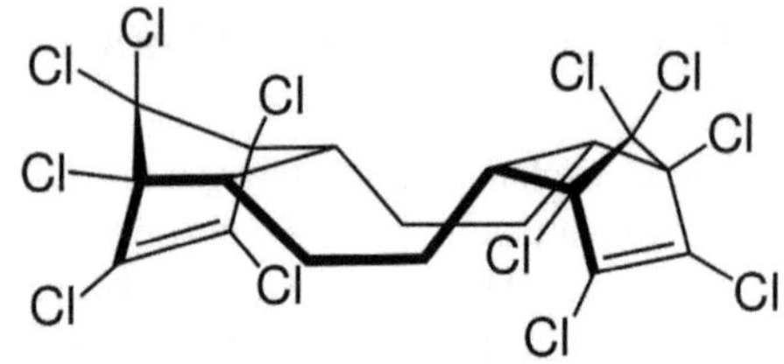

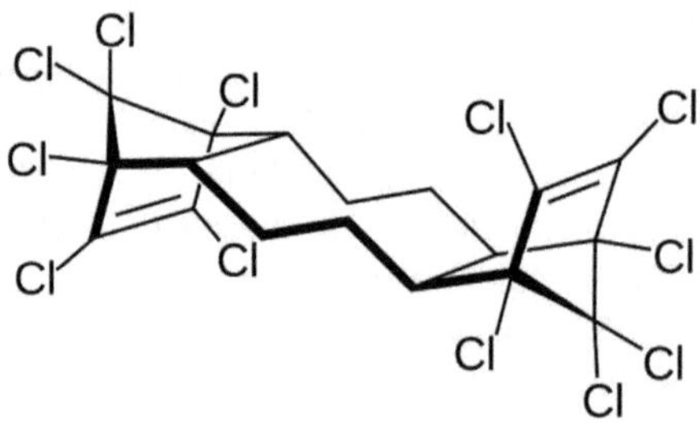

Table 5.15 Name, abbreviation, formula, and CAS number of organophosphate flame retardants

Name	Abbreviation	Formula	CAS No
Trimethyl phosphate	TMP	$C_3H_9O_4P$	512-56-1
Triethyl phosphate	TEP	$C_6H_{15}O_4P$	78-40-0
Tripropyl phosphate	TPrP	$C_9H_{21}O_4P$	513-08-06
Triisobutyl phosphate	TiBP	$C_{12}H_{27}O_4P$	126-71-6
Tris-n-butyl phosphate	TnBP	$C_{12}H_{27}O_4P$	126-73-8
Tris(2-chloroethyl) phosphate	TCEP	$C_6H_{12}Cl_3O_4P$	115-96-8
Tris(2-chloroisopropyl) phosphate	TCIPP	$C_9H_{18}Cl_3O_4P$	13674-84-5
Tris(1,3-dichloroisopropyl) phosphate	TDICPP	$C_9H_{15}Cl_6O_4P$	13674-87-8
Tricresyl phosphate (m-, p-, o-isomers)	TCrP	$C_{21}H_{21}O_4P$	1330-78-5
Tris(2-butoxyethyl) phosphate	TBOEP (TBEP)	$C_{18}H_{39}O_7P$	78-51-3
Triphenyl phosphate	TPhP	$C_{18}H_{15}O_4P$	115-86-6
2-Ethylhexyl diphenyl phosphate	EHDPHP (DPEHP)	$C_{20}H_{27}O_4P$	1241-94-7
Tris(2-ethylhexyl) phosphate	TEHP	$C_{24}H_{51}O_4P$	78-42-2

5.5.1 Health Aspects

Various comprehensive reviews of health effects are available (ATSDR 2012, Van der Veen and de Boer 2012; Wei et al. 2015; Hou et al. 2016; Saillenfait et al. 2018; ECHA 2018e, Chupeau et al. 2020; Yao et al. 2021).

Like other organophosphorus acid esters used as pesticides, chlorinated OPFRs particularly inhibit the activity of plasma cholinesterase and the so-called neurotoxic esterase in the brain at high doses, such as in cases of accidental poisoning. This inhibition increases the likelihood of convulsions, primarily due to excess acetylcholine. Additionally, neuropsychological effects following chronic exposure to organophosphates are currently under discussion. Occupational studies on TCrP and TPhP have revealed only minor health effects. In animal studies, changes were observed in the liver after repeated administration of TCEP and TBOEP, in the kidneys for TCEP and TDCIPP, and in the bladder for TnBP. Low toxicity is assumed for TPhP and TCIPP. While TnBP and TDCIPP do not significantly impact reproduction, TCrP impairs reproductive parameters in rats and mice and causes morphological changes in the testes. Fertility and fetal developmental disturbances have been reported in animal experiments for TCEP. Moreover, the scientific literature has discussed the endocrine-disrupting effects of OPFRs like TPhP and TDCIPP.

5.5.2 *Toxicokinetics*

According to available results, OPFRs are quickly absorbed by experimental animals following oral ingestion and inhalation, whereas skin absorption is restricted.

Information on the metabolic transformation of OPFRs comes from animal studies and in vitro studies. The first step involves rapid hydrolytic cleavage, resulting in the formation of diesters and monoesters, which can be further converted into a variety of hydroxylated metabolites and conjugated with glucuronide and sulfate (Wang et al. 2021a). Generally, alkyl OPFRs such as TnBP and TBOEP tend to be metabolized more quickly than aryl and chlorinated OPFRs. Studies of selected OPFRs indicate that most are primarily excreted in urine, with some differences noted between species and between male and female animals. In rats, half-lives of around 100 h after oral administration are reported in the literature. Wang et al. (2020b) estimate the half-lives for humans to be 3.8 days for TBOEP, 4.8 days for TnBP, 9.6 days for TPhP, 15.2 days for TCIPP, 17.4 days for TCEP, and 53.8 days for TDCIPP (Wang et al. 2020b). In contrast, after administering a single oral dose of TBOEP to six volunteers, 8.3% of the administered dose was excreted as BBOEHEP, with a significantly lower mean half-life of 3.9 h (Völkel et al. 2018).

5.5.3 *Occurrence in Urine*

Due to the complex metabolism of OPFRs in humans, various parameters such as triesters, diesters, and hydroxylated compounds are generally available for biomonitoring. Some triesters, including TCEP, TDCIPP, and TEHP, can be detected in urine at higher percentages. In an analysis of 20 pooled samples from Australia, He et al. (2018) reported that the substances mentioned above are found in 45% of the samples, albeit at low concentrations. The ranges in this study are < 0.03–0.90 μg/l for TCEP, < 0.02–0.07 μg/l for TDCIPP, and < 0.04–0.61 μg/l for TEHP. Given the very low levels of triesters in urine and the risk of contamination, diesters and hydroxylated compounds serve as sensible and frequently used alternatives. Table 5.16 provides an overview of the triesters and their most important metabolites. It is important to note that diesters are not always specific to a single triester. For instance, DPhP is not only a metabolite of TPhP but also of other OPFRs and is commercially used, thus occurring in the environment (Saillenfait et al. 2018).

Tables 5.17 and 5.18 summarize the results of recent studies on the occurrence of diesters and hydroxylated metabolites in the urine of adults (or the general population) and children and adolescents. Overall, the concentrations vary significantly among individuals and between populations. Yang et al. (2022) compared the metabolite composition profiles across different countries. In China, BCEP, BDCIPP, and DPhP represented the highest proportions of total concentrations, whereas in urine samples from the USA, Japan, Australia, and Europe, BDCIPP, BCIPP, and DPhP

Table 5.16 Organophosphate esters and their metabolites frequently measured

Parent substances	Abbreviation	Metabolite	Abbreviation
Tri-n-butyl phosphate	TnBP	Di-n-butyl phosphate	DnBP
Tris(2-chloroethyl) phosphate	TCEP	Di-(2-chloroethyl) phosphate	DCEP
Tris(chloroisopropyl) phosphate	TClPP	Bis(1-chloro-2-propyl) phosphate 1-Hydroxy-2-propyl bis(1-chloro-2-propyl) phosphate	BClPP BCIPHIPP
Tris(1,3-dichloroisopropyl) phosphate	TDClPP	Bis(1,3-dichloro-isopropyl) phosphate	BDCIPP
Tri(2-butoxyethyl) phosphate	TBOEP	Bis(2-butoxyethyl) phosphate 2-Hydroxyethyl bis(2-butoxyethyl) phosphate (3-Hydroxy)-2-butoxyethyl bis(2-butoxyethyl) phosphate	BBOEP BBOEHEP 3-OH-TBOEP
Triphenyl phosphate	TPhP	Diphenyl phosphate 4-Hydroxyphenyl phenyl phosphate 4-Hydroxyphenyl diphenyl phosphate 3-Hydroxyphenyl diphenyl phosphate	DPhP 4-OH-DPhP 4-OH-TPhP 3-OH-TPhP
Diphenyl(2-ethylhexyl) phosphate	EHDPHP	2-Ethylhexyl phenyl phosphate 2-Ethyl-5-hydroxyhexyl diphenyl phosphate	EHPHP 5-HO-EHDPHP
Tri(2-ethylhexyl) phosphate	TEHP	Bis(2-ethylhexyl) phosphate	BEHP
Tricresyl phosphate-Isomers	TCrP	Di-o-cresyl phosphate Di-m-cresyl phosphate Di-p-cresyl phosphate	DoCP DmCP DpCP

were predominant. The authors attribute the high BCEP levels in China to the lack of regulation for TCEP compared to other countries/regions. Significantly fewer results exist for other metabolites. For EHPHP and 5-OH-EHDPHP, the literature presents medians of < LOQ to 4.0 μg/l and 0.004 to 0.23 μg/l, respectively (Bastiaensen et al. 2021; Yang et al. 2022). DoCP and DpCP were identified in 70% of 52 urine samples from Beijing, China, with medians of 0.003 μg/l (< 0.001–0.09 μg/l) and 0.008 μg/l (< 0.001–1.13 μg/l), respectively (Hou et al. 2020). In southern China, a higher median of 0.2 μg/l (< 0.06–0.24 μg/l) was found for the combined sum of DoCP and DpCP in 411 samples (Chen et al. 2018c). The three cresyl isomers in Germany were not detected above the LOQ of 0.1 μg/l in 312 children aged 22 to 80 months (Fromme et al. 2014). In the representative US NHANES from 2013–2014, DoCP and DpCP were measured in only a few samples, with 95th percentiles of < 0.05 μg/l and 0.11 μg/l among 2316 samples (CDC 2023). This study highlighted the highest DpCP levels within the 6–11 age group, with a 95th percentile of 0.22 μg/l. Regarding BEHP, a metabolite of TEHP, recent results show low medians (ranges): In the USA, < 0.02 μg/l (< 0.02–3.48 μg/l) (Hernandez-Castro et al. 2024), and from two Chinese studies, 0.15 μg/l (< 0.07–1.91 μg/l) and 0.09 μg/l (< 0.0005–0.24 μg/l) (Hou et al. 2021; Sun et al. 2018). Conversely, Hou et al. (2020) reported unusually high contamination with a median of 6.7 μg/l (0.16–13.6 μg/l) in their analysis of 52 urine samples from Beijing, China, in 2018.

Hydroxylated metabolites of OPFRs have seldom been measured because their observed concentrations are lower than those of the diesters and often approach or fall below the detection limit of the available analytical methods, with commercial standards being limited (Yang et al. 2022). One exception is 1-hydroxy-2-propyl bis(1-chloro-2-propyl) phosphate (BCIPHIPP), the hydroxylated metabolite of TCIPP. For example, median/geometric mean values of 0.72 μg/l, 1.29 μg/l, and 0.87 μg/l were found in 61 Norwegian urine samples, 203 samples from the USA, and 46 Chinese urine samples, respectively (Xu et al. 2019; Phillips et al. 2018; Li et al. 2020). Furthermore, Hammel et al. (2021) noted higher levels in urine samples collected from 111 children in the US between 2016 and 2018, with a median of 4.12 μg/l.

5.5.4 Occurrence in Breast Milk

Only limited information is available about the occurrence of OPFRs in breast milk. The results to date are summarized in Table 5.19. Overall, the wide variation in the results of individual samples across several orders of magnitude is striking. In the two European studies, high medians of 11 and 45 ng/g lw were found for TnBP, 45 and 13 ng/g lw for TCIPP, and 4.7 and 14.8 ng/g lw for TBOEP. In Japan and Vietnam, the median OPFR levels, except for TPhP, were low, while in the Philippines, high median levels of TCEP at 42 ng/g lw and TPhP at 19 ng/g lw were observed.

To date, three studies from the USA have been published that report volume-based results. Overall, the concentrations are low. While TCIPP was the highest in the survey conducted in Washington State, with a median of 1.47 μg/l (Zheng et al.

Table 5.17 Parent compounds of OPFR and median levels of their metabolites in the urine of adults in µg/l

References	N	TCEP	TPhP	TDCIPP	TCIPP		TnBP	TBOEP		
		DCEP	DPhP	BDCIPP	BCIPHIPP	BCIPP	DnBP	BBOEHEP	BBOEP	Country, year
Tang et al. (2025)	796	–	0.15	–	–	–	<0.09	–	<0.27	China, 2021–24
Hou et al. (2021)	352	< 0.63	0.09	0.12	–	< 0.04	0.05	< 0.02	< 0.03	China, 2018/19
Hou et al. (2020)	52	2.57	0.18	0.29	–	0.15	0.23	–	2.65	China, 2018
Hernandez-Castro et al. (2023)	204	–	0.83	1.26	–	0.04	0.17	–	0.04	USA, 2017–19
CDC (2023)	~ 2800	0.34	0.91	1.24	–	< 0.10	< 0.10	–	–	USA, 2017/18
Bommarito et al. (2023)	54	0.16[a]	0.83[a]	0.74[a]	–	0.21[a]	0.10[a]	–	–	USA, 2017/18
Li et al. (2020)	46	< 0.01*	0.33	0.08	0.68	< 0.40	< 0.05	< 0.003	< 0.03	China, 2017
Sun et al. (2018)	180	< 0.003	0.06	< 0.004	–	< 0.02	0.003	–	0.07	China, 2016/17
Guo et al. (2023)	104	0.67	0.33	0.09	–	< 0.03	0.16	–	< 0.07	China, 2016
Jayatilaka et al. (2017)	76	< 0.08	0.89	0.69	–	< 0.10	< 0.05	–	–	USA, 2015
Hu et al. (2019)	15	0.67	2.47	0.09	0.49	< 0.25	0.16	< 0.06	0.12	China

(continued)

Table 5.17 (continued)

References	N	TCEP	TPhP	TDCIPP	TCIPP		TnBP	TBOEP		
		DCEP	DPhP	BDCIPP	BCIPHIPP	BCIPP	DnBP	BBOEHEP	BBOEP	Country, year
Chen et al. (2019b)	84	1.32	0.26	0.25	–	0.14	0.18	–	0.06	China, 2013–15
Carignan et al. (2018)	276	–	0.57	0.46	–	< 0.30	–	–	–	USA, 2005–15
Cequier et al. (2015)	48	–	0.51	0.12	–	–	< 0.40	–	< 0.60	Norway, 2012

[a]Geometric mean; *Calculated as TCEP

Table 5.18 Parent compounds of OPFR and median levels of their metabolites in the urine of children and adolescents in µg/l

References	Age (years)	N	TCEP	TPhP	TDCIPP	TCIPP		TnBP	TBOEP		
			DCEP	DPhP	BDCIPP	BCIPHIPP	BCIPP	DnBP	BBOEHEP	BBOEP	Country, year
Chen et al. (2023b)	0–18	135	0.68	0.67	0.02	–	0.18	0.02	–	0.38	Taiwan, 2020–22
Den Ouden et al. (2026)	4–12		<0.04	1.84	0.51	0.53	<1.0	0.16	<0.01	<0.05	Belgium, 2014–23
Zeng et al. (2023)	9–12	427	< 0.10*	0.27	0.11	0.20	< 0.40	< 0.20	0.06	0.07	Japan, 2017–20
Bastiaensen et al. (2021)	14–15	582	< 0.04*	1.37	0.34	0.71	< 1.0	< 0.15	0,04	< 0.05	Belgium, 2017/18
CDC (2023)	6–11	330	0.49	–	3.28	–	0.12	0.12	–	–	USA, 2017/18
Hammel et al. (2020)	6 weeks	100	–	0.31	0.26	0.26	1.83	–	–	–	USA, 2016–18
Hammel et al. (2020)	1	63	–	1.88	0.64	0.16	0.28	–	–	–	USA, 2016–18
Ait Bamai et al. (2029)	7	400	< 0.05*	0.64	0.10	0,40	< 1.00	< 0.15	0.20	0.10	Japan, 2017
Zhang et al. (2018)	0–5	227	0.85	0.47	0.08	–	0.69	0.004	–	0.06	China, 2016
Chen et al. (2018c)	6–14	411	1.04	0.21	0.05	–	0.15	0.12	–	0.05	China, 2015
He et al. (2018)	0–5	20 [b]	< 0.01	25 [a]	2.60 [a]	0.43 [a]	0.85 [a]	0.18 [a]	–	0.32 [a]	Australia; 2014/15

(continued)

Table 5.18 (continued)

References	Age (years)	N	TCEP	TPhP	TDCIPP	TCIPP		TnBP	TBOEP		
			DCEP	DPhP	BDCIPP	BCIPHIPP	BCIPP	DnBP	BBOEHEP	BBOEP	Country, year
Bastiaensen et al. (2019)	7–12	128	0.05*	0.32	0.07	0.18	< 1.0	< 0.40	0.36	0.20	Japan, 2009/10
Cequier et al. (2015)	6–12	54	–	1.10	0.23	–	–	< 0.40	–	< 0.60	Norway, 2012
Fromme et al. (2014)	22–80 months	312	0.20	0.80	–	–	< 0.20	0.20	–	–	Germany, 2011/12

[a] Arithmetic mean; [b] Pooled samples; *Calculated as TCEP

Table 5.19 Medians (ranges) of organophosphate flame retardants in breast milk

References	N	TEP	TnBP	TCEP	TCIPP	TDCIPP	TBOEP	TPhP	EHDPHP	TEHP	Country, year
in ng/g lw											
Beser et al. (2019)	20	3.4 (< 0.02–7.8)	10.9 (< 0.03–11)	–	12.5 (< 0.13–13)	–	14.8 (< 0.03–14.8)	9.9 (< 0.12–17.3)	–	–	Spain, 2015
Kim et al. (2014)	20	< 0.02 (< 0.02–15)	0.39 (< 0.01–25)	0.14 (< 0.02–20)	–	< 0.11 (< 0.11–162)	0.24 (< 0.04–127)	1.4 (< 0.02–15)	< 0.03 (< 0.03–0.35)	–	Japan, 2009–11
Kim et al. (2014)	41	< 0.01 (< 0.01–1.5)	1.5 (< 0.01–21))	42 (< 0.01–512)	–	< 0.08)	< 0.03 (< 0.03–206)	19 (1.82–140)	< 0.02– (< 0.02–102)	–	Philippines, 2008
Kim et al. (2014)	26	< 0.01 (< 0.01–0.80)	2.0 (< 0.01–8.4)	< 0.01 (< 0.01–18)	–	(< 0.08)	< 0.03 (< 0.03–1.6)	4.9 (< 0.01–29)	< 0.02 (< 0.02–8.2)	–	Vietnam, 2009
Sundkvist et al. (2010)	6 [a]	–	45 (22–85)	4.9 (2.1–8.2)	45 (22–82)	4.3 (1.6–5.3)	4.7 (< LOQ–63)	8.5 (3.2–11)	6.5 (3.5–13)	–	Sweden, 1997–2006
in μg/l											
Wu et al. (2024)	26	0.10 (0.03–17.9)	0.12 (0.03–0.34)	0.08 (< 0.003–0.22)	0.12 (0.004–19.8)	0.01 (< 0.012–0.71)	0.22 (0.008–2.84)	0.10 (0.004–20.0)	0.54 (0.05–7.13)	–	China
Ma et al. (2019)	100	< 0.15 (< 0.15–2.17)	0.53 (< 0.30–1.82)	< 0.002 (< 0.002–0.80)	< 0.13 (< 0.13–2.51)	–	1.41 (< 0.01–4.0)	n.a. (< 0.03–0.76)	< 0.002 (< 0.002–0.34)	< 0.06 (< 0.06–1.23)	USA, 2009–12
Zheng et al. (2021)	50	< 0.18 (< 0.18–1.11)	0.44 (< 0.14–5.15)	< 0.01 (< 0.01)	1.47 (< 0.46–14.8)	–	0.17 (< 0.13–3.66)	0.05 (< 0.06–0.24)	< 0.001 (< 0.001–0.34)	0.78 (< 0.006–3.35)	USA, 2019
He et al. (2018)	3	–	(0.26–2.1)	(< 0.13–0.47)	(<2.4–14)	(< 0.05–0.14)	(< 0.12–3.0)	(0.34–0.84)	(< 0.84)	(1.2–6.2)	USA, 2015

[a]Pooled samples; *LOQ* Limit of quantification

2021), TBOEP was the highest at 1.4 μg/l in a study of breast milk samples from six states (Ma et al. 2019). Assuming an average fat content of breast milk of 3.4 g/ 100 ml, the medians of the two US studies would be approximately 41 ng/g fat, placing TCIPP in a similar range as in Europe, but higher for TBOEP.

To date, only two studies have investigated OPFR metabolites in breast milk. Chen et al. (2022) could not detect BCEP, BCIPP, BCIPHPP, BDCIPP, and 5-OH-EHDPHP in any of the 33 samples collected from the Netherlands during 2011–2013. The most frequently detected metabolite in this study was BBOEP, found at low levels ranging from < 0.19 to 1.47 μg/l, followed by DPhP and BBOEHEP, which were present in ranges of < 0.17 to 0.09 μg/l and < 0.07 to 0.027 μg/l, respectively. DnBP was the most common metabolite, with a median value of 7.44 μg/l (<3.8–41 μg/l), followed by DPhP at 0.13 μg/l (< 0.10–1.4 μg/l) in 50 breast milk samples from Seattle, USA, in 2019 (Zheng et al. 2021). BBOEP was observed in only 36% of the samples at levels ranging from < 0.01 to 1.31 μg/l.

5.6 Dechloranes

Hexachlorobornene-based flame retardants, commonly known as dechloranes, are polychlorinated compounds. They were initially developed as a substitute for the pesticide Mirex, which was banned in the 1970s. Of particular technical importance is Dechlorane Plus (DDC-CO) (CAS No. 13560-89-9; IUPAC name: 1, 2, 3, 4, 7, 8, 9, 10, 13, 13, 14, 14-dodecachloro-1, 4, 4a, 5, 6, 6a, 7, 10, 10a, 11, 12, 12a-dodecahydro-1, 4, 7,10-dimethanodibenzo[a,e]cyclooctene), a commercially available product that consists of two stereoisomers: syn-DDC-CO (CAS No. 135821-03-3) and anti-DDC-CO (CAS No. 135821-74-8). The technical product comprises approximately 25% syn-DDC-CO and 75% anti-DDC-CO (Sverko et al. 2011). The chemical structures of the two stereoisomers are shown in Fig. 5.9. It was first produced in the 1960s by OxyChem (Niagara Falls, USA) and later in China. It was primarily used as an additive flame retardant in electronic cables and wires, plastic roofing, hard plastic compounds in televisions and computer monitors, wire coatings, and furniture. The annual global production volume is estimated to be around 5000 tons (Wang et al. 2016a). DDC-CO can be released during production, processing, and use, as well as during disposal and recycling. It has a long residence time in the environment and is considered persistent. In the atmosphere, it is predominantly bound to air particles and is only photochemically degraded to a limited extent. It can now be detected ubiquitously in many environmental media and biological matrices. Additionally, the dechloranes DDC-DBF (Dec 602, CAS No. 31107-44-5), DDC-Ant (Dec 603, CAS No. 13560-92-4), and Dec 604 (CAS No. 34571-16-9) are also used technically. However, no data on the production or use of these substances are available.

In 2018, DDC-CO, including its *syn-* and *anti-*isomers, was recognized in the EU as a substance of very high concern due to its persistent and bioaccumulative properties. DDC-CO was also listed in Annex A (parties must take measures to eliminate the production and use) of the Stockholm Convention on Persistent Organic Pollutants, albeit with specific exceptions.

Several scientific reviews concerning general substance information and effects, based on animal and epidemiological studies, have been published (US EPA 2011; ECHA 2018f; ECCC/HC 2019; ECHA 2021, 2022; UNEP 2022). These serve as the basis for the following chapters.

5.6.1 Health Aspects

Data on health effects are currently limited. Acute toxicity from oral, inhalation, and dermal exposure is estimated to be low. With repeated exposure, the liver is the main target organ, particularly impacted by the syn-isomer. DDC-CO is not genotoxic or toxic to reproduction in vitro or in vivo; however, there are no studies available on chronic toxicity or carcinogenicity in experimental animals. Epidemiological studies in environments with high exposures suggest possible associations between DDC-CO serum concentrations and effects on thyroid and sex hormones. Nonetheless, co-exposure to other flame retardants should be considered.

5.6.2 Toxicokinetics

Following oral administration, only 5–20% of DDC-CO was absorbed into the body during animal experiments and was primarily excreted unchanged in the feces. The labeled substance was detected after oral administration, predominantly in the liver and ovaries, at lower concentrations, as well as in the blood, kidneys, and lungs. Less than 0.1% of the administered dose was excreted in the urine. The elimination half-lives for both isomers were 24–25 days in serum, 44–54 days in muscle, and 179 days for syn-DDC-CO in the liver.

5.6.3 Occurrence in the Blood

Table 5.20 summarizes the previous results of biomonitoring dechloranes in the serum/plasma of the general, nonspecifically exposed population.

DDC-CO was first detected in the serum of Chinese individuals in 2005 (Ren et al. 2009). At that time, median values of 5.1 ng/g lw and 8.6 ng/g lw were recorded for the syn- and anti-isomers, respectively. In a region studied simultaneously, about 80 percent of families engaged in recycling activities using primitive methods or without protective measures; the median values were 3.4 times higher for syn-DDC-CO and 2.5 times higher for anti-DDC-CO. More recent studies from China, utilizing samples collected between 2010 and 2016, reported lower median values overall. On

Table 5.20 Median (range) dechloranes in blood samples of the general population in ng/g lw

References	N	DDC-DBF	DDC-Ant	*syn*-DDC-CO	*anti*–DDC-CO	f_{anti} (median/mean)	Country, year
Fromme et al. (2020)	30	< 0.1 (< 0.1–0.4)	< 0.1 (< 0.1)	0.6 (0.4/-3.7)	1.0 (< 0.1–6.6)	0.58/0.55	Germany, 2017
Kim et al. (2019)	170[d]	0.05 (< 0.02–2.34)	< 0.04 (< 0.04–25)	–	–	–	Canada, 2014–16
Kim et al. (2019)	166[e]	0.29 (< 0.02–5.11)	< 0.04 (< 0.04–23)	–	–	–	Canada, 2014–16
Ma et al. (2017)	20[c]	–	–	0.86[b] (0.38–1.4)	2.2[b] (1.4–3.5)	– /0.71	China, 2014/15
Bao et al. (2022a, b)	33	–	–	0.58	3.56	0.86/ –	China, 2012–16
Salah et al. (2026)	102	–	–	0.011	0.028	0.71	Finland, 2012–14
Qiao et al. (2018)	43	–	–	< 0.14 (< 0.14)	< 0.02 (< 0.02–1.2)	–	China, 2014
Fromme et al. (2015b)	42	0.23 (< 0.17–1.45)	0.09 (< 0.04–4.13)	0.77 (< 0.09–29.8)	1.23 (< 0.1–21.2)	0.57/0.63	Germany 2013/14
Tay et al. (2019)	61	–	–	< 0.80 (< 0.80–15)	< 2.1 (<2.1–41)	–	Norway, 2013
Kim et al. (2016)	61	0.67 (< 0.02–3.05)	0.13 (< 0.1–0.52)	0.21 (< 0.02–0.86)	0.52 (0.10–1.31)	0.71/0.73	Korea, 2013
Cequier et al. (2015)	46	0.18 (<LOD-0.75)	0.06 (<LOD-0.38)	0.45 (<LOD-6.7)	0.85 (LOD-25)	0.67/0.67	Norway, 2012
Cequier et al. (2013)	10	0.5 (< 0.34–1.5)	0.21 (< 0.01–0.99	0.23 (< 0.11–6.0)	0.94 (0.52–13)	0.77/0.79	Norway

(continued)

Table 5.20 (continued)

References	N	DDC-DBF	DDC-Ant	*syn*-DDC-CO	*anti*–DDC-CO	f_{anti} (median/mean)	Country, year
He et al. (2013)	10[c]	–	–	2.5[b] (0.75–9.2)	1.0[b] (0.62–2.0)	– /0.35	China 2011
Ben et al. (2014)	24	–	–	0.8 (0.35–3.2)	2.8 (1.3–8.5)	0.78/0.75	China, 2010/11
Brasseur et al. (2014)	48	0.44 (0.15–4.2)	2.0 (< 0.4–12.1)	0.22 (< 0.08–2.3)	0.89 (< 0.16–5.1)	0.89/0.78	France, 2003–5
Zhou et al. (2014b)	102	0.53 (< 0.03–5.7)	0.11 (< 0.02–2.2)	0.49 (< 0.04–15)	1.9 (< 0.12–67)	– /0.81	Canada, 2007–09
Ren et al. (2009)	20[a]	–	–	5.1 (0.35–17.6)	8.6 (0.54–32.9)	0.62/0.63	China 2005

[a]Control group; [b]Mean; [c]Pooled samples; [d]Children; [e]Adults; f_{anti} Defined as the concentration of anti-DDC-CO divided by the concentration of the sum of *syn*- and *anti*-DDC-CO

the contrary, in 60 serum samples collected 2022 from Chinese patients with thyroid diseases, higher levels with a median of 1.3 and 6.9 ng/g lw for *syn*- and *anti*-DDC-CO were reported (Wu et al. 2025).

The serum samples analyzed in Canada from women recruited between 2007 and 2009 showed a DDC-CO exposure level comparable to recent Chinese studies (Zhou et al. 2014b). Additionally, 170 plasma samples from children aged 9 to 14 were collected in Canada during 2019–2022, with dechloranes detected in only 10 to 23% of the samples. The mean levels were 1.04 ng/g lw for syn-DDC-CO, 8.18 ng/g lw for anti-DDC-CO, 0.28 ng/g lw for DDC-DBF, and 0.14 ng/g lw for DDC-Ant (Martinez et al. 2023).

Studies have been conducted across several European countries. In France, for example, 48 samples from individuals living near a municipal waste incineration plant between 2003 and 2005 were analyzed (Brasseur et al. 2014). Furthermore, results were reported from a Norwegian pilot study involving five residents in Oslo and five near Lake Mjøsa, north of Oslo (Cequier et al. 2013). The same research group also studied 46 Norwegian women recruited for a cohort study in 2012 (Cequier et al. 2015) and 61 adults from the Oslo region in 2013 (Tay et al. 2019). Fromme et al. (2015b) also analyzed 42 plasma samples from blood donors in Munich, Germany, during 2013–2014. The median concentrations from the aforementioned European studies were within a comparable range. They did not significantly differ from the exposure situations found in studies conducted simultaneously in China and Canada. In some studies, no significant differences in DDC CO exposure between women and men were observed (He et al. 2013; Brasseur et al. 2014; Fromme et al. 2015b; Kim et al. 2016), while Fromme et al. (2020) reported significantly higher concentrations for men compared to women.

To estimate a time trend, a total of 180 archived plasma samples from the German Environmental Specimen Bank, collected at six points between 1995 and 2017, were examined in Germany (Fromme et al. 2020). The medians of the sum of anti-DDC-CO and syn-DDC-CO remained relatively stable during this period, while the 95th percentiles in 1995, 1999, and 2004 fluctuated from 4.2 ng/g lw to 2.9 ng/g lw and then to 3.3 ng/g lw, followed by increases to 3.7 ng/g lw in 2008, 5.4 ng/g lw in 2013, and 5.9 ng/g lw in 2017. In comparison, the levels of DDC-DBF and DDC-Ant significantly decreased during the study period.

DDC-DBF and DDC-Ant have been measured significantly less in human serum samples. While in most studies, the levels of DDC-DBF exceeded those of DDC-Ant and were overall lower than DDC-CO, Brasseur et al. (2014) found very high median levels of DDC-Ant, well above all other measured dichlorides. In most studies, the distribution pattern of the medians is *anti*-DDC-CO > *syn*-DDC-CO > DDC-DBF > DDC-Ant. Only in the aforementioned French study were the highest medians of all dechloranes found for DDC-Ant. In Canada, a total of 414 plasma samples from 200 families were examined between 2014 and 2016, with the medians of DDC-DBF being highest in the fathers' samples at 0.38 ng/g lw, followed by the mothers' samples at 0.29 ng/g lw and lowest in the children's samples at only 0.05 ng/g lw (Kim et al. 2019). The concentrations correlated significantly between the mothers and fathers, but not between the children and their parents, which indicates a different exposure situation for the children.

The parameter f_{anti} is frequently calculated for DDC-CO, which signifies the prevalence of the most stable isomer in a technical mixture. It is defined as the concentration of the anti-isomer divided by the total concentration of the *syn*- and *anti*-isomers. Reports of f_{anti} values for technical mixtures range between 0.65 and 0.75 (Sverko et al. 2011; Tomy et al. 2007). Many studies show that f_{anti} values in Table 5.20 fall between 0.62 and 0.80. Lower values, such as those in the study by He et al. (2013), indicate a higher external exposure, for example, to workers and residents in areas with high exposure and regions where e-waste is dismantled.

5.6.4 Occurrence in Breast Milk

Only limited test results are available for this medium, indicating a low overall exposure level.

In two Canadian locations, where 84 breast milk samples were collected in 2003–2004 and 2008–2009, median values of 0.20 ng/g lw for syn-DDC-CO and *anti*-DDC-CO were found (< LOQ-3.0 ng/g lw), while 0.43 ng/g lw (< LOQ-7.1 ng/g lw) were reported (Siddique et al. 2012). In another Canadian study, 105 samples from one of the two previously investigated sites were analyzed, revealing lower medians of < 0.01 ng/g lw (< 0.01–3.0 ng/g lw) and 0.02 ng/g lw (< 0.02–12 ng/g lw) for *syn*-DDC-CO and *anti*-DDC-CO (Zhou et al. 2014b). In this study, three other dechloranes were also measured, with DDC-DBF found in 78% of the samples and DDC-Ant in 72% of the samples, presenting medians of 0.23 ng/g lw (< 0.03–19 ng/g lw) and 0.06 ng/g lw (< 0.01–5.2 ng/g lw). Dec 604 was detected in only 4.8% of the samples, with a maximum of 1.9 ng/g lw. Additionally, as part of the pan-Canadian Maternal-Infant Research on Environmental Chemicals (MIREC) study, 541 breast milk samples were analyzed between 2008 and 2011, reporting medians of 0.0042 ng/g lw (< 0.0001–2.38 ng/g lw) for *syn*-DDC-CO and 0.0086 ng/g lw (< 0.0001–0.53 ng/g lw) for *anti*-DDC-CO (Rawn et al. 2023). A positive but weak correlation was observed between the concentrations of ΣDDC-CO in breast milk and the mother's age.

In Europe, Čechová et al. (2017a) analyzed 305 breast milk samples from Norway between 2003–2006, 120 from the Netherlands between 2011–2014, and 37 from Slovakia collected between 2011–2012. The mean values for *syn*-DDC-CO and *anti*-DDC-CO were 0.36 and 0.06 ng/g lw in Norway, 0.28 and 0.16 ng/g lw in the Netherlands, and 0.11 and 0.06 ng/g lw in Slovakia. Significantly lower results were found in 40 Danish samples collected in 2007 (Hammel et al. 2024). Here, medians for *syn*-DDC-CO and *anti*-DDC-CO of < 0.008 ng/g lw (< 0.008–0.44 ng/g lw) and 0.073 ng/g lw (< 0.022–1.85 ng/g lw) were reported, respectively.

In China, during 2011–2012, medians of 0.58 ng/g lw (0.08–20 ng/g lw) were observed for *syn*-DDC-CO and 1.64 ng/g lw (0.31–32 ng/g lw) for *anti*-DDC-CO in 54 breast milk samples from a region known for a high number of plants recycling electrical and electronic products (Pan et al. 2020). Ben et al. (2013) had previously reported comparable medians in Chinese women who lived in a polluted area for a shorter period, while those who resided there for 20–40 years exhibited even higher medians of 1.33 ng/g lw for *syn*-DDC-CO and 3.32 ng/g lw for *anti*-DDC-CO compared to those who lived there for less than 3 years, which had medians of 0.50 ng/g lw and 1.58 ng/g lw, respectively.

References

Abdallah MA, Harrad S (2014) Polybrominated diphenyl ethers in UK human milk: implications for infant exposure and relationship to external exposure. Environ Int 63:130–136

Abdallah MAE, Harrad S (2011) Tetrabromobisphenol-A, hexabromocyclododecane and its degradation products in UK human milk: relationship to external exposure. Environ Int 37:443–448

Abasilim C, Persky V, Sargis RM, Day T, Tsintsifas K, Daviglus M, Cai J, Freels S, Grieco A, Peters BA, Isasi CR, Talavera GA, Thyagarajan B, Davis M, Jones R, Sjodin A, Turyk ME (2025) Associations between persistent organic pollutants and thyroid-related hormones and homeostasis parameters in middle-aged to older men and postmenopausal women: The HCHS/SOL. Environ Res 285(4):122584

Aerts R, Van Overmeire I, Colles A, Andjelković M, Malarvannan G, Poma G, Den Hond E, Van de Mieroop E, Dewolf MC, Charlet F, Van Nieuwenhuyse A, Van Loco J, Covaci A (2019) Determinants of persistent organic pollutant (POP) concentrations in the human breast milk of a cross-sectional sample of primiparous mothers in Belgium. Environ Int 131:104979

Ait Bamai Y, Bastiaensen M, Araki A, Goudarzi H, Konno S, Ito S, Miyashita C, Yao Y, Covaci A, Kishi R (2029) Multiple exposures to organophosphate flame retardants alter urinary oxidative stress biomarkers among children: the Hokkaido Study. Environ Int 131:105003

Anderson HA, Imm P, Knobeloch L, Turyk M, Mathew J, Buelow C, Persky V (2008) Polybrominated diphenyl ethers (PBDE) in serum: findings from a US cohort of consumers of sport-caught fish. Chemosphere 73:187–194

Antignac JP, Cariou R, Maume D, Marchand P, Monteau F, Zalko D, Berrebi A, Cravedi JP, Andre F, Le Bizec B (2008) Exposure assessment of fetus and newborn to brominated flame retardants in France: preliminary data. Mol Nutr Food Res 52:258–265

Antignac JP, Main KM, Virtanen HE, Boquien CY, Marchand P, Venisseau A, Guiffard I, Bichon E, Wohlfahrt-Veje C, Legrand A, Boscher C, Skakkebæk NE, Toppari J, Le Bizec B (2016) Country-specific chemical signatures of persistent organic pollutants (POPs) in breast milk of French, Danish and Finnish women. Environ Pollut 218:728–738

ATSDR (Agency for Toxic Substances and Disease Registry) (2004) Toxicological profile for polybrominated biphenyls and polybrominated diphenyl ethers.

ATSDR (2012) Toxicological profile for phosphate ester flame retardants. https://www.atsdr.cdc.gov/toxprofiles/tp202.pdf

ATSDR (2017) Toxicological profile for polybrominated diphenyl ethers (PBDEs). Online: https://www.atsdr.cdc.gov/ToxProfiles/tp.asp?id=901&tid=183

Bao J, Hou L, Wang Y, Zhang Y, Qiao J, Su M, Jin J, Chen L (2022a) Effect of affinity between dechlorane plus and human serum albumin on its serum concentration in a general population. Environ Sci Technol 56:2511–2518

Bao J, Jin H, Wang Y, Jin J, Chen L (2022b) New emerging polybromobiphenyls in serum of general population and their disruption on thyroid hormone receptor β1. Environ Int 166:107390

Bastiaensen M, Bamai YA, Araki A, Van den Eede N, Kawai T, Tsuboi T, Kishi R, Covaci A (2019) Biomonitoring of organophosphate flame retardants and plasticizers in children: Associations with house dust and housing characteristics in Japan. Environ Res 172:543–551

Bastiaensen M, Gys C, Colles A, Verheyen V, Koppen G, Govarts E, Bruckers L, Morrens B, Loots I, De Decker A, Nelen V, Nawrot T, De Henauw S, Van Larebeke N, Schoeters G, Covaci A (2021) Exposure levels, determinants and risk assessment of organophosphate flame retardants and plasticizers in adolescents (14–15 years) from the Flemish Environment and Health Study. Environ Int 147:106368

Ben Hassine S, Ameur WB, Gandoura N, Driss MR (2012) Determination of chlorinated pesticides, polychlorinated biphenyls, and polybrominated diphenyl ethers in human milk from Bizerte (Tunisia) in 2010. Chemosphere 89:369–377

Ben YJ, Li XH, Yang YL, Li L, Di JP, Wang WY, Zhou RF, Xiao K, Zheng MY, Tian Y, Xu XB (2013) Dechlorane Plus and its dechlorinated analogs from an e-waste recycling center in maternal serum and breast milk of women in Wenling, China. Environ Pollut 173:176–181

Ben Y-J, Li X-H, Yang J-L, Li L, Zheng M-Y, Wang W-Y, Xu X-B (2014) Placental transfer of dechlorane plus in mother—infant pairs in an e waste recycling area (Wenling, China). Environ Sci Technol 48:5187–5193

Bergman Å, Rydén A, Law RJ, de Boer J, Covaci A, Alaee M, Birnbaum L, Petreas M, Rose M, Sakai S, van den Eede N, van der Veen I (2012) A novel abbreviation standard for organobromine, organochlorine and organophosphorus flame retardants and some characteristics of the chemicals. Environ Int 49:57–82

Beser MI, Pardo O, Beltrán J, Yusà V (2019) Determination of 21 perfluoroalkyl substances and organophosphorus compounds in breast milk by liquid chromatography coupled to orbitrap high-resolution mass spectrometry. Anal Chim Acta 1049:123–132

Besis A, Samara C (2012) Polybrominated diphenyl ethers (PBDEs) in the indoor and outdoor environments - a review on occurrence and human exposure. Environ Pollut 169:217–229

Bi X, Qu W, Sheng G, Zhang W, Mai B, Chen D, Yu L, Fu J (2006) Polybrominated diphenyl ethers in South China maternal and fetal blood and breast milk. Environ Pollut 144:1024–1030

Bjermo H, Aune M, Cantillana T, Glynn A, Lind PM, Ridefelt P, Darnerud PO (2017) Serum levels of brominated flame retardants (BFRs: PBDE, HBCDD) and influence of dietary factors in a population-based study on Swedish adults. Chemosphere 167:485–491

Bommarito PA, Friedman A, Welch BM, Cantonwine DE, Ospina M, Calafat AM, Meeker JD, McElrath TF, Ferguson KK (2023) Temporal trends and predictors of gestational exposure to organophosphate ester flame retardants and plasticizers. Environ Int 180:108194

Bradman A, Castorina R, Sjödin A, Fenster L, Jones RS, Harley KG, Chevrier J, Holland NT, Eskenazi B (2012) Factors associated with serum polybrominated diphenyl ether (PBDE) levels among school-age children in the CHAMACOS cohort. Environ Sci Technol 46:7373–7381

Bramwell L, Fernandes A, Rose M, Harrad S, Pless-Mulloli T (2014) PBDEs and PBBs in human serum and breast milk from cohabiting UK couples. Chemosphere 116:67–74

Brasseur C, Pirard C, Scholl G, De Pauw E, Viel J-F, Shen L, Reiner EJ, Focant J-F (2014) Levels of dechloranes and polybrominated diphenyl ethers (PBDEs) in human serum from France. Environ Int 65:33–40

BSEF (Bromine Science and Environmental Forum) (2000) An introduction to bromine. http://www.cefic-efra.com/images/stories/factsheet/IntroBromineFactsheets-1_00.pdf.

Butt CM, Congleton J, Hoffman K, Fang M, Stapleton HM (2014) Metabolites of organophosphate flame retardants and 2-ethylhexyl tetrabromobenzoate in urine from paired mothers and toddlers. Environ Sci Technol 48:10432–10438

Butt CM, Miranda ML, Stapleton HM (2016) Development of an analytical method to quantify PBDEs, OH-BDEs, HBCDs, 2,4,6-TBP, EH-TBB, and BEH-TEBP in human serum. Anal Bioanal Chem 408:2449–2459

Carignan CC, Abdallah MAE, Wu N, Heiger-Bernays W, McClean M, Harrad S, Webster TF (2012) Predictors of tetrabromobisphenol-A (TBBP-A) and hexabromocyclododecanes (HBCDD) in milk from Boston mothers. Environ Sci Technol 46:12146–12153

Carignan CC, Minguez-Alarcon L, Williams PL, Meeker JD, Stapleton HM, Butt CM, Toth TL, Ford JB, Hauser R, Team ES (2018) Maternal urinary concentrations of organophosphate flame retardant metabolites, fertility measures, and pregnancy outcomes among couples undergoing in vitro fertilization. Environ Int 111:232–238

Cariou R, Antignac JP, Zalko D, Berrebi A, Cravedi JP, Maume D, Marchand P, Monteau F, Riu DA, André F, Le Bizec B (2008) Exposure assessment of French women and their newborns to tetrabromobisphenol-A: occurrence measurements in maternal adipose tissue, serum, breast milk and cord serum. Chemosphere 73:1036–1041

Carrizo D, Grimalt JO (2007) Influence of breastfeeding in the accumulation of polybromdiphenyl ethers during the first years of child growth. Environ Sci Techol 41:4907–4912

CDC (2023) National Report on Human Exposure to Environmental Chemicals. Bomonitoring Data Tables for Environmental Chemicals https://www.cdc.gov/exposurereport/data_tables.html

Čechová E, Vojta Š, Kukučka P, Kočan A, Trnovec T, Murínová ĽP, de Cock M, van de Bor M, Askevold J, Eggesbø M, Scheringer M (2017a) Legacy and alternative halogenated flame retardants in human milk in Europe: Implications for children's health. Environ Int 108:137–145

Čechová E, Seifertová M, Kukučka P, Vojta Š, Quaak I, de Cock M, van de Bor M, Kočan A (2017b) An effective clean-up technique for GC/EI-HRMS determination of developmental neurotoxicants in human breast milk. Anal Bioanal Chem 409:1311–1322

Cequier E, Marcé RM, Becher G, Thomsen C (2013) Determination of emerging halogenated flame retardants and polybrominated diphenyl ethers in serum by gas chromatographymass spectrometry. J Chromatogr A 1310:126–132

Cequier E, Marcé RM, Becher G, Thomsen C (2015) Comparing human exposure to emerging and legacy flame retardants from the indoor environment and diet with concentrations measured in serum. Environ Int 74:54–59

Charles D, Berg V, Nøst TH, Wilsgaard T, Bergdahl IA, Huber S, Ayotte P, Averina M, Sandanger T, Rylander C (2023) Polybrominated diphenyl ethers in type 2 diabetes mellitus cases and controls: repeated measurements prior to and after diagnosis. Int J Hyg Environ Health 249:114148

Chang CJ, Terrell ML, Marcus M, Marder ME, Panuwet P, Ryan PB, Pearson M, Barton H, Barr DB (2020) Serum concentrations of polybrominated biphenyls (PBBs), polychlorinated biphenyls (PCBs) and polybrominated diphenyl ethers (PBDEs) in the Michigan PBB Registry 40 years after the PBB contamination incident. Environ Int 137:105526

Chbihi K, Menouni A, Hardy E, Creta M, Grova N, Van Nieuwenhuyse A, Godderis L, El Jaafari S, Duca RC (2024) Exposure of children to brominated flame retardants and heavy metals in Morocco: urine and blood levels in association with global cytosine and adenine methylation. Environ Int 183:108409

Chen A, Park JS, Linderholm L, Rhee A, Petreas M, DeFranco EA, Dietrich KN, Ho SM (2013) Hydroxylated polybrominated diphenyl ethers in paired maternal and cord sera. Environ Sci Technol 47:3902–3908

Chen ZJ, Liu HY, Ho KL, Huang HB, Liu Q, Man YB, Lam MH, Du J, Wong MH, Wang HS (2016) Hydroxylated polybrominated diphenyl ethers (OH-PBDEs) in paired maternal and neonatal samples from South China: placental transfer and potential risks. Environ Res 148:72–78

Chen L, Wang C, Zhang Y, Zhou Y, Shi R, Cui C, Gao Y, Tian Y (2018a) Polybrominated diphenyl ethers in cord blood and perinatal outcomes from Laizhou Wan Birth Cohort, China. Environ Sci Pollut Res Int 25:20802–20808

Chen MW, Castillo BAA, Lin DY, Chao HR, Tayo LL, Gou YY, Chen FA, Huang KL (2018b) Levels of PCDD/Fs, PBDEs, and PBDD/Fs in breast milk from Southern Taiwan. Bull Environ Contam Toxicol 100:369–375

Chen Y, Fang J, Ren L, Fan R, Zhang J, Liu G, Zhou L, Chen D, Yu Y, Lu S (2018c) Urinary metabolites of organophosphate esters in children in South China: Concentrations, profiles and estimated daily intakeEnviron Pollut 235:358–364

Chen T, Huang M, Li J, Li J, Shi Z (2019a) Polybrominated diphenyl ethers and novel brominated flame retardants in human milk from the general population in Beijing, China: Occurrence, temporal trends, nursing infants' exposure and risk assessment. Sci Total Environ 689:278–286

Chen Y, Jiang L, Lu S, Kang L, Luo X, Liu G, Cui X, Yu Y (2019b) Organophosphate ester and phthalate ester metabolites in urine from primiparas in Shenzhen, China: Implications for health risks. Environ Pollut 247:944–952

Chen M, Koekkoek J, Lamoree M (2022) Organophosphate ester metabolites in human breast milk determined by online solid phase extraction coupled to high pressure liquid chromatography tandem mass spectrometry. Environ Int 159:107049

Chen L, Yin Q, Xu L, Hua M, Zhang Z, Xu Y, Xia W, Qian H, Hong J, Jin J (2023a) Serum polybrominated diphenyl ether exposure and influence factors in blood donors of Wuxi adults from 2013 to 2016. Environ Sci Pollut Res Int 30:63932–63940

Chen FS, Chen CC, Tsai CC, Lu JH, You HL, Chen CM, Huang WT, Tsai KF, Cheng FJ, Kung CT, Li SH, Wang CC, Ou YC, Lee WC, Chang YT, Hashim F, Chao HR, Wang LJ (2023b) Urinary levels of organophosphate flame retardants metabolites in a young population from Southern Taiwan and potential health effects. Front Endocrinol (Lausanne) 14:1173449

Choi G, Kim S, Kim S, Kim S, Choi Y, Kim HJ, Lee JJ, Kim SY, Lee S, Moon HB, Choi S, Choi K, Park J (2014) Occurrences of major polybrominated diphenyl ethers (PBDEs) in maternal and fetal cord blood sera in Korea. Sci Total Environ 491–492:219–226

Chupeau Z, Bonvallot N, Mercier F, Le Bot B, Chevrier C, Glorennec P (2020) Organophosphorus flame retardants: a global review of indoor contamination and human exposure in Europe and epidemiological evidence. Int J Environ Res Public Health 17:6713

Coakley JD, Harrad SJ, Goosey E, Ali N, Dirtu AC, Van den Eede N, Covaci A, Douwes J, Mannetje A' (2013) Concentrations of polybrominated diphenyl ethers in matched samples of indoor dust and breast milk in New Zealand. Environ Int 59:255 61.

Colnot T, Kacew S, Dekant W (2014) Mammalian toxicology and human exposures to the flame retardant 2,2′,6,6′-tetrabromo-4,4′-isopropylidenediphenol (TBBPA): implications for risk assessment. Arch Toxicol 88:553–573

Covaci A, Harrad S, Abdallah MAE, Ali N, Law RJ, Herzke D, de Wit CA (2011) Novel brominated flame retardants: a review of their analysis, environmental fate and behaviour. Environ Int 37:532–556

Cowell WJ, Sjödin A, Jones R, Wang Y, Wang S, Herbstman JB (2019) Temporal trends and developmental patterns of plasma polybrominated diphenyl ether concentrations over a 15-year period between 1998 and 2013. J Expo Sci Environ Epidemiol 29:49–60

CPC (Clariant Plastics and Coatings Germany GmbH (2020) Flame retardants online. https://www.flameretardants-online.com/flame-retardants/market

Croes KA, Colles A, Koppen G, Govarts E, Brucker L, Van de Mieroop E, Nelen V, Covaci A, Dirtu AC, Thomsen C, Haug LS, Becher G, Mampaey M, Schoeters G, Van Larebeke N, Baeyens W (2012) Persistent organic pollutants (POPs) in human milk: a biomonitoring study in rural areas of Flanders (Belgium). Chemosphere 89:988–994

Daniels JL, Pan I-J, Jones R, Anderson S, Patterson DG Jr, Needham LL, Sjödin A (2010) Individual characteristics associated with PBDE levels in U.S. human milk samples. Environ Health Perspect 118:155–160

Danish-EPA (Danish Environmental Protection Agency) (2013) Survey oft brominated flame retardants. Report. Copenhagen, Denmark

Darnerud PO, Eriksen GS, Jóhannesson T, Larsen PB, Viluksela M (2001) Polybrominated diphenyl ethers: occurrence, dietary exposure, and toxicology. Environ Health Perspect 109(Suppl. 1):49–68

Darnerud PO, Aune M, Larsson L, Lignell S, Mutshatshi T, Okonkwo J, Botha B, Agyei N (2011) Levels of brominated flame retardants and other pesistent organic pollutants in breast milk samples from Limpopo Province, South Africa. Sci Total Environ 409:4048–4053

Darnerud PO, Lignell S, Aune M, Isaksson M, Cantillana T, Redeby J, Glynn A (2015) Time trends of polybrominated diphenylether (PBDE) congeners in serum of Swedish mothers and comparisons to breast milk data. Environ Res 138:352–360

Darrow LA, Jacobson MH, Preston EV, Lee GE, Panuwet P, Hunter RE Jr, Marder ME, Marcus M, Barr DB (2017) Predictors of serum polybrominated diphenyl ether (PBDE) concentrations among children aged 1–5 years. Environ Sci Technol 51:645–654

den Ouden F, Cseresznye A, Engelen L, Maris E, Bosschaerts S, Gys C, Cleys P, Ait Bamai Y, Pero-Gascon R, Hemeryck LY, De Paepe E, Proost S, Vich Vila A, Raes J, Vanhaecke L, De Boevre M, De Saeger S, Nawrot TS, Poma G, Covaci A (2026) Chemical exposure in childhood: A study on organophosphate flame retardants and plasticizers in a Flemish birth cohort. Environ Res 288(Pt 1): 123275

Dereumeaux C, Saoudi A, Pecheux M, Berat B, de Crouy-Chanel P, Zaros C, Brunel S, Delamaire C, le Tertre A, Lefranc A, Vandentorren S, Guldner L (2016) Biomarkers of exposure to environmental contaminants in French pregnant women from the Elfe cohort in 2011. Environ Int 97:56–67

Devanathan G, Subramanian A, Sudaryanto A, Takahashi S, Isobe T, Tanabe S (2012) Brominated flame retardants and polychlorinated biphenyls in human breast milk from several locations in India: Potential contaminant sources in a municipal dumping site. Environ Int 39:87–95

Ding G, Yu J, Chen L, Wang C, Zhou Y, Hu Y, Shi R, Zhang Y, Cui C, Gao Y, Tian Y, Liu F (2017) Polybrominated diphenyl ethers (PBDEs) and thyroid hormones in cord blood. Environ Pollut 229:489–495

Dirtu AC, Roosens L, Geens T, Gheorghe A, Neels H, Covaci A (2008) Simultaneous determination of bisphenol A, triclosan, and tetrabromobisphenol A in human serum using solid-phase extraction and gas chromatography-electron capture negative-ionization mass spectrometry-Anal Bioanal Chem 391:1175–1181

Drage DS, Mueller JF, Hobson P, Harden FA, Toms LL (2017) Demographic and temporal trends of hexabromocyclododecanes (HBCDDD) in an Australien population. Environ Res 152:192–198

Drage DS, Harden FA, Jeffery T, Mueller JF, Hobson P, Toms LL (2019) Human biomonitoring in Australian children: Brominated flame retardants decrease from 2006 to 2015. Environ Int 122:363–368

Drobná B, Fabišiková A, Čonka K, Gago F, Oravcová P, Wimmerová S, Oktapodas Feiler M, Šovčíková E (2019) PBDE serum concentration and preschool maturity of children from Slovaki. Chemosphere 233:387–395

Dufour P, Pirard C, Charlier C (2017) Determination of phenolic organohalogens in human serum from a Belgian population and assessment of parameters affecting the human contamination. Sci Total Environ 599–600:1856–1866

Dunn RL, Huwe JK, Car GB (2010) Biomonitoring polybrominated diphenyl ethers in human milk as a function of environment, dietary intake, and demographics in New Hampshire. Chemosphere 80:1175–1182

Dvoršćak M, Jagić K, Jakovljević I, Smoljo I, Klinčić D (2024) Polybrominated diphenyl ethers in human milk of Croatian first-time mothers: 2010 Versus 2020. Arch Environ Contam Toxicol Jan 20

ECCC/HC (Environment and Climate Change Canada, Health Canada) (2019) Screening Assessment. Certain Organic Flame Retardants Substance Grouping. 1,4:7,10-Dimethanodibenzo[a,e]cyclooctene,1,2,3,4,7,8,9,10,13,13,14,14-dodecachloro-1,4,4a,5,6,6a,7,10,10a,11,12,12a-dodecahydro-Dechlorane Plus (DP). Online: https://www.canada.ca/en/environment-climate-change/services/evaluating-existing-substances/certain-organic-flame-retardants-grouping-risk-management-approach-dimethanodibenzo-cyclooctene-dodecachloro-dodecahydro-dechlorane-plus-dp.html

ECHA (European Chemicals Agency) (2011–2017) Registered substances. 1,1′-(Ethane-1,2-diyl)bis[pentabromobenzene]. Online: http://echa.europa.eu/de/information-on-chemicals/registered-substances

ECHA (2018) Screening report—an assessment of wether the use of TCEP, TCPP and TDCP in articles should be restricted. Helsinki, Finland. Online: https://echa.europa.eu/documents/10162/13641/screening_report_tcep_tcpp_td-cp_en.pdf/e0960aa7-f703-499c-24ff-fba627060698

ECHA (2021) Annex XV restriction report. Proposal for a restriction. https://echa.europa.eu/documents/10162/e226337f-0f0d-de37-b1bc-0687817f2615

ECHA (2022) Dechlorane Plus. Second draft risk management evaluation. https://echa.europa.eu/documents/10162/62e193d0-c2fa-0869-ee48-63db33e95073

ECHA (2023) Regulatory strategy for flame retardants. https://echa.europa.eu/documents/10162/2082415/flame_retardants_strategy_en.pdf/9dd56b7e-4b62-e31b-712f-16cc51d0e724?t=1679045593845

EFSA (European Food Safety Authority) (2010) Scientific Opinion on Polybrominated Biphenyls (PBBs) in Food. EFSA J 8: 1789. https://www.efsa.europa.eu/en/efsajournal/pub/1789

EFSA (2012a) Scientific opinion on emerging and novel brominated flame retardants (BFRs) in Food. EFSA Panel on Contaminants in the Food Chain (CONTAM). EFSA J 10:2908

EFSA (2012b) Scientific opinion on brominated flame retardants (BFRs) in food: Brominated phenols and their derivatives. EFSA Panel on Contaminants in the Food Chain (CONTAM). EFSA J 10:2634

EFSA (2021) Update of the risk assessment of hexabromocyclododecanes (HBCDDs) in food. EFSA J 19:6421. https://www.efsa.europa.eu/en/efsajournal/pub/6421

EFSA (2024a) Draft Update of the risk assessment of polybrominated diphenyl ethers (PBDEs) in food. file:///C:/Users/herma/Downloads/EFSA%20Journal%20-%202024%20-%20-%20Update%20of%20the%20risk%20assessment%20of%20polybrominated%20diphenyl%20ethers%20PBDEs%20in%20food-1.pdf

EFSA (2024b) Update of the scientific opinion on tetrabromobisphenol A (TBBPA) and its derivatives in food. https://www.efsa.europa.eu/en/efsajournal/pub/8859

Eick SM, Hom Thepaksorn EK, Izano MA, Cushing LJ, Wang Y, Smith SC, Gao S, Park J-S, Padula AM, DeMicco E, Valeri L, Woodruff TJ, Morello-Frosch R (2020) Associations between prenatal maternal exposure to per- and polyfluoroalkyl substances (PFAS) and polybrominated diphenyl ethers (PBDEs) and birth outcomes among pregnant women in San Francisco. Environ Health: A Global Access Science Source 19:100

Eljarrat E, Guerra P, Martinez E, Farré M, Alvarez JG, Lopez-Teijon M, Barcelo D (2009) Hexabromocyclododecane in human breast milk: levels and enantiomeric patterns. Environ Sci Technol 43:1940–1946

Erratico CA, Szeitz A, Bandiera SM (2013) Biotransformation of 2,2′,4,4′-tetrabromodiphenyl ether (BDE-47) by human liver microsomes: identification of cytochrome P450 2B6 as the major enzyme involved. Chem Res Toxicol 26:721–731

Estill CF, Mayer AC, Chen IC, Slone J, LaGuardia MJ, Jayatilaka N, Ospina M, Sjodin A, Calafat AM (2024) Biomarkers of organophosphate and polybrominated diphenyl ether (PBDE) flame retardants of american workers and associations with inhalation and dermal exposures. Environ Sci Technol 58(19):8417–8431

Fängström B, Athanassiadis I, Odsjö T, Norén K, Bergman A (2008) Temporal trends of polybrominated diphenyl ethers and hexabromocyclododecane in milk from Stockholm mothers, 1980–2004. Mol Nutr Food Res 52:187–193

Fängström B, Hovander L, Bignert A, Athanassiadis I, Linderholm L, Grandjean P, Weihe P, Bergman A (2005) Concentrations of polybrominated diphenyl ethers, polychlorinated biphenyls, and polychlorobiphenylols in serum from pregnant Faroese women and their children 7 years later. Environ Sci Technol 39:9457–9463

Fillol C, Oleko A, Saoudi A, Zeghnoun A, Balicco A, Gane J, Rambaud L, Leblanc A, Gaudreau É, Marchand P, Le Bizec B, Bouchart V, Le Gléau F, Durand G, Denys S (2021) Exposure of the French population to bisphenols, phthalates, parabens, glycol ethers, brominated flame retardants, and perfluorinated compounds in 2014–2016: Results from the Esteban study. Environ Int 147:106340

Fisher M, Arbuckle TE, Liang CL, LeBlanc A, Gaudreau E, Foster WG, Haines D, Davis K, Fraser WD (2016) Concentrations of persistent organic pollutants in maternal and cord blood from the maternal-infant research on environmental chemicals (MIREC) cohort study. Environ Health 15:59

Foster WG, Gregorovich S, Morrison KM, Atkinson SA, Kubwabo C, Stewart B, Teo K (2011) Human maternal and umbilical cord blood concentrations of polybrominated diphenyl ethers. Chemosphere 84:1301–1309

Frederiksen M, Vorkamp K, Thomsen M, Knudsen LE (2009) Human internal and external exposure to PBDEs - A review of levels and sources. Int J Hyg Environ Health 212:109–134

Frederiksen M, Thomsen C, Frøshaug M, Vorkamp K, Thomsen M, Becher G, Knudsen LE (2010) Polybrominated diphenyl ethers in paired samples of maternal and umbilical cord blood plasma and associations with house dust in a Danish cohort. Int J Hyg Environ Health 213:233–242

Frederiksen M, Vorkamp K, Jensen NM, Sørensen JA, Knudsen LE, Sørensen LS, Webster TF, Nielsen JB (2016) Dermal uptake and percutaneous penetration of ten flame retardants in a human skin ex vivo model. Chemosphere 162:308–314

Fromme H, Körner W, Shahin N, Wanner A, Albrecht M, Boehmer S, Parlar H, Mayer R, Liebl B, Bolte G (2009) Human exposure to polybrominated diphenyl ethers (PBDE), as evidenced by data from a duplicate diet study, indoor air, house dust, and biomonitoring in Germany. Environ Int 35:1125–1135

Fromme H, Lahrz T, Kraft M, Fembacher L, Mach C, Dietrich S, Burkardt R, Völkel W, Göen T (2014) Organophosphate flame retardants and plasticizers in the air and dust in German daycare centers and human biomonitoring in visiting children (LUPE 3). Environ Int 71:158–163

Fromme H, Albrecht M, Appel M, Hilger B, Völkel W, Liebl B, Roscher E (2015a) PCBs, PCDD/Fs, and PBDEs in blood samples of a rural population in South Germany. Int J Hyg Environ Health 218:41–46

Fromme H, Cequier E, Kim JT, Hanssen L, Hilger B, Thomsen C, Chang YS, Völkel W (2015b) Persistent and emerging pollutants in the blood of German adults: occurrence of dechloranes, polychlorinated naphthalenes, and siloxanes. Environ Int 85:292–298

Fromme H, Becher G, Hilger B, Völkel W (2016a) Brominated flame retardants—exposure and risk assessment for the general population. Int J Hyg Environ Health 219:1–23

Fromme H, Hilger B, Albrecht M, Gries W, Leng G, Völkel W (2016b) Occurrence of chlorinated and brominated dioxins/furans, PCBs, and brominated flame retardants in blood of German adults. Int J Hyg Environ Health 219:380–388

Fromme H, Roscher E (2018) In: Handbook of Environmental Medicine. In: Wichmann H-E, Fromme H (eds) VI-4 Organic compounds—polybrominated diphenyl ether (PBDE). Ecomed Verlag, Landsberg, Germany. [in German]

Fromme H, Roscher E (2020) In: Handbook of Environmental Medicine (Eds. Wichmann H-E, Fromme H): VI-4 Organic compounds – Important brominated flame retardants (except PBDE). Ecomed Verlag, Landsberg, Germany. [in German]

Fromme H, Thomsen C, Aschenbrenner B, Haug LS, Weber T, Kolossa-Gehring M, Völkel W, Schober W (2020) Time trend of exposure to dechloranes: plasma samples of German young adults from the environmental specimen bank collected from 1995 to 2017. Int J Hyg Environ Health 229:113593

Fromme H, Fuchs V, Albrecht M, Aschenbrenner B, Röhl C, Janitzki N, Herber-Jonat S, Wöckner M, Völkel W, Flemmer AW, Schober W (2022) Polychlorinated dioxins and dibenzofurans (PCDD/F), polybrominated dioxins and dibenzofurans (PBDD/F), polychlorinated biphenyls (PCB), polybrominated diphenyl ethers (PBDE), and per- and polyfluoroalkyl substances (PFAS) in German breast milk samples (LUPE 8). Sci Total Environ 825:154066

Fujii Y, Harada KH, Hitomi T, Kobayashi H, Koizumi A, Haraguchi K (2014a) Temporal trend and age-dependent serum concentration of phenolic organohalogen contaminants in Japanese men during 1989–2010. Environ Pollut 185:228–233

Fujii Y, Nishimura E, Kato Y, Harada KH, Koizumi A, Haraguchi K (2014b) Dietary exposure to phenolic and methoxylated organohalogen contaminants in relation to their concentrations in breast milk and serum in Japan. Environ Int 63:19–25

Fujii Y, Kato Y, Masuda N, Harada KH, Koizumi A, Haraguchi K (2018) Contamination trends and factors affecting the transfer of hexabromocyclododecane diastereomers, tetrabromobisphenol A, and 2,4,6-tribromophenol to breast milk in Japan. Environ Pollut 237:936–943

Fürst P (2006) Dioxins, polychlorinated biphenyls and other organohalogen compounds in human milk. Levels, correlations, trends and exposure through breastfeeding. Mol Nutr Food Res 50:922–933

Gao L, Li J, Wu Y, Yu M, Chen T, Shi Z, Zhou X, Sun Z (2016) Determination of novel brominated flame retardants and polybrominated diphenyl ethers in serum using gas chromatography-mass spectrometry with two simplified sample preparation procedures. Anal Bioanal Chem 408:7835–7844

García-Villarino M, Riaño-Galán I, Rodríguez-Dehli AC, Freire C, Vizcaíno E, Grimalt JO, Tardón A, Fernández-Somoano A (2020) Association between pre/perinatal exposure to POPs and children's anogenital distance at age 4 years: a study from the INMA-Asturias cohort. Int J Hyg Environ Health 229:113563

Geyer HJ, Schramm K-W, Darnerud PO, Aune M, Feicht EA, Fried KW, Henkelmann B, Lenoir D, Schmid P, McDonald TA (2004) Terminal elimination half-lives of the brominated flame retardants TBBPA, HBCD, and lower brominated PBDEs in humans. Organohal Comp 66:3820–3825

Glynn A, Lignell S, Darnerud PO, Aune M, Ankarberg EH, Bergdahl IA, Barregård L, Bensryd I (2011) Regional differences in levels of chlorinated and brominated pollutants in mother's milk from primiparous women in Sweden. Environ Int 37:71–79

Gómara B, Herrero L, Ramos JJ, Mateo JR, Fernández MA, García JF, González MJ (2007) Distribution of polybrominated diphenyl ethers in human umbilical cord serum, paternal serum, maternal serum, placentas, and breast milk from Madrid population, Spain. Environ Sci Technol 41:6961–6968

Gomes J, Begum M, Kumarathasan P (2024) Polybrominated diphenyl ether (PBDE) exposure and adverse maternal and infant health outcomes: systematic review. Chemosphere 347:140367

Guo W, Holden A, Smith SC, Gephart R, Petreas M, Park JS (2016) PBDE levels in breast milk are decreasing in California. Chemosphere 150:505–513

Guo LC, Xiao J, Zhang Y, Yu S, Lin H, Su G, Liu T, Li X, Lv S, Rutherford S, Ma W (2018) Association between serum polybrominated diphenyl ethers, new flame retardants and thyroid hormone levels for school students near a petrochemical complex, South China. Chemosphere 202:476–482

Guo Y, Chen M, Liao M, Su S, Sun W, Gan Z (2023) Organophosphorus flame retardants and their metabolites in paired human blood and urine. Ecotoxicol Environ Saf 268:115696

Gyllenhammar I, Aune M, Fridén U, Cantillana T, Bignert A, Lignell S, Glynn A (2021) Are temporal trends of some persistent organochlorine and organobromine compounds in Swedish breast milk slowing down? Environ Res 197:111117

Gyllenhammar I, Näslund M, Cantillana T, Fridén U, Hedvall Kallerman P, Lignell S, Halldin Ankarberg E, Aune M (2023) Levels of PCB, PCDD/F and BFR in pooled serum samples from 4-, 8- and 12-year-old children in Uppsala, sampled 2008–2021 (POPUP), and PCB and PCDD/F in 12-year-old children from Sweden, sampled 2016–2017 (Riksmaten Adolescents). https://www.diva-portal.org/smash/get/diva2:1737451/FULLTEXT01.pdf

Hakk H, Letcher RJ (2003) Metabolism in the toxicokinetics and fate of brominated flame retardants - a review. Environ Int 29:801–828

Hakk H, Larsen G, Bowers J (2004) Metabolism, tissue disposition, and excretion of 1,2-bis(2,4,6-tribromophenoxy)ethane (BTBPE) in male Sprague-Dawley rats. Chemosphere 54:1367–1374

Hammel SC, Zhang S, Lorenzo AM, Eichner B, Stapleton HM, Hoffman K (2020) Young infants' exposure to organophosphate esters: Breast milk as a potential source of exposure. Environ Int 143:106009

Hammel SC, Stapleton HM, Eichner B, Hoffman K (2021) Reconsidering an Appropriate Urinary Biomarker for Flame Retardant Tris(1-chloro-2-propyl) Phosphate (TCIPP) Exposure in Children. Environ Sci Technol Lett 8:80−85

Hammel SC, Vorkamp K, Nielsen JB, Sørensen LS, Knudsen LE, Frederiksen M (2024) Novel and legacy brominated flame retardants in human breast milk and house dust from Denmark. J Environ Expo Assess 3:8

Hanioka N, Isobe T, Saito K, Nagaoka K, Mori Y, Jinno H, Ohkawara S, Tanaka-Kagawa T (2024) Hepatic glucuronidation of tetrabromobisphenol A and tetrachlorobisphenol A: interspecies differences in humans and laboratory animals and responsible UDP-glucuronosyltransferase isoforms in humans. Arch Toxicol 98(3):837–848

Haraguchi K, Ito Y, Takagi M, Fujii Y, Harada KH, Koizumi A (2016) Levels, profiles and dietary sources of hydroxylated PCBs and hydroxylated and methoxylated PBDEs in Japanese women serum samples. Environ Int 97:155–162

Harrad S, Abdallah MA (2015) Concentrations of polybrominated diphenyl ethers, hexabromocyclododecanes and tetrabromobisphenol-A in breast milk from United Kingdom. Women do not decrease over twelve months of lactation. Environ Sci Technol 49:13899–13903

Hartmann C, Kaiser AM, Moche W, Weiss S, Raffesberg W, Scharf S, Graf-Rohrmeister K, Thanhaeuser M, Haiden N, Uhl M (2024) Persistent organic pollutants in Austrian human breast milk collected between 2013 and 2016. J Xenobiot 14(1):247–266

Hays SM, Kirman CR (2017) Risk assessment and biomonitoring equivalent for 2-ethylhexyl-2,3,4,5 tetrabromobenzoate (TBB) and tetrabromobenzoic acid (TBBA). Regul Toxicol Pharmacol 89:186–192

HC (2011) Proposed risk management approach for hexabromocyclododecane (HBCD). Chemical Abstracts Service Registry Number (CAS RN): 3194–55–6. https://www.ec.gc.ca/ese-ees/default.asp?lang=En&n=5F5A32FB-1

He SJ, Li M, Jin J, Wang Y, Bu Y, Xu M, Yang X, Liu A (2013) Concentrations and trends of halogenated flame retardants in the pooled serum of residents of Laizhou, China. Environ Toxicol Chem 32:1242–1247

He C, Toms LML, Thai P, Van den Eede N, Wang XY, Li Y, Baduel C, Harden FA, Heffernan AL, Hobson P, Covaci A, Mueller JF (2018) Urinary metabolites of organophosphate esters: concentrations and age trends in Australian children. Environ Int 111:124–130

Hedvall Kallerman P, Cantillana T, Aune M, Gyllenhammar I (2023) Concentrations of brominated flame retardants (PBDEs, HBCDD, HBB, PBEB, BTBPE, and DBDPE) in blood serum from first-time mothers in Uppsala 1996–2021. Report to the Swedish EPA. https://www.diva-portal.org/smash/get/diva2:1741263/FULLTEXT01.pdf

Herbstman JB, Sjödin A, Apelberg BJ, Witter FR, Patterson DG Jr, Halden RU, Jones RS, Park A, Zhang Y, Heidler J, Needham LL, Goldman LR (2007) Determinants of prenatal exposure to polychlorinated biphenyls (PCBs) and polybrominated diphenyl ethers (PBDEs) in an urban population. Environ Health Perspect 115:1794–1800

Herbstman JB, Sjödin A, Kurzon M, Lederman SA, Jones RS, Rauh V, Needham LL, Tang D, Niedzwiecki M, Wang RY, Perera F (2010) Prenatal exposure to PBDEs and neurodevelopment. Environ Health Perspect 118:712–719

Herceg Romanić S, Milićević T, Jovanović G, Matek Sarić M, Mendaš G, Fingler S, Jakšić G, Popović A, Relić D (2023) Persistent organic pollutants in Croatian breast milk: an overview of pollutant levels and infant health risk assessment from 1976 to the present. Food Chem Toxicol 179:113990

Hernandez-Castro I, Eckel SP, Chen X, Yang T, Vigil MJ, Foley HB, Kannan K, Robinson M, Grubbs B, Lerner D, Lurvey N, Al-Marayati L, Habre R, Dunton GF, Farzan SF, Aung MT, Breton CV, Bastain TM (2024) Prenatal exposures to organophosphate ester metabolites and early motor development in the MADRES cohort. Environ Pollut 342:123131

Hernik A, Góralczyk K, Struciński P, Czaja K, Korcz W, Minorczyk M, Ludwicki JK (2013) Polybrominated diphenyl ethers and polychlorinated biphenyls in cord blood from women in Poland. Chemosphere 93:526–531

Ho KL, Yau MS, Murphy MB, Wan Y, Fong BM, Tam S, Giesy JP, Leung KS, Lam MH (2015) Urinary bromophenol glucuronide and sulfate conjugates: potential human exposure molecular markers for polybrominated diphenyl ethers. Chemosphere 133:6–12

Ho KL, Yuen KK, Yau MS, Murphy MB, Wan Y, Fong BM, Tam S, Giesy JP, Leung KS, Lam MH (2017) Glucuronide and sulfate conjugates of tetrabromobisphenol A (TBBPA): chemical synthesis and correlation between their urinary levels and plasma TBBPA content in voluntary human donors. Environ Int 98:46–53

Hoffman K, Fang M, Horman B, Patisaul HB, Garantziotis S, Birnbaum LS, Stapleton HM (2014) Urinary tetrabromobenzoic acid (TBBA) as a biomarker of exposure to the flame retardant mixture Firemaster® 550. Environ Health Perspect 122:963–969

Hoffman K, Tang X, Cooper EM, Hammel SC, Sjodin A, Phillips AL, Webster TF, Stapleton HM (2024) Children's exposure to brominated flame retardants in the home: the TESIE study. Environ Pollut 352:124110

Hood RB, Terrell ML, Smith AK, Curtis S, Conneely K, Pearson M, Barton H, Barr DB, Marder EM, Marcus M (2023) Elimination of PBB-153; findings from a cohort of Michigan adults. Environ Res 220:115146

Hoopmann M, Albrecht UV, Gierden E, Huppmann R, Suchenwirth R (2012) Time trends and individual characteristics associated with polybrominated diphenyl ethers in breast milk samples 2006–2009 in Lower Saxony, Germany. Int J Hyg Environ Health 215:352–359

Hou R, Xu Y, Wang Z (2016) Review of OPFRs in animals and humans: absorption, bioaccumulation, metabolism, and internal exposure research. Chemosphere 153:78–90

Hou M, Shi Y, Jin Q, Cai Y (2020) Organophosphate esters and their metabolites in paired human whole blood, serum, and urine as biomarkers of exposure. Environ Int 139:105698

Hou M, Fang J, Shi Y, Tang S, Dong H, Liu Y, Deng F, Giesy JP, Godri Pollitt KJ, Cai Y, Shi X (2021) Exposure to organophosphate esters in elderly people: relationships of OPE body burdens with indoor air and dust concentrations and food consumption. Environ Int 157:106803

Hu LQ, Tao Y, Luo D, Feng JW, Wang LM, Yu M, Li YP, Covaci A, Mei SR (2019) Simultaneous biomonitoring of 15 organophosphate flame retardants metabolites in urine samples by solvent induced phase transition extraction coupled with ultra-performance liquid chromatography-tandem mass spectrometry. Chemosphere 233:724–732

Huang M, Li J, Xiao Z, Shi Z (2020) Tetrabromobisphenol A and hexabromocyclododecane isomers in breast milk from the general population in Beijing, China: Contamination levels, temporal trends, nursing infant's daily intake, and risk assessment. Chemosphere 244:125524

Huwe JK, Hakk H, Smith DJ, Diliberto JJ, Richardson V, Stapleton HM, Birnbaum LS (2008) Comparative absorption and bioaccumulation of polychlorinated diphenyl ethers following ingestion via dust and oil in male rats. Environ Sci Technol 42:2694–2700

IARC (International Agency for Research on Cancer) (1999) Re-evaluation of some organic chemicals, hydrazine and hydrogen peroxide. Vol. 71. Online: http://monographs.iarc.fr/ENG/Monographs/vol71/index.php

IARC (2015) IARC Working Group on the Evaluation of Carcinogenic Risks to Humans Polychlorinated biphenyls and polybrominated biphenyls. Volume 107, Lyon, France.

Ingle ME, Mínguez-Alarcón L, Carignan CC, Stapleton HM, Williams PL, Ford JB, Moravek MB, O'Neill MS, Wang L, Hauser R, Meeker JD (2021) Reproductive outcomes associated with flame retardants among couples seeking fertility treatment: a paternal perspective. Environ Res 192:110226

Inoue K, Harada K, Takenaka K, Uehara S, Kono M, Shimizu T, Takasuga T, Senthilkumar K, Yamashita F, Koizumi A (2006) Levels and concentration ratios of polychlorinated biphenyls and polybrominated diphenyl ethers in serum and breast milk in Japanese mothers. Environ Health Perspect 114:1179–1185

Iszatt N, Janssen S, Lenters V, Dahl C, Stigum H, Knight R, Mandal S, Peddada S, González A, Midtvedt T, Eggesbø M (2019) Environmental toxicants in breast milk of Norwegian mothers and gut bacteria composition and metabolites in their infants at 1 month. Microbiome 27:34

Inthavong C, Hommet F, Bordet F, Rigourd V, Guérin T, Dragacci S (2017) Simultaneous liquid chromatography-tandem mass spectrometry analysis of brominated flame retardants (tetrabromobisphenol A and hexabromocyclododecane diastereoisomers) in French breast milk-Chemosphere 186: 762–769.

ITU (International Telecommunication Union) (2024) Global E-waste Monitor 2020. https://www.itu.int/en/ITU-D/Environment/Pages/Spotlight/Global-Ewaste-Monitor-2020.aspx

Jacobson MH, Barr DB, Marcus M, Muir AB, Lyles RH, Howards PP, Pardo L, Darrow LA (2016) Serum polybrominated diphenyl ether concentrations and thyroid function in young children. Environ Res 149:222–230

Jagić K, Dvoršćak M, Tariba Lovaković B, Klinčić D (2024) Polybrominated diphenyl ethers in paired dust-breast milk samples: Levels, predictors of contamination, and health risk assessment for infants and mothers. Environ Toxicol Pharmacol 111:104547

Jakobsson K, Thuresson K, Sjödin A, Hagmar L, Bergman A (2003) A summary of exposures to polybrominated diphenyl ethers (PBDEs) in Swedish workers, and determination of half-lives of PBDEs. Organohalogen Compd 61:17–20

Jakobsson K, Fång J, Athanasiadou M, Rignell-Hydbom A, Bergman A (2012) Polybrominated diphenyl ethers in maternal serum, umbilical cord serum, colostrum and mature breast milk. Insights from a pilot study and the literature. Environ Int 47:121–130

Jayatilaka NK, Restrepo P, Williams L, Ospina M, Valentin-Blasini L, Calafat AM (2017) Quantification of three chlorinated dialkyl phosphates, diphenyl phosphate, 2,3,4,5-tetrabromobenzoic acid, and four other organophosphates in human urine by solid phase extraction-high performance liquid chromatography-tandem mass spectrometry. Anal Bioanal Chem 409:1323–1332

Jin YT, Deng XK, Zhao YY, Li JL, Song Q, Zhang YH, Yang Q, Chen SQ (2020) Concentrations of polybrominated diphenyl ethers in maternal blood, placental size, and risk for fetal growth restriction: a nested case-control study. Biomed Environ Sci: BES 33:821–828

Jin W, Yang S, Yin L, Jia Y, Li X, An K, Li J, Zhang Y (2024) Determination of nine bisphenol analogues in human urine by high-throughput solid-phase extraction and ultra-high performance liquid chromatography-tandem mass spectrometry analysis. J Chromatogr A 1730:465096

Kakimoto K, Akutsu K, Konishi Y, Tanaka Y (2008) Time trend of hexabromocyclododecane in the breast milk of Japanese women. Chemosphere 71:1110–1114

Kakimoto K, Akutsu K, Konishi Y, Toriba A, Hayakawa K (2012) Hexabromocyclododecane levels in foodstuff and human breast milk in Osaka, Japan. In: Interdisciplinary Studies on Environmental Chemistry—Environmental Pollution and Ecotoxicology. (Eds.: Kawaguchi et al. Page 253–259.

Kalantzi OI, Geens T, Covaci A, Siskos PA (2011) Distribution of polybrominated diphenyl ethers (PBDEs) and other persistent organic pollutants in human serum from Greece. Environ Int 37:349–353

Karlsson M, Julander A, van Bavel B, Hardell L (2007) Levels of brominated flame retardants in blood in relation to levels in household air and dust. Environ Int 33:62–69

Kawashiro Y, Fukata H, Omori-Inoue M, Kubonoya K, Jotaki T, Takigami H, Sakai SI, Mori C (2008) Perinatal exposure to brominated flame retardants and polychlorinated biphenyls in Japan. Endocr J 55:1071–1084

Kim TH, Lee YJ, Lee E, Patra N, Lee J, Kwack SJ, Kim KB, Chung KK, Han SY, Han JY, Lee BM, Kim HS (2009) Exposure assessment of polybrominated diphenyl ethers (PBDE) in umbilical cord blood of Korean infants. J Toxicol Environ Health A 72:1318–1326

Kim UJ, Kim MY, Hong YH, Lee DH, Oh JE (2012) Assessment of impact of internal exposure to PBDEs on human thyroid function–comparison between congenital hypothyroidism and normal paired blood. Environ Sci Technol 46:6261–6268

Kim UJ, Oh JE (2014) Tetrabromobisphenol A and hexabromocyclododecane flame retardants in infant-mother paired serum samples, and their relationships with thyroid hormones and environmental factors. Environ Pollut 184:193–200

Kim JW, Isobe T, Muto M, Tue NM, Katsura K, Malarvannan G, Sudaryanto A, Chang KH, Prudente M, Viet PH, Takahashi S, Tanabe S (2014) Organophosphorus flame retardants (PFRs) in human breast milk from several Asian countries. Chemosphere 116:91–97

Kim S, Park J, Kim HJ, Lee JJ, Choi G, Choi S, Kim S, Kim SY, Moon HB, Kim S, Choi K (2015) Association between several persistent organic pollutants and thyroid hormone levels in cord blood serum and bloodspot of the newborn infants of Korea. PLoS ONE 10:e0125213

Kim J, Son MH, Shin ES, Choi SD, Chang YS (2016) Occurrence of Dechlorane compounds and polybrominated diphenyl ethers (PBDEs) in the Korean general population. Environ Pollut 212:330–336

Kim J, Staedelin L, Takser L, Abdelouahab N, Zhu J (2019) Assessment of selected chlorinated and brominated flame retardants in human plasma samples among co-residing family members. Environ Pollut 252(Pt B):1035–1041

Kim JT, Oh D, Choi SD, Chang YS (2021) Factors associated with partitioning behavior of persistent organic pollutants in a feto-maternal system: a multiple linear regression approach. Chemosphere 263:128247

Klinčić D, Dvoršćak M, Jagić K, Mendaš G, Herceg Romanić S (2020) Levels and distribution of polybrominated diphenyl ethers in humans and environmental compartments: a comprehensive review of the last five years of research. Environ Sci Pollut Res Int 27:5744–5758

Knudsen GA, Sanders JM, Sadik AM, Birnbaum LS (2014) Disposition and kinetics of tetrabromobisphenol A in female Wistar Han rats. Toxicol Rep 1:214–223

Knudsen GA, Hughes MF, Sanders JM, Hall SM, Birnbaum LS (2016) Estimation of human percutaneous bioavailability for two novel brominated flame retardants, 2-ethylhexyl 2,3,4,5-tetrabromobenzoate (EH-TBB) and bis(2-ethylhexyl) tetrabromophthalate (BEH-TEBP). Toxicol Appl Pharmacol 311:117–127

Knudsen LE, Hansen PW, Mizrak S, Hansen HK, Mørck TA, Nielsen F, Siersma V, Mathiesen L (2017a) Biomonitoring of Danish school children and mothers including biomarkers of PBDE and glyphosate. Rev Environ Health 32:279–290

Knudsen GA, Sanders JM, Hughes MF, Hull EP, Birnbaum LS (2017b) The biological fate of decabromodiphenyl ethane following oral, dermal or intravenous administration. Xenobiotica 47:894–902

Koh TW, Chen SC-C, Chang-Chien GP, Lin DY, Chen F-U, Chao H-R (2010) Breast-milk levels of polybrominated diphenyl ether flame retardants in relation to women's age and pre-pregnant body mass index. Int J Hyg Environ Health 213:59–65

Kuester RK, Solyom AM, Rodriguez VP, Sipes IG (2007) The effects of dose, route, and repeated dosing on the disposition and kinetics of tetrabromobisphenol A in male F-344 rats. Toxicol Sci 96:237–245

Landrigan PJ, Wilcox KR Jr, Silva J Jr, Humphrey HE, Kauffman C, Heath CW Jr (1979) Cohort study of Michigan residents exposed to polybrominated biphenyls: epidemiologic and immunologic findings. Ann N Y Acad Sci 320:284–294

Lankova D, Lacina O, Pulkrabova J, Hajslova J (2013) The determination of perfluoroalkyl substances, brominated flame retardants and their metabolites in human breast milk and infant formula. Talanta 117:318–325

Lee S, Kim S, Kim E, Lee IS, Choi G, Kim HJ, Park J, Lee JJ, Choi S, Kim SY, Kim S, Kim S, Choi K, Moon HB (2013) Polybrominated diphenyl ethers (PBDEs) in breast milk of Korea in 2011: current contamination, time course variation, influencing factors and health risks. Environ Res 126:76–83

Leijs MM, van Teunenbroek T, Olie K, Koppe JG, ten Tusscher GW, van Aalderen WMC, de Voogt P (2008) Assessment of current serum levels of PCDD/Fs, dl-PCBs and PBDEs in a Dutch cohort with known perinatal PCDD/F exposure. Chemosphere 73:176–181

Lenters V, Thomsen C, Smit LA, Jönsson BA, Pedersen HS, Ludwicki JK, Zviezdai V, Piersma AH, Toft G, Bonde JP, Becher G, Vermeulen R, Heederik D (2013) Serum concentrations of polybrominated diphenyl ethers (PBDEs) and a polybrominated biphenyl (PBB) in men from Greenland, Poland and Ukraine. Environ Int 61:8–16

Lenters V, Iszatt N, Forns J, Čechová E, Kočan A, Legler J, Leonards P, Stigum H, Eggesbø M (2019) Early-life exposure to persistent organic pollutants (OCPs, PBDEs, PCBs, PFASs) and attention-deficit/hyperactivity disorder: A multi-pollutant analysis of a Norwegian birth cohort. Environ Int 125:33–42

Li P, Jin J, Wang Y, Hu J, Xu M, Sun Y, Ma Y (2017) Concentrations of organophosphorus, polybromobenzene, and polybrominated diphenyl ether flame retardants in human serum, and relationships between concentrations and donor ages. Chemosphere 171:654–660

Li MQ, Yao YM, Wang Y, Bastiaensen M, Covaci A, Sun HW (2020) Organophosphate ester flame retardants and plasticizers in a Chinese population: significance of hydroxylated metabolites and implication for human exposure. Environ Pollut 257:113633

Li Z, Zhang J, Miao W, Qi X, Dai Y, Wang Z, Guo J, Chang X, Wu C, Zhou Z (2024) Associations of cord serum polybrominated diphenyl ether (PBDE) mixture with birth outcomes and mediating role of thyroid function: evidence from the Sheyang Mini Birth Cohort Study. Environ Res 251(Pt 2):118605

Liang S, Xu F, Tang W, Zhang Z, Zhang W, Liu L, Wang J, Lin K (2016) Brominated flame retardants in the hair and serum samples from an e-waste recycling area in southeastern China: the possibility of using hair for biomonitoring. Environ Sci Pollut Res Int 23:14889–14897

Lignell S, Glynn A, Darnerud PO, Aune M (2008) Temporal trends of PBDEs in mother's milk from Uppsala county, Sweden. Organohal Comp 70:2123–2126

Lignell S, Aune M, Glynn A, Cantillana T, Fridén U (2012) Levels of persistent halogenated organic pollutants (POP) in mother's milk from first-time mothers in Uppsala, Sweden – results from 2008/2010 and temporal trends 1996–2010. Report for the Swedish EPA.

Lin M, Ma S, Tang J, Yu Y, Li G, Fan R, Zhang G, Mai B, An T (2023) Polybrominated diphenyl ethers and bromophenols in paired serum, hair, and urine samples of e-waste dismantlers: insights into hair as an indicator of endogenous exposure. Sci Total Environ 871:161980

Link B, Gabrio T, Mann V, Schilling B, Maisner V, König M, Flicker-Kleine A, Zöllner I, Fischer G (2012) Polybrominated diphenyl ethers (PBDE) in blood of children in Baden-Württemberg between 2002/03 and 2008/09. Int J Hyg Environ Health 215:224–228

Liu S, Zhao G, Li J, Zhao H, Wang Y, Chen J, Zhao H (2017) Association of polybrominated diphenylethers (PBDEs) and hydroxylated metabolites (OH-PBDEs) serum levels with thyroid function in thyroid cancer patients. Environ Res 159:1–8

Liu Y, Gong S, Ye L, Li J, Liu C, Chen D, Fang M, Letcher RJ, Su G (2021) Organophosphate (OP) diesters and a review of sources, chemical properties, environmental occurrence, adverse effects, and future directions. Environ Int 155:106691

Liu M, Li A, Meng L, Zhang G, Guan X, Zhu J, Li Y, Zhang Q, Jiang G (2022) Exposure to novel brominated flame retardants and organophosphate esters and associations with thyroid cancer risk: a case-control study in Eastern China. Environ Sci Technol 56(24):17825–17835

Long M, Sonne C, Dietz R, Bossi R, Jørgensen N, Olsen TI, Bonefeld-Jørgensen EC (2023) Diet, lifestyle and contaminants in three east Greenland Inuit municipalities. Chemosphere 344:140368

Lopez-Espinosa MJ, Costa O, Vizcaino E, Murcia M, Fernandez-Somoano A, Iñiguez C, Llop S, Grimalt JO, Ballester F, Tardon A (2015) Prenatal exposure to polybrominated flame retardants and fetal growth in the INMA cohort (Spain). Environ Sci Technol 49:10108–10116

Lunder S, Hovander L, Athanassiadis I, Bergman A (2010) Significantly higher polybrominated diphenyl ether levels in young U.S. children than in their mothers. Environ Sci Technol 44:5256–5262

Lyche JL, Rosseland C, Berge G, Polder A (2015) Human health risk associated with brominated flame-retardants (BFRs). Environ Int 74:170–180

Ma Y, Li P, Jin J, Wang Y, Wang Q (2017) Current halogenated flame retardant concentrations in serum from residents of Shandong Province, China, and temporal changes in the concentrations. Environ Res 155:116–122

Ma J, Zhu H, Kannan K (2019) Organophosphorus flame retardants and plasticizers in breast milk from the united states. Environ Sci Technol Lett 6:525–531

Ma J, Li Y, Qian L, Geng S, Yao X, Gao X, Yan Y, Wen J (2023) Serum levels of polychlorinated biphenyls and polybrominated diphenyl ethers in early pregnancy and their associations with gestational diabetes mellitus. Chemosphere 339:139640

Makey CM, McClean MD, Braverman LE, Pearce EN, He XM, Sjödin A, Weinberg JM, Webster TF (2016) Polybrominated diphenyl ether exposure and thyroid function tests in North American adults. Environ Health Perspect 124:420–425

Malmberg T, Athanasiadou M, Marsh G, Brandt I, Bergman A (2005) Identification of hydroxylated polybrominated diphenyl ether metabolites in blood plasma from polybrominated diphenyl ether exposed rats. Environ Sci Technol 39:5342–5348

Malmvärn A, Marsh G, Kautsky L, Athanasiadou M, Bergman A, Asplund L (2005) Hydroxylated and methoxylated brominated diphenyl ethers in the red algae Ceramium tenuicorne and blue mussels from the Baltic Sea. Environ Sci Technol 39:2990–2997

Mancini FR, Cano-Sancho G, Mohamed O, Cervenka I, Omichessan H, Marchand P, Boutron-Ruault MC, Arveux P, Severi G, Antignac JP, Kvaskoff M (2020) Plasma concentration of brominated flame retardants and postmenopausal breast cancer risk: a nested case-control study in the French E3N cohort. Environ Health 19:54

Martinez G, Zhu J, Takser L, Baccarelli AA, Bellenger JP (2023) Complementarity of plasma and stool for the characterization of children's exposure to halogenated flame retardants: Update on analytical methods and application to a Canadian cohort. Chemosphere 344:140222

Martinez G, Zhu J, Takser L, Baccarelli AA, Bellenger JP (2024) Indoor environment, physiological factors, and diet as predictors of halogenated flame retardant levels in stool and plasma of children from a Canadian cohort. Chemosphere 352:141443

Matovu H, Li ZM, Henkelmann B, Bernhöft S, De Angelis M, Schramm KW, Sillanpää M, Kato CD, Ssebugere P (2021) Multiple persistent organic pollutants in mothers' breastmilk: Implications for infant dietary exposure and maternal thyroid hormone homeostasis in Uganda. East Africa. Sci Total Environ 770:145262

Mazdai A, Dodder NG, Abernathy MP, Hites RA, Bigsby RM (2003) Polybrominated diphenyl ethers in maternal and fetal blood samples. Environ Health Perspect 111:1249–1252

McGrath TJ, Ball AS, Clarke BO (2017) Critical review of soil contamination by polybrominated diphenyl ethers (PBDEs) and novel brominated flame retardants (NBFRs); concentrations, sources and congener profiles. Environ Pollut 230:741–757

Meijer L, Weiss J, Van Velzen M, Brouwer A, Bergman A, Sauer PJJ (2008) Serum concentrations of neutral and phenolic organohalogens in pregnant women and some of their infants in the Netherlands. Environ Sci Technol 42:3428–3433

Meironyté DM, Norén K, Bergman A (1999) Analysis of Polybrominated Diphenyl Ethers in Swedish human milk. a time-related trend-study, 1972–1997. J Toxicol Environ Health Part A 58:329–341

Meironyté-Guvenius DM, Aronsson A, Ekman-Ordeberg G, Bergman A, Norén K (2003) Human prenatal and postnatal exposure to polybrominated diphenyl ethers, polychlorinated biphenyls, polychlorobiphenylols, and pentachlorophenol. Environ Health Perspect 111:1235–1241

Meng T, Cheng J, Tang Z, Yin H, Zhang M (2021) Global distribution and trends of polybrominated diphenyl ethers in human blood and breast milk: a quantitative meta-analysis of studies published in the period 2000–2019. J Environ Manage 280:111696

Milić J, Lugonja N, Knudsen TŠ, Marinković V, Avdalović J, Ilić M, Nakano T (2025) Polychlorinated biphenyls and polybrominated diphenyl ethers in infant food: occurrence and exposure assessment. Sci Total Environ 958:178011

Nakao T, Akiyama E, Kakutani H, Mizuno A, Aozasa O, Akai Y, Ohta S (2015) Levels of tetrabromobisphenol A, tribromobisphenol A, dibromobisphenol A, monobromobisphenol A, and bisphenol A in Japanese breast milk. Chem Res Toxicol 28:722–728

Nguyen KH, Abou-Elwafa Abdallah M, Moehring T, Harrad S (2017) Biotransformation of the flame retardant 1,2-dibromo-4-(1,2-dibromoethyl)cyclohexane (TBECH) in vitro by human liver microsomes. Environ Sci Technol 51:10511–10518

NICNAS (National Industrial Chemicals Notification and Assessment Scheme (2007) Interim public health risk assessment of certain PBDE congeners. Online: https://www.nicnas.gov.au/__data/assets/word_doc/0020/39305/PBDE.docx.

NICNAS (2012) Hexabromocyclododecane. Priority Existing Chemical Assessment Report No. 34. Online: https://www.nicnas.gov.au/__data/assets/word_doc/0009/34848/PEC34-HBCD.docx

Nomeir AA, Markham PM, Ghanayem BI, Chadwick M (1993) Disposition of the flame retardant 1,2-bis(2,4,6-tribromophenoxy)ethane in rats following administration in the diet. Drug Metab Dispos 21:209–214

NTP (2016) Technical report on the toxicology studies of a pentabromodiphenyl ether mixture [DE-71 (technical grade)] (CAS NO. 32534–81–9) in F344/N rats and B6C3F1/N mice and toxicology and carcinogenesis studies of a pentabromodiphenyl ether mixture [DE-71 (technical grade)] in Wistar Han [Crl:WI(Han)] rats and B6C3F1/N mice (gavage studies). Online: https://ntp.niehs.nih.gov/go/764686

Oulhote Y, Chevrier J, Bouchard MF (2016) Exposure to polybrominated diphenyl ethers (PBDEs) and hypothyroidism in Canadian women. J Clin Endocrinol Metab 101:590–598

Pan HY, Li JF, Li XH, Yang YL, Qin ZF, Li JB, Li YY (2020) Transfer of dechlorane plus between human breast milk and adipose tissue and comparison with legacy lipophilic compounds. Environ Pollut 265(Pt A):115096

Parizek O, Gramblicka T, Parizkova D, Polachova A, Bechynska K, Dvorakova D, Stupak M, Dusek J, Pavlikova J, Topinka J, Sram RJ, Pulkrabova J (2023) Assessment of organohalogenated pollutants in breast milk from the Czech Republic. Sci Total Environ 871:161938

Peters RJB (2004) Man-made chemicals in human blood. TNO Report R2004/493. Prepared for Greenpeace Nederland, Nov 2004. http://www.greenpeace.org/eu-unit/Global/eu-unit/reports-briefings/2009/3/man-made-chemicals-inhuman-bl.pdf

Phillips AL, Hammel SC, Hoffman K, Lorenzo AM, Chen A, Webster TF, Stapleton HM (2018) Children's residential exposure to organophosphate ester flame retardants and plasticizers: Investigating exposure pathways in the TESIE study. Environ Int 116:176–185

Plattard N, Venisse N, Carato P, Dupuis A, Haddad S (2022) Hepatic metabolism of chlorinated derivatives of bisphenol A (ClxBPA) and interspecies differences between rats and humans. Arch Toxicol 96(3):783–792

Polder A, Gabrielsen GW, Odland JO, Savinova TN, Tkachev A, Loken KB, Skaare JU (2008) Spatial and temporal changes of chlorinated pesticides, PCBs, dioxins (PCDDs/PCDFs) and brominated flame retardants in human breast milk from Northern Russia. Sci Total Environ 391:41–54

Pratt I, Anderson W, Crowley D, Daly S, Evans R, Fernandes A, Fitzgerald M, Geary M, Keane D, Morrison JJ, Reilly A, Tlustos C (2013) Brominated and fluorinated organic pollutants in the breast milk of first-time Irish mothers: is there a relationship to levels in food? Food Addit Contam Part A 30:1788–1798

Przybyla J, Kile ML, Smit E, Houseman EA (2016) Cross-sectional study of polybrominated flame retardants and self-reported attention deficit hyperactivity disorder in US youth aged 12–15 (NHANES 2003–2004). J Environ Public Health 2392045.

Qiao L, Zheng XB, Yan X, Wang MH, Zheng J, Chen SJ, Yang ZY, Mai BX (2018) Brominated flame retardant (BFRs) and Dechlorane Plus (DP) in paired human serum and segmented hair. Ecotoxicol Environ Saf 147:803–808

Qin YY, Leung CKM, Lin CK, Leung AOW, Wang HS, Giesy JP, Wong MH (2011) Halogenated POPs and PAHs in blood plasma of Hong Kong residents. Environ Sci Technol 45:1630–1637

Qiu X, Bigsby RM, Hites RA (2009) Hydroxylated metabolites of polybrominated diphenyl ethers in human blood samples from the United States. Environ Health Perspect 117:93–98

Raab U, Preiss U, Albrecht M, Shahin N, Parlar H, Fromme H (2008) Concentrations of polybrominated diphenyl ethers, organochlorine compounds and nitro musks in mother's milk from Germany (Bavaria). Chemosphere 72:87–94

Rawn DF, Ryan JJ, Sadler AR, Sun WF, Weber D, Laffey P, Haines D, Macey K, Van Oostdam J (2014) Brominated flame retardant concentrations in sera from the Canadian Health Measures Survey (CHMS) from 2007 to 2009. Environ Int 63:26–34

Rawn DFK, Quade SC, Corrigan C, Ménard C, Sun WF, Breton F, Arbuckle TE, Fraser WD (2023) Differences in mirex [dechlorane] and dechlorane plus [syn- and anti-] concentrations observed in Canadian human milk. Chemosphere 316:137784

Rawn DGK, Sadler AR, Casey VA, Breton F, Sun W-F, Yu Feng S, Arbuckle TE (2024) Legacy halogenated flame retardants in Canadian human milk from the maternal-infant research on environmental chemicals study. J Environ Expo Assess 3:16

Ren G, Yu Z, Ma S, Li H, Peng P, Sheng G, Fu J (2009) Determination of Dechlorane Plus in serum from electronics dismantling workers in south China. Environ Sci Technol 43:9453–9457

Renzelli V, Gallo M, Morviducci L, Marino G, Ragni A, Tuveri E, Faggiano A, Mazzilli R, Natalicchio A, Zatelli MC, Montagnani M, Fogli S, Giuffrida D, Argentiero A, Danesi R, D'Oronzo S, Gori S, Franchina T, Russo A, Monami M, Sciacca L, Cinieri S, Colao A, Avogaro A, Di Cianni G, Giorgino F, Silvestris N (2023) Polybrominated diphenyl ethers (PBDEs) and human health: effects on metabolism, diabetes and cancer. Cancers (Basel) 15:4237

Rodríguez-Gómez R, Jiménez-Díaz I, Zafra-Gómez A, Ballesteros O, Navalón A (2014) A multiresidue method for the determination of selected endocrine disrupting chemicals in human breast milk based on a simple extraction procedure. Talanta. 130:561–570

Roosens L, Abdallah MA, Harrad S, Neels H, Covaci A (2009) Factors influencing concentrations of polybrominated diphenyl ethers (PBDEs) in students from Antwerp, Belgium. Environ Sci Technol 43:3535–3541

Roosens L, Cornelis C, D'Hollander W, Bervoets L, Reynders H, Van Campenhout K, Van Den Heuvel R, Neels H, Covaci A (2010) Exposure of the flemish population to brominated flame retardants: model and risk assessment. Environ Int 36:368–376

Rose M, Bennet DH, Bergman A, Fängström B, Pessah IN, Hertz-Picciotto I (2010) PBDEs in 2–5 year-old children from California and associations with diet and indoor environment. Environ Sci Technol 44:2648–2653

Rovira J, Martínez MÁ, Mari M, Cunha SC, Fernandes JO, Marmelo I, Marques A, Haug LS, Thomsen C, Nadal M, Domingo JL, Schuhmacher M (2022) Mixture of environmental pollutants in breast milk from a Spanish cohort of nursing mothers. Environ Int 166:107375

Ryan JJ, Rawn DF (2014) The brominated flame retardants, PBDEs and HBCDD, in Canadian human milk samples collected from 1992 to 2005; concentrations and trends. Environ Int 70:1–8

Sahlström LM, Sellström U, de Wit CA, Lignell S, Darnerud PO (2014) Brominated flame retardants in matched serum samples from Swedish first-time mothers and their toddlers. Environ Sci Technol 48:7584–7592

Saillenfait A-M, Ndaw S, Robert A, Sabat´e JP (2018) Recent biomonitoring reports on phosphate ester flame retardants: a short review. Arch. Toxicol 92: 2749–2778.

Salah S, Myllynen P, Prribyloá P, Kukuccka P, Laatio L, Sieppi E, Kummua M, Palaniswamy S, Abass K (2026) Monitoring and assessing the association of polybrominated diphenyl ethers (PBDEs) and flame retardants (Dechlorane Plus and DBDPE) with parental lifestyle factors in a mother–child cohort of the Finnish population (NUGEN). Emerg Contam 12:100596

Sales C, Poma G, Malarvannan G, Portolés T, Beltrán J, Covaci A (2017) Simultaneous determination of dechloranes, polybrominated diphenyl ethers and novel brominated flame retardants in food and serum. Anal Bioanal Chem 409:4507–4515

SC (Stockholm Convention) (2019) Stockholm Convention on Persistent Organic Pollutants. List of substances. https://www.pops.int/TheConvention/ThePOPs/AllPOPs/tabid/2509/Default.aspx

Schächtele A, Malisch R, Hardebusch B, Rolaf van Leeuwen F-X, Moy G, Tritscher A, van Duursen M, van den Berg M, Šebková K, Klánová J, Kalina J (2023) WHO-and UNEP-Coordinated Exposure Studies 2000–2019: Findings of Polybrominated Substances (PBDE, HBCDD, PBB 153, PBDD/PBDF). In: (Malisch et al. (Eds.): Persistent Organic Pollutants in Human Milk. Springer Nature

Schauer UM, Völkel W, Dekant W (2006) Toxicokinetics of tetrabromobisphenol A in humans and rats after oral administration. Toxicol Sci 91:49–58

Schecter A, Päpke O, Tung KC, Joseph J, Harris TR, Dahlgren J (2005) Polybrominated diphenyl ether flame retardants in the US population: current levels, temporal trends, and comparison with dioxins, dibenzofurans, and polychlorinated biphenyls. J Occup Environ Med 47:199–211

Schecter A, Päpke O, Harris TR, Tung KC (2006) Partitioning of polybrominated diphenyl ether (PBDE) congeners in human blood and milk. Toxicol Environ Chem 88:319–324

Schreder E, Zheng G, Sathyanarayana S, Gunaje N, Hu M, Salamova A (2023) Brominated flame retardants in breast milk from the United States: first detection of bromophenols in U.S. breast milk. Environ Pollut 334:122028.

Schultz IR, Kuo LJ, Cullinan V, Cade S (2020) Occupational and dietary differences in hydroxylated and methoxylated PBDEs and metals in plasma from Puget Sound, Washington. USA Region Volunteers. Sci Total Environ 714:136566

Sharkey M, Harrad S, Abou-Elwafa Abdallah M, Drage DS, Berresheim H (2020) Phasing-out of legacy brominated flame retardants: the UNEP stockholm convention and other legislative action worldwide. Environ Int 144:106041

Shen H, Ding G, Han G, Wang X, Xu X, Han J, Lou X, Xu C, Cai D, Song Y, Lu W (2010) Distribution of PCDD/Fs, PCBs, PBDEs and organochlorine residues in children's blood from Zhejiang, China. Chemosphere 80:170–175

Shi ZX, Wu YN, Li JG, Zhao YF, Feng JF (2009) Dietary exposure assessment of Chinese adults and nursing infants to tetrabromobisphenol-A and hexabromocyclododecanes: occurrence measurements in foods and human milk. Environ Sci Technol 43:4314–4319

Shi Z, Jiao Y, Hu Y, Sun Z, Zhou X, Feng J, Li J, Wu Y (2013) Levels of tetrabromobisphenol A, hexabromocyclododecanes and polybrominated diphenyl ethers in human milk from the general population in Beijing, China. Sci Total Environ 452–453:10–18

Shin MY, Lee S, Choi H, Jeong DI, Moon HB, Kim S (2017) Placental and lactational transfer of decabromodiphenyl ether and 2,2′,4,4′-tetrabromodiphenyl ether in dam-offspring pairs of Sprague-Dawley rats. Food Chem Toxicol 102:198–203

Siddique S, Xian Q, Abdelouahab N, Takser L, Phillips SP, Feng Y-L, Wang B, Zhu J (2012) Levels of dechlorane plus and polybrominated diphenylethers in human milk in two Canadian cities. Environ Int 39:50–55

Silva MJ, Hilton D, Furr J, Gray LE, Preau JL, Calafat AM, Ye X (2016) Quantification of tetrabromo benzoic acid and tetrabromo phthalic acid in rats exposed to the flame retardant Uniplex FPR-45. Arch Toxicol 90:551–557

Sjödin A, Jones RS, Focant JF, Lapeza C, Wang RY, McGahee EE 3rd, Zhang Y, Turner WE, Slazyk B, Needham LL, Patterson DG Jr (2004) Retrospective time-trend study of polybrominated diphenyl ether and polybrominated and polychlorinated biphenyl levels in human serum from the United States. Environ Health Perspect 112:654–658

Sjödin A, Wong LY, Jones RS, Park A, Zhang Y, Hodge C, Dipietro E, McClure C, Turner W, Needham LL, Patterson DG Jr (2008) Serum concentrations of polybrominated diphenyl ethers (PBDEs) and polybrominated biphenyls (PBB) in the United States population: 2003–2004. Environ Sci Technol 42:1377–1384

Sjödin A, Jones RS, Wong LY, Caudill SP, Calafat AM (2019) Polybrominated diphenyl ethers and biphenyl in serum: time trend study from the national health and nutrition examination survey for years 2005/06 through 2013/14. Environ Sci Technol 53:6018–6024

Sun Y, Gong X, Lin W, Liu Y, Wang Y, Wu M, Kannan K, Ma J (2018) Metabolites of organophosphate ester flame retardants in urine from Shanghai, China. Environ Res 164:507–515

Sunday OE, Bin H, Guanghua M, Yao C, Zhengjia Z, Xian Q, Xiangyang W, Weiwei F (2022) Review of the environmental occurrence, analytical techniques, degradation and toxicity of TBBPA and its derivatives. Environ Res 206:112594

Sundkvist AM, Olofsson U, Haglund P (2010) Organophosphorus flame retardants and plasticizers in marine and fresh water biota and in human milk. J Environ Monit 12:943–951

Sverko E, Tomy GT, Reiner EJ, Li Y-F, McCarry BE, Arnot JA, Law RL, Hites RA (2011) Dechlorane Plus and related compounds in the environment: a review. Environ Sci Technol 45:5088–5098

t Mannetje A, Coakley J, Mueller JF, Harden F, Toms L-M, Douwes J (2012) Partitioning of persistent organic pollutants (POPs) between human serum and breast milk: a literature review. Chemosphere 89:911–918

Tang W, Liu X, Cao Y, Liu J, Zhao L, Qiu J, Zhang J, Chen X, Ao J, Zhang Q (2025) Exposure to individual organophosphate esters and their mixture in relation to semen quality in Chinese men. Environ Int 205:109907

Tao F, Abdallah MA, Ashworth DC, Douglas P, Toledano MB, Harrad S (2017) Emerging and legacy flame retardants in UK human milk and food suggest slow response to restrictions on use of PBDEs and HBCDD. Environ Int 105:95–104

Tay JH, Sellström U, Papadopoulou E, Padilla-Sánchez JA, Haug LS, de Wit CA (2019) Serum concentrations of legacy and emerging halogenated flame retardants in a Norwegian cohort: Relationship to external exposure. Environ Res 178:108731

Terry P, Towers CV, Liu LY, Peverly AA, Chen J, Salamova A (2017) Polybrominated diphenyl ethers (flame retardants) in mother-infant pairs in the Southeastern U.S. Int J Environ Health Res 27:205–214

Thomsen C, Lundanes E, Becher G (2002a) Brominated flame retardants in archived serum samples from Norway: a study on temporal trends and the role of age. Environ Sci Technol 36:1414–1418

Thomsen C, Leknes H, Lundanes E, Becher G (2002b) A new method for determination of halogenated flame retardants in human milk using solid-phase extraction. J Anal Toxicol 26:129–137

Thomsen C, Liane VH, Becher G (2007) Automated solid-phase extraction for the determination of polybrominated diphenyl ethers and polychlorinated biphenyls in serum—application on archived Norwegian samples from 1977 to 2003. J Chromatogr B 846:252–263

Thomsen C, Knutsen HK, Liane VH, Frøshaug M, Kvalem HE, Haugen M, Meltzer HM, Alexander J, Becher G (2008) Consumption of fish from a contaminated lake strongly affects the concentrations of polybrominated diphenyl ethers and hexabromocyclododecane in serum. Mol Nutr Food Res 52:228–237

Thomsen C, Stigum H, Frøshaug M, Broadwell SL, Becher G, Eggesbø M (2010) Determinants of brominated flame retardants in breast milk from a large scale Norwegian study. Environ Int 36:68–74

Toms LM, Harden F, Päpke O, Hobson P, Ryan JJ, Müller JF (2008) Higher accumulation of polybrominated diphenyl ethers in infants than in adults. Environ Sci Technol 42:7510–7515

Toms LM, Sjödin A, Harden F, Hobson P, Jones R, Edenfield E, Mueller JF (2009) Serum polybrominated diphenyl ether (PBDE) levels are higher in children (2–5 years of age) than in infants and adults. Environ Health Perspect 117:1461–1465

Toms LM, Guerra P, Eljarrat E, Barceló D, Harden FA, Hobson P, Sjödin A, Ryan E, Müller JF (2012) Brominated flame retardants in the Australian population: 1993–2009. Chemosphere 89:398–403

Toms LM, Mazaheri M, Brommer S, Clifford S, Drage D, Mueller JF, Thai P, Harrad S, Morawska L, Harden FA (2015) Polybrominated diphenyl ethers (PBDEs) in dust from primary schools in South East Queensland, Australia. Environ Res 142:135–140

Tomy GT, Pleskach K, Ismail N, Whittle DM, Helm PA, Sverko E, Zaruk D, Marvin CH (2007) Isomers of Dechlorane Plus in Lake Winnipeg and Lake Ontario food webs. Environ Sci Technol 41:2249–2254

Torres-Moreno AC, Mejia-Grau K, Puente-DelaCruz L, Codling G, Villa AL, Ríos-Marquez O, Patequiva-Chauta L, Cobo M, Johnson-Restrepo B (2023) Polybrominated diphenyl ethers (PBDEs), polychlorinated biphenyls (PCBs), organochlorine pesticides (OCPs) in human breast milk from Colombia: a probabilistic risk assessment approach. Chemosphere 339:139597

Trudel D, Scheringer M, von Goetz N, Hungerbühler K (2011) Total consumer exposure to polybrominated diphenyl ethers in North America and Europe. Environ Sci Technol 45:2391–2397

Tsai MH, Chao HR, Hsu WL, Tsai CC, Lin CW, Chen CH (2021) Analysis of polybrominated diphenyl ethers and lipid composition in human breast milk and their correlation with infant neurodevelopment. Int J Environ Res Public Health 18:11501

Turyk ME, Anderson HA, Steenport D, Buelow C, Imm P, Knobeloch L (2010) Longitudinal biomonitoring for polybrominated diphenyl ethers (PBDEs) in residents of the Great Lakes basin. Chemosphere 81:517–522

UNEP (2022) Report of the Persistent Organic Pollutants Review Committee on the work of its seventeenth meeting. Addendum. Draft Risk Management Evaluation on Dechlorane Plus. Geneva, Switzerland. https://www.meti.go.jp/shingikai/kagakubusshitsu/shinsa/pdf/229_s07_00.pdf

US EPA (US Environmental Protection Agency) (2008) IRIS (Integrated Risk Information System). BDE-47, BDE-99, BDE-153, BDE-209 (Summaries, Technical reports). Online: über http://www.epa.gov/iris/

US EPA (2011) Screening-level hazard characterization. Dechlorane Plus®.

US EPA (2014). An alternatives assessment for the flame retardant decabromodiphenyl ether (decaBDE). Online: https://www.epa.gov/sites/production/files/2014-05/documents/decabde_final.pdf

US EPA (2015). High Production Volume Information System (HPVIS). Detailed chemical results: 1,2-Benzenedicarboxylic acid, 3,4,5,6-tetrabromo-, bis(2-ethylhexyl) ester. Online: http://ofmpub.epa.gov/oppthpv/quicksearch.display?pChem=102247

US EPA (2020) Risk Evaluation for Cyclic Aliphatic Bromide Cluster (HBCD). https://www.epa.gov/sites/default/files/2020-09/documents/1._risk_evaluation_for_cyclic_aliphatic_bromide_cluster_hbcd_casrn25637-99-4_casrn_3194-5_casrn_3194-57-8.pdf

van der Schyff V, Kalina J, Govarts E, Gilles L, Schoeters G, Castaño A, Esteban-López M, Kohoutek J, Kukučka P, Covaci A, Koppen G, Andrýsková L, Piler P, Klánová J, Jensen TK, Rambaud L, Riou M, Lamoree M, Kolossa-Gehring M, Vogel N, Weber T, Göen T, Gabriel C, Sarigiannis DA, Sakhi AK, Haug LS, Murinova LP, Fabelova L, Tratnik JS, Mazej D, Melymuk L (2023) Exposure to flame retardants in European children—results from the HBM4EU aligned studies. Int J Hyg Environ Health 247:114070

Van der Veen I, de Boer J (2012) Phosphorus flame retardants: properties, production, environmental occurrence, toxicity and analysis. Chemosphere 88:1119–1153

Vieth B, Rüdiger T, Ostermann B, Mielke H (2005) Esidues of flame retardants in breast milk from Germany with special consideration of polybrominated diphenyl ethers (PBDE). Final report UFOPLAN 202 61 218/03 on behalf of the Federal Environment Agency. http://bfr.bund.de/cm/343/rueckstaende_von_flammschutzmitteln_in_frauenmilch_aus_deutschland_abschlussbericht.pdf. [in German]

Vizcaino E, Grimalt JO, Lopez-Espinosa MJ, Llop S, Rebagliato M, Ballester F (2011) Polybromodiphenyl ethers in mothers and their newborns from a non-occupationally exposed population (Valencia, Spain). Environ Int 37:152–157

Völkel W, Fuchs V, Wöckner M, Fromme H (2018) Toxicokinetic of tris(2-butoxyethyl) phosphate (TBOEP) in humans following single oral administration Arch Toxicol 92:651–660

Vorkamp K, Nielsen F, Kyhl HB, Husby S, Nielsen LB, Barington T, Andersson AM, Jensen TK (2014) Polybrominated diphenyl ethers and perfluoroalkyl substances in serum of pregnant women: levels, correlations, and potential health implications. Arch Environ Contam Toxicol 67:9–20

Wan Y, Choi K, Kim S, Ji K, Chang H, Wiseman S, Jones PD, Khim JS, Park S, Park J, Lam MH, Giesy JP (2010) Hydroxylated polybrominated diphenyl ethers and bisphenol A in pregnant women and their matching fetuses: placental transfer and potential risks. Environ Sci Technol 44:5233–5239

Wang F, Wang J, Dai J, Hu G, Wang J, Luo X, Mai B (2010) Comparative tissue distribution, biotransformation and associated biological effects by decabromodiphenyl ethane and decabrominated diphenyl ether in male rats after a 90-day oral exposure study. Environ Sci Technol 44:5655–5660

Wang HS, Chen ZJ, Ho KL, Ge LC, Du J, Lam MH, Giesy JP, Wong MH, Wong CK (2012) Hydroxylated and methoxylated polybrominated diphenyl ethers in blood plasma of humans in Hong Kong. Environ Int 47:66–72

Wang P, Zhang Q, Zhang H, Wang T, Sun H, Zheng S, Li Y, Liang Y, Jiang G (2016a) Sources and environmental behaviors of dechlorane plus and related compounds—a review. Environ Int 88:206–220

Wang Y, Liu S, Zhao H, Zhao G, Chen J, Zhai G, Zhao H (2016b) Polybrominated diphenylethers (PBDEs) and their hydroxylated metabolites (OH-PBDEs) in female serum from Dalian, China. Int J Hyg Environ Health 219:816–822

Wang X, Zhu Q, Yan X, Wang Y, Liao C, Jiang G (2020a) A review of organophosphate flame retardants and plasticizers in the environment: analysis, occurrence and risk assessment. Sci Total Environ 731:139071

Wang X, Liu Q, Zhong W, Yang L, Yang J, Covaci A, Zhu L (2020b) Estimating renal and hepatic clearance rates of organophosphate esters in humans: Impacts of intrinsic metabolism and binding affinity with plasma proteins. Environ Int 134:105321

Wang S, Jin J, Guo C, Li Z, Wang Y, Wei Y, Jin J (2021a) Previously identified and unidentified polybrominated biphenyl congeners in serum from people living in an electronic waste dismantling area in China. Chemosphere 279:130478

Wang Y, Zhang W, Li A, Song M (2021b) Tetrachlorobisphenol A induced immunosuppression and uterine injury in mice. Ecotoxicol Environ Saf 207:111527

Wang S, Jin J, Zhang Z, Guo C, Li Z, Hu J, Wang Y, Wei Y, Jin J (2024a) Polybrominated biphenyl exposure and thyroid function tests in electronic waste dismantling and general population of Chinese adults. Emerging Contam 10:100272

Wang Q, Wang Y, Chen C, Zhang L, Wang C, Lou X, Chen D, Jin J, Fan Y, Sui S, Liu Z (2024b) Study on association between human plasma polybrominated diphenyl ethers (PBDEs) and thyroid homeostasis and carcinoma. Environ Geochem Health 6(12):502

Wattigney WA, Irvin-Barnwell E, Li Z, Ragin-Wilson A (2022) Biomonitoring of toxic metals, organochlorine pesticides, and polybrominated biphenyl 153 in Michigan urban anglers. Environ Res 203:111851

Wei G-L, Li D-Q, Zhuo M-N, Liao Y-S, Xie Z-X, Guo T-L, Li J-J, Zhang S-Y, Liang Z-Q (2015) Organophosphorus flame retardants and plasticizers: Sources, occurrence, toxicity and human exposure. Environ Pollut 196:29–46

Weiss J, Wallin E, Axmon A, Jönsson BAG, Akesson H, Janak K, Hagmar L, Bergman A (2006) Hydroxy-PCBs, PBDEs, and HBCDDDs in Serum from an Elderly Population of Swedish Fishermen's Wives and Associations with Bone Density. Environ Sci Technol 40:6282–6289

Wemken N, Drage DS, Cellarius C, Cleere K, Morrison JJ, Daly S, Abdallah MA, Tlustos C, Harrad S, Coggins MA (2020) Emerging and legacy brominated flame retardants in the breast milk of first time Irish mothers suggest positive response to restrictions on use of HBCDD and Penta- and Octa-BDE formulations. Environ Res 180:108805

Wong F, Cousins IT, Macleod M (2013) Bounding uncertainties in intrinsic human elimination half-lives and intake of polybrominated diphenyl ethers in the North American population. Environ Int 59:168–174

Woods MM, Lanphear BP, Braun JM, McCandless LC (2017) Gestational exposure to endocrine disrupting chemicals in relation to infant birth weight: a Bayesian analysis of the HOME study. Environ Health: A Global Access Science Source 16:115

Wu XM, Bennett DH, Moran RE, Sjödin A, Jones RS, Tancredi DJ, Tulve NS, Clifton MS, Colón M, Weathers W, Hertz-Picciotto I (2015) Polybrominated diphenyl ether serum concentrations in a Californian population of children, their parents, and older adults: an exposure assessment study. Environ Health 14:23

Wu Z, He C, Han W, Song J, Li H, Zhang Y, Jing X, Wu W (2020) Exposure pathways, levels and toxicity of polybrominated diphenyl ethers in humans: a review. Fast determination of organophosphate esters and their metabolites. Environ Res 187:109531.

Wu J, Yao S, Wang Y, Fan S, Cui X, Chen D, Shi Z (2024) Fast determination of organophosphate esters and their metabolites in human matrices by a straightforward cold-induced strategy coupled with HPLC-MS/MS. Anal Bioanal Chem 416(28):6663–6677

Wu Y, Fan Y, Lou X, Chen D, Jin J, Wang Y, Yang T, Liu Z (2025) Serum Dechlorane Plus (DPs) levels in thyroid disease patients: Influence of gender, age, and metastasis. Ecotoxicol Environ Saf 299:118410

Xiong P, Yan X, Zhu Q, Qu G, Shi J, Liao C, Jiang G (2019) A review of environmental occurrence, fate, and toxicity of novel brominated flame retardants. Environ Sci Technol 53:13551–13569

Xu X, Yekeen TA, Xiao Q, Wang Y, Lu F, Huo X (2013) Placental IGF-1 and IGFBP-3 expression correlate with umbilical cord blood PAH and PBDE levels from prenatal exposure to electronic waste. Environ Pollut 182:63–69

Xu B, Wu M, Wang M, Pan C, Qiu W, Tang L, Xu G (2018) Polybrominated diphenyl ethers (PBDEs) and hydroxylated PBDEs in human serum from Shanghai, China: a study on their presence and correlations. Environ Sci Pollut Res Int 25:3518–3526

Xu FC, Eulaers I, Alves A, Papadopoulou E, Padilla-Sanchez JA, Lai FY, Haug LS, Voorspoels S, Neels H, Covaci A (2019) Human exposure pathways to organophosphate flame retardants: associations between human biomonitoring and external exposure. Environ Int 127:462–472

Xu L, Hu Y, Zhu Q, Liao C, Jiang G (2022) Several typical endocrine-disrupting chemicals in human urine from general population in China: regional and demographic-related differences in exposure risk. J Hazard Mater 424(Pt B):127489

Xue J, Xiao Q, Zhang M, Li D, Wang X (2023) Toxic effects and mechanisms of polybrominated diphenyl ethers. Int J Mol Sci 24:13487

Yang Q, Qiu X, Li R, Liu S, Li K, Wang F, Zhu P, Li G, Zhu T (2013) Exposure to typical persistent organic pollutants from an electronic waste recycling site in Northern China. Chemosphere 91(2):205–211

Yang L, Lu Y, Wang L, Chang F, Zhang J, Liu Y (2016) Levels and profiles of polybrominated diphenyl ethers in breast milk during different nursing durations. Bull Environ Contam Toxicol 97:510–516

Yang Y, Chen P, Ma S, Lu S, Yu Y, An T (2022) A critical review of human internal exposure and the health risks of organophosphate ester flame retardants and their metabolites. Critical Reviews in Environ Sci Technol 52:1528–1560

Yao C, Yang H, Li Y (2021) A review on organophosphate flame retardants in the environment: Occurrence, accumulation, metabolism and toxicity. Sci Total Environ 15:148837

Yu YJ, Lin BG, Chen XC, Qiao J, Li LZ, Liang Y, Zhang GZ, Jia Y, Zhou XQ, Chen CR, Kan HD (2019) Polybrominated diphenyl ethers in human serum, semen and indoor dust: Effects on hormones balance and semen quality. Sci Total Environ 671:1017–1025

Yuan KY, Gu YH, Pei YH, Yu SY, Li TZ, Feng T, Liu Y, Tian J, Miao X, Xiong J, Hu M, Yuan BF (2025) Comprehensive analysis of transplacental transfer of environmental pollutants detected in paired maternal and cord serums. J Hazard Mater 486:136970

Zeng Y, Goudarzi H, Ait Bamai Y, Ketema RM, Roggeman M, den Ouden F, Gys C, Miyashita C, Ito S, Konno S, Covaci A, Kishi R, Ikeda-Araki A (2023) Exposure to organophosphate flame retardants and plasticizers is positively associated with wheeze and FeNO and eosinophil levels among school-aged children: the Hokkaido study. Environ Int 181:108278

Zhang L, Yin S, Zhao Y, Shi Z, Li J, Wu Y (2017) Polybrominated diphenyl ethers and indicator polychlorinated biphenyls in human milk from China under the Stockholm Convention. Chemosphere 189:32–38

Zhang B, Lu S, Huang M, Zhou M, Zhou Z, Zheng H, Jiang Y, Bai X, Zhang T (2018) Urinary metabolites of organophosphate flame retardants in 0–5-year-old children: Potential exposure risk for inpatients and home-stay infants. Environ Pollut 243:318–325

Zhang Z, Hu M, Xuan D, Wu L, Zhang Y, He G, Zhou Y (2022) Physiologically based pharmacokinetic (PBPK) modeling of BDE-209 following oral exposure in Chinese population. Food Chem Toxicol 169:113416

Zhao Y, Song Q, Ge W, Jin Y, Chen S, Zhao Y, Xiao X, Zhang Y (2019) Associations between in utero exposure to polybrominated diphenyl ethers, pathophysiological state of fetal growth and placental DNA methylation changes. Environ Int 133(Pt B):105255

Zhao X, Shi Z (2021) Legacy brominated flame retardants in human milk from the general population in Beijing, China: Biomonitoring, temporal trends from 2011 to 2018, and nursing infant's exposure assessment. Chemosphere 285:131533

Zhao X, Zhang L, Lyu B, Li J, Zhao Y, Wu Y, Shi Z (2025) Integrated risk assessment of dietary exposure and body burden of three brominated flame retardants in childbearing-aged Chinese women and breastfeeding infants. Environ Res 269:120881

Zheng G, Schreder E, Dempsey JC, Uding N, Chu V, Andres G, Sathyanarayana S, Salamova A (2021) Organophosphate esters and their metabolites in breast milk from the United States: Breastfeeding Is an important exposure pathway for infants. Environ Sci Technol Lett 8:224–230

Zhou SN, Buchar A, Siddique S, Takser L, Abdelouahab N, Zhu J (2014a) Measurements of selected brominated flame retardants in nursing women: implications for human exposure. Environ Sci Technol 48:8873–8880

Zhou SN, Siddique S, Lavoie L, Takser L, Abdelouahab N, Zhu J (2014b) Hexachloronorbornene-based flame retardants in humans: Levels in maternal serum and milk. Environ Int 66:11–17

Zota AR, Linderholm L, Park JS, Petreas M, Guo T, Privalsky ML, Zoeller RT, Woodruff TJ (2013) Temporal comparison of PBDEs, OH-PBDEs, PCBs, and OH-PCBs in the serum of second trimester pregnant women recruited from San Francisco General Hospital, California. Environ Sci Technol 47:11776–11784

Zota AR, Geller RJ, Romano LE, Coleman-Phox K, Adler NE, Parry E, Wang M, Park J-S, Elmi AF, Laraia BA, Epel ES (2018) Association between persistent endocrine-disrupting chemicals (PBDEs, OH-PBDEs, PCBs, and PFASs) and biomarkers of inflammation and cellular aging during pregnancy and postpartum. Environ Int 115:9–20

Chapter 6
Other Industrial Chemicals and Microplastics

Abstract This chapter discusses common industrial chemicals such as surfactants, copolymers, and consumer products, including textiles. Due to their chemical differences, these substances have varying potential health effects. It offers a concise overview of their chemical and physical properties, human exposure, primary health risks, and toxicokinetics, with references to additional scientific literature. The main goal is to provide detailed information on their presence in human samples like blood, urine, and breast milk. Specifically, it covers aromatic amines, diisocyanate, perchlorate, thiocyanate, nonyl- and octylphenols, cyclic volatile methylsiloxanes, and chloro- and bromophenols. Additionally, the increasingly important topic of microplastics is introduced, and in this context, health aspects, toxicokinetics, and the concentrations found to date in human test materials are discussed.

Keywords Aromatic amines · Diisocyanate · Perchlorate · Nonylphenol · Cyclic volatile methylsiloxanes · Chlorophenols · Microplastics · Micro particles

6.1 Aromatic Amines

Amines are organic compounds in which one, two, or three organic groups, such as alkyl or aryl, surround the nitrogen atom. A distinction is made between aliphatic amines, where a carbon bridge binds the nitrogen atom; heterocyclic amines, in which one or more nitrogen atoms are directly incorporated into a ring system; and aromatic amines, where the nitrogen is bound to an aromatic ring. Some amines occur naturally or can be formed through enzymatic processes in the organism.

This chapter focuses on aromatic amines (arylamines). The prominent representatives of this group, frequently used in biomonitoring studies, are listed in Table 6.1. Additionally, the chemical structures of some substances are illustrated in Fig. 6.1. They have been produced in the chemical industry since the early nineteenth century as raw materials for manufacturing azo dyes, synthesizing pesticides and pharmaceuticals, and for use in the rubber and plastics industry. For instance, aniline, as a high-production volume chemical, had a production volume of approximately 7

© The Author(s), under exclusive license to Springer Nature Switzerland AG 2026
H. Fromme, *Human Biomonitoring II*, https://doi.org/10.1007/978-3-032-10036-8_6

million tons worldwide in 2022, with 74% utilized in isocyanate production (4,4′-methylene diphenyl diisocyanate), 10% in rubber-processing chemicals, 8% in dyes and pigments, and 8% in other applications (Chemanalyst 2023). Another significant source of arylamine exposure is tobacco smoke due to the formation of amines during smoking. Dobaradaran et al. (2022) found significantly higher mean concentrations in freshly smoked cigarette butts compared to unsmoked cigarettes, with concentrations of 1349 versus 73 ng/g for ortho-anisidine, 740 versus 71 ng/g for 2,6-dimethylaniline, 549 versus 20 ng/g for the sum of toluidines, 332 versus 83 ng/g for aniline, and 172 versus 36 ng/g for 2-naphthylamine. Concentrations significantly higher than in mainstream smoke were observed in the sidestream smoke of nine different cigarette brands, indicating the importance of amine exposure for passive smokers (Goniewicz and Czogala 2005). Further exposure is possible through tattoos, as aromatic amines may be unintentionally introduced with the ink or formed in the skin through various processes, such as the reductive cleavage of organic azo dyes (Negi et al. 2022).

6.1.1 Health Aspects

Regarding aniline, the simplest aromatic amine in terms of structure, symptoms such as cyanosis, lacrimation, tremors, tachypnea, weakness, disorientation, dizziness, impaired gait, lethargy, drowsiness, convulsions, and loss of consciousness have been particularly observed in cases of acute poisoning. The most critical effects noted in animal studies after repeated oral administration included methemoglobin formation, hemolytic anemia due to erythrocyte damage, hyperplasia, fibrosis, and tumors in the spleen (IARC 2021). Methemoglobin formation has also been documented during the exposure of human volunteers and at polluted workplaces. Consequently, the conversion of oxyhemoglobin to methemoglobin impairs oxygen transport in the body.

The primary health endpoint of aromatic amines is carcinogenicity, which is well-documented for various individual substances. As early as 1895, German surgeon Ludwig Rehn described the development of bladder tumors in workers exposed to the dye fuchsin, attributing this to the long-term inhalation of aniline vapors. In animal experiments, aniline, ortho-toluidine, benzidine, ortho-anisidine, 4-aminobiphenyl, ortho-nitroanisol, and 2-naphthylamine showed increased incidences of various tumor forms, including sarcomas, hemangiosarcomas, hepatocellular carcinomas, adenomas, fibromas, and bladder carcinomas (IARC 2009, 2021). In humans, case reports and cohort studies indicated that exposure to o-toluidine, benzidine, 4-aminobiphenyl, and 2-naphthylamine caused bladder tumors.

The WHO International Agency for Research on Cancer classifications of various aromatic amines are presented in Table 6.1. Additionally, two related industrial processes (auramine production and magenta production) have been classified as carcinogenic to humans (Group 1). In contrast, auramine and magenta themselves are classified as possibly carcinogenic to humans (Group 2B). Furthermore, the

Table 6.1 Common amines and their IARC classification

Name	Abbreviation	CAS No	IARC classification*
Aniline	A	62-53-3	2A
Ortho-toluidine	o-TD	95-53-4	1
Meta-toluidine	m-TD	108-44-1	
Para-toluidine	p-TD	106-49-0	
Ortho-anisidine	o-ANSD	90-04-0	2A
Para-anisidine	p-ANSD	104-94-9	3
2-chloroaniline	2-CA	95-51-2	
3-chloroaniline	3-CA	108-42-9	
4-chloroaniline	4-CA	106-47-8	2B
2,4-Dichloroaniline	24-DCA	554-00-7	
3,4-dichloroaniline	34-DCA	95-76-1	
3,5-dichloroaniline	35-DCA	626-43-7	
2-aminobiphenyl	2-ABP	90-41-5	
4-aminobiphenyl	4-ABP	92-67-1	1
1-aminonaphthalin	1-ANP	134-32-7	3
2-aminonaphthalin	2-ANP	91-59-8	1
2,4-dimethylaniline	2,4-DMA	95-68-1	3
2,6-dimethylaniline	2,6-DMA	87-62-7	
Para-cresidine	p-CD	120-71-8	2B
2,4,5-trimethylaniline	2,4,5-TMA	137-17-7	3
2,4,6-trimethylaniline	2,4,6-TMA	88-05-1	3
Benzidine	BD	92-87-5	1
2,4-diaminotoluene	2,4 DAT	95-80-7	2B
1,5-Naphthalenediamine	1,5-NDA	2243-62-1	3
ortho-Nitroanisole	o-NAS	91-23-6	2A
2,4-diaminoanisole	2,4-DAAS	615-05-4	2B
3,3′-dimethylbenzidine	3,3′-DMBD	119-93-7	2B
4,4′-methylenedianiline	4,4′-MDA	101-77-9	2B
4,4′-methylenedi-o-toluidine	4,4’-MTD	838-88-0	2B
4,4′-oxydianiline	4,4-OD	101-80-4	2B
4-chloro-o-toluidine	4-CTD	95-69-2	2A
5-chloro-o-toluidine	5-CTD	95-79-4	3

*Group 1: the substance is carcinogenic to humans; Group 2A: the substance is probably carcinogenic to humans; Group 2B: the substance is possibly carcinogenic to humans; Group 3: the substance is not classifiable as to its carcinogenicity to humans (nach IARC 2025)

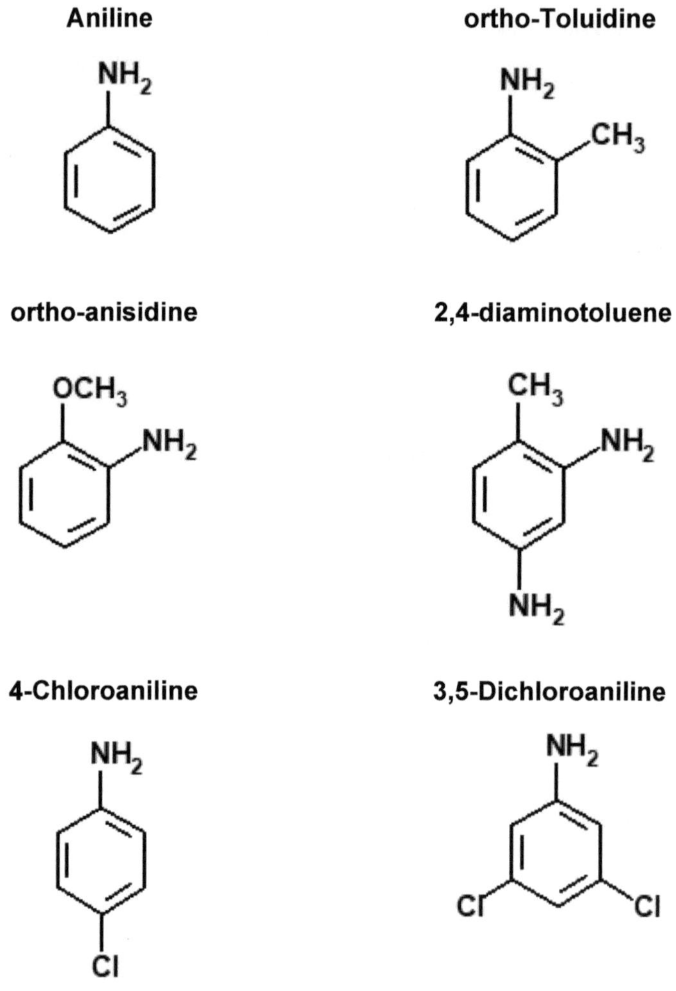

Fig. 6.1 Structure of some common aromatic amines

three azo dyes Direct Black 38, Direct Blue 6, and Direct Brown 95, which can be metabolized to benzidine, are classified as carcinogenic to humans in Group 1.

6.1.2 Toxicokinetics

Aromatic amines (arylamines) like aniline and o-toluidine are well absorbed through dermal, inhalation, and oral exposure, with the dermal route significantly contributing to overall workplace exposure. They are then rapidly distributed throughout the organism. Three primary metabolic pathways have been observed in laboratory animals and humans. First, N-acetylation and ring hydroxylation are detoxifying reactions, whereas nitrogen hydroxylation is the initial step in bioactivation. In the latter case, the N-hydroxylarylamine is formed first, which in a subsequent step releases the arylnitrenium ion, capable of forming adducts with DNA and is considered crucial for chemical carcinogenesis (Turesky and Le Marchand 2011; Wang et al. 2019a; IARC 2021; Gheni et al. 2024). Due to genetic polymorphisms of N-acetyltransferase, significant differences in acetylation reaction rates can occur among individuals.

As an example, the metabolism of aniline is depicted schematically in Fig. 6.2. Aniline is primarily converted to para-aminophenol through ring hydroxylation and is further acetylated to N-acetyl-4-aminophenol or converted to N-acetyl-4-aminophenol via acetanilide. N-acetyl-4-aminophenol is subsequently excreted as a sulfated or glucuronidated conjugate or mercapturic acid. In contrast, acetylation at the nitrogen atom of aniline leads, via phenylhydroxylamine in erythrocytes, to the formation of hemoglobin adducts and methemoglobin. Furthermore, it is noted that phenylhydroxylamine could also result in N-acetoxyaniline, which may lead to DNA adducts through the formation of phenylnitrenium ion (IARC 2021). Additionally, other pathways are discussed for aromatic amines that could affect the process of carcinogenesis via the generation of radicals and oxidative stress (Wang et al. 2019a).

Some toxicokinetic studies in experimental animals and human subjects exist. For instance, Käfferlein et al. (2014) monitored the excretion of total aniline in the urine of 19 subjects after continuous whole-body exposure to 2 ppm aniline for 6 h. They estimated the biological half-life in urine to be about 5 h. Modick et al. (2016) administered isotope-labeled aniline orally to four subjects on a single occasion. They discovered that, on average, 65.2% was excreted as N-acetyl-4-aminophenol, 4.0% as a mercapturic acid conjugate of N-acetyl-4-aminophenol, 0.26% as acetanilide, and 0.24% as aniline. The mean elimination half-lives based on the creatinine-corrected urine concentrations in this study were 3.9 h for N-acetyl-4-aminophenol, 4.8 h for mercapturic acid conjugate, 1.4 h for acetanilide, and 0.9 h for aniline.

6.1.3 Occurrence in Urine

In principle, the total content of individual arylamines in urine, following the cleavage of conjugates or adducts with hemoglobin, can be determined as part of biomonitoring studies of the general population. The results of studies on the exposure of the general population, which is occupationally unexposed, are summarized in Table 6.2. They show that the level of contamination from exposure varies considerably among the different groups examined and that there is evidence of general internal exposure to this class of substances.

Smoking tobacco cigarettes can significantly influence internal exposure. Grimmer et al. (2000) found higher mean excretions of 1-ANP (507 vs. 69 ng/24 h), 2-ABP (45 vs. 67 ng/24 h), and 4-ABP (79 vs. 68 ng/24 h) in 12 smokers compared

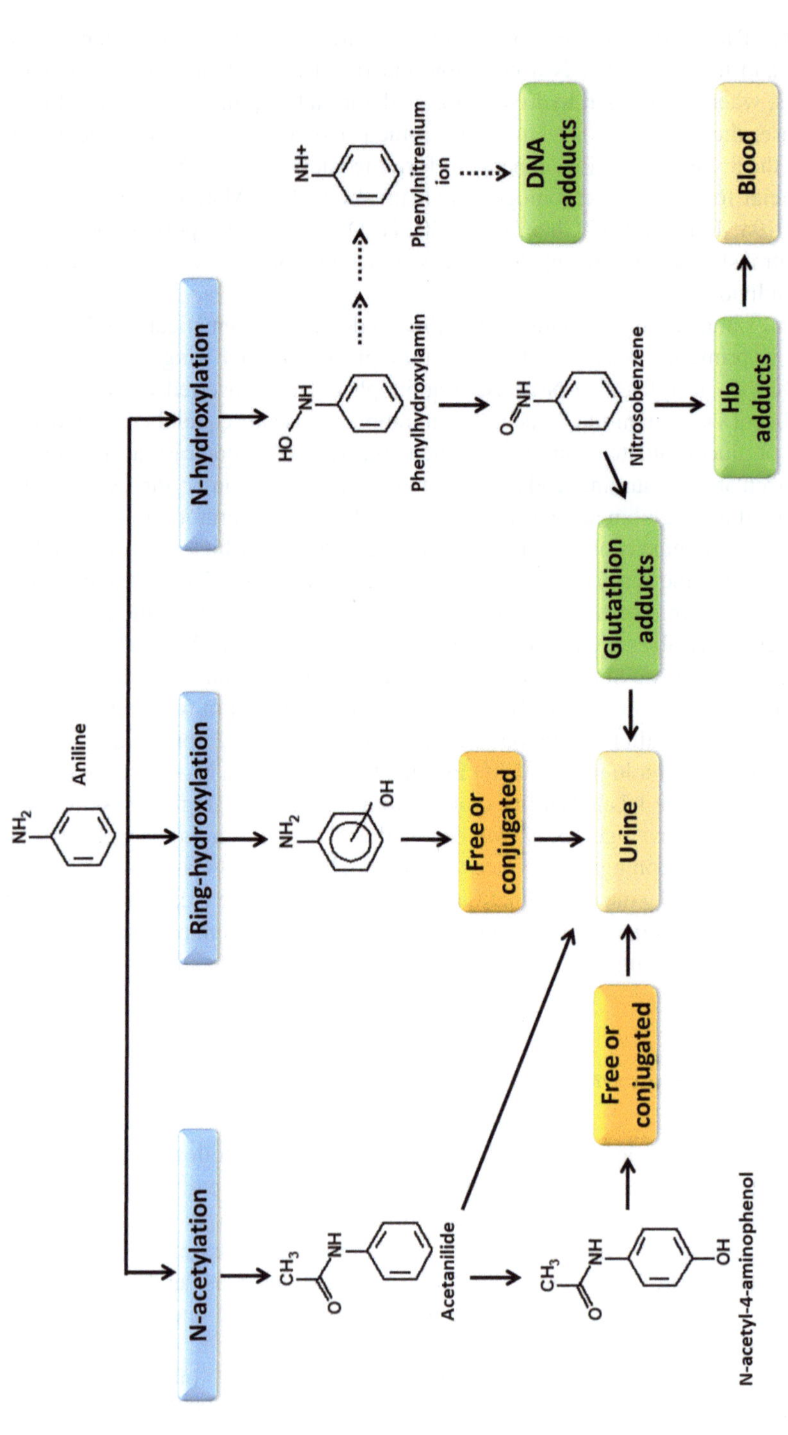

Fig. 6.2 Proposed metabolism of aniline in experimental animals (according to IARC 2021, HBC 2011)

Table 6.2 Median (95th percentile) concentration of aromatic amines in the urine of the general population in μg/l

Name	Weiss and Angerer (2002)	Kütting et al. (2009)	Choi et al. (2022)	CDC (2023)	Choi et al. (2024)	Chinthakindi and Kannan (2022)	Kütting et al. (2009)	CDC 2023
Aniline	3.5 (7.9)	2.99 (14.5)	0.74 (34[c])	–	11.3 (1380[c])	6.0 (13.7)	3.3 (13.5)	–
Ortho-toluidine	0.12 (2.7)	< 0.05 (0.19)	0.77 (7.1)	0.27 (1.46)	1.82 (98[c])	0.10 (0.47)	< 0.05 (0.41)	0.47 (1.96)
Meta-toluidine	0.17 (2.2)	< 0.05 (0.24)		–			0.11 (0.57)	–
Para-toluidine	0.11 (0.43)	< 0.05 (1.23)	0.04 (0.38)	–	< 0.03 (44[c])	0.12 (0.51)	< 0.05 (1.51)	–
Ortho-anisidine	0.22 (0.68)	0.21 (1.04)	0.04 (0.48)	0.05 (0.22)	0.41 (98[c])	< 0.03	0.29 (1.24)	0.06 (0.24)
Para-anisidine	–	–	< 0.05 (29)	–	< 0.05 (59[c])	< 0.03	–	–
3-chloroaniline	< 0.05 (0.55)	< 0.05 (0.98)	0.09 (8.6)	–	< 0.05 (252[c])	< 0.05	< 0.05 (0.40)	–
4-chloroaniline	0.11 (0.57)	< 0.05 (0.98)	0.12 (67)	–	0.87 (333[c])	0.09 (1.3)	< 0.05 (0.61)	–
3,4-dichloroaniline	< 0.05 (0.12)	< 0.05 (0.42)	0.20 (5.7)	–	< 0.10 (68[c])	–	< 0.05 (0.50)	–
3,5-dichloroaniline	0.18 (1.5)	0.16 (4.30)	–	–	–	–	0.29 (6.10)	–
2-aminobiphenyl		–	< 0.05	–	< 0.05 (15[c])	< 0.03	–	–
4-aminobiphenyl	< 0.05	–	< 0.05	0.005 (0.016)	< 0.05 (18[c])	< 0.03	–	0.02 (0.08)
1-aminonaphthalin	< 0.15	–	–	0.002 (0.010)	–	–	–	0.04 (0.19)
2-aminonaphthalin	< 0.08	–	< 0.03	0.004 (0.014)	< 0.03 (29[c])	0.18 (1.1)	–	0.015 (0.078)

(continued)

Table 6.2 (continued)

Name	Weiss and Angerer (2002)	Kütting et al. (2009)	Choi et al. (2022)	CDC (2023)	Choi et al. (2024)	Chinthakindi and Kannan (2022)	Kütting et al. (2009)	CDC 2023
2,4-dimethylaniline	–	–	< 0.05 (0.31)	–	< 0.05 (11[c])	–	–	–
2,6-dimethylaniline	–	–	< 0.05	0.02 (0.33)	< 0.05 (46[c])	3.8 (13.3)	–	0.04 (0.57)
Para-cresidine	–	–	< 0.03 (0.66)	–	< 0.03 (23[c])	0.14 (0.59)	–	–
2,4,5-trimethylaniline	–	–	< 0.05 (0.18)	–	< 0.05 (0.27[c])	< 0.03	–	–
Benzidine	–	–	< 0.05	–	< 0.05 (21[c])	< 0.03	–	–
2,4-diaminotoluene	–	–	0.69 (21)	–	< 0.10 (50[c])	< 0.05	–	–
Smoke status	Nonsmoker	Nonsmoker		Nonsmoker		Nonsmoker	Smoker	Smoker
Number	20	856	171	~ 1200 [d]	1231	213 [b]	145	~ 370 [e]
Country, year	Germany, 1998	Germany, 2003/04	USA, 2008–20	USA, 2013/14	USA, 2014–20	USA, 2018	Germany, 2003/04	USA, 2013/14

[a]Mean; [b]Collected from 15 participants; [c]Maximum; [d]Aged $\geq$ 20 years; [e]Aged $\geq$ 18 years

to 14 nonsmokers, while 2-ANP was higher in nonsmokers. In 22 passive smokers, the excretion in this study was not significantly different from that of nonsmokers. In another study, 10 smokers excreted substantially higher amounts of o-toluidine (204 vs. 104 ng/24 h), 2-ANP (20.8 vs. 10.7 ng/24 h), and 4-ABP (15.3 vs. 9.6 ng/24 h) than nonsmokers (Riedel et al. 2006). A Chinese study also observed similar differences between 20 smokers and 7 nonsmokers, with mean values of 19 versus 40 pg/mg crea for 1-ANP, 416 versus 66 pg/mg crea for 2-ANP, 172 versus < LOQ pg/mg crea for 3-ABP, and 419 versus16 pg/mg crea for 4-ABP (Yu et al. 2020). In the United States, the sample-weighted geometric means of the concentrations in 675 adults who participated in the representative NHANES study 2013–15 and smoked 10–19 cigarettes daily compared to adult nonsmokers ($n = 520$) were 30 times higher for 1-ANP (59 vs. 2.17 ng/g crea) and 4 to 6 times higher for 2-ANP (20.1 vs. 4.9 ng/g crea) and 4-ABP (30.5 vs. 4.8 ng/g crea) (Seyler et al. 2023). In this study, passive smokers were not significantly more exposed than nonsmokers. Similarly, in 13 e-cigarette users, the mean concentrations of 2-ANP and o-TD were slightly higher at 2.33 and 1.46 μg/l than in a control group, which had values of 1.00 and 1.13 μg/l (Fuller et al. 2018).

Dierkes et al. (2014) analyzed the urine of 31 individuals with known aniline exposure and 6 individuals in a control group without known exposure. They reported significantly higher medians in the exposed group compared to the control group. The median urinary concentrations were 5720 μg/l versus 80 μg/l for N-acetyl-4-aminophenol and 78.7 μg/l versus < 0.09 μg/l (for all nonexposed subjects) for acetanilide. In two individuals who had taken the medication paracetamol (chemically N-acetyl-4-aminophenol) the day before the experiment, concentrations of 159 and 275 mg/l of N-acetyl-4-aminophenol were detected in the urine. At the same time, acetanilide could not be measured above the LOQ of 0.09 μg/l in these individuals.

Hemoglobin adducts correlate well with DNA adducts, making them useful as a surrogate marker for the biologically active dose and for assessing longer term exposure (Richter and Branner 2002). Numerous workplace studies are available, primarily concerning 4-ABP, and less frequently addressing other aromatic amines. Research on the general population has largely focused on differences between smokers and nonsmokers. Richter and Branner (2002) summarized results from about 23 studies conducted during this period, reporting that Hb adduct levels of 4-ABP are 3–8 times higher in smokers than in nonsmokers. In an earlier study, significant differences emerged between 43 and 18 smokers (two types of tobacco) and 25 nonsmokers, with mean 4-ABP Hb adduct levels of 176 and 288 pg/g Hb, compared to 51 pg/g Hb (Bryant et al. 1988). Among children aged between 6.3 and 9.3 years from three different regions in Germany, mean Hb adduct levels were observed at 19–30 pg/g Hb for 4-ABP, 558–621 pg/g Hb for o-TD, and 256–275 pg/g Hb for o-ANSD (Richter et al. 2001). Among children exposed to secondhand smoke, 4-ABP adduct levels were only slightly elevated, though not significantly. The aforementioned review paper noted significantly higher Hb adduct levels for methyl- and dimethylaniline compared to 4-ABP, even though differences between smokers and nonsmokers were minimal (Richter and Branner 2002). A later review by Sabbioni (2017) summarized more recent biomonitoring data.

6.1.4 Occurrence in Breast Milk

Research on the presence of aromatic amines in breast milk has been quite limited thus far. In principle, although limited, transfer from the mother to the milk appears possible. LaVoie et al. (1989) administered ^{14}C-labeled 4-ABP intravenously to rats 11 days postpartum and noted that only 0.2% of the dose was passed to each of the five pups in the litters.

DeBruin et al. (1999) analyzed the breast milk of 31 subjects in Canada and found aniline and o-toluidine with geometric mean values of 0.15 ppb and 0.03 ppb in the 24 nonsmokers and 0.12 ppb and 0.03 ppb in the 7 smokers. While aniline was detected in all samples, o-toluidine was only found in five smokers and 19 nonsmokers.

6.2 Diisocyanates

Isocyanates are organic chemicals with the chemical formula R–N=C=O that contain one or more functional groups (R). In particular, diisocyanates are significant with regard to human exposure, especially in certain workplaces. In their case, R can be an aliphatic, cycloaliphatic, or aromatic group. Monoisocyanates contain only one NCO group in the molecule (e.g., methyl isocyanate) and are primarily used for chemical synthesis. They can form during the thermal decomposition of plastics (AGS 2009). Di- or tri-isocyanates have two or three NCO groups in the molecule and can thereby form polymer chains or cross-linked molecules. Diisocyanates exist as solids or liquids and exhibit high reactivity. Airborne isocyanates occur in various forms, such as vapors, aerosols, or adsorbed on suspended particles. While monoisocyanates are more volatile, diisocyanates have low vapor pressure.

The four most technically significant diisocyanates, which represent about 94% of the amounts commercially produced or imported in Europe, are toluene diisocyanates (TDIs), 4,4′-methylenediphenyl diisocyanate (4,4′-MDI), 2,4-toluene diisocyanate (2,4-TDI), and 1,6-hexamethylene diisocyanate (1,6-HDI) (ECHA 2019). Among these, TDI (as a mixture of 2,4-TDI and 2,6-TDI) constitutes 48%, 4,4′-MDI makes up 29%, 2,4-TDI accounts for 12%, and 1,6-HDI comprises 4.3% of the total. Commercial TDI available on the market is either the pure 2,4-TDI isomer or a mixture of 2,4-TDI and 2,6-TDI in ratios ranging from 80:20 to 65:35 (Schupp and Plehiers 2022). In the United States, TDI, MDI, and their polyisocyanates comprise more than 90% of the commercial market (ATSDR 2018). Table 6.3 displays the chemical structures of the aforementioned substances. Other commonly utilized diisocyanates include hydrogenated MDI (H12MDI or 4,4′-diisocyanato dicyclohexylmethane) and isophorone diisocyanate (IPDI).

Globally, approximately 5 million tons of MDI and 2 million tons of TDI are produced each year and utilized in the manufacture of polyurethanes (ACS 2023). Polyurethanes (PU) are created by reacting diisocyanates with various polyols (alcohols containing more than two reactive hydroxyl groups per molecule). The properties

Table 6.3 Chemical structures of certain diisocyanates and their metabolites (according to Sabbioni et al. 2022)

Parent compounds, acronym, CAS No	Metabolites, acronym, CAS No	Structure
4,4′-Methylenediphenyl diisocyanate (4,4′-MDI) CAS: 101-68-8	4,4′-Methylenedianiline (4,4′-MDA) CAS:101-77-9	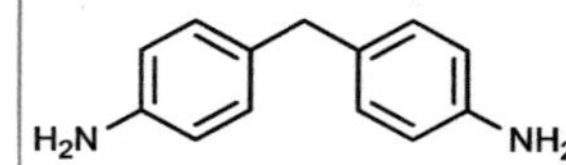
	N-Acetyl-4,4′-methylenedianiline (AcMDA) CAS: 24367-94-0	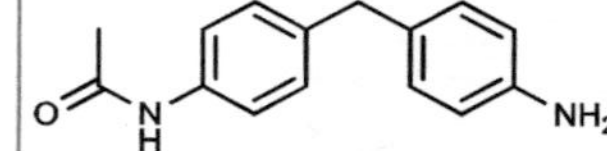
	N6-[[[4-[(4-Aminophenyl)methyl]phenyl]amino]carbonyl]-L-lysine (MDI-Lys) CAS: 1200446-89-4	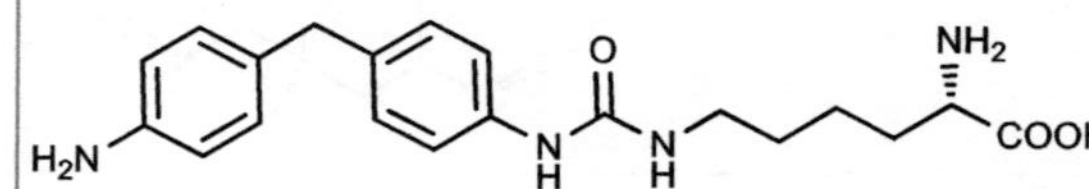
	3-[4-[(4-Aminophenyl)methyl]phenyl]-5-(1-methylethyl)-2,4-imidazolidinedione (MDI-Val-Hyd) CAS:264285-90-7	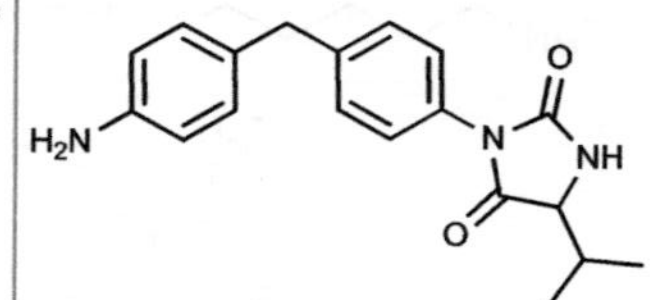
2,4-Toluene diisocyanate (2,4-TDI) CAS: 584-84-9	2,4-toluenediamine (2,4-TDA) CAS: 95-80-7	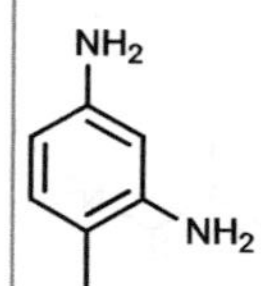

(continued)

Table 6.3 (continued)

Parent compounds, acronym, CAS No	Metabolites, acronym, CAS No	Structure
2,6-Toluene diisocyanate (2,6-TDI) CAS: 91-08-7	2,6-toluenediamine (2,6-TDA) CAS: 823-40-5	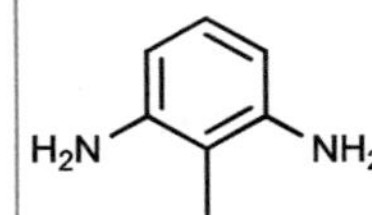
1,5-Naphthylene diisocyanate (1,5-NDI) CAS: 3173-72-6	1,5-Naphthalenediamine (1,5-NDA) CAS:2243-62-1	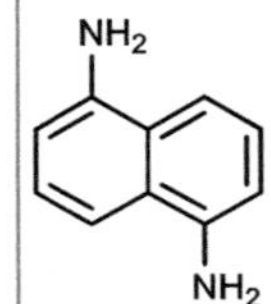
1,6-Hexamethylene diisocyanate (1,6-HDI) Cas: 822-06-0	1,6-Hexanediamine (1,6-HDA) CAS:124-09-4	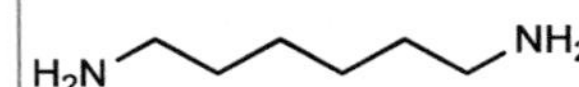
Isophorone diisocyanate (IPDI) CAS: 4098-71-9	Isophorone diamine (IPDA) CAS: 2855-13-2	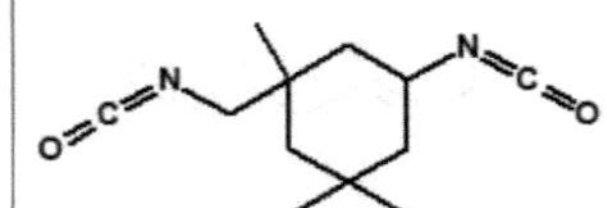

of PU can be adjusted by varying the number and quantity of ingredients, resulting in a spectrum of densities and molecular weights, which aids in maintaining the flexibility or rigidity of the final product (ISOPA 2020). These polyurethane chemicals are used in various continuous or batch processes. They are frequently combined with other materials to produce polyurethane foams (both flexible and rigid foams), assembly foams (e.g., insulation boards), foundry cores (molds), coating materials such as paints, lacquers, and varnishes, adhesives and sealants, glues, elastomers, prepolymers for chemical synthesis, engineering plastics, and polyurethane fibers or composite materials (ECHA 2019). Liang et al. (2021a, b) present a high-resolution material flow analysis of PU within the US economy and report that around 3700 kt were consumed in the US in 2016.

The European Commission has amended Annex XVII of the REACH regulation with Regulation (EU) 2020/1149, which came into force in August 2020. It states that after August 2023, diisocyanates may not be used, either as substances in their own right or as constituents of other substances or in mixtures, for industrial and commercial purposes, unless a) the concentration of individual diisocyanates, both individually and in combination, is below 0.1% by weight, or b) the employer or self-employed person ensures that industrial or commercial users have completed training on the safe use of diisocyanates before using the substances or mixtures (BAUA 2023).

Several review articles have thoroughly examined the occurrence, biological effects, and health impacts of diisocyanates (ATSDR 2018; DECOS 2018; ECHA 2019; Scholten et al. 2020; Schupp and Plehiers 2022; Sabbioni et al. 2022; Hartwig et al. 2023). These articles serve as the foundation for the next two subchapters.

6.2.1 Health Aspects

Numerous epidemiological and experimental studies have examined the toxic effects of 2,4- and 2,6-TDI, as well as their mixture, on the respiratory tract and skin. Acute exposure studies indicate that local respiratory irritation is the primary effect at low concentrations, while high concentrations result in more severe respiratory symptoms and possibly asthma. Furthermore, long-term occupational exposure can lead to bronchial asthma, asthma-like respiratory symptoms, and a deterioration of lung function (ATSDR 2018). The respiratory toxicity of MDI has not been researched as thoroughly as that of TDI, but the effects appear to be similar. Animal studies with various species and human studies demonstrate that DI can cause respiratory sensitization. The primary clinical manifestation of respiratory sensitization following multiple exposures is occupational asthma. Additionally, there is evidence that dermal exposure is a risk factor for the induction phase of respiratory sensitization.

In animal experiments, no increased tumor rates were observed following chronic inhalation exposure to TDI, although such effects occurred with oral administration. A chronic inhalation study in rats involving polymeric MDI showed an increase in lung adenomas. Several cohort studies have found a significantly higher lung cancer risk for women in PU-producing plants. The WHO's International Agency for Research on Cancer concluded that there is inadequate evidence for the carcinogenicity of TDI in humans. However, there is sufficient evidence for its carcinogenicity in experimental animals, classifying TDI as possibly carcinogenic to humans in group 2B (IARC 1986, 1999). For 4,4′-MDI or polymeric 4,4′-MDI, IARC found inadequate evidence for its carcinogenicity in humans and limited evidence in experimental animals, classifying it in group 3 (not classifiable as to its carcinogenicity to humans) (IARC 1999).

6.2.2 *Toxicokinetics*

Due to their high reactivity, diisocyanates can react with the surface molecules of the airways after inhalation and be hydrolyzed in the respiratory tract. Even after oral administration, they are readily hydrolyzed to the corresponding amines in the acidic environment of the gastrointestinal tract, which can then be further metabolized. After absorption, diisocyanates are almost exclusively bound to blood albumin. The organism metabolizes them through hydrolysis to the corresponding diamine, which is excreted in the urine, mainly conjugated to glutathione. For example, 4,4′-MDI is metabolized to 4,4′-MDA and AcMDA. However, these metabolites are not specific for MDI exposure since diamines are also widely used as industrial chemicals. The elimination half-lives of 4,4′-MDA, the two TDA isomers, and HDA are approximately 2–5 h (Scholten et al. 2020). In addition, diisocyanates form albumin adducts in the blood (e.g., MDI-Lys), which are specific for MDI exposure and are used as markers in occupational settings. The half-life of these adducts is assumed to be 20–25 days. After the further reaction of diamines to N-hydroxyarylamines, diisocyanate-specific hemoglobin adducts (MDI-Val-Hyd) with a half-life of about 120 days occur, and the formation of DNA adducts is also possible.

Various studies on test subjects have shown that, based on the excretion of diamines in urine, the uptake after inhalation of 2,4- and 2,6-TDI, both alone and in a mixture, ranges from approximately 14% to 20%. Dermal systemic uptake of diisocyanates is estimated to be lower and slower than inhalation uptake. In a diagnostic study, Budnik et al. (2011) exposed 121 patients to HDI, TDI, MDI, NDI, or IPDI at concentrations between 0.5 and 30 ppb for 0 to 120 min. They determined the corresponding diamines in the urine for up to 24 h after exposure. The peak concentrations in urine were found after 2 h for 1,6-HDA, 4.1 h for 2,4-TDA, 4.8 h for 2,6-TDA, 14 h for 4,4′-MDA, 6 h for 1,5-NDA, and 5.6 h for IPDA.

6.2.3 *Occurrence in the Blood and Urine*

For decades, biomonitoring of diisocyanates has been employed to manage contaminated work areas. Scholten et al. (2020) published a comprehensive review of biomonitoring data in occupational settings. Thus far, measurements from the general population not exposed to work have yielded limited results.

Sennbro et al. (2005) reported a median MDA level of 0.2 μg/l (<0.05–3 μg/l) and a 95th percentile of 0.3 μg/l in 97% of urine samples from 120 adults in Sweden who were not occupationally exposed. 2,4-TDA and 2,6-TDA were detected in urine samples in only 7% and 15% of cases, respectively. The ranges for 2,4-TDA and 2,6-TDA were < 0,1–0,4 μg/l and < 0,1–0,2 μg/l, with 95th percentiles of 0.1 μg/l in both cases. 1,5-NDA was found above the LOD of 0.1 μg/l in only 3% of samples, with a maximum value of 0.2 μg/l. In plasma, only MDA was detected in a larger proportion of samples, with median and maximum values of 0.2 and 0.4 μg/l, respectively. In a further study, Robert et al. (2007) examined the urine of 120 control subjects in France without known MDI exposure, finding a mean value of 0.14 μg/l (or 0.11 μg/g crea) and a range of < 0.10–0.87 μg/l (or < 0.10–0.88 μg/g crea). In the representative US National Health and Nutrition Examination Survey (NHANES) from 2015–2016, 2828 participants had median (95th percentile) levels of 0.06 μg/l (0.38 μg/l) for MDA, < 0.6 μg/l (0.79 μg/l) for 2,4-TDA, < 0.6 μg/l (<0.6 μg/l) for 2,6-TDA, and < 1.0 μg/l (<1.0 μg/l) for 1,5-NDA (CDC 2023). In comparison, Chinthakindi and Kannan (2022), who analyzed 231 urine samples from 15 participants in the USA over 5 weeks in 2018, reported a median MDA level of 0.08 μg/l and a 95th percentile of 0.31 μg/l. Additionally, 171 urine samples were examined as part of the US-ECHO cohort from 2008 to 2020, revealing a median (range) of 0.12 μg/l (<0.03–1.3 μg/l) (Choi et al. 2022).

Another nonselective biomarker is the mercapturic acid N-acetyl-S-(N-methyl carbamoyl)-L-cysteine (AMCC), a metabolite of both N,N-dimethylformamide and MDI. Kenwood et al. (2021) analyzed urine samples from the representative US NHANES for the years 2005–2006 and 2011–2016, involving 8272 individuals aged 12 and older. This study found that cigarette smokers had a significantly higher level of exposure than nonsmokers, with a median AMCC concentration of 517 μg/g crea compared to 127 μg/g crea.

Gries and Leng (2013) analyzed blood samples from 40 individuals with no known exposure to 4,4′-MDI. They were unable to detect the hemoglobin adduct 5-isopropyl-3-[4 (4-aminobenzyl)phenyl]hydantoin (ABP-Val-Hyd) above the LOQ of 0.02 ng/g globin (0.062 pmol/g globin).

6.3 Perchlorate and Thiocyanate

Perchlorates (ClO_4; CAS No. 14797-73-0) are the salts of perchloric acid ($HClO_4$). Perchloric acid (CAS No. 7601-90-3) is a colorless liquid that fumes in air and has a pungent odor. Specifically, sodium, ammonium, magnesium, and lithium perchlorates hold significant technical importance. Some perchlorates occur naturally in the soil, but they primarily originate from anthropogenic sources, such as the use of nitrate fertilizers. Notably, ammonium perchlorate (NH_4ClO_4; CAS No. 77-90-9) is widely used as an oxidizing agent in explosives and propellants, as well as in fireworks, matches, analytical chemistry, and various industrial processes like paper finishing (Charnley 2008). Additionally, perchlorate can form in the atmosphere through natural processes and be deposited in the soil via precipitation. As an environmental contaminant, perchlorate was first detected in groundwater near military sites and in the vicinity of military, aerospace, and explosives industries. Due to the chemical and physical characteristics of this substance, absorption of perchlorates is presumed, particularly after ingestion. Dermal uptake is minimal, and inhalation exposure is rather unlikely due to its low vapor pressure.

Inorganic thiocyanates are substances known to dissociate in solution, forming the thiocyanate anion (SCN-). Humans are exposed to SCN- both endogenously, through the microbiome's activity, and exogenously, for instance, through the conversion of glycosylates in plants (Willemin and Lumen 2017). The primary exposure route to thiocyanates for the general population appears to be through the consumption of foods that naturally contain thiocyanate, such as cabbage, kale, spinach, and kohlrabi. However, additional sources include cyanide-containing foods, cigarette smoke, the combustion of nitrogenous polymers, and environments contaminated by industrial processes (ATSDR 2024). Thiocyanates like ammonium thiocyanate (CAS No. 1762-95-4) serve various purposes, including the production of metal thiocyanates and pharmaceuticals, in freezing solutions, for pickling and printing, as a corrosion inhibitor, in photography, in fertilizers, in electroplating, and in analytical chemistry. Potential sources of occupational exposure to ammonium thiocyanate include manufacturing electronic computing equipment, research and development laboratories, commercial printing products like newspapers, general medical and surgical hospitals, production of adhesives and sealants, and industries such as construction and furniture (ATSDR 2024).

6.3.1 Health Aspects

Studies on animals have shown that perchlorate disrupts thyroid homeostasis in a dose-dependent manner. It acts as a competitive inhibitor of iodide transport that blocks the sodium iodide symporter without entering the thyroid gland via the iodide transporter (Lisco et al. 2020). Animal studies demonstrated a blockage of iodine uptake into the thyroid gland, leading to a decrease in thyroid hormone levels and an increase in serum TSH (thyroid-stimulating hormone) levels. In response, the organ undergoes hyperplasia and hypertrophy with signs of hypothyroidism. Epidemiological studies suggest that perchlorate impacts maternal thyroid function during pregnancy (EFSA 2024a). Various thyroid tumors (adenomas and carcinomas) have been observed after oral administration in rats and mice. Since the underlying mechanism is not deemed significant in humans and epidemiological studies have yielded inconclusive results, the carcinogenic risk is considered negligible (EFSA 2014). This mechanism of action of perchlorates is also utilized therapeutically. Sodium perchlorate (Irenat®) helps prevent iodine-induced reactions in diagnostics involving iodine-containing X-ray contrast agents. Further reviews on the effects of perchlorate are available (ATSDR 2008; Pleus and Corey 2018; Niziński et al. 2020).

Similar to perchlorate, the primary effects of thiocyanate on the thyroid gland stem from its greater binding affinity for the sodium-iodide symporter compared to iodine, the physiological ligand. However, its effect is significantly less than that of perchlorate (ATSDR 2024). Long-term studies, especially regarding carcinogenicity, are lacking; for other endpoints, they are deemed invalid (EFSA 2013). Thiocyanates were utilized as antihypertensives in the early 1900s. Symptoms of overdose include dizziness, an uncomfortable weakness in the arm and leg muscles, an enlarged thyroid gland, disorientation, confusion, delirium, hallucinations, convulsions, and ultimately, coma and death (NICNAS 2016). Watson et al. (2022) documented a case of intentional poisoning with potassium thiocyanate, exhibiting signs of cardiac and neurological toxicity along with symptoms such as delirium, tachycardia, and presumed seizure activity, as well as concurrent lactatemia and acidemia.

6.3.2 Toxicokinetics

Studies in various animal species show that perchlorate is absorbed quickly and efficiently from the gastrointestinal tract, is widely distributed in the body, and is minimally metabolized (EFSA 2014, 2024a, b). After oral exposure, excretion in humans and experimental animals primarily occurs via the kidneys. Greer et al. (2002) administered perchlorate in drinking water to 37 subjects for 14 days and calculated a serum half-life of 6.0–9.3 h (mean 8.1 h). In experiments with rats and calves, over 90% of the perchlorate dose was excreted in the urine (Seiivanova and Arefaeva 1986). In an older study involving two subjects who received potassium perchlorate orally, an elimination rate of over 90% was also reported (Eichler 1929). Braverman et al. (2006) administered two doses of perchlorate to nine subjects over six months and calculated that, on average, 65–70% of the administered dose was excreted in the urine.

Thiocyanate is rapidly absorbed following oral administration, widely distributed in the extracellular space, and does not accumulate in the body (EFSA 2013). After administering thiocyanate orally in rats, maximum plasma levels were observed after 30 min, with a plasma half-life of about 5 h (cited in Willemin and Lumen 2017). Excretion occurs quickly, primarily through urine and to a significantly lesser extent in feces and exhalation. Even when cyanide is ingested through food, the primary metabolite, thiocyanate, along with 2-amino-2-thiazoline-4-carboxylic acid to a lesser extent, is formed in further metabolism using a sulfur donor, and both are excreted in the urine (EFSA 2019). Overall, thiocyanate accounts for about 80% of the ingested cyanide dose.

6.3.3 Occurrence of Perchlorate in Urine

So far, research results are particularly available from the USA and China. Pearce et al. (2011) analyzed 107 and 134 urine samples from pregnant women in Argentina and the USA between 2004 and 2007, reporting medians (ranges) of 13.5 μg/l (1–676 μg/l) and 7.8 μg/l (0.4–284 μg/l), respectively. In Canada, however, the analysis of 150 urine samples from pregnant women during 2009–2011 revealed only a median of 3.2 μg/l (range: 0.5–48.1 μg/l) (Leung et al. 2014). In 2009–2010, a median of 2.57 μg/g crea and a 95th percentile of 8.74 μg/g crea were found in the urine of 284 pregnant women in the USA (Horton et al. 2015). In the representative NHANES 2001–2002 study in the USA, Blount et al. (2007) found perchlorate levels (95th percentile) in the urine of 5.2 μg/l (19 μg/l) in 6–11-year-olds and 3.5 μg/l (12 μg/l) in individuals aged $\geq$ 20 years. In the follow-up study to NHANES 2017–2018, slightly lower concentrations were observed in 6–11-year-olds ($n = 325$) and adults aged $\geq$ 20 years ($n = 1697$), with medians (95th percentile) of 3.42 μg/l (11.4 μg/l) and 2.33 μg/l (8.22 μg/l) (CDC 2023). Slightly higher concentrations were recorded in California during 2002–2003, with a median of 6.6 μg/l (0.23–177 μg/l) in the urine of 1880 people living in an area with elevated perchlorate concentrations in drinking water (Steinmaus et al. 2016). In urine samples from 31 subjects aged 11–66 years, collected in 2011 in New York State, USA, the perchlorate median was 2.35 μg/l (Kannan et al. 2024). For the NHANES studies conducted from 2001 to 2008, vegetables significantly contributed to urinary concentrations for adolescents and adults, while fruits and eggs were significant contributors only for adults (Lau et al. 2013).

Lee et al. (2017) reported a median of 3.1 μg/l and a range of 0.2–94.5 μg/l in 376 lactating women aged 18 years and older across three regions of the USA between 2008 and 2016. A study conducted on 64 breastfed infants aged 1 to 3 months and their mothers in the USA during 2008–2011 also found median (range) perchlorate concentrations of 3.1 μg/l (0.2–22.4 μg/l) in maternal urine and 4.7 μg/l (0.3–134.5 μg/l) in infant urine (Leung et al. 2012). Cao et al. (2010) observed significantly higher concentrations in the urine of 92 children in California during their first 12 months of life, with 28 μg/g crea in breastfed infants compared to 9.5 and 2.5 μg/g crea in those consuming cow's milk formula and soy-based formula, respectively.

In Thailand, Charatcharoenwitthaya et al. (2014) determined a median (range) of 1.9 μg/l (0.1–35.5 μg/l) for perchlorate in 200 women. Higher concentrations have been reported in China. In 178 urine samples collected between 2012 and 2013 in two regions with firework manufacturing facilities and six areas without such facilities, notable differences were observed in the medians of 52 and 136 μg/l compared to 14.1–38.3 μg/l (Zhang et al. 2015). Song et al. (2019) also confirmed these findings. In this study, 38 urine samples from China collected in 2012 revealed concentrations ranging from 19.0 to 289 μg/l in a fireworks manufacturing area, while levels in an area devoid of such industry were only 6.5 to 69.9 μg/l (Song et al. 2019). To characterize maternal exposure, Li et al. (2023) analyzed 1028 urine samples collected 2013–2018 in China and found a median (95th percentile) of 14.6 μg/l (55.4 μg/l) for perchlorate. Furthermore, King et al. (2023) reported a median of 16.7 μg/l (3.54–2411 μg/l) when examining the urine of 2441 adults in China during

2018–2019. In this case, the concentrations were significantly higher in men than in women, in those aged over 50 compared to younger individuals, and in nonsmokers compared to current smokers. Wang et al. (2022) found a median of 9.29 μg/l in Chinese adults in 2018–2019 when analyzing 184 urine samples. Moreover, a median (range) of 12 μg/l (1.4–31 μg/l) for perchlorate was reported in 43 urine samples from China (Wang et al. 2025).

In Italy and the UK, a total of 526 and 480 urine samples from pregnant women were analyzed between 2002 and 2006, respectively. Medians (range) of 5.2 μg/l (0.2–168 μg/l) and 2.6 μg/l (0.3–49 μg/l) were determined (Pearce et al. 2010). The same research group reported a median of 4.1 μg/l (0.2–119 μg/l) in 134 samples taken from pregnant women in Greece between 2008 and 2010 (Pearce et al. 2012). In Turkey, Ozpina et al. (2014) analyzed 24-h urine samples from 255 participants across three study areas in 2010 and found medians comparable to those of other European studies, ranging from 5.01 to 7.72 μg/l.

6.3.4 Occurrence of Thiocyanate in Urine

Overall, the median internal exposure to thiocyanate determined so far is comparable, but the wide range of individual results is striking. In the first representative NHANES 2001–2002 study in the USA, median (95th percentile) urinary thiocyanate levels of 1500 μg/l (10,000 μg/l) were observed in 6- to 11-year-olds and 1400 μg/l (12,000 μg/l) in adults aged $\geq$ 20 years. In the follow-up NHANES 2017–2018 study, decreased medians (95th percentiles) of 855 μg/l (2960 μg/l) and 1150 μg/l (8810 μg/l) were noted (CDC 2023). Comparable thiocyanate concentrations were found in California during 2002–2003, with a median of 898 μg/l (20–16,200 μg/l) in the urine of 1880 individuals (Steinmaus et al. 2016).

In 2009–2010, a median of 672 μg/g crea and a 95th percentile of 2853 μg/g crea were reported in the urine of 284 pregnant women in the USA (Horton et al. 2015). In Canada, a median of 351 μg/l (28–1195 μg/l) was observed from the analysis of 150 urine samples from pregnant women in 2009–2011. Lee et al. (2017) found only a median of 514 μg/l and a range of 12.3–4250 μg/l in 376 lactating women aged 18 years and older in three regions of the USA during 2008–2016. In a study involving 64 breastfed infants aged 1–3 months and their mothers in the USA from 2008–2011, comparable medians (range) of 374 μg/l (31.1–2420 μg/l) in maternal urine and 193 μg/l (21.7–1880 μg/l) in infant urine were reported (Leung et al. 2012). Cao et al. (2010) observed significantly higher concentrations in the urine of 92 children in California during the first 12 months of life, with 1000 μg/g crea in breast-fed infants compared to 620 and 200 μg/g crea in cow's milk formula and soy formula, respectively.

In Thailand, Charatcharoenwitthaya et al. (2014) determined a median (range) of 511 μg/l (68–3525 μg/l) for thiocyanate in 200 women. Li et al. (2023) analyzed 1028 urine samples 2013-2018 in China and found a median (95th percentile) of 390 μg/l (751 μg/l). King et al. (2023) analyzed the urine of 2441 adults in China from 2018 to 2019 and observed a slightly median of 717 μg/l (106–76,152 μg/l) for thiocyanate. Concentrations were significantly higher in nonsmokers compared to current smokers, while no differences were found regarding gender or age.

In Italy and the UK, a total of 526 and 480 urine samples from pregnant women were analyzed from 2002 to 2006, respectively, with median (range) levels of 373 μg/l (132–6650 μg/l) and 471 μg/l (34–3100 μg/l) reported (Pearce et al. 2010). The same research group found a similar median of 413 μg/l (range: 0.6–1635 μg/l) in 134 samples from pregnant women in Greece during 2008–2010 (Pearce et al. 2012). In Turkey, Ozpina et al. (2014) analyzed 24-h urine samples from 255 participants across three study areas and found medians ranging from 273 to 378 μg/l. In this study, thiocyanate levels in urine increased significantly with smoking status, from 268 μg/l in nonsmokers to 1110 μg/l in light smokers, and up to 2410 μg/l in heavy smokers.

6.3.5 Occurrence in Breast Milk

Pearce et al. (2007) detected perchlorate in milk samples collected from 49 mothers in Boston, USA, from 2002 to 2006, with a median concentration of 9.1 μg/l and a range of 1.3 to 411 μg/l. In a study involving 64 mothers in the USA from 2008 to 2011, slightly lower medians (range) of 4.4 μg/l (0.5 to 29.5 μg/l) were found in breast milk (Leung et al. 2012). This study showed that breast milk perchlorate levels were significantly correlated with maternal urine levels. Similar findings were reported by Shelor et al. (2012), who analyzed 43 breast milk samples from 13 mothers in the USA, discovering concentrations between 4.5 and 68.2 μg/l. Kirk et al. (2007) collected 6 to 18 breast milk samples from 10 mothers over a period of 2–14 days. Across all 147 samples, the median concentration was 4.0 μg/l, with a range of 0.5–39.5 μg/l. The individual perchlorate concentrations fluctuated significantly over time, as well as between individuals. The nationwide Maternal-Infant Research on Environmental Chemicals study conducted in Canada from 2008 to 2011 also found a similar range in 439 breast milk samples, with a median of 4.56 μg/kg (<0.3–676 μg/kg) (Wang et al. 2019b). The highest concentrations reported to date have come from Chile, where three regions were examined (Téllez et al. 2005). In a region known for high drinking water contamination, an average of 95.6 μg/l was measured in 25 milk samples. In contrast, the other two areas with lower contamination had averages of 81.6 μg/l ($n = 6$) and 18.3 μg/l ($n = 23$). In these locations, the authors suspected that another source, aside from drinking water, must be responsible for the elevated breast milk levels.

In the third National Breast Milk Monitoring Program conducted between 2016 and 2019 in China, 100 pooled breast milk samples from 100 cities were analyzed, and perchlorate was detected in all samples, showing a high median of 40.5 μg/kg and ranging from 4.03 to 411 μg/kg (Chen et al. 2023). No significant differences in concentration were observed between samples from urban and rural areas. Zhang et al. (2016a) also reported a high median of 20 μg/l (<0.2–204 μg/l) in 30 breast milk samples from regions with firework manufacturing facilities. Again, comparable results were published by Wang et al. (2025) analyzing 15 breast milk samples from Beijing with a median of 20 μg/l (3.4–50 μg/l). In contrast, in southern China in

2020, when examining 62 breast milk samples, perchlorate was noted with a median (range) of only 0.65 μg/l (0.18–7.23 μg/l) (Li et al. 2022).

On the contrary, thiocyanate has rarely been analyzed in breast milk to date. For example, Kirk et al. (2007) detected thiocyanate in 6 to 18 breast milk samples from 10 mothers collected over a period of 2 to 14 days among all 147 samples. The median in this case was 5.6 μg/l, with a range of 0.4 to 228 μg/l. Individual thiocyanate levels fluctuated considerably over time and among individuals. In a study of 64 mothers in the USA conducted from 2008 to 2011, higher medians (range) of 46.5 μg/l (2.9–1080 μg/l) were found in breast milk (Leung et al. 2012). In this study, breast milk levels were not significantly correlated with maternal urine levels.

6.4 Nonyl- and Octylphenols

Alkylphenols are a significant group of organic compounds created through the alkylation of phenols. Nonyl- and octylphenol, along with their derivatives, hold particular technical importance.

Nonylphenols (NP) are synthetic substances featuring a straight or branched alkyl group of nine carbon atoms, which may be positioned at various points in the molecule. They are chemically produced from phenols through alkylation processes and comprise a mixture of isomers. This mixture includes highly branched nonylphenols (br-NP; CAS No 84852-15-3), predominantly mono-substituted in the para-position, alongside small quantities of ortho- and di-substituted nonylphenols (US EPA 2010). Much of the literature discusses linear nonylphenol (n-NP; CAS No. 25152-52-3), often focusing on the specific para-isomer, namely 4-nonylphenol (4-NP; CAS No. 104-40-5), the chemical structure of which is illustrated in Fig. 6.3. Adding ethoxyl groups to this basic structure yields nonylphenol ethoxylates (NPE). NP is primarily employed on a large scale in industry as an intermediate for producing NPE. NPEs are categorized as nonionic surfactants, which are utilized in numerous applications, including detergents, cleaners, degreasers, dry cleaning aids, petroleum dispersants, emulsifiers, wetting agents, adhesives, agrochemicals, indoor pesticides, cosmetics, paper and textile processing formulations, prewash spotters, metalworking fluids, oilfield chemicals, paints and coatings, and dust control agents (US EPA 2010; IMAP 2020a). Aside from point sources in the production process, environmental exposure mainly occurs through wastewater (EC 2022). In the environment, NPs biodegrade into NPs.

The technically produced octylphenols (OP) are the reaction products of phenol and octene, resulting in a mixture of various linear and branched OP isomers. The most significant octylphenol in terms of quantity is 4-tert-octylphenol (4-t-OP; 4-(1,1,3,3-tetramethylbutyl)phenol; CAS No. 140–66-9), which serves as an intermediate in the production of octylphenol ethoxylates (OEPs), phenol/formaldehyde resins, polymers, antioxidants, and rubber and plastic products (IMAP 2020b). The

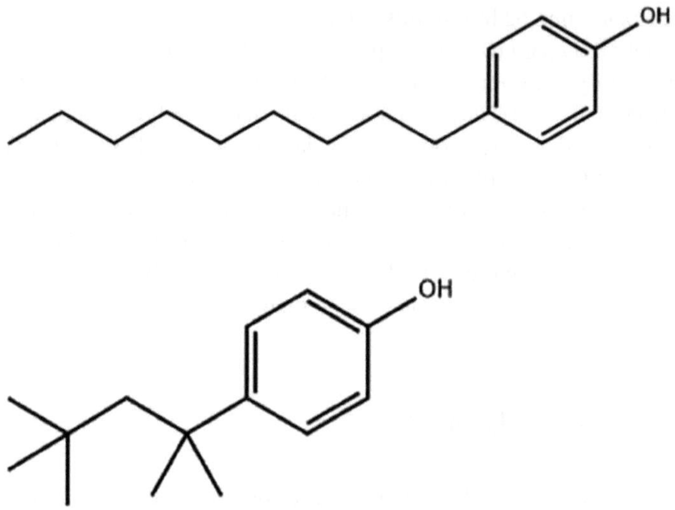

Fig. 6.3 Chemical structure of 4-nonylphenol (above) and 4-*tert*-octylphenol (below)

chemical structure of 4-t-OP is shown in Fig. 6.3. The OPEs, which also do not occur naturally, are nonionic surfactants commonly used in paints, emulsions, and wetting agents (EC/HC 2022). In addition, they are used as additives in surfactants like emulsifiers and polymer stabilizers, as coatings, in agriculture as emulsifiers or dispersants, and to a limited extent in cleaning agents. OPEs can also gradually break down into OP in the environment.

6.4.1 Health Aspects

After repeated oral administration, the liver was the target organ, showing slight histopathological changes at high doses in animal experiments (ECB 2002). Nonylphenol has exhibited slight estrogenic activity in several in vitro and in vivo assays. A multigenerational study in rats demonstrated that NP can cause minor perturbations in the reproductive system of offspring. The observed disturbances in the offspring align with the predictable or suspected effects of exogenous estrogenic activity (ECB 2002). In a subchronic NP exposure study with low doses, some male parameters were altered, and histopathological abnormalities were found in the testis and epididymis of F1 mice (Kim et al. 2019). NP was not mutagenic in various tests (Nielsen et al. 2000).

The toxicological data on OP is currently limited. Repeated dose studies showed reductions in body weight gain and changes in liver and kidney organ weights at high concentrations; however, only inconsistent histopathological changes were observed (ECHA 2011). In vitro tests did not provide any evidence of genotoxicity.

6.4.2 Toxicokinetics

In animal experiments, approximately 70% of labeled NP is excreted in the feces and 20% in the urine after oral administration (Nielsen et al. 2000). Results are available from two human studies. In the first study, Müller et al. (1998) administered $^{13}C_6$-labeled NP orally to one male subject and intravenously to another. The half-life of elimination from the blood was 2–3 h for both routes of administration, and the oral bioavailability was determined to be about 20%. The total dose of free and conjugated NP excreted in the urine was approximately 10% of the administered dose. In the second study, Ringbeck et al. (2021a) orally administered a single dose of $^{13}C_6$-labeled NP diluted in ethanol with water to three male subjects. They monitored the urinary excretion of the metabolites for 48 h. Alkyl chain hydroxylation (6OH-NP; 4-(6-hydroxy-1,1-dimethylheptyl)phenol) constituted the major metabolic pathway, with mean values of 43.7% or 62.2% (depending on the mass transition used for quantification) of the given NP dose excreted in urine. The urinary excretion fraction for oxo-NP (4-(6-oxo-1,1-dimethylheptyl)phenol) was 6.0 or 9.3%, and for parent NP it was 6.6%. A biphasic excretion kinetics was observed, with a maximum urine concentration occurring after an average of 2.7 h and elimination half-lives of 1.1 to 1.5 h in the first phase and 5.2–6.8 h in the second phase.

After oral administration, 4-t-OP is rapidly absorbed, with bioavailability in animal studies ranging from 10 to 55% (ECHA 2011). Dermal absorption is deemed possible. OP experiences rapid first-pass metabolism in the liver through phase I and phase II reactions, and it is excreted in the urine following hydroxylation, glucuronidation, and sulfation. Toxicokinetic studies in humans are unavailable for OP.

6.4.3 Occurrence in Urine

Table 6.4 summarizes the results of measurements of 4-NP and 4-t-OP in the healthy adult population. The median concentrations of 4-NP are generally low; only in Korea and Taiwan, and partially in China, do the values exceed 1 µg/l. The concentrations of 4-t-OP are often even lower. A similar picture emerges for children and adolescents, although the available results are limited (see Table 6.5). In various studies, no difference was observed between levels in women and men (Li et al. 2013; Ihde et al. 2015; Park and Kim 2017). In China, Li et al. 2013 examined a total of 287 children and adolescents aged 3–24 years and observed that the adjusted least square geometric means of 4-NP declined significantly from ages 3–6 to 7–12 years and from ages 7–12 to 13–17 years, then increased significantly to 18–24 years.

Ringbeck et al. (2021b) developed an analytical method to determine the oxidized metabolites of NP and analyzed urine collected in 2014 from 32 adults aged 25 to 56 years. OH-NP was quantified in 94% of the samples, with a maximum concentration of 25.3 µg/l (median 1.64 µg/l), and oxo-NP was found in 47% of samples with a

Table 6.4 Concentrations of alkylphenols in the urine of adults in μg/l

References	Age*	N	4-n-nonylphenol		4-tert-octylphenol		Country, year
			> LOQ (%)	Median (range)	> LOQ (%)	Median (range)	
Li et al. (2013)	18–24	64	100	21.6 (15.2–41.3[e])	–	–	China
Peng et al. (2016)	20–40	30	0	< 1.8	–	–	China
Chen et al. (2012)	30[a]	713	30	< 0.02 (0.10[h])	12	< 0.34 (0.58[h])	China, 2005–10
Park and Kim (2017)	18–69	1865	83	8.1 (<0.05–4477)	92	0.77 (<0.05–9.89)	Korea, 2009
Lv et al. (2016)	26[a]	1100	–	–	100	0.90 (0.25–20.1)	China, 2009/10
CDC (2023)	20	1914	–	–	n.g.	< 0.20 (0.30[h])	USA, 2009/10
Liu et al. (2016)	27[a]	137	87	2.0 (<0.12–17.7)	70	0.3 (<0.24–12.9)	China, 2010–11
Tang et al. (2013)	18–45	567	14	< 0.02 (<0.02–1.02)	15	< 0.02 (<0.02–2.53)[f]	China, 2010–12
Pirard et al. (2012)	1–75	131	0	< 0.68	–	–	Belgium, 2011
Shen et al. (2013)	30–50	156	n.g	1.6[a]	n.g	0.02[a]	China, 2011–13
Wang et al. (2015b)	33.4[a]	146	n.g	3.1	–	–	Taiwan, 2014
Azzouz et al. (2016)	n.g	12	25	< 0.2 (<0.2–1.8)	17	< 0.4 (<0.4–3.0)	Spain
Huang et al. (2017)	33[a]	232	99	2.98 (<0.20–38.9)	–	–	Taiwan, 2014–16
Sukuroglu et al. (2022)	3–47	103	9	0.014 (<LOQ–0.083)[b]	37	0.008 (<LOQ–0.165)[b]	Turkey, 2015/16
Jamal et al. (2020)	25–35	189	35	< 0.25 (<0.25–0.78)	–	–	Iran, 2016
Hou et al. (2021)	29.5[a]	390	n.g	0.44 (0.12–9.12)	n.g	1.46 (0.07–9.84)[g]	China, 2017/18
Chen et al. (2021)	62[a]	60	n.g	0.87[a]	n.g	0.33[a]	Taiwan, 2017–19
Xu et al. (2022)	18–90	58[c]	74	0.21 (0.02–1.69)	–	–	China, 2019
Xu et al. (2022)	18–90	69[d]	69	0.34 (0.02–4.24)	–	–	China, 2019
Wang et al. (2020)	43[a]	34	n.g	1.63	–	–	China, 2019

(continued)

Table 6.4 (continued)

References	Age*	N	4-n-nonylphenol		4-tert-octylphenol		Country, year
			> LOQ (%)	Median (range)	> LOQ (%)	Median (range)	
Battal et al. (2021)	n.g	15	20	< LOD (<LOD–0.07[b])	73	< LOD (n.g. –0.16[b])	Turkey
Lee et al. (2023)	n.g	21	95	5.55 (<0.16–33.0)	–	–	Korea
Shin et al. (2023)	19–35	25	100	4.87 (1.45–44.4)	–	–	Korea
Schmidtkunz et al. (2023)	n.g	40	0	< 1.0	–	–	Germany

[a]Mean; [b]All values in μg/g crea; [c]Urban; [d]Rural; [e]10th and 95th percentile; [f]As 4-n-OP; [g]As 2-t-OP; [h] 95th percentile; *Years; n.g.: not given

Table 6.5 Concentrations of alkylphenols in the urine of infants/adolescents in μg/l

References	Age*	N	4-n-nonylphenol		4-tert-octylphenol		Country
			> LOQ (%)	Median (range)	> LOQ (%)	Median (range)	
Li et al. (2013)	3–6	56	100	23.9 (16.7–78.7[b])	–	–	China
Li et al. (2013)	7–12	95	100	19.1 (11.6–36.2[b])	–	–	China
Li et al. (2013)	13–17	72	100	8.43 (5.2–28.2[b])	–	–	China
CDC (2023)	6–11	415	–	–	n.g	< 0.20 (0.40[c])	USA, 2009/10
Choi et al. (2014)	6–12	58	n.g	10.9[a] (<LOQ–110)	n.g	9.0[a] (<LOQ–48.9	South Korea, 2011
Hou et al. (2015)	6.5–15	308	97	3.91 (<LOQ–29.0)	–	–	Taiwan, 2012/13
Ihde et al. (2015)	4–8	50	72	0.11 (<0.23–6.25)	–	–	USA, 2012–14
Yu et al. (2016)	4–15	110	96	3.76 (<1.6–18.4)	–	–	Taiwan
Zhang et al. (2016b)	3–6	20	–	–	85	0.08 (<0.04–0.34)	China
Kuraoka et al. (2024)	4	~3400	4	9.9[a]	1.4	2.5[a]	Japan, 2017–19

[a]Mean; [b]10th and 95th percentile; [c]95th percentile; n.g.: not given; *LOQ* Limit of quantification: *years

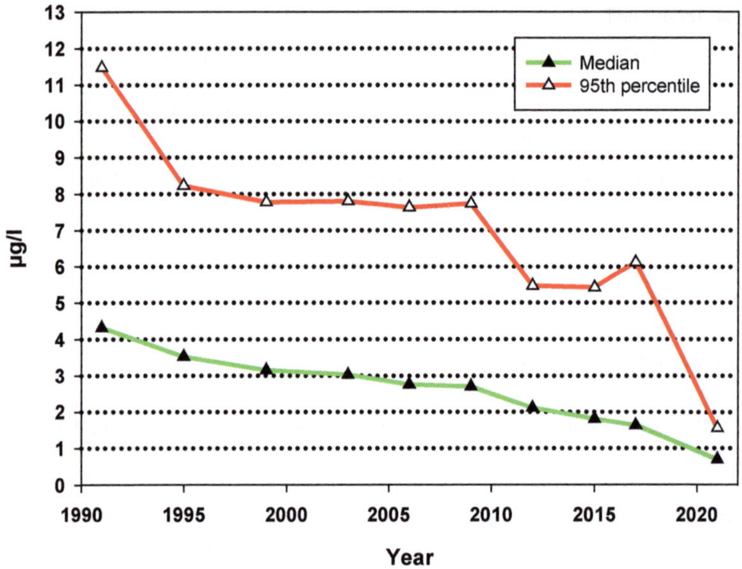

Fig. 6.4 Time-trend of hydroxy nonylphenol (OH–NP) concentrations in 24-h urine samples from Germany in µg/l (after Ringbeck et al. 2022c)

maximum concentration of 4.3 µg/l. In 180 urine samples collected from 7-year-old Japanese children between 2012 and 2017, OH-NP and oxo-NP were detected in 100% and 66% of the samples, respectively, with median (95th percentile) levels of 2.69 µg/l (11.0 µg/l) and 0.36 µg/l (1.2 µg/l) (Ringbeck et al. 2022a). Additionally, the same research group reported an OH-NP median (95th percentile) of 8.12 µg/l (25.3 µg/l) in 89 urine samples from Indonesia, 8.57 µg/l (27.3 µg/l) in 108 samples from Saudi Arabia, and 1.07 µg/l (7.12 µg/l) in 104 samples from Thailand (Ringbeck et al. 2022b). For oxo-NP, the medians in these countries were 0.95 µg/l, 1.82 µg/l, and < 0.25 µg/l.

Again, Ringbeck et al. (2022c) analyzed 660 24-h urine samples from the German Environmental Specimen Bank, collected over eleven sampling years from 1991 to 2021, involving young adults aged 20 to 29. Median OH-NP concentrations significantly decreased from 4.32 µg/l in 1991 to 0.70 µg/l in 2021. Figure 6.4 illustrates the temporal development of the medians and 95th percentiles.

6.4.4 Occurrence in Breast Milk

In initial studies, Guenther et al. (2002) analyzed a breast milk sample from Germany, while Otaka et al. (2003) analyzed three samples from Japan. They found low NP levels of 0.3 µg/kg and between 0.65 and 1.4 µg/kg, respectively. An analysis of

89 breast milk samples from Slovakia also showed a low mean value of 1.16 μg/l (0.67–2.27 μg/l) (Repková et al. 2023). In contrast, significantly higher concentrations were found in 10 Italian human mature milk samples, with a mean NP value of 32 μg/l (13.4–56.3 μg/l) (Ademollo et al. 2008). The authors attributed this to different dietary habits, observing a correlation between NP concentrations and fish consumption. Lin et al. (2009) also determined 4-NPs in 19 of 20 breast milk samples from Taiwan at concentrations of 1.7–11.6 μg/kg (mean 4.4 μg/kg). In a further study from Taiwan, which analyzed 59 samples, a comparable geometric mean value of 2.26 μg/kg was found for 4-NP, with the NP content significantly associated with the consumption of fish oil capsules and processed fish products (Chen et al. 2010). Even higher geometric mean values of 23.5 μg/l and 57.3 μg/l were found in 16 and 15 breast milk samples taken in Taiwan in the first and third months after birth (Huang et al. 2014). In Turkey, NP was detected in 100 samples with a geometric mean of 5.0 μg/l (0.26–47.5 μg/l), and a significant correlation was observed here with fish consumption and the use of cleaning products (Sise and Uguz 2017). NP was only detected in 2 out of 6 breast milk samples from Spain, with concentrations of 1.2 and 5.6 μg/l, both above the limit of detection of 1.1 μg/l (Azzouz et al. 2016).

In the USA, 4-tOP media of 2.6 μg/l (2.6–7.6 μg/l) was found in only 5 of 20 breast milk samples (Ye et al. 2006). Ademollo et al. (2008) found very low mean OP levels of 0.12 μg/l (0.01–0.21 μg/l) in 10 Italian human mature milk samples, with the two octylphenol ethoxylates, namely OP1EO and OP2EO, detected in 27% and 30% of the samples, with mean values of 0.07 and 0.16 μg/l, respectively. Lin et al. (2009) also detected 4-t-OP in 8 of 20 breast milk samples from Taiwan at concentrations ranging from 0.4 to 1.1 μg/kg. In another study from Taiwan, which analyzed 59 samples, an even lower geometric mean value of 0.02 μg/kg was found for 4-t-OP, with the 4-t-OP levels significantly associated with the consumption of cooking oil and the intake of fish oil capsules (Chen et al. 2010).

6.5 Cyclic Volatile Methylsiloxanes

Siloxanes (or silicones) are a group of aliphatic and cyclic oxygen-containing silicon compounds characterized by the basic structure Si-O-Si, with various molecular groups such as methyl, ethyl, or phenyl attached. Depending on their molecular size and structure, siloxanes are categorized into three groups:

- Volatile methylsiloxanes (VMS).
- cyclic VMS (cVMS).
- linear VMS (lVMS).
- polydimethylsiloxanes (PDMS).
- polyethermethylsiloxanes (PEMS)

Cyclic volatile methyl siloxanes (cVMS) are among the most widely used siloxanes and will be presented in this chapter. Generally, cVMS are based on the chemical structure [-Si-$(CH_3)_2$O-]n, with four units forming a ring in D4, five in D5, and six in D6 (see Table 6.6). Cyclomethicone, a mixture of different cyclic dimethylsiloxanes with varying proportions of D4, D5, and D6 (as well as D3 and D7), is listed under CAS number 69430–24-6. Cyclomethicone may also refer to pure D4 or D5 in cosmetic ingredient lists. cVMS are clear, volatile, low-viscosity liquids at room temperature that are poorly soluble in water, possess low molecular weight, and exhibit high vapor pressure. Due to the Si–O and Si–C bonds, siloxanes have relatively high thermal stability. There are no known natural sources of cVMS.

According to the Global Silicones Council, worldwide sales of silicones in 2013 reached 2,122,000 metric tons (GSC 2021). Figure 6.5 illustrates the percentage distribution of the primary applications of silicones. Cyclic siloxanes are mainly employed as raw materials for producing higher-molecular polydimethylsiloxanes and copolymers. These are utilized in industry and in a diverse range of consumer

Table 6.6 CAS No., names, structures of volatile cyclic methyl siloxanes

CAS No	IUPAC name	Abbreviation, synonyms	Sum formula	Chemical structure
541-05-9	Hexamethyl cyclotrisiloxane	D3, Cyclotrisiloxane	$C_6H_{18}O_3Si_3$	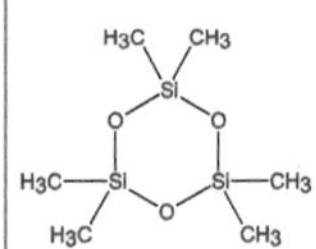
556-67-2	Octamethyl cyclotetrasiloxane	D4, Cyclotetrasiloxane	$C_8H_{24}O_4Si_4$	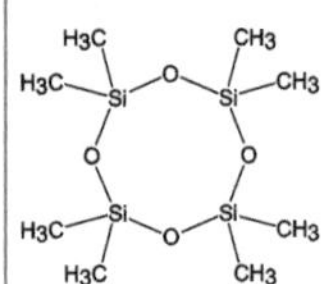
541-02-6	Decamethyl cyclopentasiloxane	D5, Cyclopentasiloxane	$C_{10}H_{30}O_5Si_5$	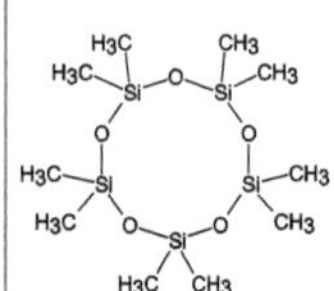
540-97-6	Dodecamethyl cyclohexasiloxane	D6, Cyclohexasiloxane	$C_{12}H_{36}O_6Si_6$	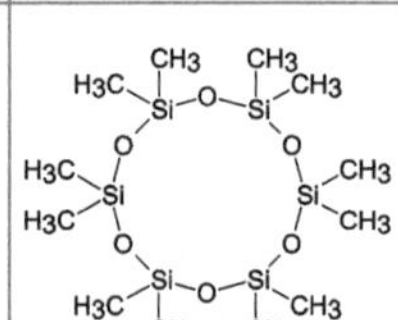
107–50-6	Tetradecamethyl cycloheptasiloxane	Siloxan D7, Cycloheptasiloxane	$C_{14}H_{42}O_7Si_7$	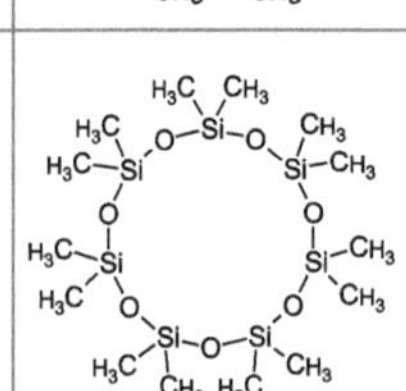

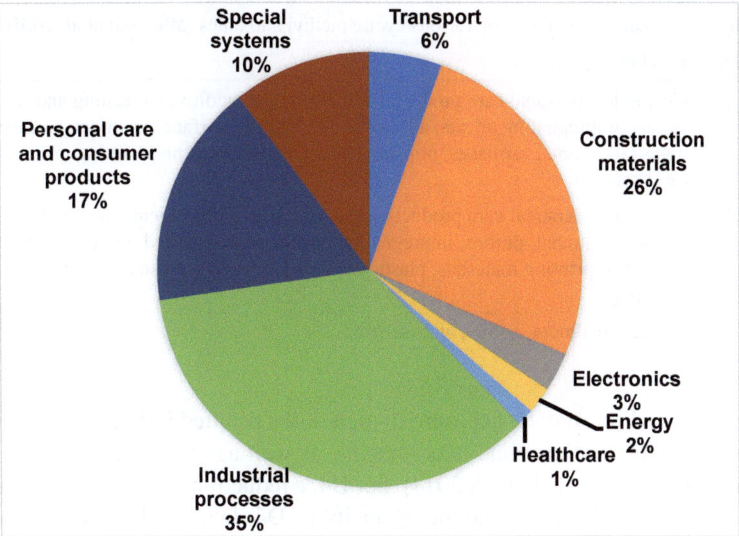

Fig. 6.5 Downstream applications of silicones in tons in 2013 (after GSC 2021)

products, including topical or orally applied pharmaceutical formulations, and as anti-foaming agents in food or medical devices (e.g., breast implants, catheters, infusion sets) (CES 2016). Due to their volatility, low surface tension, transparency, hydrophobicity, and odorlessness, siloxanes are incorporated in cosmetics and personal care products such as shampoos, conditioners, antiperspirants, soaps, moisturizers, and body lotions, as well as in decorative cosmetics (make-up, mascara, lipsticks). The highest proportions of D4 and D5 are found in hair and skin care products and in antiperspirants (EC/HC 2008a; b). A specialized application for D5 is chemical cleaning (CES 2016). Table 6.7 offers an overview of the various uses of cVMS. In cosmetics, siloxanes act as solvents, moisturizers, and antifoams for viscosity control and structure enhancement. Moreover, silicone-containing formulations exhibit excellent spreading ability on the skin. Siloxanes in deodorants and antiperspirants help keep the skin feeling dry and soft. In hair care products, siloxanes provide an antistatic effect, impart a beautiful shine to the hair, and speed up the drying process during blow-drying (SEHSC 2009).

6.5.1 Health Aspects

No epidemiological studies are reported in the scientific literature. However, numerous animal and experimental studies on human subjects exist. The results indicated that cVMS has very low acute toxicity, and no irritation to the eyes or skin was observed.

Table 6.7 Applications and uses of various cyclic methyl siloxanes (after Kaj et al. 2005b)

Siloxane	Field of application
D4	Cosmetics, personal care products, antifoams, fuel additives, cleaning and washing agents, impregnating agents, adhesives and binders, surface treatment, construction materials, paints, varnishes, polishes, fillers, nonstick agents, printing technology (anti-set-off)
D5	Cosmetics, personal care products, cleaning and washing agents, fuel additives, surface treatment, putties, impregnating agents, adhesives and binders, paints and varnishes, printing materials, plasticizers, surface-active substances, chemical cleaning
D6	Surface treatment, colors, and varnishes

In repeated-dose studies, exposure to high doses resulted in liver effects, such as increased relative and absolute liver weights, as well as effects on the lungs in the experimental animals (Dan EPA 2014). For D4, a reduction in fertility was observed in female animals. Tests for the mutagenicity of D4, D5, and D6 yielded negative results. In a 2-year inhalation study in rats, D4 was found to increase the incidence of cystic endometrial hyperplasia and uterine adenomas. The assumed underlying mechanism for the uterine effects is the influence of prolactin secretion through interactions with dopamine receptors in the nervous system of the animals, which is not relevant for humans (SCCS 2010, EC/HC 2008b). Additionally, the adenocarcinomas of the uterus observed in the long-term inhalation study in rats after D5 exposure are considered unlikely to be transferable to humans (SCCS 2015; Klaunig et al. 2016). Some studies have indicated that D4, but not D5, may exhibit very weak estrogenic and anti-estrogenic activity in the rat uterotrophic assay.

6.5.2 *Toxicokinetics*

In various studies with rats, resorption rates of approximately 4.4–8.8% for D4 and 1–3% were observed after inhalation, regardless of sex and dose (Plotzke et al. 2000, SEHSC 2002a, b, Tobin et al. 2008, Schmitt et al. 2023). In all studies, the majority of the administered cVMS was exhaled immediately. Studies involving 12 subjects who inhaled D4 for 1 h showed an average uptake of approximately 13% of the administered dose after a single exposure and around 10% after multiple exposures at rest and 7% and 4% during physical activity, respectively (Utell et al. 1998).

Many orally administered siloxanes remain in the gastrointestinal tract as poorly absorbable microemulsions and are ultimately excreted unchanged in the feces (SEHSC 2004). Depending on the carrier substance, animal experiments showed an uptake of between 12 and 51% for D4 and 12% for D6 (SEHSC 1998, 2004).

Overall, the dermal absorption of cVMS demonstrated a low uptake through the skin (Clewell et al. 2024). Animal studies with labeled individual substances reported an absorption of approximately 1% (Jovanovic et al. 2008; Zareba et al. 2002), while

D5 showed an absorption rate of 0.1–0.8% (Jovanovic et al. 2008; SEHSC 2000) of the administered dose. In six subjects, 13C-labeled D4 and D5 were applied to the axillae, and only systemic absorption rates of 0.1–0.3% and 0.05% were observed (Reddy et al. 2007). For D6, no penetration through the skin was detected (SEHSC 2003).

The absorbed portion of the siloxanes is generally well distributed across all tissues. The highest concentrations in plasma and tissues are found immediately after exposure ends, particularly in the lungs, ovaries, and especially in fatty tissue.

Two studies have been conducted in humans, wherein urine was analyzed after inhalation exposure to labeled D4 (Utell et al. 1998; Reddy et al. 2003). Dimethylsilanediol and methylsilanetriol were identified as major metabolites, accounting for approximately 27% and 70% of the total observed metabolites, respectively. A schematic representation of the proposed metabolism of D4 in humans can be found in Fig. 6.6. It is assumed that biotransformation via a reaction catalyzed by the cytochrome P450 system leads to oxidative demethylation and ring opening. Further hydrolysis and/or oxidation steps ultimately form various short-chain, linear methylsiloxanes, which are then excreted in the urine (Johnson et al. 2011). A comparable mechanism is assumed for the metabolism of D5. Upon inhalation, cVMS are primarily excreted through the breath and urine, with only a minor proportion being excreted in the feces. Elimination from plasma and tissues (except fat) is generally very rapid and multiphase, with most of the dose being excreted within the first 24 h (SEHSC 1996; Plotzke et al. 2000). However, elimination from fatty tissue is slow, with a half-life of over 100 h (D5) in some cases (Tobin et al. 2008).

6.5.3 Occurrence in the Blood

In principle, determining internal exposure to siloxanes is challenging because measuring the starting substances poses a significant risk of contamination. Overall, the studies available to date indicate a low level of exposure for the general population that is not occupationally exposed. Human biomonitoring in urine is not feasible because the observed metabolites are not specific to cVMS exposure.

Data on the exposure of the general population are limited. Hansen et al. (2013) presented initial results from two Norwegian cohorts, from which 94 and 17 plasma samples were taken from women in 2005 and 2009, respectively. In the first study, medians (% > LOQ, maximum) reported for D4, D5, and D6 were 4.8 μg/l (85%, 12.7 μg/l) for D4, 1.9 μg/l (18%, 3.9 μg/l) for D5, and 2.9 μg/l (5%, 3.2 μg/l) for D6. In the second study, only D4 was found above the LOQ of < 1.4–2.7 μg/l in 18% of the samples, while D5 and D6 could not be determined above the LOQs of 1.7 and 1.5 μg/l. There was no significant correlation between the concentrations of D4 and the reported total body cream use. In a German study of 42 plasma samples conducted in 2013/14, D4, D5, and D6 were only quantified above the LOQ in 31%,

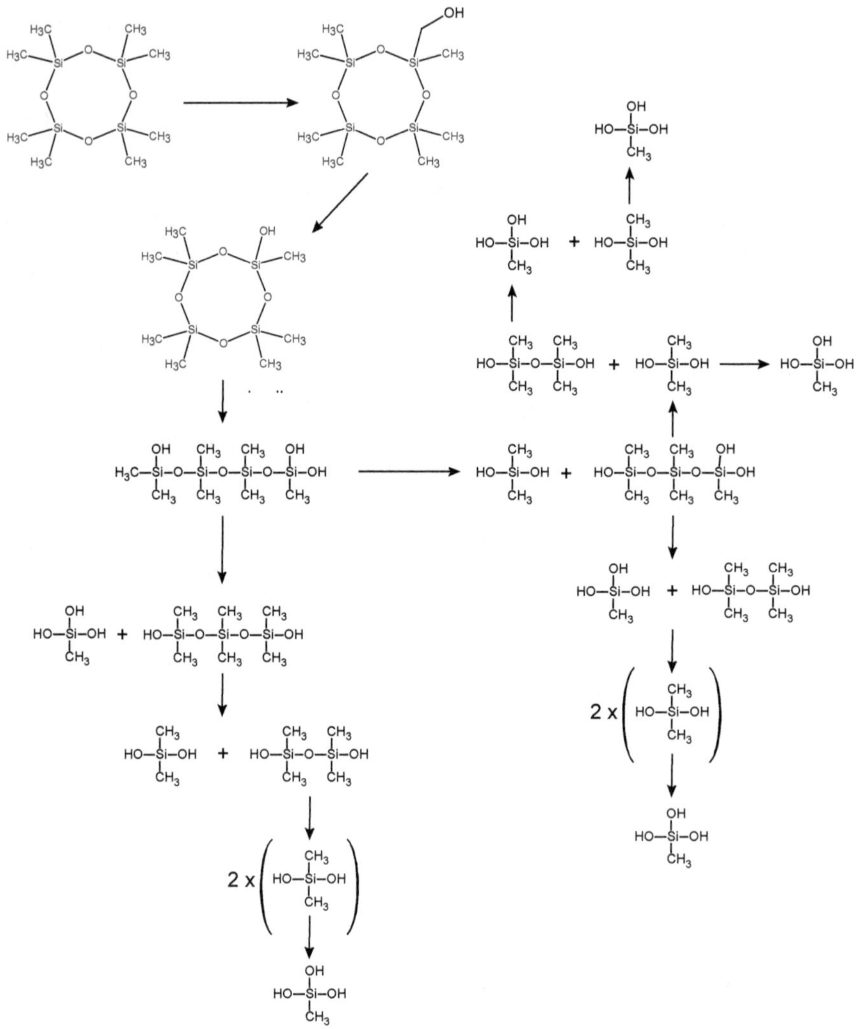

Fig. 6.6 Schematic representation of the proposed metabolism of D4 in humans (modified from Reddy et al. 2003)

12%, and 12% of the samples, respectively (Fromme et al. 2015). The concentrations ranged in a very low range of < 0.18–0.73 μg/l for D4, < 0.29–0.48 μg/l for D5, and < 0.44–0.79 μg/l for D6.

In a Chinese study conducted in 2012, the levels of D4 in the blood plasma of 519 individuals in the general population were < 0.8–5.95 μg/l (3.7% > LOQ), for D5 < 1.0–6.22 μg/l (3.7% > LOQ), and D6 < 1.0–7.50 μg/l (1.7% > LOQ) (Xu et al. 2015). Additionally, among the 80 plasma samples from Chinese residents (ages 1–65) who had no known exposure collected in 2018, cVMS were detected in only eight females, with concentrations ranging from < 0,4–7.5 µg/l (10% > LOD), < 0,5–8.3 µg/l (10% > LOD), and < 0.4–6.3 µg/l (8.8% > LOD) for D4, D5, and D6, respectively (Guo et al. 2020). In another Chinese study involving 100 children up to 12 years of age, cVMS were detected in only nine children living in residential areas, with levels of < 0.4–11 µg/l (4% > LOD) for D4, < 0.5–17 µg/l (4–5% > LOD) for D5, and < 0.4–9.5 µg/l (3–4% > LOD) for D6 (Guo et al. 2021). Additionally, Guo et al. (2022) analyzed 50 blood samples from Chinese students in 2018 and reported ranges from <-19 µg/l for D4, <-22 µg/l for D5, and <-17 µg/l for D6. Levels were higher in female than in male participants. The concentrations of cVMS in children from industrial areas were one to two orders of magnitude higher than those from residential areas. Jiang et al. (2023) investigated the internal exposure of 3–7 individuals living near a large Chinese petroleum refinery. In downwind areas, the detection rates of D4 and D5 in male and female plasma were 100% at sites 200, 600, and 800 m from the refinery, while the rates were 0% and 33–100% for males and females, respectively. At downwind sites (compared to upwind sites), the mean concentrations in male and female plasma samples were < LOD–8.0 µg/l (all < LOD) and 0.8–8.2 µg/l (<LOD–0.7 µg/l) for D4, < LOD–6.8 µg/l (<LOD–1.0 µg/l) and 2.2–7.4 µg/l (<LOD–2.0 µg/l) for D5, and < LOD–3.5 µg/l (all < LOD) for D6. In another Chinese study conducted in 2017/18, only medians of 0.2 µg/l for D4, 0.3 µg/l for D5, and 0.2 µg/l for D6 were observed in 142 adults without occupational exposure who lived in a nonindustrialized area (Zhang et al. 2023).

6.5.4 Occurrence in Women with Breast Implants

The commonly used breast implants consist of a shell made of silicone elastomer, into which silicone gel is incorporated. It is known that diffusion processes from intact implants (so-called bleeding), and in particular, ruptures in the gel cushion, can expose surrounding tissue and possibly even the entire organism to substances from the gel (Lykissa et al. 1997; Birkefeld et al. 2003; Semple 2007). The rupture rates range from 4 to 30% (Wazir et al. 2015).

In four women with silicone breast implants, the concentrations of D3 in plasma ranged from 2 to 12 µg/l, and for D4, from 2–36 µg/l, while D5 and D6 were not found above the LOQ of 2 µg/l (Flassbeck et al. 2001). In the control group (n = 2), cVMS could not be detected above the LOQ. Occasionally, siloxanes were detected in the subjects' blood even several years after a ruptured or burst implant was removed. In a more recent study, the internal exposure of 11 women with ruptured breast implants and 13 women with intact implants was determined (Rosendahl et al. 2016). D5 could only be quantified in one person from each group in this study. D4 was not detected above the LOQ of 0.18 ng/g whole blood in any individual in the group with intact implants, while in the group with defective implants, 7 of 11 samples had concentrations above the LOQ, ranging from 0.19 to 0.57 ng/g whole blood. Similar conditions were observed for D6, which was found in 7 of 11 samples from the group with damaged implants at concentrations of approximately 0.13 to 0.16 ng/g whole blood above the LOQ.

6.5.5 Occurrence in Breast Milk

It can be assumed that the highly lipophilic siloxanes can also pass into breast milk. However, no valid exposure data are available to date. In a national screening program conducted in 2004 in Sweden, 49 breast milk samples were examined for their cVMS content (Kaj et al. 2005a). At least one siloxane was detected in 11 of the 49 samples. The maximum concentrations for D4, D5, and D6 were 10 μg/l, 4.5 μg/l, and 4.8 μg/l, with a detection limit of 2 μg/l. However, the results should be viewed cautiously, as siloxanes were not originally intended to be analyzed; the selected processing could lead to a reduction of the analytes, and only external calibration was carried out without internal standards.

6.6 Chlorophenols

The most important chlorophenols, along with their abbreviations and chemical structures, are presented in Table 6.8. Chlorophenols are considered moderately persistent, exhibiting resistance to biological degradation, a tendency toward lipid solubility, and increased bioconcentration as the chlorine content rises (ATSDR 2022b). Pentachlorophenol (PCP) has been extensively used since around 1930 as a fungicide, bactericide, herbicide, molluscicide, algicide, and insecticide in agriculture and forestry, as well as in textiles, antifouling paints, and oil drilling (ATSDR 2022a). It has served as a wood preservative dissolved in hydrocarbon diluents. Due to its widespread use, it can be detected in water and soil, and under certain conditions, it may be released from these environments. Volatilization should also be anticipated following application as a liquid formulation in wood preservation (ATSDR 2022a). The other chlorophenols, with lower degrees of chlorination, are mainly employed in the manufacture of agricultural chemicals, such as herbicides, and, to a lesser extent, in the production of pharmaceuticals, biocides, and dyes (ATSDR 2022b).

Beginning in the 1980s, the use was restricted in many countries. Furthermore, PCP and its salts and esters are included in Annex A of the Stockholm Convention on Persistent Organic Pollutants, under which parties must take measures to halt production and use unless an exemption has been requested (SC 2025).

Extensive reviews on chlorophenols serve as the foundation for the next two paragraphs (IARC 2019; ATSR 2022a, b).

6.6.1 Health Aspects

The most sensitive PCP effects observed in animal experiments and human studies were those on the liver, including hepatomegaly, increases in liver enzymes, alterations in porphyrin excretion, and centrilobular degeneration. Animal studies also

Table 6.8 Names, abbreviations, CAS No. and the chemical structure of common chlorophenols

Name	Abbreviation	CAS No	Structure
2-Chlorophenol	2-MCP	95-57-8	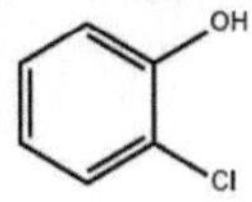
4-Chlorophenol	4-MCP	106-48-9	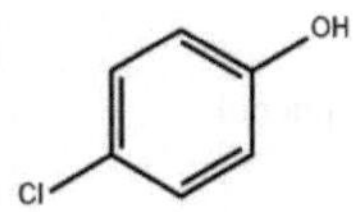
2,4-Dichlorophenol	2,4-DCP	120-83-2	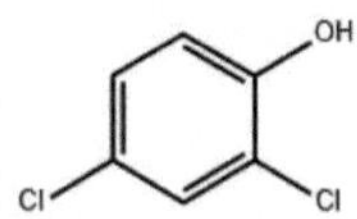
2,5-Dichlorophenol	2,5-DCP	583-78-8	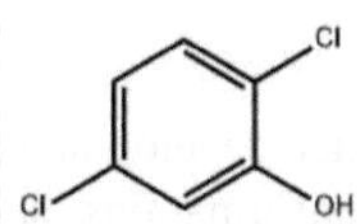
2,3,4-Trichlorophenol	2,3,4-TCP	15950-66-0	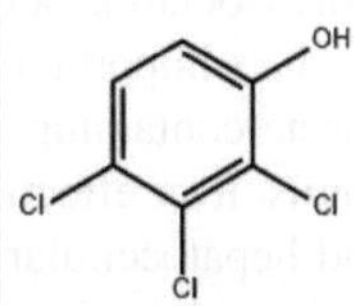
2,4,5-Trichlorophenol	2,4,5-TCP	95-95-4	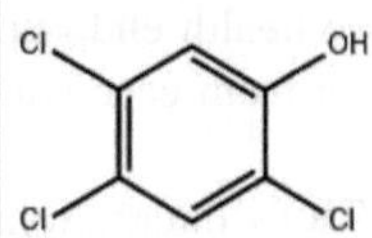
2,4,6-Trichlorophenol	2,4,6-TCP	88-06-2	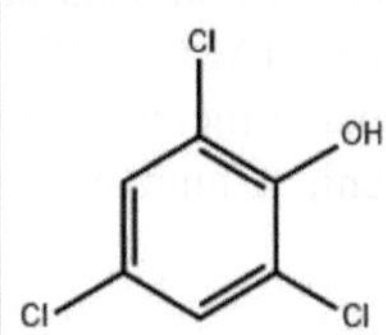

(continued)

Table 6.8 (continued)

Name	Abbreviation	CAS No	Structure
2,3,4,6-Tetrachlorophenol	2,3,4,6-TeCP	58-90-2	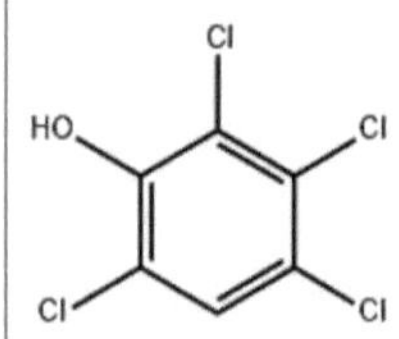
Pentachlorophenol	PCP	87-86-5	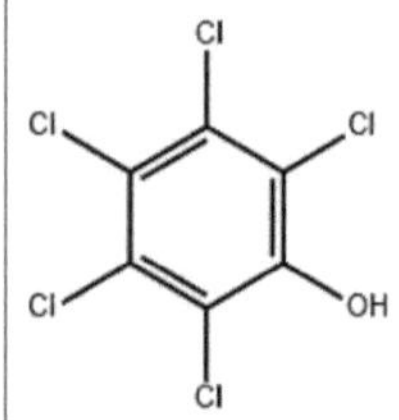

demonstrated developmental effects, including increases in fetal/neonatal mortality, skeletal malformations, and decreases in growth. Various epidemiological studies indicated an association between PCP exposure and an increased risk of non-Hodgkin lymphoma. It is important to note that technical PCP is always contaminated with other chlorine-containing substances, such as dioxins and furans. For the other chlorophenols, liver effects, including changes in clinical chemistry, increased liver weight, and hepatocellular hypertrophy, as well as impacts on the central nervous system, body weight, immune system, and reproductive function, were observed as sensitive health endpoints in animal experiments following oral administration. Information from case studies and epidemiological studies of these compounds is limited.

The WHO's International Agency for Research on Cancer has classified PCP as carcinogenic to humans in Group 1, based on sufficient evidence in both humans and animals (IARC 2019). For 2,4,6-TCP, the IARC found inadequate evidence in humans but sufficient evidence in experimental animals, classifying it as possibly carcinogenic to humans in Group 2B.

6.6.2 *Toxicokinetics*

To date, toxicokinetic data are unavailable for all chlorophenols. In animal experiments, absorption from inhalation, oral, and dermal routes is rapid and nearly complete. Chlorophenols are widely distributed throughout the body, with the highest concentrations found in the liver, kidneys, and spleen. The binding to plasma proteins is high and increases with the degree of chlorination. Chlorophenols are rapidly metabolized, particularly in the liver, and, depending on the species, dose, and site of application, are predominantly excreted in urine in a glucuronidated or sulfated form. Furthermore, reactive quinones and semiquinones are produced during metabolism, particularly with di- and tetrachlorophenols. Oxidative dechlorination to tetrachloro-p-hydroquinone (TCHQ) has been noted for PCP. In animal studies, the substances are primarily excreted in urine, with half-lives ranging from a few hours to several days. Research involving volunteers has shown that approximately 80–90% of PCP is excreted unchanged in urine, with the remainder primarily as glucuronide conjugates. TCPs and TeCPs are mostly excreted as conjugates, particularly sulfate conjugates. In a study of five sawmill workers exposed to 2,4,6-TCP, 2,3,4,6-TeCP, and PCP through inhalation and dermal contact, the elimination half-lives were found to be 18 h, 4.2 days, and 16 days, respectively (Pekari et al. 1991).

For humans, polluted air and contaminated drinking water represent the main exposure pathways, while intake through food is regarded as secondary.

6.6.3 Occurrence of PCP in the Blood

Currently, results of measurements of PCP in blood are unavailable, but were conducted prior to the corresponding bans or restrictions on its use. Zeng et al. (2011) summarized findings from various studies on blood contamination in the general population from 1967 to 2003. They calculated midpoints for the USA and Germany for specific periods, noting a decrease from 123 μg/l in the USA and 61 μg/l in Germany between 1978 and 1986 to 1.36 μg/l and 3.14 μg/l, respectively, between 1995 and 2001. The authors observed an overall global mean annual decrease of 18% during this time, with decreases of 26% and 20% in the USA and Germany, respectively. In Japan, Fujii et al. (2014a) analyzed serum samples from 20 adults in 1989, 1999, and 2010. Mean PCP values declined during this period from 0.65 ng/g ww to 0.30 ng/g ww and then to 0.14 ng/g ww.

The PCP results of the most recent studies are summarized in Table 6.9. Glynn et al. (2011) repeatedly analyzed serum from 10 Swedish women for PCP during pregnancy and found PCP medians of 2.9 ng/g ww, 1.5 ng/g ww, and 1.5 ng/g ww in the 9th-13th, 23rd-26th, and 35th-36th weeks of pregnancy, respectively. There were no significant changes in levels during pregnancy; however, notable inter-individual variation in PCP concentrations was observed, with a 20-fold increase in early pregnancy (weeks 9–16) and a threefold increase in late pregnancy.

Several studies have simultaneously investigated PCP in maternal blood and cord blood. For example, Meijer et al. (2008) observed a cord blood to maternal serum concentration ratio (C/M ratio) of 0.9 (0.6–1.6) in nine mother–child pairs in the Netherlands between 2000 and 2001. In Slovakia, Park et al. (2008) analyzed 92 maternal serum and cord blood samples collected immediately after delivery and found a significant correlation between the concentrations in the two matrices. Considering the median values, the C/M ratio in this study was 1.1. Dallaire et al.

Table 6.9 PCP concentrations in serum/plasma of the general population in ng/g fw or ww

References	N	Median (range)	Country, year
Adults			
Fujii et al. (2014a)	20	0.14 (0.04–0.85)	Japan, 2010
Dirtu et al. (2010)	52	< 0.50 (<0.50–1.20)[b]	Romania, 2006/07
Park et al. (2008)	92	0.65 (0.33–1.40[d])	Slovakia, 2002–04
Guvenius et al. (2003)	15	2.83 (1.36–13.2)	Sweden, 2000/01
Dirtu et al. (2010)	20	6.30 (1.29–20.3)[b]	Belgium, 2000
Glynn et al. (2011)	10[e]	2.9 (0.6–8.1)	Sweden, 1996–99
	10[e]	1.5 (1.4–9.5)	
	10[e]	1.5 (1.0–2.9)	
Meijer et al. (2008)	90	0.97 (0.30–8.50)	Netherlands, 2001/02
Dallaire et al. (2009a)	25	0.93[a] (0.24–2.89)	Canada, 1995–2001
Heudorf et al. (2000)	485	1.50 (<LOQ-59.3)[b]	Germany, 1998
Sandau et al. (2000)	13	2.02[a] (0.56–7.78)	Canada, 1992
Children/adolescents			
Carrizo et al. (2008)	66	6.40[c] (1.5–35)[b]	Spain, 1997–99
Carrizo et al. (2008)	131	0.61[c] (<0.30–4.7)[b]	Spain, 1997–99
Heudorf et al. (2000)	138	2.50 (<LOQ-44.7)[b]	Germany, 1998
Cord serum			
Park et al. (2008)	92	0.69 (0.37–1.64[d])	Slovakia, 2000–04
Guvenius et al. (2003)	15	1.96 (0.82–7.58)	Sweden, 2000/01
Meijer et al. (2008)	9	1.50 (0.42–1.90)	Netherlands, 2001/02
Dallaire et al. (2009a)	41	1.08[a] (0.30–2.91)	Canada, 1995–2001
Sandau et al. (2002)	30	1.67[a] (0.63–7.68)	Canada, 1993–96

[a]Geometric mean; [b]ng/ml; [c]Mean; [d]10th and 90th percentile; [e]Three diffent sampling times during pregnancy

(2009a) also identified a significant correlation between maternal PCP concentrations and those in the umbilical cord from Canadian samples. If the median levels of the 25 maternal serum samples and the 41 cord blood samples examined in this study are used, this case would yield a C/M ratio of 1.2.

6.6.4 *Occurrence of PCP in Urine*

For PCP, results from various studies on its occurrence in urine are available, with the findings of more recent studies compiled in Table 6.10. It is important to note that PCP itself can be a metabolite of other environmental contaminants such as hexachlorobenzene, pentachlorobenzene, or pentachloronitrobenzene, so its detection may also be due to exposure to these substances. Overall, the concentrations found in urine samples from both adults and children/adolescents are currently low. Nevertheless, Sun et al. (2023) still found significant differences in PCP concentrations among various age groups, with a higher median of 0.72 μg/l in participants aged 20–45 years compared to those aged 45–60 years (0.42 μg/l) and over 60 years (0.26 μg/l), respectively. In another Chinese study conducted in six cities in 2018/19, significant differences were observed in the urine of 425 adults among the study locations, ranging from a median of 0.17 μg/l to 0.93 μg/l (Li et al. 2025a, b). Additionally, significantly higher concentrations were found in younger individuals compared to older people, in urban residents compared to rural residents, and in those with a higher education level compared to those with a lower level.

Zeng et al. (2011) calculated a midpoint for the USA and Germany for specific periods based on various studies examining the contamination of urine in the general population conducted between approximately 1967 and 2003 (see Fig. 6.7). Globally, there was an average annual decrease of 11% during this time, with declines of 8% in the USA and 18% in Germany. For comparison, the figure also displays the median and geometric mean from a US study conducted from 2010 to 2012 and a German study from 2014 to 2017 (limited to subjects aged 14 to 17 years). Additionally, Schmied-Tobies et al. (2021) reported significantly lower concentrations in a representative German survey conducted in 1990/91 ($N = 678$) compared to a more recent study carried out in 2014/15 involving 6- to 14-year-old subjects. The geometric mean urinary PCP concentration of 0.18 μg/l in the later study was roughly 8% of that found in the 1990/1991 investigation.

6.6.5 Occurrence of chlorophenols in Breast Milk

To date, chlorophenols have been infrequently analyzed in breast milk. In 2000, breast milk samples were collected from 11 women in a Chinese study 3–7 days after birth (Hong et al. 2005). The PCP levels in this study ranged from 0.32 to 12.8 ng/g fw, with an average of 2.15 ng/g fw. Concentrations increased significantly with the mothers' age, which ranged from 24 to 33 years. Hong et al. (2005) also report on another, less widely available Chinese study by Zhang et al. (1995), in which a mean PCP concentration of 3.63 ng/g fw was found in breast milk samples. In Sweden, milk samples from 15 mothers were collected in 2000–2001, revealing significantly lower mean PCP levels of 0.020 ng/g fw (0.010–0.57 ng/g fw) (Guvenius et al. 2003). In the mothers' blood plasma samples, measured in parallel, significantly higher concentrations were observed, with a median of 2.83 ng/g fw (1.36–3.20 ng/g fw).

Ye et al. (2006) reported that 2,4-DCP, 2,4,5-TCP, and 2,4,6-TCP were each detected in 5% of 20 pooled breast milk samples, with detection limits of 0.16 μg/l, 0.10 μg/l, and 1.22 μg/l, respectively. In contrast, 2,5-DCP was not detected above the limit of detection of 0.42 μg/l. No results were reported in this study.

Table 6.10 Concentration of PCP in the urine of the general population

References	Age (year)	N	DF (%)	Median (range)	Country, year
Adults					
Luo et al. (2025a, b)	20–50	656	97	0.70 (< 0.03–31.5)	China, 2023
Li et al. (2025a, b)	≥ 20	425	99.6	0.35 (< 0.04–0.99[b])	China, 2018/19
Yang et al. (2021)	47[a]	297		0.41 (nr-0.81[b])	China, 2017–19
Sun et al. (2023)	≥ 20	129	100	0.44 (nr-14.2)	China, 2018–19
Tagne-Fotso et al. (2023)	18–47	900	17	< 0.15 (<0.15–0.23[c])	France, 2014–16
Huo et al. (2022)	23–44	879	100	0.70 (0.10–3.99[d])	China, 2013–17
Honda and Kannan (2018)	23–75	30	16.7	< 0.06 (<0.06–0.14)	Greece, 2012–14
Honda and Kannan (2018)	3–87	30	56.7	0.08 (<0.06–0.54)	Saudi Arabia, 2010–14
Honda and Kannan (2018)	2–67	47	72.3	0.08 (<0.06–7.64)	China, 2010–12
Honda and Kannan (2018)	24–75	41	58.5	0.10 (<0.06–7.56)	India, 2010–12
Honda and Kannan (2018)	2–59	36	19.4	< 0.06 (<0.06–0.13)	Japan, 2010–12
Honda and Kannan (2018)	6–73	40	90.0	0.20 (<0.06–0.78)	Kuwait, 2010–12
Honda and Kannan (2018)	19–83	19	78.9	0.15 (<0.06–0.33)	Vietnam, 2010–12
Honda and Kannan (2018)	11–66	31	71.0	0.17 (<0.06–0.82)	USA, 2010–12
Lu et al. (2015)	Adults	18	100	0.10 (nr-0.59[c])	China
Chen et al. (2012)	Adults	325		< 0.41 (nr-1.92[c])	China
Guo et al. (2019)	26	377	100	0.27 (0.03–15.2)	China, 2009/10
Hardy et al. (2021)	18–35	93	100	0.36 (0.08–11.3)	France, 2007
Honda and Kannan (2018)	30	26	61.5	0.07 (<0.06–0.24)	Korea, 2006/07
Children/adolescents					
Ding et al. (2024)	~ 8–10	624	100	0.42 (0.05–1.56[c])	China, 2017–20
Schmied-Tobies et al. (2021)	6–10	158	90	0.20 (<0.1–6.7)	Germany, 2014–17

(continued)

Table 6.10 (continued)

References	Age (year)	N	DF (%)	Median (range)	Country, year
Schmied-Tobies et al. (2021)	14–17	143	86	0.20 (<0.1–3.5)	Germany, 2014–17
Tagne-Fotso et al. (2023)	6–17	500	15	< 0.15 (<0.15–0.34[c])	France, 2014–16
Ding et al. (2024)	~ 3–7	821	99.5	0.08 (<0.003–0.57[c])	China, 2011–17
Guo et al. (2019)	3	377	98.7	0.07 (<LOD-6.31)	China, 2012/13
Ding et al. (2024)	~ 1–2	586	100	0.21 (0.05–0.60[c])	China, 2010–12
Lu et al. (2015)	0.3–4.1	11	100	0.09 (nr-0.35[c])	China
Becker et al. (2007)	3–14	599	49	< 0.60 (<0.60–9.71)	Germany, 2003/04

[a]Mean; [b]75th percentile; [c]95th percentile; [d]5th and 95th percentile; *DF* detection frequency

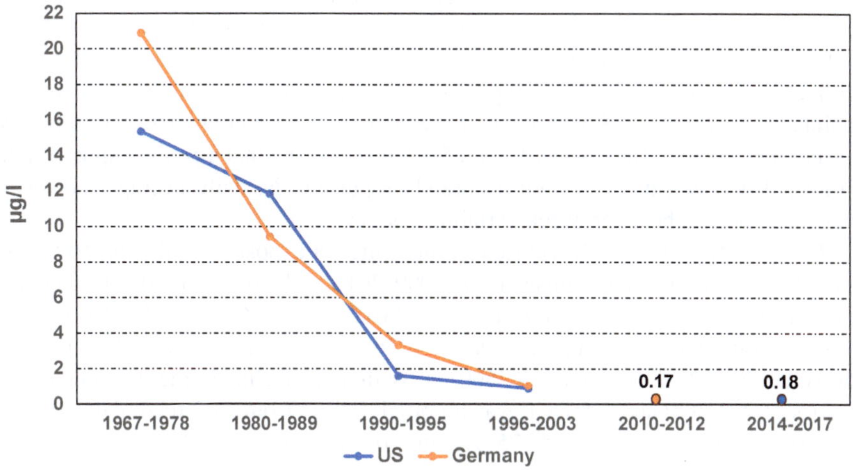

Fig. 6.7 Time trend of PCP in urine. The data from 1967 to 2003 were obtained from various studies on the exposure of the general population, which were then transformed and summarized by Zheng et al. (2011) into a midpoint value. For comparison, data from 2010 to 2012 represent the median from a US study (Hong and Kannan 2018); data from 2014 to 2017 represent the geometric mean of a German study (Schmied-Tobies et al. 2021)

6.6.6 *Occurrence of Other Chlorophenols in Urine*

As of now, no reliable biomarkers exist for mono- to tetrachlorophenols because measuring both the parent compounds and the dechlorinated derivatives is not specific to chlorophenol exposure. These compounds can form from the degradation of chlorobenzenes and insecticides/herbicides such as lindane, 2,4-dichlorophenoxyacetic acid, triclosan, prochloraz, fenchlorfos, fenoprop, and 2,4,5-trichlorophenoxyacetic acid (Schmied-Tobies et al. 2021).

Table 6.11 presents the results of chlorophenol measurements that have been frequently examined. Overall, the medians are in a low range, but significant individual fluctuations in concentrations occur across several orders of magnitude. The highest medians were observed for 2,5-DCP. Notably high 2,5-DCP medians were reported in earlier studies from the USA and China, with concentrations of 19.1 μg/l and 11.2 μg/l in adults, and 16.5 μg/l in the urine of children in China (Pollack et al. 2018; Lu et al. 2015). Following this, 2,4-DCP was detected with medians ranging from < 0.15 to 3.45 μg/l. For trichlorophenols, only very low concentrations, if any, above the limits of quantification have been reported so far. As part of the representative Korean National Human Biomonitoring 2009, urine samples were collected from 186 Koreans aged 18–69 years (Park and Kim 2018). There were no significant differences regarding gender or age.

Lassen et al. (2013) collected urine samples from 33 young men in Denmark over a 3-month period. They report that the osmolality-adjusted median concentrations of the sum of 2,4-DCP and 2,5-DCP were significantly lower in 24-h urine samples compared to first-morning urine samples (median values: < 0.06 μg/l vs. 0.64 μg/l) from the same participants. In eight French pregnant women, urine samples were collected over one week, averaging 60 samples per week per woman, at three time points (15th, 24th, and 32nd weeks of pregnancy) (Vernet et al. 2018). Within-day variability was high for 2,4-DCP and 2,5-DCP, indicating that a random spot sample collected within a day does not accurately represent the daily average; however, between-day and between-week variability is low.

Frederiksen et al. (2020) reported a time course of exposure, analyzing 100 urine samples each from young Danish men in 2009, 2013, and 2017. A significant decrease of 63% and 80% was observed from 2009 to 2017 for 2,4-DCP and 2,5-DCP, respectively. Additionally, 2,4,5-TCP decreased by 36% during this period. In the representative US NHANES (n = each ~ 2500), a continuous decline in the 95th percentile values of 2,4-DCP and 2,5-DCP was reported, falling from 21.3 μg/l and 705 μg/l in 2003–2004 to 7.70 μg/l and 175 μg/l in 2015–2016 (CDC 2023).

Other chlorophenols have been examined in urine samples to a lesser extent so far. In two German studies conducted from 2003 to 2006 and from 2014 to 2017, 2-MCP was found to have a median (95th percentile) of 1.70 μg/l (6.78 μg/l) and 0.2 μg/l (1.8 μg/l), while 4-MCP had values of 4.58 μg/l (15.3 μg/l) and 1.4 μg/l (4.5 μg/l) (Becker et al. 2007; Schmied-Tobies et al. 2021). In the period from 2014 to 2016, 4-MCP was also reported in a French study, showing medians (95th percentiles) of < 0.15 μg/l (0.21 μg/l) in adults and < 0.15 μg/l (0.21 μg/l) in children (Tagne-Fotso et al. 2023). 2,3,4,6-TeCP was observed by Becker et al. (2007) in 10% of the samples, with a 95th percentile of 0.43 μg/l. Furthermore, for 2,3,4,6-TeCP, 95th percentiles of 0.3 μg/l (in 44% of the samples) were reported in Germany, and < 0.15 μg/l in French adults (in 4% of the samples) and children (in 1.6%) (Schmied-Tobies et al. 2021; Tagne-Fotso et al. 2023).

Table 6.11 Median (range) concentrations of dichlorophenols (DCPs) and trichlorophenols (TCPs) in urine in μg/l

References	N	2,4-DCP	2,5-DCP	2,3,4-TCP	2,4,5-TCP	2,4,6-TCP	Country, year
Adults/general population							
Yang et al. (2021)	297	0.50	–	–	–	0.24	China, 2017–19
Dodson et al. (2020)	726	0.24 (<0.10–9.75)	0.81 (<0.20–242)	–	–	–	USA, 2017/18
Frederiksen et al. (2020)	100	0.21 (<0.04–1.97)	0.17 (<0.04–10.0)	–	0.03 (<0.01–0.33)	–	Denmark, 2017
CDC (2023)	1690	0.50 (7.40[b])	1.9 (164[b])	–	–	–	USA, 2015/16
Tagne–Fotso et al. (2023)	900	< 0.15 (nr)	< 0.15 (nr)	< 0.15	< 0.15	< 0.15	France, 2014–16
Zhou et al. (2014)	100	0.73 (<0.10–165)	6.02 (0.18–1843)	–	0.07 (<0.10–0.57)	0.28 (<0.10–6.11)	USA, 2009–12
HC (2013)	~ 2500	1.0 (12.0[c])	5.0 (77.0[c])	–	< 0.50 (<0.50[c])	< 1.0(<1.0[c])	Canada, 2009–11
Guo et al. (2019)	377	0.81 (0.14–185)	2.32 (0.06–130	–	< LOQ (<LOQ–158)	1.14 (0.52–6.93)	China, 2009/10
Lee et al. (2019)	1874	0.14[b]	0.42[b]	–	0.08[b]	0.37[b]	Korea, 2009/10
Lu et al. (2015)	18	0.57 (0.8[b])	11.2 (79.7[b])	–	< 0.006	–	China
Park et al. (2018)	1865	0.08 (<0.05–83.6)	0.40 (<0.05–63.0	–	–	–	Korea, 2009
Forde et al. (2015)	150	0.68 (<0.20–26)	1.80 (<0.20–1400)	–	< 0.50 (<0.50[b])	< 1.0 (<1.0[b])	Caribbean islands, 2008–11
Lassen et al. (2013)	97	–	–	–	< 0.06(<0.06–2.09)	–	Denmark, 2008
Chen et al. (2012)	325	–	–	< 0.28 (0.55[c])	< 0.15 (0.39[c])	–	China

(continued)

Table 6.11 (continued)

References	N	2,4-DCP	2,5-DCP	2,3,4-TCP	2,4,5-TCP	2,4,6-TCP	Country, year
Philippat et al. (2013)	71	0.6 (5.7[c])	7.5 (178[c])	–	–	–	USA, 2005–08
Pollack et al. (2018)	143	0.86 (<0.5–38.6)	19.1 (<0.5–804)	–	0.08 (<0.08–2.8)	0.27 (0.03–8.5)	USA, 2005–07
Children/adolescents							
Seo et al. (2021)	165	0.40 (0.88[a])	0.32 (1.40[a])	–	< 0.04 (0.20[a])	< 0.08 (0.75[a])	Korea, 2019
Parastar et al. (2018)	242	2.0 (7.5[c])	7.0 (14.0[c])	–	0.81 (4.6[c])	0.49 (6.8[c])	Iran, 2016
CDC (2023)	415	1.30 (8.50[b])	2.70 (196[b])	–	–	–	USA, 2015/16
Schmied–Tobies et al. (2021)	485	0.2 (–15.2)	0.2 (<0.10–357)	< 0.10 (<0.10–0.5)	< 0.10 (<0.10–0.9)	0.1 (<0.10–2.8)	Germany, 2014–17
Tagne-Fotso et al. (2023)	500	< 0.15 (<0.15[c])	< 0.15 (<0.15[c])	< 0.15	< 0.15	< 0.15	France, 2014–16
Guo et al. (2019)	377	3.45 (<LOQ-167)	4.64 (<LOQ-141)	–	0.02 (<LOQ-8.29)	0.07 (<LOQ-6.31)	China, 2009/10
Lu et al. (2015)	11	2.02 (5.8[c])	16.5 (73.4[c])	–	< 0.006 (0.19[c])	–	China
Becker et al. (2007)	599	0.33 (<0.1–19.3)	0.71 (<0.1–517)	< 0.1 (<0.1c)	0.15 (<0.1–4.55)	0.11 (<0.1–13.7)	Germany, 2003–06

[a]90th percentile; [b]Geometric mean; [c]95th percentile; nr: not reported; *LOQ* limit of quantification

6.7 Bromophenols

Brominated phenols are a class of various brominated substances. The most important representatives are listed in Table 6.12. They have been used, starting with 4-BP in the 1920s, as flame retardants and for applications including the synthesis of other BFRs, pigments, herbicides, germicides, and antifungal agents (EFSA 2024b). 2,4-DBP, 2,4,6-TBP, and PBP are the most commonly used bromophenols on a larger scale (Michałowicz et al. 2022). For example, 2,4,6-TBP is marketed under trade names such as Bromkal Pur 3, Bromol, Flammex 3BP, and NSC 2136. Due to its fungicidal effect, it was also used as a wood preservative in the past (EFSA 2024b). Brominated phenols can form naturally and are also transformation products of other brominated flame retardants, such as tetrabromobisphenol-A (TBBPA) and polybrominated diphenyl ethers (PBDEs), through UV photolysis, thermal treatment, and biological processes (Michałowicz et al. 2022).

Various reviews are available in the scientific literature and were also used as a basis for the next two chapters (EFSA 2012, 2024b; Michałowicz et al. 2022).

6.7.1 Health Effects

Data from human studies are not yet available for this class of substances, and experimental data from animals are extremely limited. In rats, the target organs of 2,4,6-TBP after oral administration are the liver and kidneys, which show increased organ weights and histopathological changes. At very high doses, this substance has been

Table 6.12 Names, abbreviations, and CAS No. of common bromophenols

Name	Abbreviation	CAS No
3-bromophenol	3-MBP	591-20-8
4-bromophenol	4-MBP	106-41-2
2,3-dibromophenol	2,3-DBP	57,383-80-9
2,4-dibromophenol	2,4-DBP	615-58-7
2,5-dibromophenol	2,5-DBP	28165-52-8
2,6-dibromophenol	2,6-DBP	608-33-3
2,3,6-tribromophenol	2,3,6-TBP	28165-57-3
2,4,5-tribromophenol	2,4,5-TBP	14401-61-7
2,4,6-tribromophenol	2,4,6-TBP	118-79-6
2,3,5,6-tetrabromophenol	2,3,5,6-TeBP	na
2,3,4,6-tetrabromophenol	2,3,4,6-TeBP	14400-94-3
Pentabromophenol	PBP	608-71-9

na: not available

observed to affect reproduction and development. Genotoxicity is not considered likely. There is evidence that 2,4,6-TBP disrupts thyroid hormone homeostasis.

6.7.2 Toxicokinetics

To date, only limited toxicokinetic data are available, primarily regarding 2,4,6-TBP. In rats, this substance was rapidly absorbed following oral administration and was primarily distributed to the kidneys, lungs, and liver. Seventy to ninety percent were excreted in the urine within 48 h, and the half-life in the blood was 2 h. After both oral and intravenous administration of labeled 2,4,6-TBP to rats, Knudsen et al. (2019) reported a bioavailability of 30% for this substance.

Ho et al. (2015) analyzed 100 urine samples in China collected in 2010 and found that the glucuronide and sulfate conjugates had geometric mean values of 0.32 and 0.11 μg/g crea for 2,4-DBP, and 0.87 and 0.10 μg/g crea for 2,4,6-TBP. They observed no age dependency; however, a significant difference was noted in the urine concentrations of 2,4-DBP-glucuronide between males and females, with higher concentrations found in men.

6.7.3 Occurrence in Urine

Table 6.13 presents the results of studies that analyzed bromophenols in urine samples. The number of analyzed samples is sometimes small because these measurements are often taken from actual samples during method development. The most frequently detected substance at the highest concentrations was 2,4,6-TBP, followed by 2,4-DBP. Based on the limited studies available, a European study found lower concentrations than those in China, although levels in China vary greatly. For MBPs, 3-MBP was often not detected above the LOQ of 8.8 or 80 ng/l in China (Feng et al. 2016; Lin et al. 2020). In Spain, it was only found in one of three samples at 110 ng/l. Additionally, Table 6.13 shows significantly higher concentrations for 4-MBP. Furthermore, 2,3-DBP, 2,4-DBP, and 2,6-DBP could not be found in any Chinese urine sample (lowest LOQ of 4.7–5.1 ng/l) (Feng et al. 2016; Lin et al. 2020), while Pena et al. (2024) report only positive results between 100 and 220 ng/l in one of three subjects examined. PBP has also not been found in any urine samples above the LOQs from various studies (Feng et al. 2016; Lin et al. 2020, Pena et al. 2024).

Table 6.13 Mean (range) concentrations of bromophenols in urine and Serum/plasma samples of the general population

References	N	4-MBP	2,4-DBP	2,3,6-TBP	2,4,5-TBP	2,4,6-TBP	2,3,4,6-TeBP	Country, year
Urine (ng/l)								
Peña et al. (2024)	3	– (<7.2–150)	– (<6.0–110)	–	< 12	– (50–460)	< 17	Spain
Lin et al. (2020)	10	270 (<27–480)	90 (<86–530)	< 68	< 13 (<13–150)	2210 (240–7390)	< 23	China
Chen et al. (2016)	30		44.5[a] (0.65–916)	–	23.9[a] (1.47–239)	9.74[a] (0.30–103)	–	China, 2012
Feng et al. (2016)	10	7550 (-)	650 (–)	< 8.5 (–)	< 4.5	5570 (–)	< 16.1	China
Blood (ng/l or ng/kg ww)								
Dufour et al. (2020)	160	–	–	< 2.4	< 5.0	74.6 (<50–1277)	< 4.1	Belgium, 2015–18
Dufour et al. (2016)	274	–	–	< 2.4	< 5.0	81.2 (<50–1277)	< 4.1 (<4.1–51)	Belgium, 2015
Eguchi et al. (2015)	34	–	–	–	< 0.5	220 (100–400)	< 0.5 (<0.5–7.9)	Vietnam, 2010/11
Fujii et al. (2014a)	20	–	–	–	–	350 (62–960)	–	Japan, 2010
Eguchi et al. (2012)	20	–	–	–	2.4 (<0.3–7.1)	360 (72–1200)	0.24 (<0.3–2.0)	India, 2007
Kawashiro et al. (2008)	6	–	–	–	–	22 (<60–130)	–	Japan, 2005/06
Dallaire et al. (2009b)	112–722	–	148[b] (<50–4100)	–	–	58[b] (<5.0–280)	7.0[b] (<1.0–1395)	Canada, 2004

(continued)

Table 6.13 (continued)

References	N	4-MBP	2,4-DBP	2,3,6-TBP	2,4,5-TBP	2,4,6-TBP	2,3,4,6-TeBP	Country, year
Blood (ng/g lw)								
Lin et al. (2023)	20	1.92 (<LOQ–6.75)	7.49 (<LOQ-47.4)	–	0.02 (<LOQ-0.46)	146 (<LOQ-381)	1.85 (<LOQ-4.35)	China, 2017/18
Ali et al. (2013)	17	–	–	–	–	3.5 (<5.0–15)	–	Pakistan, 2012
Butt et al. (2016)	34		–	–	< 1.2	19.2^b (5.10–63.4)	–	USA, 2008–10
Qiu et al. (2009)	4	–	1.5 (–)	–	0.2 (–)	0.8 (–)	–	USA, 2003/04
Thomsen et al. (2001)	5	–	–	–	–	11 (0.17–28)	–	Norway

[a]Median; [b]Geometric mean; *LOQ* Limit of quantification; *LOQ = 1.4–2.5 ng/g lw

6.7.4 Occurrence in the Blood

The study results in the blood are summarized in Table 6.13. As observed in the urine samples, 2,4,6-TBP is the most frequently found substance with the highest concentrations, followed by 2,4-DBP. The median levels of 2,4,6-TBP are generally higher in Asia than in Europe and the USA. The other bromophenols listed in the table appear at very low concentrations. 2,6-DBP and PBP have thus far only been detected in 8% of samples in Canada at concentrations exceeding the LOQ of 5.0 and 1.0 ng/l blood, respectively, with maximum values of 580 and 250 ng/l (Dallaire et al. 2009b). In Vietnam, PBP could not be detected in any sample above 0.5 ng/kg ww (Eguchi et al. 2015). Ali et al. (2013) analyzed serum samples from 17 mothers and their 2–10-year-old children from Pakistan and found 2,4,6-TBP with medians of < 5.0 in all samples from urban areas, and 5 ng/g lw in mothers and 11 ng/g lw in children from rural areas. Detection rates in this study ranged between 50 and 88%. Other bromophenols have rarely been analyzed. Eguchi et al. (2012) reported 2,3,5,6-TeBP with a median of 0.32 pg/g ww (<0.3–1.3 pg/g ww) in 20 Indian serum samples and not above the LOQ of 0.5 pg/g ww in 35 Vietnamese samples (Eguchi et al. 2015). Again, Lin et al. (2023) found 2,3,5,6-TeBP in 60% of the samples and reported a mean value of 0.20 ng/g lw (<LOQ-1.01 ng/g lw) in 20 Chinese serum samples.

To date, only one study from Japan has investigated the time course of internal exposure. Fujii et al. (2014a) analyzed serum samples from 20 adults in 1989, 1999, and 2010, observing increasing mean values of 2,4,6-TBP from 150 pg/g ww to 240 pg/g ww and 350 pg/g ww.

Various studies have determined fetal exposure. In the USA from 2003 to 2004, 16 cord blood samples were taken along with 4 maternal serum samples (Qiu et al. 2009). The mean values in the cord blood samples were higher than those in the maternal blood, with 20 versus 1.5 ng/g lw for 2,4 DBP, 7.9 versus 0.2 ng/g lw for 2,4,5-TBP, and 6.6 versus 0.8 ng/g lw for 2,4,6-TBP. In an analysis of six maternal blood samples and six cord blood samples conducted in Japan from 2005 to 2006, Kawashiro et al. (2008) also found mean 2,4,6-TBP levels of 22 pg/g ww (<60–130 pg/g ww) in 17% of the samples and 37 pg/g ww (<50–110 pg/g ww) in 33% of the samples. Furthermore, Chen et al. (2016) collected a total of 30 cord blood samples in China in 2012 and reported medians (range) of 6.79 pg/g ww (0.45–177 pg/g ww) for 2,4-DBP, 16.5 pg/g ww (0.74–297 pg/g ww) for 2,4,5-TBP, and 6.67 pg/g ww (0.50–95.0 pg/g ww) for 2,4,6-TBP.

6.7.5 Occurrence in Breast Milk

So far, only a very limited number of measurement results are available for this medium. In 2001, 2,4,6-TBP was found in six pooled breast milk samples from

Norwegian mothers at 627 pg/g lw, while PBP was below the limit of quantification at 0.3 pg/g milk (Thomsen et al. 2002). Chen et al. (2016) collected a total of 30 breast milk samples in China in 2012, reporting medians (range) of 98.4 pg/g ww (3.41–522 pg/g ww) for 2,4-DBP, 27.5 pg/g ww (3.41–70.2 pg/g ww) for 2,4,5-TBP, and 7.76 pg/g ww (0.80–67.0 pg/g ww) for 2,4,6-TBP. In southern Japan, Fujii et al. (2014b) analyzed nine milk samples taken between 2005 and 2006 and found a median of 1060 pg/g lw (130–2730 pg/g lw) for 2,4,6-TBP. The authors discussed whether marine food contamination is responsible for these levels. Comparable concentrations are reported by Fujii et al. (2018), who found 2,4,6-TBP in 94% of 64 randomly selected breast milk samples from six Japanese prefectures between 2008 and 2010, with a median of 750 pg/g lw (<31–3400 pg/g lw). No significant correlation was found between the levels of phenolic brominated flame retardants in milk and the mothers' age, workplace, or drinking/smoking habits in this study. For 50 breast milk samples from the USA in 2019, medians of < 90 pg/g lw (<90–25,000 pg/g lw) for 2,6-DBP, 94 pg/g lw (<30–1060 pg/g lw) for 2,3,5-TBP, 326 pg/g lw (<30–10,400 pg/g lw) for 2,4,6-TBP, and < 370 pg/g lw (<30–1060 pg/g lw) for 2,3,6-TBP were reported (Schreder et al. 2023). 3-BP + 4-BP and 2,3,4 + 2,4,5-TBP could not be separated into distinct samples, but could only be determined in 4% of the samples. For 2,4,6-TBP, the values were comparable to those in the Japanese studies.

6.8 Microplastic (MP)

Plastics are widely used in our modern world. Global plastic consumption is expected to increase from 435 million tons in 2020 to 736 million tons by 2040 (OECD 2024). Of this, 32% will be used for packaging, 15% in building and construction, 14% in transportation, and 10% each in textiles and consumer and institutional products. Additionally, the plastic lifecycle in 2020 emitted about 1.8 gigatonnes of carbon dioxide equivalent of greenhouse gases. The use also generated 81 million tonnes of mismanaged waste and resulted in the leakage of 20 million tonnes of plastics into the environment in 2020 (OECD 2024). Table 6.14 lists the most important polymers and their main application areas, and Fig. 6.8 shows the share of individual polymers in global consumption.

So far, there is no universally accepted definition of microplastics, but the European Food Safety Authority describes them as heterogeneous mixtures of differently shaped materials such as fragments, fibers, spheroids, granules, pellets, flakes, or beads measuring from 0.1 to 5000 μm (EFSA 2016). Their sources are varied and include natural, organic particles from everyday objects like clothing and shoes, as well as particles created by abrasion or aging of plastics from packaging, toys, construction materials, and personal care products. Tire abrasion particles also contribute (RIVM 2021). While primary microplastics are manufactured at this small size, secondary microplastics come from the fragmentation and breakdown of larger plastic items. Conversely, nanoplastics are defined as materials with any external

Table 6.14 Plastic categories and their main applications (modified after UNEP 2023)

Abbreviation	Name	Main applications
ABS	Acrylonitrile butadiene styrene	Drain-waste-vent pipe systems, musical instruments (recorders, plastic clarinets, piano movements), golf club heads, automotive trim components, automotive bumper bars, medical devices for blood access, enclosures for electrical and electronic assemblies, protective headgear, whitewater canoes, buffer edging for furniture and joinery panels, luggage and protective carrying cases, small kitchen appliances, toys (including Lego and Kre-O bricks), household and consumer goods
ASA	Acrylonitrile styrene acrylate	Weather-resistant panel in the construction industry, electrical appliances subject to high heat loads in the household, in the sports and leisure sector, such as a surfboard, for example, or in the exterior of cars
HDPE	High-density polyethylene	Packaging, construction, and various consumer products
LDPE	Low-density polyethylene	Packaging, including films, bags, and containers, and also finds applications in consumer goods, industrial products, and agriculture
LLDPE	Linear low-density polyethylene	Packaging films, bags, and containers, as well as various consumer goods and industrial applications
PA	Polyamides	Textiles, automotive industry, carpets, kitchen utensils, sportswear
PC	Polycarbonate	Electronic components, construction materials, data storage, automotive, aircraft, railway, security components, medical applications, phones
PET	Polyethylene terephthalate	Plastic bottles, tape applications, packaging, rigid cosmetic jars, microwavable containers, transparent films
PP	Polypropylene	Packaging, consumer goods, automotive applications
PS	Polystyrene	protective packaging (such as packing peanuts and CD and DVD cases), containers, lids, bottles, trays, tumblers, disposable cutlery, in the making of models
PUR	Polyurethane	Cleaning products, furniture, automobile seats, hoses, sculptures, decorations, filling of spaces and cavities

(continued)

Table 6.14 (continued)

Abbreviation	Name	Main applications
PVC	Polyvinyl chloride	Building and construction, health care, electronics, automobile and other sectors, piping and siding, blood bags and tubing, wire and cable insulation, windshield system components
SAN	Styrene acrylonitrile	Food containers, water bottles, kitchenware, computer products, packaging material, battery cases, plastic optical fibers

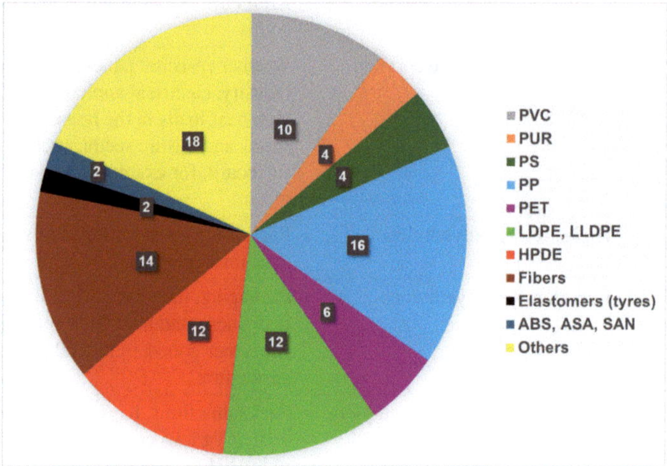

Fig. 6.8 Global plastic use by polymer in percent (for abbreviations see Table 6.14) (OECD 2024)

dimension in the nanoscale or with internal or surface structures in the nanoscale range from 0.001 to 0.1 μm (EFSA 2016). Nanoplastics can result from further breaking down of microplastic waste or originate from industrial materials used in various processes.

The EU defined synthetic polymer microparticle (SPM) or microplastic as polymers that (a) are solid and contained in particles and constitute at least 1% by weight of those particles, or build a continuous surface coating on particles; and (b) at least 1% by weight of the particles referred to in point (a) fulfil either of the following conditions: (i) all dimensions of the particles are equal to or less than 5 mm; (ii) the length of the particles is equal to or less than 15 mm and their length to diameter ratio is greater than 3 (EC 2023). Since October 2023, SPM shall not be placed on the EU market, with some exceptions, as substances on their own or, where the synthetic polymer microparticles are present to confer a sought-after characteristic, in mixtures in a concentration equal to or greater than 0,01% by weight (EC 2023).

The following end dates (each on October 16 of the year) have been established: microbeads in 2023; rinse-off cosmetics in 2027; detergents, fertilisers, and agricultural products that are not plant protection products or biocides in 2028; leave-on cosmetic products, fragrance encapsulates used in any application, and medical devices in 2029; granular infill on sports fields in 2031: and make up, lip and nail leave on cosmetic products in 2035. Again in the US, the Microbead-Free Waters Act of 2015 prohibited the manufacture, packaging, and distribution of cosmetics containing plastic microbeads, a type of microplastic intentionally added to products like rinse-off cosmetics.

It is now believed that MPs are widely present in all environmental media, and they can also be found in food, bottled and drinking water (Beaurepaire et al. 2021; Zhang et al. 2020; Yee et al. 2021; Ageel et al. 2022; WHO 2022; Wu et al. 2022; Ramsperger et al. 2023; Khu et al. 2025). In particular, contamination of the aquatic ecosystem and its food products is well documented.

6.8.1 Health Aspects

Currently, there are no studies on the effects of environmental exposure to MPs on human health. Data on the relationship between airborne MPs and disease are only available from workplaces in the synthetic textile industry, the flock industry, and the vinyl chloride (VC) and polyvinyl chloride (PVC) sectors. Workplace studies have reported cases of pneumoconiosis, interstitial lung disease, chronic bronchitis, and various other acute and chronic symptoms, including cough and dyspnea, at very high exposure levels (Prata 2018; Wieland et al. 2022; WHO 2022).

Studies on human- and animal-derived cell cultures, especially involving polystyrene polymers and spherical particles, have shown that MPs increasingly trigger inflammatory reactions and oxidative stress. This results in the upregulation of proinflammatory cytokine gene expression and demonstrates cytotoxic effects (Li et al. 2025a, b, Sniadach et al. 2025, Zhao et al. 2024, Sutkar et al. 2025). However, the effects vary greatly depending on the polymers and cell lines used and were not seen in all studies at low doses. In animal studies, effects on immune function, microbiota imbalance, and disruptions in intestinal barrier integrity were discussed following oral administration of PS particles in the gastrointestinal tract. Lim et al. (2021) reported no concentration–response relationship for several endpoints, including physiological, biochemical, hematological, and respiratory markers, after a 24-day exposure of rats to 0.1 μm polystyrene microparticles; however, they did observe an increase in inflammatory proteins in lung tissue. Similarly, Lu et al. (2021) examined the effects of nasal instillation of green fluorescent spherical polymer particles (diameter ranging from 1 to 5 μm). They observed respiratory effects, including infiltration of inflammatory cells and increased cytokine levels in the bronchoalveolar lavage fluid, in both normal and asthmatic mice.

Chemicals linked to microplastics, such as plasticizers, flame retardants, and metals, can have additional effects on human health when they enter the body. However, the specific contribution of substances delivered by MPs to potential effects, compared to direct exposure from the environment and food, remains unclear.

Facciolà et al. (2021) and Ageel et al. (2022) provided an overview of the current state of knowledge on health effects. Overall, there is a great need for further research to come to a valid health risk assessment of MPs (Wright and Kelly 2017; Barbosa et al. 2020).

6.8.2 *Toxicokinetics*

In principle, human exposure to MPs can happen through inhalation and ingestion, while dermal contact with MPs is also possible; however, uptake into the body is insignificant.

In principle, MPs can be deposited in the respiratory tract depending on their size and shape, with those having an aerodynamic diameter of < 10 µm also reaching deeper areas of the respiratory system. Measurements of MPs in lung tissue support these findings. As demonstrated for ultrafine particles in environmental aerosols, it must also be assumed that nanoscale plastics are absorbed after inhalation and that MPs are likely to be removed from the lungs through mucociliary clearance and uptake by macrophages.

Various mechanisms for uptake from the gastrointestinal tract have been discussed; however, paracellular transport through the intestinal wall is unlikely due to the small pore size of the tight junction channels, which are in the lower nanometer range. Generally, particles smaller than 500 nm are typically transported through cells via endocytosis, while particles larger than 500 nm are believed to be internalized by phagocytosis (Yoo et al. 2011). Although these constitute only about 1% of the intestinal surface, significant absorption seems to occur via the so-called M cells of Peyer's patches, a collection of lymph follicles that are part of the gut-associated lymphatic tissue in the small intestinal mucosa (Zheng et al. 2024). In an animal experiment, the highest uptake rate was observed for standard, nonionic, plain polystyrene particles, 2 µm in diameter, compared to negatively charged carboxylated microspheres and the paramagnetic form with a dense iron core (Carr et al. 2012). The same research group also found a low percentage of uptake of 2 µm latex particles, ranging from 0.04 to 0.32%, after oral administration in animals of different species and ages. Overall, several factors, including type, size, shape, composition, surface charge, hydrophilicity, and presence of a biofilm or biocorona (a layer of adsorbed biomolecules), affected absorption. Additionally, after oral administration of 50 nm PS particles with different surface modifications in rats, the estimated bioavailability of various nanoparticles ranged only from 0.2 to 1.7% (Walczak et al. 2015). Moreover, Keinänen et al. (2021) administered orally radiolabeled PS particles of several sizes (20 nm, 220 nm, 1 µm, and 6 µm) to mice, and most particles remained mainly in the gastrointestinal tract, with only

0.10–0.36% of the dose being absorbed, primarily into the gallbladder. Liang et al. (2021a, b) observed a clear size dependence in bioavailability, with values of 6.2, 1.5, and 0.46% after a single oral dose of PS particles measuring 50 nm, 500 nm, and 5 μm, respectively. It is important to consider that after oral intake, MPs interact with proteins, lipids, carbohydrates, nucleic acids, ions, and water, which can significantly influence absorption rates. Hong et al. (2024) administered fluorescent nano and micro PS with diameters of 100 nm, 3 μm, and 10 μm via oral gavage, finding rapid absorption of the nanoscale particles from the gastrointestinal tract and distribution across various organs, whereas the microscale particles were barely or not absorbed at all and were excreted in the feces.

6.8.3 Occurrence in Human Specimens

Analyzing MPs is still complicated and requires various methods and strict quality control in sampling, sample preparation, and instrumental testing. Several recent reviews discuss the analytical techniques and limitations used for measuring MPs in biological samples and their presence in human tissues (Noventa et al. 2021; Zuri et al. 2023; Roslan et al. 2024; Vdovchenko and Resmini 2024; Thompson et al. 2024; Sutkar et al. 2025; Luo et al. 2025a, b). Overall, they see an urgent need for more validation and harmonization, mainly because the measured concentrations are often low. In their review, Vdovchenko and Resmini (2024) find that PET, PA, PUR, and PP can be detected in human test samples at concentrations of 39%, 17%, 15%, and 9.4%, respectively.

Limited results on MPs in urine have been published so far. Pironti et al. (2022) examined urine samples from six volunteers living in different southern Italian cities. They found only four pigmented microplastic fragments (4–15 μm) in four samples, made of polyethylene vinyl acetate, PVC, PP, and PE. Similarly, Massardo et al. (2024) detected 23 particles, ranging from 3 to 13 μm in size, in 10 urine samples collected in Italy, with PE and PS being the most common polymers. In China, an average of 0.06 particles/ml was reported in the urine of 86 adults (Zhu et al. 2024). Additionally, in the UK, 123 MNPs, representing 22 polymer types, were identified in 17 out of 29 samples from healthy individuals; in contrast, 232 MPs of 16 polymers were found in 12 out of 19 samples from participants with endometriosis (Rotchell et al. 2024). The average was 2.6 particles/ml in healthy subjects and 4.7 particles/ml in those with the condition. PE, PS, and PP were the most prevalent polymers in healthy donors, while polytetrafluoroethylene and PE dominated samples from endometriosis patients.

When analyzing pyrolysis products in the blood of 22 individuals, PET, PE, and styrene polymers were detected, with an average total concentration of plastic particles in blood of 1.6 μg/ml (Leslie et al. 2022). However, it is important to note that duplicate measurements of the blood samples yielded different polymer types and concentrations in each sample. Lee et al. (2024) identified MNPs in 36 blood samples, mainly composed of PS, PP, and PE, with an average concentration of 4.2 particles/

ml. Additionally, in 26 thrombi collected from patients undergoing cardiovascular surgery, 16 contained 87 particles identified, ranging in size from 2.1 to 26.0 μm; only one microplastic particle was found (Wu et al. 2023). Furthermore, Marfella et al. (2024) examined plaques from carotid endarterectomy in 304 patients and found PE in 58% of samples, with an average level of 21.7 μg/mg plaque, and PVC in 12% of samples, with an average of 5.2 μg/ml. Patients with carotid artery plaque containing MPs had a higher risk of experiencing a composite event of myocardial infarction, stroke, or death from any cause over 34 months of follow-up compared to those without detectable MPs. In thrombus samples from 30 patients who underwent thrombectomy procedures, MPs, mainly PA, PVC, and PE, were detected in 80% of the samples, using pyrolysis products for quantification (Wang et al. 2024). The overall mean concentrations were 106 μg/g in thrombi from ischemic stroke, 142 μg/g from myocardial infarction, and 66 μg/g from deep vein thrombosis.

Table 6.15 summarizes the results published to date on measurements in stool, which have mainly been analyzed using Raman spectroscopy or Fourier transform infrared spectroscopy. PP, PET, and PE particles were very frequently detected, mostly in low concentrations. High median concentrations are reported from Nanjing and Hong Kong, China, with higher levels found in Nanjing, particularly in participants who ate takeaway food compared to those who ate home-cooked meals, and in Hong Kong, where there was a high proportion of seafood in the diet (Ho et al. 2022; Yan et al. 2022). Again, Yan et al. (2022) analyzed samples from healthy subjects and inflammatory bowel disease (IBD) patients and reported significantly lower concentrations in healthy individuals. Moreover, in this study, fecal microplastic concentration is positively correlated with the severity of IBD, and plastic packaging materials for drinking water and food, as well as dust exposure under working and living conditions, are important sources of human exposure. Additionally, in China, a moderate correlation was observed between microplastic abundance in feces and intake of packaged water and beverages, but not tap water, milk and dairy products, or beer in 26 young men (Zhang et al. 2021). Conversely, Refosco et al. (2025) reported no significant association between microplastic levels and seafood consumption or other food groups, such as chocolate and chips, in 18 Norwegian adults. In a study of 18 adults from Austria, the influence of different plastic use and food consumption scenarios—ranging from a regular diet to high consumption of packaged foods and beverages in plastic containers—was tested (Hartmann et al. 2024). Overall, oral MP intake showed no significant differences between the scenarios, but specific scenarios revealed significant differences, highlighting the role of food processing degree. In another Chinese study, participants from suburban areas had significantly higher microplastic levels in feces compared to those from rural and urban areas (Zhu et al. 2024).

In 18 Chinese breast milk samples, the median number of MPs (20–500 μm, 16 polymers) was 20.2 particles/g, with PU and PA being the most abundant at 9.9 and 3.1 particles/g, respectively (Liu et al. 2023). The authors reported that water intake and the use of scrub cleansers or toothpaste could be sources of exposure for pregnant women and breastfeeding individuals. In another study, Ragusa et al.

Table 6.15 Microparticles in human stool samples (in particles/g)

References	N	Size range (μm)	Positive (%)	Median (range)	Main polymers (abundance)	Country, year
Ho et al. (2022)	8	30–180	100	36.4 (20.4–139)	PS > PP > PE > PET	China, 2020
Zhang et al. (2021)	23	20–800	96	7 (0–36)	PP > PET > PS > PE	China
Wibowo et al. (2021)	11	n.g	64	– (0–16.5)[c]	PP > HDPE > PET > PS	Indonesia
Schwabl et al. (2019)	8	50–500	100	2.0 (0.8–41.6)	PP > PET > PS > PE	8 countries*
Refosco et al. (2025)	18	10–5000	94	6.5 (0–158)	PP > PE > PS	Norway
Yan et al. (2022)	50[a]	~ 4–400	100	28.0	PET > PA > PP > PE	China
Yan et al. (2022)	52[b]	~ 4–400	98	41.8	PET > PA > PVC > PP > PE	China
Hartmann et al. (2024)	15	50–500	95	3.5 (0.26–49)	PE > PET > PA > PP > PS	Austria
Zhu et al. (2024)	86	> 20	n.g	6.9[d]	PVC > PP > PE	China

[a]Healthy subjects; [b]Patients with inflammatory bowel disease; [c]In μg/g; [d]Mean; *Japan, Russia, the Netherlands, the United Kingdom, Italy, Poland, Finland, and Austria; *Japan, Russia, the Netherlands, the United Kingdom, Italy, Poland, Finland, and Austria; n.g.; not given

(2021) analyzed MPs (2–12 μm) in 26 of 34 breast milk samples in Italy, finding PE (38%), PVC (21%), and PP (17%) to be the most prevalent.

Some information is available on the occurrence of MPs in different human organs. Dzierżyński et al. (2025) examined seven organs post-mortem and found concentrations ranging from 5.2 particles/g in the chest muscle to 40.4 particles/g in the thyroid gland, especially in PET, PS, and polyacrylonitrile polymers.

To date, two studies have explored the presence of MPs in human lung tissue, with polymer particles and fibers found in 13 out of 20 samples and 11 out of 13 samples, respectively (Amato-Lourenco et al. 2021, Jenner et al. 2021). Jenner et al. (2022) reported an average of 1.42 particles/g, with higher levels observed in tissue from the upper lung.

Braun et al. (2021) collected placenta and meconium samples from two women during cesarean delivery and analyzed MPs in the range of 50–500 μm (10 polymers). The screening in placental tissue was positive for PE, PP, and PU, while the meconium samples tested positive for PE. Ragusa et al. (2021) analyzed MPs (5–10 μm) in four placentas in Italy and found 12 fragments, all of which were pigmented, with three identified as stained PP. Additionally, in 10 placenta and amniotic fluid samples from the Czech Republic, the number of MNPs (20–50 μm) ranged from 0 to 20 particles/g and 0 to 1.6 particles/ml, respectively (Halfar et al. 2023). These were primarily chlorinated. In 18 Chinese placenta and meconium samples, the median number of MPs (20–500 μm, 16 polymers) was 18.0 and 51.4 particles/g, respectively, with PU and PA being the most abundant (Liu et al. 2023). In a second Chinese study, Zhang et al. (2025) collected placentas from 1121 women and reported lower median levels of 1.5 particles/g, with equal concentrations of 0.5 particles/g for PVC, PP, and polybutylene succinate.

Additionally, more results are available for samples from the kidney (Massardo et al. 2024), the intestine (Ibrahim et al. 2020; Cetin et al. 2023), the heart (Yang et al. 2023), and the testes (Zhao et al. 2023).

References

ACS (American Chemistry Council) (2023) Applications & benefits. https://www.americanchemistry.com/industry-groups/center-for-the-polyurethanes-industry-cpi/applications-benefits

Ademollo N, Ferrara F, Delise M, Fabietti F, Funari E (2008) Nonylphenol and octylphenol in human breast milk. Environ Int 34:984–987

Ageel HK, Harrad S, Abdallah MA (2022) Occurrence, human exposure, and risk of microplastics in the indoor environment. Environ Sci Process Impacts 24(1):17–31

AGS (Committee on Hazardous Substances) (2009) Technical Rules for Hazardous Substances. Isocyanates – Risk assessment and protective measures (TRGS 43) [in German]. Online: file:///C:/Users/herma/Downloads/TRGS-430-1.pdf

Ali N, Eqani SA, Malik RN, Neels H, Covaci A (2013) Organohalogenated contaminants (OHCs) in human serum of mothers and children from Pakistan with urban and rural residential settings. Sci Total Environ 461–462:655–662

Amato-Lourenço LF, Carvalho-Oliveira R, Júnior GR, Dos Santos GL, Ando RA, Mauad T (2021) Presence of airborne microplastics in human lung tissue. J Hazard Mater 416:126124

ATSDR (2008) Toxicological Profile for perchlorates. https://wwwn.cdc.gov/TSP/ToxProfiles/ToxProfiles.aspx?id=895&tid=181

ATSDR (2018) Toxicological profile for toluene diisocyanate and methylenediphenyl diisocyanate. https://www.atsdr.cdc.gov/ToxProfiles/tp206-p.pdf

ATSDR (2022a) Toxicological profile for pentachlorophenol. https://www.atsdr.cdc.gov/toxprofiles/tp51.pdf

ATSDR (2022b) Toxicological profile for vhlorophenols. https://www.atsdr.cdc.gov/toxprofiles/tp107.pdf

ATSDR (2024) Toxicological profile for cyanide. https://wwwn.cdc.gov/TSP/ToxProfiles/ToxProfiles.aspx?id=72&tid=19

Azzouz A, Rascón AJ, Ballesteros E (2016) Simultaneous determination of parabens, alkylphenols,phenylphenols, bisphenol A and triclosan in human urine, blood and breast milk by continuous solid-phase extraction and gas chromatography–mass spectrometry. J Pharm Biomed Anal 119:16–26

Barbosa F, Adeyemi JA, Bocato MZ, Comas A, Campiglia A (2020) A critical viewpoint on current issues, limitations, and future research needs on micro- and nanoplastic studies: from the detection to the toxicological assessment. Environ Res 182:109089

Battal D, Sukuroglu AA, Kocadal K, Cok I, Unlusayin I (2021) Establishment of rapid, sensitive, and quantitative liquid chromatography-electrospray ionization-tandem mass spectrometry method coupled with liquid-liquid extraction for measurement of urinary bisphenol A, 4-t-octylphenol, and 4-nonylphenol. Rapid Commun Mass Spectrom 35(12):e9084

BAUA (German Federal Institute for Occupational Safety and Health) (2023) Regulation on the restriction of diisocyanates. https://www.baua.de/EN/Topics/Work-design/Hazardous-substances/REACH-assessment-unit/Diisocyanates.html

Beaurepaire M, Dris R, Gasperi J, Tassin B (2021) Microplastics in the atmospheric compartment: a comprehensive review on methods, results on their occurrence and determining factors. Curr Opin Food Sci 41:159–168

Becker K, Müssig-Zufika M, Conrad A, Lüdecke A, Schulz C, Seiwert M, Kolossa-Gehring M (2007) German environmental survey for children 2003/06 - KUS - Human biomonitoring of children in Germany. [in German] https://www.umweltbundesamt.de/en/publikationen/kinder-umwelt-survey-200306-human-biomonitoring

Birkefeld AB, Bertermann R, Eckert H, Pfleiderer B (2003) Liquid- and solid-state high-resolution NMR methods for the investigation of aging processes of silicone breast implants. Biomaterials 24(1):35–46

Blount BC, Valentin-Blasini L, Osterloh JD, Mauldin JP, Pirkle JL (2007) Perchlorate exposure of the US Population, 2001–2002. J Expo Sci Environ Epidemiol 17:400–407

Braun T, Ehrlich L, Henrich W, Koeppel S, Lomako I, Schwabl P, Liebmann B (2021) Detection of microplastic in human placenta and meconium in a clinical setting. Pharmaceutics 13(7):921

Braverman LE, Pearce EN, He X, Pino S, Seeley M, Beck B, Magnani B, Blount BC, Firek A (2006) Effects of six months of daily low-dose perchlorate exposure on thyroid function in healthy volunteers. J Clin Endocrinol Metab 91:2721–2724

Bryant MS, Vineis P, Skipper PL, Tannenbaum SR (1988) Hemoglobin adducts of aromatic amines: associations with smoking status and type of tobacco. Proc Natl Acad Sci U S A 85(24):9788–9791

Budnik LT, Nowak D, Merget R, Lemiere C, Baur X (2011) Elimination kinetics of diisocyanates after specific inhalative challenges in humans: mass spectrometry analysis, as a basis for biomonitoring strategies. J Occup Med Toxicol 6:9

Butt CM, Miranda ML, Stapleton HM (2016) Development of an analytical method to quantify PBDEs, OH-BDEs, HBCDs, 2,4,6-TBP, EH-TBB, and BEH-TEBP in human serum. Anal Bioanal Chem 408(10):2449–2459

Cao Y, Blount BC, Valentin-Blasini L, Bernbaum JC, Phillips TM, Rogan WJ (2010) Goitrogenic anions, thyroid-stimulating hormone, and thyroid hormone in infants. Environ Health Perspect 118:1332–1337

Carr KE, Smyth SH, McCullough MT, Morris JF, Moyes SM (2012) Morphological aspects of interactions between microparticles and mammalian cells: intestinal uptake and onward movement. Prog Histochem Cytochem 46:185–252

Carrizo D, Grimalt JO, Ribas-Fito N, Torrent M, Sunyer J (2008) Pentachlorobenzene, hexachlorobenzene, and pentachlorophenol in children's serum from industrial and rural populations after restricted use. Ecotoxicol Environ Saf 71(1):260–266

CDC (2023) National report on human exposure to environmental chemicals. Bomonitoring data tables for environmental chemicals https://www.cdc.gov/exposurereport/data_tables.html

CES (2016) The socioeconomic impact of the silicones industry in Europe. London, UK.

Cetin M, Demirkaya Miloglu F, Kilic Baygutalp N, Ceylan O, Yildirim S, Eser G, Gul HI (2023) Higher number of microplastics in tumoral colon tissues from patients with colorectal adenocarcinoma. Environ Chem Lett 21:639–646

Charatcharoenwitthaya N, Ongphiphadhanakul B, Pearce EN, Somprasit C, Chanthasenanont A, He X, Chailurkit L, Braverman LE (2014) The association between perchlorate and thiocyanate exposure and thyroid function in first-trimester pregnant Thai women. J Clin Endocrinol Metab 99:2365–2371

Charnley G (2008) Perchlorate: overview of risks and regulation. Food Chem Toxicol 46:2307–2315

Chemanalyst (2023) Aniline market analysis. https://www.chemanalyst.com/industry-report/aniline-market-282
Chen CY, Sun CY, Hsu HJ, Wu IW, Chen YC, Lee CC (2021) Xenoestrogen exposure and kidney function in the general population: Results of a community-based study by laboratory tests and questionnaire-based interviewing. Environ Int 155:106585
Chen D, Li H, Li J, Liu Z, Lyu B, Li J, Zhao Y, Zhong Y, Wu Y (2023) Comprehensive evaluation of perchlorate dietary exposure and health risks for Chinese adult males and breastfed infants. Sci Total Environ 891:164696
Chen GW, Ding WH, Ku HY, Chao HR, Chen HY, Huang MC, Wang SL (2010) Alkylphenols in human milk and their relations to dietary habits in central Taiwan. Food Chem Toxicol 48(7):1939–1944
Chen M, Zhu P, Xu B, Zhao R, Qiao S, Chen X, Tang R, Wu D, Song L, Wang S, Xia Y, Wang X (2012) Determination of nine environmental phenols in urine by ultra-high-performance liquid chromatography-tandem mass spectrometry. J Anal Toxicol 36(9):608–615
Chen Z-J, Liu H-J, Ho K-L, Huang H-B, Liu Q, Mane Y-B, Hon-Wah Lam M, Du J, Wong M-H, Wang H-S (2016) Hydroxylated polybrominated diphenylethers (OH-PBDEs) inpaired maternal and neonatal samples from South China: placental transfer and potential risks. Environ Res 148:72–78
Chinthakindi S, Kannan K (2022) Variability in urinary concentrations of primary aromatic amines. Sci Total Environ 831:154768
Choi G, Kuiper JR, Bennett DH, Barrett ES, Bastain TM, Breton CV, Chinthakindi S, Dunlop AL, Farzan SF, Herbstman JB, Karagas MR, Marsit CJ, Meeker JD, Morello-Frosch R, O'Connor TG, Pellizzari ED, Romano ME, Sathyanarayana S, Schantz S, Schmidt RJ, Watkins DJ, Zhu H, Kannan K, Buckley JP, Woodruff TJ (2022) Exposure to melamine and its derivatives and aromatic amines among pregnant women in the United States: The ECHO Program. Chemosphere 307(Pt 2):135599
Choi G, Xun X, Bennett DH, Meeker JD, Morello-Frosch R, Sathyanarayana S, Schantz SL, Trasande L, Watkins D, Pellizzari ED, Li W, Kannan K, Woodruff TJ, Buckley JP (2024) Associations of prenatal urinary melamine, melamine analogues, and aromatic amines with gestational duration and fetal growth in the ECHO Cohort. Environ Int 195:109227
Choi J, Eom J, Kim J, Lee S, Kim Y (2014) Association between some endocrine-disrupting chemicals and childhood obesity in biological samples of young girls: a cross-sectional study. Environ Toxicol Pharmacol 38(1):51–57
Clewell H, Greene T, Gentry R (2024) Dermal absorption of cyclic and linear siloxanes: a review. J Toxicol Environ Health B Crit Rev 27(3):106–129
Dallaire R, Ayotte P, Pereg D, Déry S, Dumas P, Langlois E, Dewailly E (2009a) Determinants of plasma concentrations of perfluorooctanesulfonate and brominated organic compounds in Nunavik Inuit adults (Canada). Environ Sci Technol 43:5130–5136
Dallaire R, Muckle G, Dewailly E, Jacobson SW, Jacobson JL, Sandanger TM, Sandau CD, Ayotte P (2009b) Thyroid hormone levels of pregnant inuit women and their infants exposed to environmental contaminants. Environ Health Perspect 117(6):1014–1020
Dan EPA (Danish Environmental Protection Agency) (2014) Siloxanes (D3, D4, D5, D6, HMDS). Evaluation of health hazards and proposal of a health-based quality criterion for ambient air. Environmental Project No. 1531. https://www2.mst.dk/udgiv/publications/2014/01/978-87-93026-85-8.pdf
DeBruin LS, Pawliszyn JB, Josephy PD (1999) Detection of monocyclic aromatic amines, possible mammary carcinogens, in human milk. Chem Res Toxicol 12(1):78–82
DECOS (2018). Di- and triisocyanates. Health-based recommendation on occupational exposure limits. Report to the State Secretary of Social Affairs and Employment. No. 2018/20. The Hague, Health Council of the Netherlands.
Dierkes G, Weiss T, Modick H, Käfferlein HU, Brüning T, Koch HM (2014) N-Acetyl-4-aminophenol (paracetamol), N-acetyl-2-aminophenol and acetanilide in urine samples from

the general population, individuals exposed to aniline and paracetamol users. Int J Hyg Environ Health 217(4–5):592–599

Ding J, Dai Y, Zhang L, Wang Z, Zhang B, Guo J, Qi X, Lu D, Chang X, Wu C, Zhang J, Zhou Z (2024) Identifying childhood pesticide exposure trajectories and critical window associated with behavioral problems at 10 years of age: findings from SMBCS. Environ Int 193:109079

Dirtu AC, Jaspers VL, Cernat R, Neels H, Covaci A (2010) Distribution of PCBs, their hydroxylated metabolites, and other phenolic contaminants in human serum from two European countries. Environ Sci Technol 44(8):2876–2883

Dobaradaran S, Mutke XAM, Schmidt TC, Swiderski P, De-la-Torre GE, Jochmann MA (2022) Aromatic amines contents of cigarette butts: fresh and aged cigarette butts vs unsmoked cigarette. Chemosphere 301:134735

Dodson RE, Boronow KE, Susmann H, Udesky JO, Rodgers KM, Weller D, Woudneh M, Brody JG, Rudel RA (2020) Consumer behavior and exposure to parabens, bisphenols, triclosan, dichlorophenols, and benzophenone-3: results from a crowdsourced biomonitoring study. Int J Hyg Environ Health: 113624

Dufour P, Pirard C, Charlier C (2016) Validation of a novel and rapid method for the simultaneous determination of some phenolic organohalogens in human serum by GC- Dufour P, Pirard C, Charlier C (2017) Determination of phenolic organohalogens in human serum from a Belgian population and assessment of pa rameters affecting the human contamination. Sci Total Environ 599–600:1856–1866

Dufour P, Pirard C, Petrossians P, Beckers A, Charlier C (2020) Association between mixture of persistent organic pollutants and thyroid patholo gies in a Belgian population. Environ Res 181:108922

Dzierżyński E, Blicharz-Grabias E, Komaniecka I, Panek R, Forma A, Gawlik PJ, Puźniak D, Flieger W, Choma A, Suśniak K, Teresiński G, Baj J, Kupisz K, Flieger J (2025) Post-mortem evidence of microplastic bioaccumulation in human organs: insights from advanced imaging and spectroscopic analysis. Arch Toxicol Jun 25

EC (Environment and Climate Change Canada) (2022) Performance Measurement Evaluation for Risk Management of Nonylphenol and Its Ethoxylates, eco-component. https://publications.gc.ca/site/eng/9.911095/publication.html?wbdisable=false

EC (European Commission) (2023) Commission regulation (EU) 2023/2055 of 25 September 2023 amending Annex XVII to Regulation (EC) No 1907/2006 of the European Parliament and of the Council concerning the Registration, Evaluation, Authorisation and Restriction of Chemicals (REACH) as regards synthetic polymer microparticles. Off J EU L 238/67. https://eur-lex.europa.eu/legal-content/EN/TXT/PDF/?uri=CELEX:32023R2055

EC/HC (2008a) Environment / Health Canada. Screening Assessment for the Challenge Decamethylcyclopentasiloxane (D5) Chemical Abstracts Service Registry Number 541–02–6. http://www.chemicalsubstanceschimiques.gc.ca/challenge-defi/batch-lot_2_e.html.

EC/HC (2008b) Environment / Health Canada. Screening Assessment for the Challenge Octamethylcyclotetrasiloxane (D4) Chemical Abstracts Service Registry Number 556–67–2. http://www.chemicalsubstanceschimiques.gc.ca/challenge-defi/batch-lot_2_e.html

EC/HC (2022) Screening Assessment Poly(alkoxylates/ethers) Group. https://www.canada.ca/content/dam/eccc/documents/pdf/pded/poly(alkoxylates-ethers)/Screening-assessment-poly-alkoxylates-ethers.pdf

ECB (European Chemical Bureau) (2002) 4-Nonylphenol (branched) and nonylphenol. Summary risk assessment report. https://www.bfr.bund.de/cm/343/4_nonylphenol_und_nonylphenol.pdf

ECHA (2019) Scientific report for evaluation of limit values for diisocyanates at the workplace. https://echa.europa.eu/documents/10162/db15bfdf-eec8-c10a-67c4-f65166c5110a

ECHA (European Chemicals Agency) (2011) Annex XV dossier for 4-(1,1,3,3-tetramethylbutyl)phenol

EFSA (2012) Scientific Opinion on Brominated Flame Retardants (BFRs) in Food: Brominated Phenols and their Derivatives. EFSA J 10(4): 2634. https://www.efsa.europa.eu/en/efsajournal/pub/2634

EFSA (2013) Conclusion on the peer review of the pesticide risk assessment of the active substance potassium thiocyanate. EFSA J 11(6):2922. https://www.efsa.europa.eu/en/efsajournal/pub/2922

EFSA (2014) Scientific Opinion on the risks to public health related to the presence of perchlorate in food, in particular fruits and vegetables. EFSA J 12(10):3869. https://www.efsa.europa.eu/en/efsajournal/pub/3869

EFSA (2016) Presence of microplastics and nanoplastics in food, with particular focus on seafood. EFSA J 14(6):4501. https://www.efsa.europa.eu/en/efsajournal/pub/4501

EFSA (2019) Evaluation of the health risks related to the presence of cyanogenic glycosides in foods other than raw apricot kernels. EFSA J 17(4):5662. https://www.efsa.europa.eu/en/efsajournal/pub/5662

EFSA (2024a) Update of the scientific opinion on the risks for human health related to the presence of perchlorate in food. https://www.efsa.europa.eu/en/news/perchlorate-food-draft-scientific-opinion-public-consultation

EFSA (2024b) Update of the risk assessment of brominated phenols and their derivatives in food. EFSA J 22:e9034. https://www.efsa.europa.eu/en/efsajournal/pub/9034

Eguchi A, Nomiyama K, Devanathan G, Subramanian A, Bulbule KA, Parthasarathy P, Takahashi S, Tanabe S (2012) Different profiles of anthropogenic and naturally produced organohalogen compounds in serum from residents living near a coastal area and e-waste recycling workers in India. Environ Int 47:8–16

Eguchi A, Nomiyama K, Minh Tue N, Trang PT, Hung Viet P, Takahashi S, Tanabe S (2015) Residue profiles of organohalogen compounds in human serum from e- waste recycling sites in North Vietnam: association with thyroid hormone levels. Environ Res 137:440–449

Eichler O (1929) Zur Pharmakologie der Perchloratwirkung. Naunyn-Schmiedeberg's Arch Exp Pathol Pharmakol 144:251–260

Facciolà A, Visalli G, Pruiti Ciarello M, Di Pietro A (2021) Newly emerging airborne pollutants: current knowledge of health impact of micro and nanoplastics. Int J Environ Res Public Health 18(6):2997

Feng C, Xu Q, Jin Y, Lin Y, Qiu X, Lu D, Wang G (2016) Determination of urinary bromophenols (BrPs) as potential biomarkers for human exposure to polybrominated diphenyl ethers (PBDEs) using gas chromatography-tandem mass spectrometry (GC-MS/MS). J Chromatogr B 1022:70–74

Flassbeck D, Pfleiderer B, Grumping R, Hirner AV (2001) Determination of low molecular weight silicones in plasma and blood of women after exposure to silicone breast implants by GC/MS. Anal Chem 73:606–611

Forde MS, Robertson L, Laouan Sidi EA, Côté S, Gaudreau E, Drescher O, Ayotte P (2015) Evaluation of exposure to organophosphate, carbamate, phenoxy acid, and chlorophenol pesticides in pregnant women from 10 Caribbean countries. Environ Sci Process Impacts 17(9):1661–1671

Frederiksen H, Nielsen O, Koch HM, Skakkebaek NE, Juul A, Jørgensen N, Andersson AM (2020) Changes in urinary excretion of phthalates, phthalate substitutes, bisphenols and other polychlorinated and phenolic substances in young Danish men; 2009–2017. Int J Hyg Environ Health 223(1):93–105

Fromme H, Cequier E, Kim JT, Hanssen L, Hilger B, Thomsen C, Chang YS, Völkel W (2015) Persistent and emerging pollutants in the blood of German adults: occurrence of dechloranes, polychlorinated naphthalenes, and siloxanes. Environ Int 85:292–298

Fujii Y, Harada KH, Hitomi T, Kobayashi H, Koizumi A (2014Feb) Haraguchi K (2014a) Temporal trend and age-dependent serum concentration of phenolic organohalogen contaminants in Japanese men during 1989–2010. Environ Pollut 185:228–233

Fujii Y, Kato Y, Masuda N, Harada KH, Koizumi A, Haraguchi K (2018) Contamination trends and factors affecting the transfer of hexabromocyclododecane diastereomers, tetrabromobisphenol A, and 2,4,6- tribromophenol to breast milk in Japan. Environ Pollut 237:936–943

Fujii Y, Nishimura E, Kato Y, Harada KH, Koizumi A, Haraguchi K (2014) Dietary exposure to phenolic and methoxylated organohalogen contaminants in relation to their concentrations in breast milk and serum in Japan. Environ Int 63:19–25

Gheni SA, Ali MM, Ta GC, Harbin HJ, Awad SA (2024) Toxicity, hazards, and safe handling of primary aromatic amines. ACS Chem Health Saf 31:8–21

Glynn A, Larsdotter M, Aune M, Darnerud PO, Bjerselius R, Bergman A (2011) Changes in serum concentrations of polychlorinated biphenyls (PCBs), hydroxylated PCB metabolites and pentachlorophenol during pregnancy. Chemosphere 83(2):144–151

Goniewicz MŁ, Czogała J (2005) Exposure of active and passive smokers to aromatic amines present in tobacco smoke. Toxicol Mech Methods 15:235–245

Greer MA, Goodman G, Pleus RC, Greer SE (2002) Health effects assessment for environmental perchlorate contamination: the dose response for inhibition of thyroidal radioiodine uptake in humans. Environ Health Perspect 110:927–937

Gries W, Leng G (2013) Analytical determination of specific 4,4′-methylene diphenyl diisocyanate hemoglobin adducts in human blood. Anal Bioanal Chem 405:7205–7213

Grimmer G, Dettbarn G, Seidel A, Jacob J (2000) Detection of carcinogenic aromatic amines in the urine of non-smokers. Sci Total Environ 247:81–90

GSC (Global Silicone Council) (2021) Silicon research-an industry commitment. https://globalsilicones.org/wp-content/uploads/2020/10/Silicone-Monitoring-initiatives.pdf

Guenther K, Heinke V, Thiele B, Kleist E, Prast H, Raecker T (2002) Endocrine disrupting nonylphenols are ubiquitous in food. Environ Sci Technol 36:1676–1680

Guo J, Wu C, Zhang J, Jiang S, Lv S, Lu D, Qi X, Feng C, Liang W, Chang X, Zhang Y, Xu H, Cao Y, Wang G, Zhou Z (2019) Anthropometric measures at age 3 years in associations with prenatal and postnatal exposures to chlorophenols. Chemosphere 228:204–211

Guo J, Zhou Y, Sun M, Cui J, Zhang B, Zhang J (2020) Methylsiloxanes in plasma from potentially exposed populations and an assessment of the associated inhalation exposure risk. Environ Int 143:105931

Guo J, Zhou Y, Wang Y, Zhang B, Zhang J (2021) Assessment of internal exposure to methylsiloxanes in children and associated non-dietary exposure risk. Environ Int 154:106672

Guo J, Zhou Y, Wang Y, Chen Y, Zhang B, Zhang J (2022) Methylsiloxanes risk assessment combining external and internal exposure for college students. Sci Total Environ 845:157379

Guvenius DM, Aronsson A, Ekman-Ordeberg G, Bergman A, Norén K (2003) Human prenatal and postnatal exposure to polybrominated diphenyl ethers, polychlorinated biphenyls, polychlorobiphenylols, and pentachlorophenol. Environ Health Perspect 111(9):1235–1241

Halfar J, Čabanová K, Vávra K, Delongová P, Motyka O, Špaček R, Kukutschová J, Šimetka O, Heviánková S (2023) Microplastics and additives in patients with preterm birth: the first evidence of their presence in both human amniotic fluid and placenta. Chemosphere 343:140301

Hanssen L, Warner NA, Braathen T, Odland JØ, Lund E, Nieboer E, Sandanger TM (2013) Plasma concentrations of cyclic volatile methylsiloxanes (cVMS) in pregnant and postmenopausal Norwegian women and self-reported use of personal care products (PCPs). Environ Int 51:82–87

Hardy EM, Dereumeaux C, Guldner L, Briand O, Vandentorren S, Oleko A, Zaros C, Appenzeller BMR (2021) Hair versus urine for the biomonitoring of pesticide exposure: results from a pilot cohort study on pregnant women. Environ Int 152:106481

Hartmann C, Lomako I, Schachner C, El Said E, Abert J, Satrapa V, Kaiser AM, Walch H, Koppel S (2024) Assessment of microplastics in human stool: a pilot study investigating the potential impact of diet-associated scenarios on oral microplastics exposure. Sci Total Environ 951:175825

Hartwig A, Arand M (2023) The permanent senate commission for the investigation of health hazards of chemical compounds in the work area (2023) Toluene diisocyanates. The MAK Collection for Occupational Health and Safety, 8(4):Doc081. https://doi.org/10.34865/mb58484isme8_4ad

HBC (Human Biomonitoring Commission) (2011) Substance monograph and reference values for monocyclic aminoaromatics in urine. Bundesgesundheitsbl 54:650–663. [in German]

HC (Health Canada) (2013) Second Report on Human Biomonitoring of Environmental Chemicals in Canada. https://www.canada.ca/content/dam/hc-sc/migration/hc-sc/ewh-semt/alt_formats/pdf/pubs/contaminants/chms-ecms-cycle2/chms-ecms-cycle2-eng.pdf
Heudorf U, Letzel S, Peters M, Angerer J (2000) PCP in the blood plasma: current exposure of the population in Germany, based on data obtained in 1998. Int J Hyg Environ Health 203(2):135–139
Ho KL, Yau MS, Murphy MB, Wan Y, Fong BMW, Tam S, Giesy JP, Leung KSY, Lam MHW (2015) Urinary bromophenol glucuronide and sulfate conjugates: potentialhuman exposure molecular markers for polybrominated diphenyl ethers. Chemosphere 133:6–12
Ho YW, Lim JY, Yeoh YK, Chiou JC, Zhu Y, Lai KP, Li L, Chan PKS, Fang JK (2022) Preliminary findings of the high quantity of microplastics in faeces of Hong Kong residents. Toxics 10:414
Honda M, Kannan K (2018) Biomonitoring of chlorophenols in human urine from several Asian countries, Greece and the United States. Environ Pollut 232:487–493
Hong HC, Zhou HY, Luan TG, Lan CY (2005) Residue of pentachlorophenol in freshwater sediments and human breast milk collected from the Pearl River Delta. China. Environ Int 31(5):643–649
Hong T, Sun W, Deng Y, Lyu JD, Jin CH, Bai YL, Na J, Zhang R, Gao Y, Pan GW, Yang ZS, Yan LJ (2024) The uptake and distribution evidence of nano- and microplastics in vivo after a single high dose of oral exposure. Biomed Environ Sci 37(1):31–41
Horton MK, Blount BC, Valentin-Blasini L, Wapner R, Whyatt R, Gennings C, Factor-Litvak P (2015) CO-occurring exposure to perchlorate, nitrate and thiocyanate alters thyroid function in healthy pregnant women. Environ Res 143:1–9
Hou JW, Lin CL, Tsai YA, Chang CH, Liao KW, Yu CJ, Yang W, Lee MJ, Huang PC, Sun CW, Wang YH, Lin FR, Wu WC, Lee MC, Pan WH, Chen BH, Wu MT, Chen CC, Wang SL, Lee CC, Hsiung CA, Chen ML (2015) The effects of phthalate and nonylphenol exposure on body size and secondary sexual characteristics during puberty. Int J Hyg Environ Health 218(7):603–615
Hou Y, Li S, Xia L, Yang Q, Zhang L, Zhang X, Liu H, Huo R, Cao G, Huang C, Tian X, Sun L, Cao D, Zhang M, Zhang Q, Tang N (2021) Associations of urinary phenolic environmental estrogens exposure with blood glucose levels and gestational diabetes mellitus in Chinese pregnant women. Sci Total Environ 754:142085
Huang YF, Wang PW, Huang LW, Lai CH, Yang W, Wu KY, Lu CA, Chen HC, Chen ML (2017) Prenatal nonylphenol and bisphenol a exposures and inflammation are determinants of oxidative/nitrative stress: a taiwanese cohort study. Environ Sci Technol 51(11):6422–6429
Huang YF, Wang PW, Huang LW, Yang W, Yu CJ, Yang SH, Chiu HH, Chen ML (2014) Nonylphenol in pregnant women and their matching fetuses: placental transfer and potential risks of infants. Environ Res 134:143–148
Huo Y, Wan Y, Huang Q, Wang A, Mahai G, He Z, Xu S, Xia W (2022) Pentachlorophenol exposure in early pregnancy and gestational diabetes mellitus: a nested case-control study. Sci Total Environ 831:154889
IARC (1999) Re-evaluation of some organic chemicals, hydrazine and hydrogen peroxide (Part 1, Part 2, Part 3). IARC monographs on the evaluation of carcinogenic risks to humans volume 71. http://monographs.iarc.fr/ENG/Monographs/vol71/mono71-47.pdf
IARC (2009) A review of human carcinogens. Part F: Chemical agents and related occupations/ IARC Working Group on the Evaluation of Carcinogenic Risks to Humans. Lyon, France
IARC (2019) Pentachlorophenol and some related compounds. IARC working group on the evaluation of carcinogenic risks to humans. IARC monographs on the evaluation of carcinogenic risks to humans. Vol. 117. file:///C:/Users/herma/Downloads/117–008-Pentachlorophenol.pdf
IARC (2021) Some aromatic amines and related compounds. IARC Monographs on the Identification of Carcinogenic Hazards to Humans Volume 127. https://publications.iarc.fr/599
IARC (2025) List of classifications https://monographs.iarc.who.int/list-of-classifications

IARC (International Agency for Research on Cancer) (1986) Some Chemicals Used in Plastics and Elastomers. IARC Monographs on the Evaluation of the Carcinogenic Risk of Chemicals to Humans Volume 39. https://publications.iarc.fr/Book-And-Report-Series/Iarc-Monographs-On-The-Identification-Of-Carcinogenic-Hazards-To-Humans/Some-Chemicals-Used-In-Plastics-And-Elastomers-1986

Ibrahim YS, Tuan Anuar S, Azmi AA, Wan Mohd Khalik WMA, Lehata S, Hamzah SR, Ismail D, Ma ZF, Dzulkarnaen A, Zakaria Z, Mustaffa N, Tuan Sharif SE, Lee YY (2020) Detection of microplastics in human colectomy specimens. JGH Open 5(1):116–121

Ihde ES, Loh JM, Rosen L (2015) Association of environmental chemicals and estrogen metabolites in children. BMC Endocr Disord 15:83

IMAP (2020b) Octylphenols: human health tier II assessment. https://www.industrialchemicals.gov.au/sites/default/files/Octylphenols_Human%20health%20tier%20II%20assessment.pdf

IMAP (Inventory Multitiered Assessment and Prioritisation) (2020a) Nonylphenols: human health tier II assessment. https://www.industrialchemicals.gov.au/sites/default/files/Nonylphenols_Human%20health%20tier%20II%20assessment.pdf

ISOPA (European Diisocyanate and Polyol Producers Association) (2020) Polyurethanes—Benefits of Polyurethanes. https://www.isopa.org/wp-content/uploads/2022/09/isopa-factsheet_benefits-of-polyurethanes_v05_a.pdf

Jamal A, Rastkari N, Dehghaniathar R, Nodehi RN, Nasseri S, Kashani H, Shamsipour M, Yunesian M (2020) Prenatal urinary concentrations of environmental phenols and birth outcomes in the mother-infant pairs of Tehran Environment and Neurodevelopmental Disorders (TEND) cohort study. Environ Res 184:109331

Jenner LC, Rotchell JM, Bennett RT, Cowen M, Tentzeris V, Sadofsky LR (2022) Detection of microplastics in human lung tissue using μFTIR spectroscopy. Sci Total Environ 831:154907

Jiang Y, Guo J, Zhou Y, Dong K, Zhang B, Han J, Wang Y, Chen Y (2023) Assessment of the internal and external exposure risks to methylsiloxanes in communities near a petroleum refinery. Sci Total Environ 900:166314

Johnson W, Bergfeld W, Belsito D, Hill RN, Klaassen C, Liebler D, Marks J, Shank R, Slaga T, Snyder P, Andersen F (2011) Safety assessment of cyclomethicone, cyclotetrasiloxane, cyclopentasiloxane, cyclohexasiloxane, and cycloheptasiloxane. Int J Toxicol 30(Suppl 3):149S-227S

Jovanovic ML, McMahon JM, McNett DA, Tobin JM, Plotzke KP (2008) In vitro and in vivo percutaneous absorption of 14C-octamethylcyclotetrasiloxane (14C–D4) and 14C-decamethylcyclopentasiloxane (14C–D5). Regul Toxicol Pharmacol 50:239–248

Käfferlein HU, Broding HC, Bünger J, Jettkant B, Koslitz S, Lehnert M, Marek EM, Blaszkewicz M, Monsé C, Weiss T, Brüning T (2014) Human exposure to airborne aniline and formation of methemoglobin: a contribution to occupational exposure limits. Arch Toxicol 88:1419–1426

Kaj L, Andersson J, Cousins AP, Remberger M, Brorström-Lundén E (2005) Results from the swedish national screening programme 2004, Subreport 4: Siloxanes. Swedish Environmental Research Institute (IVL). http://www.imm.ki.se/Datavard/PDF/B1643_siloxaner.pdf

Kaj L, Schlabach M, Andersson J, Cousins AP, Schmidbauer N, Brorström-Lundén E (2005b) Siloxanes in the nordic environment. Norden. http://www.norden.org/sv/publikationer/publikationer/2005-2593

Kannan K, Wu Q, Li Z-M (2024) Sources and pathways of human exposure to perchlorate in the United States: a comparison of environmental monitoring and biomonitoring approaches. Environ Chem Ecotoxicol 6:363–369

Kawashiro Y, Fukata H, Omori-Inoue M, Kubonoya K, Jotaki T, Takigamo H, Sakai S-I, Mori C (2008) Perinatal exposure to brominated flame retardants and polychlorinated biphenyls in Japan. Endocrine J 55:1071–1084

Keinänen O, Dayts EJ, Rodriguez C, Sarrett SM, Brennan JM, Sarparanta M, Zeglis BM (2021) Harnessing PET to track micro- and nanoplastics in vivo. Sci Rep 11:11463

Kenwood BM, Bagch P, Zhang L (2021) Characterization of US population levels of urinary methylcarbamoyl mercapturic acid, a metabolite of N,N-dimethylformamide and methyl isocyanate, in

the National Health and Nutrition Examination Survey (NHANES) 2005–2006 and 2011–2016. Environ Sci Pollut Res Int 28(13):16781–16791
Khu ST, Li F, Zhao W (2025) Microplastics in drinking water distribution systems: occurrence, environmental behavior, and human health concerns. Environ Pollut 382:126666
Kim YB, Cheon YP, Lee SH (2019) Adverse effect of nonylphenol on the reproductive system in F1 male mice: a subchronic low-dose exposure model. Dev Reprod 23(2):93–99
King L, Wang Q, Xia L, Wang P, Jiang G, Li W, Huang Y, Liang X, Peng X, Li Y, Chen L, Liu L (2023) Environmental exposure to perchlorate, nitrate and thiocyanate, and thyroid function in Chinese adults: a community-based cross-sectional study. Environ Int 171:107713
Kirk AB, Dyke JV, Martin CF, Dasgupta PK (2007) Temporal patterns in perchlorate, thiocyanate, and iodide excretion in human milk. Environ Health Perspect 115(2):182–186
Klaunig JE, Dekant W, Plotzke K, Scialli AR (2016) Biological relevance of decamethylcyclopentasiloxane (D5) induced rat uterine endometrial adenocarcinoma tumorigenesis: mode of action and relevance to humans. Regul Toxicol Pharmacol 74(Suppl):S44–S56
Knudsen GA, Trexler AW, Richards AC, Hall SM, Hughes MF, Birnbaum LS (2019) 2,4,6-Tribromophenol disposition and kinetics in rodents: effects of dose, route, sex, and species. Toxicol Sci 169:167–179
Kütting B, Göen T, Schwegler U, Fromme H, Uter W, Angerer J, Drexler H (2009) Moonoarylamines in the general population—a cross-sectional populatioon-based study including 1004 bavarian subjects. Int J Hyg Enviroon Health 212(3):298–309
Kuraoka S, Oda M, Ohba T, Mitsubuchi H, Nakamura K, Katoh T, Japan Environment and Children's Study (JECS) Group (2024) Association of phenol exposure during pregnancy and asthma development in children: the Japan environment and Children's study. Environ Pollut 361:124801
La Voie EJ, Stern SL, Burrill C, Weyand EH (1989) On the maternal transfer of 4-aminobiphenyl in rats Carcinogenesis 10(2):231–236
Lassen TH, Frederiksen H, Jensen TK, Petersen JH, Main KM, Skakkebæk NE, Jørgensen N, Kranich SK, Andersson AM (2013) Temporal variability in urinary excretion of bisphenol A and seven other phenols in spot, morning, and 24-h urine samples. Environ Res 126:164–170
Lau FK, deCastro BR, Mills-Herring L, Tao L, Valentin-Blasini L, Alwis KU, Blount BC (2013) Urinary perchlorate as a measure of dietary and drinking water exposure in a representative sample of the United States population 2001–2008. J Expo Sci Environ Epidemiol 23(2):207–214
Lee DW, Jung J, Park SA, Lee Y, Kim J, Han C, Kim HC, Lee JH, Hong YC (2024) Microplastic particles in human blood and their association with coagulation markers. Sci Rep 14(1):30419
Lee KM, Han SM, Lee HJ, Kang M, Jeong TY, Son J, Min H, Cha S, Oh HB, Oh WK, Lee J (2023) Influence of mobile phase composition on the analytical sensitivity of LC-ESI-MS/MS for the concurrent analysis of bisphenols, parabens, chlorophenols, benzophenones, and alkylphenols. Environ Res 221:115305
Lee S, Ahn RM, Kim JH, Han YD, Lee JH, Son BS, Lee K (2019) Study design, rationale and procedures for human biomonitoring of hazardous chemicals from foods and cooking in Korea. Int J Environ Res Public Health 16(14):2583
Lee SY, McCarthy AM, Stohl H, Ibrahim S, Jeong C, Braverman LE, Ma W, He X, Mestman JH, Schuller KE, Jahreis KA, Pearce EN, Leung AM (2017) Urinary Iodine, perchlorate, and thiocyanate concentrations in U.S. lactating women. Thyroid 27(12):1574–1581
Leslie HA, Van Velzen MJ, Brandsma SH, Vethaak AD, Garcia-Vallejo JJ, Lamoree MH (2022) Discovery and quantification of plastic particle pollution in human blood. Environ Int 163:107199
Leung AM, Braverman LE, He X, Schuller KE, Roussilhes A, Jahreis KA, Pearce EN (2012) Environmental perchlorate and thiocyanate exposures and infant serum thyroid function. Thyroid 22(9):938–943
Leung AM, Katz PM, He X, Feig DS, Pearce EN, Braverman LE (2014) Urinary perchlorate and thiocyanate concentrations in pregnant women from Toronto. Canada. Thyroid 24(1):175–176

Li B, Li M, Du D, Tang B, Yi W, He M, Liu R, Yu H, Yu Y, Zheng J (2025a) Characteristics and influencing factors of microplastics entering human blood through intravenous injection. Environ Int 198:109377

Li M, Xiao M, Xiao Q, Chen Y, Guo Y, Sun J, Li R, Li C, Zhu Z, Qiu H, Liu X, Lu S (2022) Perchlorate and chlorate in breast milk, infant formulas, baby supplementary food and the implications for infant exposure. Environ Int 158:106939

Li J, Tu F, Wan Y, Qian X, Mahai G, Wang A, Ma J, Yang Z, Xia W, Xu S, Zheng T, Li Y (2023) Associations of trimester-specific exposure to perchlorate, thiocyanate, and nitrate with childhood neurodevelopment: a birth cohort study in China. Environ Sci Technol 57(49): 20480–20493

Li R, Wan Y, He Z, Wang D, Xu S, Zhao X, Xia W (2025b) Exposure to organophosphates, pyrethroids, neonicotinoids, and pentachlorophenol: spatial variations in urinary biomarkers and associations with oxidative stress based on a repeated-measure study. Sci Total Environ 969:178934

Li X, Ying GG, Zhao JL, Chen ZF, Lai HJ, Su HC (2013) 4-Nonylphenol, bisphenol-A and triclosan levels in human urine of children and students in China, and the effects of drinking these bottled materials on the levels. Environ Int 52:81–86

Liang BX, Zhong YZ, Huang YJ, Lin X, Liu J, Lin L, Hu MJ, Jiang JY, Dai MZ, Wang B, Zhang BL, Meng H, Lelaka JJJ, Sui HX, Yang XF, Huang ZL (2021a) Underestimated health risks: polystyrene micro- and nanoplastics jointly induce intestinal barrier dysfunction by ROS-mediated epithelial cell apoptosis. Part Fibre Toxicol 18(1):20

Liang C, Gracida-Alvarez UR, Gallant ET, Gillis PA, Marques YA, Abramo GP, Hawkins TR, Dunn JB (2021b) Material flows of polyurethane in the United States. Environ Sci Technol 55(20):14215–14224

Lim D, Jeong J, Song KS, Sung JH, Oh SM, Choi J (2021) Inhalation toxicity of polystyrene micro(nano)plastics using modified OECD TG 412. Chemosphere 262:128330

Lin M, Ma S, Tang J, Yu Y, Li G, Fan R, Zhang G, Mai B, An T (2023) Polybrominated diphenyl ethers and bromophenols in paired serum, hair, and urine samples of e-waste dismantlers: Insights into hair as an indicator of endogenous exposure. Sci Total Environ 871:161980

Lin M, Ma S, Yu Y, Li G, Mai B, An T (2020) Simultaneous determination of multiple classes of phenolic compounds in human urine: insight into metabolic biomarkers of occupational exposure to E-waste. Environ Sci Technol Lett 7:323–329

Lin W-C, Wang S-L, Cheng C-Y, Ding W-H (2009) Determination of alkylphenol residues in breast and commercial milk by solid-phase extraction and gas chromatography–mass spectrometry. Food Chem 114:753–757

Lisco G, De Tullio A, Giagulli VA, De Pergola G, Triggiani V (2020) Interference on iodine uptake and human thyroid function by perchlorate-contaminated water and food. Nutrients 12(6):1669

Liu C, Xu X, Zhang Y, Li W, Huo X (2016) Associations between maternal phenolic exposure and cord sex hormones in male newborns. Hum Reprod 31(3):648–656

Liu S, Guo J, Liu X, Yang R, Wang H, Sun Y, Chen B, Dong R (2023) Detection of various microplastics in placentas, meconium, infant feces, breastmilk and infant formula: a pilot prospective study. Sci Total Environ 854:158699

Lu D, Feng C, Wang D, Lin Y, Ip HS, She J, Xu Q, Wu C, Wang G, Zhou Z (2015) Analysis of twenty phenolic compounds in human urine: hydrochloric acid hydrolysis, solid-phase extraction based on K_2CO_3-treated silica, and gas chromatography tandem mass spectrometry. Anal Bioanal Chem 407(14):4131–4141

Lu K, Lai KP, Stoeger T, Ji S, Lin Z, Lin X, Chan TF, Fang JK, Lo M, Gao L, Qiu C, Chen S, Chen G, Li L, Wang L (2021) Detrimental effects of microplastic exposure on normal and asthmatic pulmonary physiology. J Hazard Mater 416:126069

Luo L, Qian X, Duan Y, Luo X, Li R, Zhang X, Guo X, Xiong S, Huang G, Zeng H, Zhang Q, Wan Y, He Q (2025a) Association of pentachlorophenol in urine and follicular fluid with ovarian reserve and reproductive outcomes among women undergoing in vitro fertilization based on a prospective cohort study. Environ Res 270:120950

Luo Q, Tan H, Ye M, Jho EH, Wang P, Iqbal B, Zhao X, Shi H, Lu H, Li G (2025b) Microplastics as an emerging threat to human health: an overview of potential health impacts. J Environ Manage 387:125915

Lv S, Wu C, Lu D, Qi X, Xu H, Guo J, Liang W, Chang X, Wang G, Zhou Z (2016) Birth outcome measures and prenatal exposure to 4-tert-octylphenol. Environ Pollut 212:65–70

Lykissa ED, Kala SV, Hurley JB, Lebovitz RM (1997) Release of low molecular weight silicones and platinum from silicone breast implants. Anal Chem 69(23):4912–4916

Marfella R, Prattichizzo F, Sardu C, Fulgenzi G, Graciotti L, Spadoni T, D'Onofrio N, Scisciola L, La Grotta R, Frigé C, Pellegrini V, Municinò M, Siniscalchi M, Spinetti F, Vigliotti G, Vecchione C, Carrizzo A, Accarino G, Squillante A, Spaziano G, Mirra D, Esposito R, Altieri S, Falco G, Fenti A, Galoppo S, Canzano S, Sasso FC, Matacchione G, Olivieri F, Ferraraccio F, Panarese I, Paolisso P, Barbato E, Lubritto C, Balestrieri ML, Mauro C, Caballero AE, Rajagopalan S, Ceriello A, D'Agostino B, Iovino P, Paolisso G (2024) Microplastics and nanoplastics in atheromas and cardiovascular events. N Eng J Med 390(10):900–910

Massardo S, Verzola D, Alberti S, Caboni C, Santostefano M, Verrina EE, Artini C (2024) Micro-Raman spectroscopy detects the presence of microplastics in human urine and kidney tissue. Environ Int 184:108444

Meijer L, Weiss J, Van Velzen M, Brouwer A, Bergman A, Sauer PJ (2008) Serum concentrations of neutral and phenolic organohalogens in pregnant women and some of their infants in The Netherlands. Environ Sci Technol 42(9):3428–3433

Michałowicz J, Włuka A, Bukowska B (2022) A review on environmental occurrence, toxic effects and transformation of man-made bromophenols. Sci Total Environ 811:152289

Modick H, Weiss T, Dierkes G, Koslitz S, Käfferlein HU, Brüning T, Koch HM (2016) Human metabolism and excretion kinetics of aniline after a single oral dose. Arch Toxicol 90:1325–1333

Negi S, Bala L, Shukla S, Chopra D (2022) Tattoo inks are toxicological risks to human health: a systematic review of their ingredients, fate inside skin, toxicity due to polycyclic aromatic hydrocarbons, primary aromatic amines, metals, and overview of regulatory frameworks. Toxicol Ind Health 38:417–434

NICNAS (2016) Inventory multitiered assessment and prioritisation (IMAP) human health tier II assessment thiocyanate salts. https://www.industrialchemicals.gov.au/sites/default/files/Thiocyanate%20salts_Human%20health%20tier%20II%20assessment.pdf

Nielsen E, Østergaard G, Thorup I, Ladefoged O, Jelnes JE (2000) Toxicological evaluation and limit values for nonylphenol, nonylphenol ethoxylates, tricresyl, phosphates and benzoic acid. Environmental Project No. 512. https://www2.mst.dk/udgiv/publications/1999/87-7909-566-6/pdf/87-7909-565-8.pdf

Niziński P, Błażewicz A, Kończyk J, Michalski R (2020) Perchlorate—properties, toxicity and human health effects: an updated review. Rev Environ Health 36(2):199–222

Noventa S, Boyles MSP, Seifert A, Belluco S, Sánchez Jiménez A, Johnston HJ, Tran L, Fernandes TF, Mughini-Gras L, Orsini M, Corami F, Castro K, Mutinelli F, Boldrin M, Puntes V, Sotoudeh M, Mascarello G, Tiozzo B, McLean P, Ronchi F, Booth AM, Koelmans AA, Losasso C (2021) Paradigms to assess the human health risks of nano- and microplastics. Microplast Nanoplast 1:9

OECD (2024) Policy scenarios for eliminating plastic pollution by 2040. OECD Publishing, Paris. https://doi.org/10.1787/76400890-en

Otaka H, Yasuhara O, Morita M (2003) Determination of bisphenol A and 4-nonylphenol in human milk using alkaline digestion and cleanup by solid-phase extraction. Anal Sci 19:1663

Ozpinar A, Kelestimur F, Songur Y, Can O, Valentin L, Caldwell K, Arikan E, Unsal I, Serteser M, Inal T, Erdemgil Y, Coskun A, Bakirci N, Sezgin O, Blount B (2014) Iodine status in Turkish populations and exposure to iodide uptake inhibitors. PLoS ONE 9(2):e88206

Parastar S, Ebrahimpour K, Hashemi M, Maracy MR, Ebrahimi A, Poursafa P, Kelishadi R (2018) Association of urinary concentrations of four chlorophenol pesticides with cardiometabolic risk factors and obesity in children and adolescents. Environ Sci Pollut Res Int 25(5):4516–4523

Park H, Kim K (2017) Urinary levels of 4-nonylphenol and 4-t-octylphenol in a representative sample of the Korean adult population. Int J Environ Res Public Health 14(8):932

Park H, Kim K (2018) Concentrations of 2,4-dichlorophenol and 2,5-dichlorophenol in Urine of Korean Adults. Int J Environ Res Public Health 15(4):589

Park JS, Bergman A, Linderholm L, Athanasiadou M, Kocan A, Petrik J, Drobna B, Trnovec T, Charles MJ, Hertz-Picciotto I (2008) Placental transfer of polychlorinated biphenyls, their hydroxylated metabolites and pentachlorophenol in pregnant women from eastern Slovakia. Chemosphere 70(9):1676–1684

Pearce EN, Alexiou M, Koukkou E, Braverman LE, He X, Ilias I, Alevizaki M, Markou KB (2012) Perchlorate and thiocyanate exposure and thyroid function in first-trimester pregnant women from Greece. Clin Endocrinol (Oxf) 77:471–474

Pearce EN, Lazarus JH, Smyth PP, He X, Dall'amico D, Parkes AB, Burns R, Smith DF, Maina A, Bestwick JP, Jooman M, Leung AM, Braverman LE (2010) Perchlorate and thiocyanate exposure and thyroid function in first-trimester pregnant women. J Clin Endocrinol Metab 95:3207–3215

Pearce EN, Leung AM, Blount BC, Bazrafshan HR, He X, Pino S, Valentin-Blasini L, Braverman LE (2007) Breast milk iodine and perchlorate concentrations in lactating Boston-area women. J Clin Endocrinol Metab 92(5):1673–1677

Pearce EN, Spencer CA, Mestman JH, Lee RH, Bergoglio LM, Mereshian P, He X, Leung AM, Braverman LE (2011) Effect of environmental perchlorate on thyroid function in pregnant women from Córdoba, Argentina, and Los Angeles, California. Endocr Pract 17:412–417

Pekari K, Luotamo M, Järvisalo J, Lindroos L, Aitio A (1991) Urinary excretion of chlorinated phenols in saw-mill workers. Int Arch Occup Environ Health 63(1):57–62

Peña J, González-Mariño I, Pérez Pavón JL (2024) In-situ acetylation followed by liquid-liquid extraction and gas chromatography—mass spectrometry for the determination of bromophenols in urine. Talanta 275:126146

Peng F, Ji W, Zhu F, Peng D, Yang M, Liu R, Pu Y, Yin L (2016) A study on phthalate metabolites, bisphenol A and nonylphenol in the urine of Chinese women with unexplained recurrent spontaneous abortion. Environ Res 150:622–628

Philippat C, Wolff MS, Calafat AM, Ye X, Bausell R, Meadows M, Stone J, Slama R, Engel SM (2013) Prenatal exposure to environmental phenols: concentrations in amniotic fluid and variability in urinary concentrations during pregnancy. Environ Health Perspect 121(10):1225–1231

Pirard C, Sagot C, Deville M, Dubois N, Charlier C (2012) Urinary levels of bisphenol A, triclosan and 4-nonylphenol in a general Belgian population. Environ Int 48:78–83

Pironti C, Notarstefano V, Ricciardi M, Motta O, Giorgini E, Montano L (2022) First evidence of microplastics in human urine, a preliminary study of intake in the human body. Toxics 11:40

Pleus RC, Corey LM (2018) Environmental exposure to perchlorate: a review of toxicology and human health. Toxicol Appl Pharmacol 358:102–109

Plotzke KP, Crofoot SD, Ferdinandi ES, Beattie JG, Reitz RH, McNett DA, Meeks RG (2000) Disposition of radioactivity in fischer 344 rats after single and multiple inhalation exposure to [(14)C]Octamethylcyclotetrasiloxane ([(14)C]D(4)). Drug Metab Dispos 28:192–204

Pollack AZ, Mumford SL, Krall JR, Carmichael AE, Sjaarda LA, Perkins NJ, Kannan K, Schisterman EF (2018) Exposure to bisphenol A, chlorophenols, benzophenones, and parabens in relation to reproductive hormones in healthy women: a chemical mixture approach. Environ Int 120:137–144

Prata JC (2018) Airborne microplastics: consequences to human health? Environ Pollut 234:115–126

Qiu X, Bigsby RM, Hites RA (2009) Hydroxylated metabolites of polybrominated diphenyl ethers in human blood samples from the United States. Environ Health Perspect 117(1):93–98

Ragusa A, Svelato A, Santacroce C, Catalano P, Notarstefano V, Carnevali O, Papa F, Rongioletti MCA, Baiocco F, Draghi S, D'Amore E, Rinaldo D, Matta M, Giorgini E (2021) Plasticenta: first evidence of microplastics in human placenta. Environ Int 146:106274

Ramsperger AFRM, Bergamaschi E, Panizzolo M, Fenoglio I, Barbero F, Peters R, Undas A, Purker S, Giese B, Lalyer CR, Tamargo A, Moreno-Arribas MV, Grossart HP, Kühnel D, Dietrich J, Paulsen F, Afanou AK, Zienolddiny-Narui S, Eriksen Hammer S, Kringlen Ervik T, Graff P, Brinchmann BC, Nordby KC, Wallin H, Nassi M, Benetti F, Zanella M, Brehm J, Kress H, Löder MGJ, Laforsch C (2023) Nano- and microplastics: a comprehensive review on their exposure routes, translocation, and fate in humans. NanoImpact 29:100441

Reddy MB, Andersen ME, Morrow PE, Dobrev ID, Varaprath S, Plotzke KP, Utell MJ (2003) Physiological modeling of inhalation kinetics of octamethylcyclotetrasiloxane in humans during rest and exercise. Toxicol Sci 72:3–18

Reddy MB, Looney RJ, Utell MJ, Plotzke KP, Andersen ME (2007) Modeling of human dermal absorption of octamethylcyclotetrasiloxane (D(4)) and decamethylcyclopentasiloxane (D(5)). Toxicol Sci 99:422–431

Refosco A, Dierkes J, Kögel T, Nitter Dankel S, Laupsa-Borge J, Gomiero A (2025) Benny Daniel D (2025) Microplastics in human feces: a pilot study exploring links with dietary habits. Microplast Nanoplast 5:22

Repková A, Mišľanová C, Hrabčáková J, Masár M, Wimmerová S, Žemlička L, Slezáková Z, Valachovičová M (2023) Influence of selected lifestyle factors on nonylphenol concentration in breast milk of mothers in Slovakia. PrePrints.org

Richter E, Branner B (2002) Biomonitoring of exposure to aromatic amines: haemoglobin adducts in humans. J Chromatogr B 778:49–62

Richter E, Rösler S, Scherer G, Gostomzyk JG, Grübl A, Krämer U, Behrendt H (2001) Haemoglobin adducts from aromatic amines in children in relation to area of residence and exposure to environmental tobacco smoke. Int Arch Occup Environ Health 74:421–428

Riedel K, Scherer G, Engl J, Hagedorn H, Tricker AR (2006) Determination of three carcinogenic aromatic amines in urine of smokers and nonsmokers. J Anal Toxicol 30:187–195

Ringbeck B, Belov VN, Schmidtkunz C, Küpper K, Gries W, Weiss T, Brüning T, Hayen H, Bury D, Leng G, Koch HM (2021a) Human metabolism and urinary excretion kinetics of nonylphenol in three volunteers after a single oral dose. Chem Res Toxicol 34:2392–2403

Ringbeck B, Bury D, Hayen H, Weiss T, Brüning T, Koch HM (2021b) Determination of specific urinary nonylphenol metabolites by online-SPE-LC-MS/MS as novel human exposure biomarkers. J Chromatogr B 1177:122794

Ringbeck B, Bury D, Ikeda-Araki A, Ait Bamai Y, Ketema RM, Miyashita C, Brüning T, Kishi R, Koch HM (2022a) Nonylphenol exposure in 7-year-old Japanese children between 2012 and 2017- Estimation of daily intakes based on novel urinary metabolites. Environ Int 161: 107145.

Ringbeck B, Bury D, Lee I, Lee G, Alakeel R, Alrashed M, Tosepu R, Jayadipraja EA, Tantrakarnapa K, Kliengchuay W, Brüning T, Choi K, Koch HM (2022b) Biomarker-determined nonylphenol exposure and associated risks in children of Thailand, Indonesia, and Saudi Arabia. Environ Sci Technol 56(14):10229–10238

Ringbeck B, Weber T, Bury D, Kasper-Sonnenberg M, Pälmke C, Brüning T, Koch HM, Kolossa-Gehring M (2022c) Nonylphenol (NP) exposure in Germany between 1991 and 2021: Urinary biomarker analyses in the German environmental specimen bank (ESB). Int J Hyg Environ Health 245:114010

RIVM (2021) Microplastics in indoor air - a literature review. RIVM letter report 2021–0059. https://www.rivm.nl/publicaties/microplastics-in-indoor-air

Robert A, Ducos P, Francin JM, Marsan P (2007) Biological monitoring of workers exposed to 4,4′-methylenediphenyl diisocyanate (MDI) in 19 French polyurethane industries. Int Arch Occup Environ Health 80:412–422

Rosendahl P, Hippler J, Schmitz OJ, Hoffmann O, Rusch P (2016) Cyclic volatile methylsiloxanes in human blood as markers for ruptured silicone gel-filled breast implants. Anal Bioanal Chem 408:3309–3317

Roslan NS, Lee YY, Ibrahim YS, Tuan Anuar S, Yusof KMKK, Lai LA, Brentnall T (2024) Detection of microplastics in human tissues and organs: a scoping review. J Glob Health 14:04179

Rotchell JM, Austin C, Chapman E, Atherall CA, Liddle CR, Dunstan TS, Blackburn B, Mead A, Filart K, Beeby E, Cunningham K, Allen J, Draper H, Ba G (2024) Microplastics in human urine: Characterisation using μFTIR and sampling challenges using healthy donors and endometriosis participants. Ecotoxicol Environ Saf 274:116208

Sabbioni G (2017) Hemoglobin adducts and urinary metabolites of arylamines and nitroarenes. Chem Res Toxicol 30(10):1733–1766

Sabbioni G, Castaño A, Esteban López M, Göen T, Mol H, Riou M, Tagne-Fotso R (2022) Literature review and evaluation of biomarkers, matrices and analytical methods for chemicals selected in the research program Human Biomonitoring for the European Union (HBM4EU). Environ Int 169:107458

Sandau CD, Ayotte P, Dewailly E, Duffe J, Norstrom RJ (2000) Analysis of hydroxylated metabolites of PCBs (OH-PCBs) and other chlorinated phenolic compounds in whole blood from Canadian inuit. Environ Health Perspect 108(7):611–616

Sandau CD, Ayotte P, Dewailly E, Duffe J, Norstrom RJ (2002) Pentachlorophenol and hydroxylated polychlorinated biphenyl metabolites in umbilical cord plasma of neonates from coastal populations in Québec. Environ Health Perspect 110(4):411–417

SC (Stockholm Convention) (2025) POPs listed in the Stockholm Convention. https://www.pops.int/TheConvention/ThePOPs/AllPOPs/tabid/2509/Default.aspx

SCCS (2015) Opinion on decamethylcyclopentasiloxane (cyclopentasiloxane, D5) in cosmetic products. SCCS/1549/15

SCCS (Scientific Committee on Consumer Safety) (2010) Cyclomethicone, octamethylcyclotetrasiloxane (cyclotetrasiloxane, D4) and decamethylcyclopentasiloxane (cyclopentasiloxane, D5). SCCS/1241/10

Schmidtkunz C, Gries W, Küpper K, Leng G (2023) A "dilute-and-shoot" column-switching UHPLC-MS/MS procedure for the rapid determination of branched nonylphenol in human urine: method optimisation and some fundamental aspects of nonylphenol analysis. Anal Bioanal Chem 415(5):975–989

Schmied-Tobies MIH, Murawski A, Schmidt L, Rucic E, Schwedler G, Apel P, Göen T, Kolossa-Gehring M (2021) Pentachlorophenol and nine other chlorophenols in urine of children and adolescents in Germany - Human biomonitoring results of the German Environmental Survey 2014–2017 (GerES V). Environ Res 196:110958

Schmitt BG, Tobin J, McNett DA, Kim J, Durham J, Plotzke KP (2023) Comparative pharmacokinetic studies of 14C-octamethylcyclotetrasiloxane (14C–D4) in Fischer 344 and Sprague Dawley CD rats after single and repeated inhalation exposure. Toxicol Lett 373:13–21

Scholten B, Kenny L, Duca RC, Pronk A, Santonen T, Galea KS, Loh M, Huumonen K, Sleeuwenhoek A, Creta M, Godderis L, Jones K (2020) Biomonitoring for occupational exposure to diisocyanates: a systematic review. Ann Work Expo Health 64(6):569–585

Schreder E, Zheng G, Sathyanarayana S, Gunaje N, Hu M, Salamova A (2023) Brominated flame retardants in breast milk from the United States: first detection of bromophenols in U.S. breast milk. Environ Pollut 334:122028

Schupp T, Plehiers PM (2022) Absorption, distribution, metabolism, and excretion of methylene diphenyl diisocyanate and toluene diisocyanate: many similarities and few differences. Toxicol Ind Health 38(9):500–528

Schwabl P, Köppel S, Königshofer P, Bucsics T, Trauner M, Reiberger T, Liebmann B (2019) Detection of various microplastics in human stool. Ann Intern Med 171:453–457

SEHSC (1996) EPA DCN 86970000024. Pharmacokinetics of 14COctamethylcyclotetrasiloxane (D4) in the Rat Following Single Nose-Only Vapor Inhalation Exposure. Siloxane Research Program, http://www.sehsc.com/srp-research-reports.asp.

SEHSC (1998) EPA DCN 86980000184. An Oral Gavage Study to Compare the Absorption Potential of 14C-Octamethylcyclotetrasiloxane (D4) in Fischer 344 Rats When Delivered in Various Carriers. Siloxane Research Program, http://www.sehsc.com/srpresearch-reports.asp

SEHSC (2000) EPA DCN 86010000009. In Vivo Dermal ADME of C14-D4 in Fischer 344 Rats (to clarify expired air values). Siloxane Research Program, http://www.sehsc.com/srpresearch-reports.asp

SEHSC (2002a) EPA DCN 84030000028. Disposition of Octamethylcyclotetrasiloxane (D4) in Female Fischer 344 and Sprague-Dawley IGS Rats Following 14 Repeat Nose-Only Vapor Inhalation Exposures and a Single Exposure to 14C-D4 at 700 ppm. Siloxane Research Program, http://www.sehsc.com/srp-research-reports.asp

SEHSC (2002b) EPA DCN 84030000027. Single Exposure Pharmacokinetics and Metabolism of 14C-Decamethylcyclopentasiloxane (D5) via Nose-Only Vapor Inhalation. Siloxane Research Program, http://www.sehsc.com/srp-research-reports.asp

SEHSC (2003) EPA DCN 84030000026. Absorption of Dodecamethylcyclohexasiloxane (D6) Using the Flow-Through Diffusion Cell System for In Vitro Absorption in Human Skin. Siloxane Research Program, http://www.sehsc.com/srp-research-reports.asp

SEHSC (2004) EPA DCN 84040000016. Pharmacokinetics of 14C-D6 in the Rat Following Single Oral Exposure. Siloxane Research Program, http://www.sehsc.com/srp-researchreports.asp.

SEHSC (2009) Key Features of Silicones. http://www.sehsc.com/versatile.asp

Seiivanova LN, Arefaeva ZS (1986) The dynamics behind absorption and elimination of perchloric acid salts in laboratory animals and agricultural livestock. Chemistry P.S.X. 24(5):43–45

Semple JL (2007) Breast-feeding and silicone implants. Plast Reconstr Surg 120(Suppl 1):123S-128S

Sennbro CJ, Littorin M, Tinnerberg H, Jönsson BAG (2005) Upper reference limits for biomarkers of exposure to aromatic diisocyanates. Int Arch Occup Environ Health 78:541–546

Seo MY, Choi MH, Hong Y, Kim SH, Park MJ (2021) Association of urinary chlorophenols with central obesity in Korean girls. Environ Sci Pollut Res Int 28(2):1966–1972

Seyler T, Mazumder S, Ahamed R, Zhu W, Blount BC, Apelberg BJ, Wang L (2023) Tobacco smoke is a major source of aromatic amine exposure in U.S. adults: 2013–2014 National health and nutrition examination survey (NHANES) cancer epidemiol biomarkers Prev 17:OF1-OF9

Shelor CP, Kirk AB, Dasgupta PK, Kroll M, Campbell CA, Choudhary PK (2012) Breastfed infants metabolize perchlorate. Environ Sci Technol 46(9):5151–5159

Shen Y, Xu Q, Ren M, Feng X, Cai Y, Gao Y (2013) Measurement of phenolic environmental estrogens in women with uterine leiomyoma. PLoS ONE 8(11):e79838

Shin C, Lee SM, Kim M, Kim YS (2023) Simultaneous determination of the free and total forms of nonylphenol, nonylphenol monoethoxylate, and nonylphenol diethoxylate in human urine by gas chromatography-mass spectrometry. Anal Bioanal Chem 415(26):6583–6593

Sise S, Uguz C (2017) Nonylphenol in human breast milk in relation to sociodemographic variables, diet, obstetrics histories and lifestyle habits in a turkish population. Iran J Public Health 46:491–499

Śniadach J, Kicman A, Szymkowiak S, Waszkiewicz N (2025) The hidden threat of microplastics in traditional cigarettes: a narrative review of health and environmental risks. J Clin Med 14(11):3721

Song S, Ruan J, Bai X, Xie L, Zhang B, He Y, Zhang T (2019) One-step sample processing method for the determination of perchlorate in human urine, whole blood and breast milk using liquid chromatography tandem mass spectrometry. Ecotoxicol Environ Saf 174:175–180

Steinmaus C, Pearl M, Kharrazi M, Blount BC, Miller MD, Pearce EN, Valentin-Blasini L, DeLorenze G, Hoofnagle AN, Liaw J (2016) Thyroid hormones and moderate exposure to perchlorate during pregnancy in women in Southern California. Environ Health Perspect 124:861–867

Sukuroglu AA, Battal D, Kocadal K, Sungur MA, Cok İ, Unlusayin I (2022) Biomonitoring of bisphenol A, 4-nonylphenol, and 4-t-octylphenol in Turkish population: exposure and risk assessment. Environ Sci Pollut Res Int 29(18):26250–26262

Sun Y, Liu Z, Xia W, He Z, Wan Y (2023) Urinary pentachlorophenol in general population of central China: reproducibility, predictors, and associations with oxidative stress biomarkers. Environ Sci Pollut Res Int 30(13):37598–37606

Sutkar P, Dhulap V, Girigosavi S, Deshmukh S, Desai M (2025) Microplastics in the human body: a comprehensive review of exposure, detection, and health risks. Wastewater Purif 1(1):20250317163738

Tagne-Fotso R, Zeghnoun A, Saoudi A, Balestier A, Pecheux M, Chaperon L, Oleko A, Marchand P, Le Bizec B, Vattier L, Bouchart V, Limon G, Le Gléau F, Denys S, Fillol C (2023) Exposure of the general French population to herbicides, pyrethroids, organophosphates, organochlorines, and carbamate pesticides in 2014–2016: Results from the Esteban study. Int J Hyg Environ Health 254:114265

Tang R, Chen MJ, Ding GD, Chen XJ, Han XM, Zhou K, Chen LM, Xia YK, Tian Y, Wang XR (2013) Associations of prenatal exposure to phenols with birth outcomes. Environ Pollut 178:115–120

Tellez RT, Chacon PM, Abarca CR, Blout BC, Van Landingham CB, Crump KS, Gibbs JP (2005) Long-term environmental exposure to perchlorate through drinking water and thyroid function during pregnancy and the neonatal period. Thyroid 15:963–975

Thomsen C, Leknes H, Lundanes E, Becher G (2002) A new method for determination of halogenated flame retardants in human milk using solid-phase extraction. J Anal Toxicol 26(3):129–137

Thomsen C, Lundanes E, Becher G (2001) Brominated flame retardants in plasma samples from three different occupational groups in Norway. J Environ Monit 3(4):366–370

Thompson RC, Courtene-Jones W, Boucher J, Pahl S, Raubenheimer K, Koelmans AA (2024) Twenty years of microplastic pollution research-what have we learned? Science 386(6720):eadl2746

Tobin JM, McNett DA, Durham JA, Plotzke KP (2008) Disposition of decamethylcyclopentasiloxane in Fischer 344 rats following single or repeated inhalation exposure to 14C-decamethylcyclopentasiloxane (14C–D5). Inhal Toxicol 20:513–531

Turesky RJ, Le Marchand L (20011) Metabolism and biomarkers of heterocyclic aromatic amines in molecular epidemiology studies: lessons learned from aromatic amines. Chem Res Toxicol 24:1169–1214

UNEP (2023) United nations environment programme and secretariat of the basel, rotterdam and stockholm conventions. Chemicals in plastics: a technical report. Geneva. https://digitallibrary.un.org/record/4023718?v=pdf

US EPA (2010) Nonylphenol (NP) and Nonylphenol Ethoxylates (NPEs) Action Plan [RIN 2070-ZA09]. https://www.epa.gov/sites/default/files/2015-09/documents/rin2070-za09_np-npes_action_plan_final_2010-08-09.pdf

Utell MJ, Gelein R, Yu CP, Kenaga C, Geigel E, Torres A, Chalupa D, Gibb FR, Speers DM, Mast RW, Morrow PE (1998) Quantitative exposure of humans to an octamethylcyclotetrasiloxane (D4) vapor. Toxicol Sci 44:206–213

Vdovchenko A, Resmini M (2024) Mapping microplastics in humans: analysis of polymer types, and shapes in food and drinking water-a systematic review. Int J Mol Sci 25(13):7074

Vernet C, Philippat C, Calafat AM, Ye X, Lyon-Caen S, Siroux V, Schisterman EF, Slama R (2018) Within-day, between-day, and between-week variability of urinary concentrations of phenol biomarkers in pregnant women. Environ Health Perspect 126(3):037005

Walczak AP, Hendriksen PJ, Woutersen RA, van der Zande M, Undas AK, Helsdingen R, van den Berg HH, Rietjens IM, Bouwmeester H (2015) Bioavailability and biodistribution of differently charged polystyrene nanoparticles upon oral exposure in rats. J Nanopart Res 17(5):231

Wang H, Jiang Y, Song J, Liang H, Liu Y, Huang J, Yin P, Wu D, Zhang H, Liu X, Zhou D, Wei W, Lei L, Peng J, Zhang J (2022) The risk of perchlorate and iodine on the incidence of thyroid tumors and nodular goiter: a case-control study in southeastern China. Environ Health 21:4

Wang JM, Xu R, Di QN, Fu HW, Xu Q (2020) Determination of urinary carnitine levels as a potential indicator of uterine fibroids caused by nonylphenol exposure. J Chromatogr B 1152:122248

Wang PW, Chen ML, Huang LW, Yang W, Wu KY, Huang YF (2015) Nonylphenol exposure is associated with oxidative and nitrative stress in pregnant women. Free Radic Res 49(12):1469–1478

Wang S, Hanna D, Sugamori KS, Grant DM (2019a) Primary aromatic amines and cancer: novel mechanistic insights using 4-aminobiphenyl as a model carcinogen. Pharmacol Ther 200:179–189

Wang T, Yi Z, Liu X, Cai Y, Huang X, Fang J, Shen R, Lu W, Xiao Y, Zhuang W, Guo S (2024) Multimodal detection and analysis of microplastics in human thrombi from multiple anatomically distinct sites. EBioMedicine 103:105118

Wang Z, Sparling M, Wang KC, Arbuckle TE, Fraser W (2019b) Perchlorate in human milk samples from the maternal-infant research on environmental chemicals study (MIREC) Food Addit Contam Part A 36(12):1837–1846

Wang Z, Hu X, Guo Q, Fan M, Xing Y, Fan S, Yang Y, Zhang J, Shao B (2025) Simultaneous determination of 11 haloacetic acids and oxyhalides by UPLC-HRMS in human serum, breast milk, and urine. Talanta Open 12:100589

Watson CJ, Overbeek DL, Allegri-Machado G, Kellogg MD, Patterson A, McAlvin JB, Burns MM (2022) Thiocyanate toxicity: a teaching case. Clin Toxicol (Phila) 60(7):876–881

Wazir U, Kasem A, Mokbel K (2015) The clinical implications of poly implant prothèse breast implants: an overview. Arch Plast Surg 42:4–10

Weiss T, Angerer J (2002) Simultaneous determination various aromatic amines and metabolites of aromatic compounds in urine for low level exposure using gas chromatography-mass spectromometry. J Chromatogr B 778(1–2):179–192

WHO (2022) Dietary and inhalation exposure to nano- and microplastic particles and potential implications for human health. Geneva. https://www.who.int/publications/i/item/9789240054608

Wibowo AT, Nugrahapraja H, Wahyuono RA, Islami I, Haekal MH, Fardiansyah Y, Sugiyo PWW, Putro YK, Fauzia FN, Santoso H, Götz F, Tangahu BV, Luqman A (2021) Microplastic contamination in the human gastrointestinal tract and daily consumables associated with an indonesian farming community. Sustain Sci Pract Policy 13:12840

Wieland S, Balmes A, Bender J, Kitzinger J, Meyer F, Ramsperger AF, Roeder F, Tengelmann C, Wimmer BH, Laforsch C, Kress H (2022) From properties to toxicity: comparing microplastics to other airborne microparticles. J Hazard Mater 428:128151

Willemin ME, Lumen A (2017) Thiocyanate: a review and evaluation of the kinetics and the modes of action for thyroid hormone perturbations. Crit Rev Toxicol 47(7):537–563

Wright SL, Kelly FJ (2017) Plastic and human health: a micro issue? Environ Sci Technol 51(12):6634–6647

Wu D, Feng Y, Wang R, Jiang J, Guan Q, Yang X, Wei H, Xia Y, Luo Y (2023) Pigment microparticles and microplastics found in human thrombi based on Raman spectral evidence. J Adv Res 49:141–150

Wu P, Lin S, Cao G, Wu J, Jin H, Wang C, Wong MH, Yang Z, Cai Z (2022) Absorption, distribution, metabolism, excretion and toxicity of microplastics in the human body and health implications. J Hazard Mater 437:129361

Xu C, Ling H, Fan C, Xiang L, Zhang S, Li W, Yi C (2022) Higher levels of nonylphenol were found in human urine and drinking water from rural areas as compared to metropolitan regions of Wuhan. China. Environ Sci Pollut Res Int 29(44):66950–66959

Xu L, Shi Y, Liu N, Cai Y (2015) Methyl siloxanes in environmental matrices and human plasma/fat from both general industries and residential areas in China. Sci Total Environ 505:454–463

Yan Z, Liu Y, Zhang T, Zhang F, Ren H, Zhang Y (2022) Analysis of microplastics in human feces reveals a correlation between fecal microplastics and inflammatory bowel disease status. Environ Sci Technol 56:414–421

Yang WJ, Wu HB, Zhang C, Zhong Q, Hu MJ, He JL, Li GA, Zhu ZY, Zhu JL, Zhao HH, Zhang HS, Huang F (2021) Exposure to 2,4-dichlorophenol, 2,4,6-trichlorophenol, pentachlorophenol and risk of thyroid cancer: a case-control study in China. Environ Sci Pollut Res Int 28(43):61329–61343

Yang Y, Xie E, Du Z, Peng Z, Han Z, Li L, Zhao R, Qin Y, Xue M, Li F, Hua K, Yang X (2023) Detection of various microplastics in patients undergoing cardiac surgery. Environ Sci Technol 57:10911–10918

Yao Y, Xue J, Wang Y, Kannan K, Zhang T (2025) Exposure to carcinogenic primary aromatic amines and associated health risks among the Chinese general population: A nationwide biomonitoring study. Environ Chem Ecotoxicol 7:1624–1633

Ye X, Kuklenyik Z, Needham LL, Calafat AM (2006) Measuring environmental phenols and chlorinated organic chemicals in breast milk using automated on-line column-switching-high performance liquid chromatography-isotope dilution tandem mass spectrometry. J Chromatogr B 831(1–2):110–115

Yee MS, Hii LW, Looi CK, Lim WM, Wong SF, Kok YY, Tan BK, Wong CY, Leong CO (2021) Impact of microplastics and nanoplastics on human health. Nanomaterials (Basel) 11(2):496

Yoo JW, Doshi N, Mitragotri S (2011) Adaptive micro and nanoparticles: temporal control over carrier properties to facilitate drug delivery. Adv Drug Deliv Rev 63:1247–1256

Yu CJ, Du JC, Chiou HC, Yang SH, Liao KW, Yang W, Chung MY, Chien LC, Hwang B, Chen ML (2016) Attention deficit/hyperactivity disorder and urinary nonylphenol levels: a case-control study in taiwanese children. PLoS ONE 11(2):e0149558

Yu J, Wang B, Cai J, Yan Q, Wang S, Zhao G, Zhao J, Pan L, Liu S (2020) Selective extraction and determination of aromatic amine metabolites in urine samples by using magnetic covalent framework nanocomposites and HPLC-MS/MS RSC Adv 10:28437–28446

Zareba G, Gelein R, Morrow PE, Utell MJ (2002) Percutaneous absorption studies of octamethylcyclotetrasiloxane using the human skin/nude mouse model Skin Pharmacol. Appl Skin Physiol 15:184–194

Zhang B, Zhou Y, Guo J (2023) Association of volatile methylsiloxanes exposure with non-alcoholic fatty liver disease among Chinese adults. Environ Pollut 334:122128

Zhang N, Li Y, bin He HR, Zhang JF, Ma GS (2021) You are what you eat: microplastics in the feces of young men living in Beijing. Sci Total Environ 767:144345

Zhang Q, Xu EG, Li J, Chen Q, Ma L, Zeng EY, Shi H (2020) A review of microplastics in table salt, drinking water, and air: direct human exposure. Environ Sci Technol 54(7):3740–3751

Zhang S, Chen H, Li L, Li Z, Wang D (2025) Exposure to placental microplastic and placental and umbilical cord blood telomere length. Ecotoxicol Environ Saf 302:118536

Zhang S-H, Zhang Y-X, Ji G-X, Xu H-Z, Liu J-N, Shi L-L (2016a) Determination of bisphenol A, tetrabromobisphenol A and 4-tert-octylphenol in children and adults urine using high performance liquid chromatography-tandem mass spectrometry. Chinese J Anal Chem 44(1):19–24

Zhang T, Chen X, Wang D, Li R, Ma Y, Mo W, Sun H, Kannan K (2015) Perchlorate in indoor dust and human urine in China: contribution of indoor dust to total daily intake. Environ Sci Technol 49:2443–2450

Zhang T, Ma Y, Wang D, Li R, Chen X, Mo W, Qin X, Sun H, Kannan K (2016b) Placental transfer of and infantile exposure to perchlorate. Chemosphere 144:948–954

Zhang WS, Xu XQ, Zhou JZ, Fan WZ, Zhang PS, Zhou JA, et al. (1995) Pentachlorophenol contamination and its health effects on human being in Jintan districts. Jiangsu Prev Med (China) 1995;2:20–3

Zhao B, Rehati P, Yang Z, Cai Z, Guo C, Li Y (2024) The potential toxicity of microplastics on human health. Sci Total Environ 912:168946

Zhao Q, Zhu L, Weng J, Jin Z, Cao Y, Jiang H, Zhang Z (2023) Detection and characterization of microplastics in the human testis and semen. Sci Total Environ 877:162713

Zheng W, Wang X, Yu H, Tao X, Zhou Y, Qu W (2011) Global trends and diversity in pentachlorophenol levels in the environment and in humans: a meta-analysis. Environ Sci Technol 45(11):4668–4675

Zheng Y, Luo S, Xu M, He Q, Xie J, Wu J, Huang Y (2024) Transepithelial transport of nanoparticles in oral drug delivery: from the perspective of surface and holistic property modulation. Acta Pharm Sin B 14(9):3876–3900

Zhou X, Kramer JP, Calafat AM, Ye X (2014) Automated on-line column-switching high performance liquid chromatography isotope dilution tandem mass spectrometry method for the quantification of bisphenol A, bisphenol F, bisphenol S, and 11 other phenols in urine. J Chromatogr B 944:152–156

Zhu L, Wu Z, Dong J, Zhao S, Zhu J, Wang W, Ma F, An L (2024) Unveiling Small-Sized plastic particles hidden behind Large-Sized ones in human excretion and their potential sources. Environ Sci Technol 58(27):11901–11911

Zuri G, Karanasiou A, Lacorte S (2023) Human biomonitoring of microplastics and health implications: a review. Environ Res 237(Pt 1):116966

MIX
Papier aus verantwortungsvollen Quellen
Paper from responsible sources
FSC® C105338

If you have any concerns about our products,
you can contact us on
ProductSafety@springernature.com

In case Publisher is established outside the EU,
the EU authorized representative is:
Springer Nature Customer Service Center GmbH
Europaplatz 3, 69115 Heidelberg, Germany

Printed by Libri Plureos GmbH
in Hamburg, Germany